Are Magnus Bruaset Aslak Tveito (Eds.)

Numerical Solution of Partial Differential Equations on Parallel Computers

With 201 Figures and 42 Tables

Springer

Editors

Are Magnus Bruaset
Aslak Tveito
Simula Research Laboratory
P.O. Box 134
1325 Lysaker, Fornebu, Norway
email: arem@simula.no
aslak@simula.no

The second editor of this book has received financial support from the NFF – Norsk faglitterær forfatter- og oversetterforening

Library of Congress Control Number: 2005934453

Mathematics Subject Classification:
Primary: 65M06, 65M50, 65M55, 65M60, 65Y05, 65Y10
Secondary: 65N06, 65N30, 65N50, 65N55, 65F10, 65F50

ISBN-10 3-540-29076-1 Springer Berlin Heidelberg New York
ISBN-13 978-3-540-29076-6 Springer Berlin Heidelberg New York

Springer is a part of Springer Science+Business Media
springer.com

Printed in The Netherlands

Typesetting: by the authors and TechBooks using a Springer LaTeX macro package

Cover design: *design & production* GmbH, Heidelberg

Printed on acid-free paper SPIN: 11548843 46/TechBooks 5 4 3 2 1 0

Preface

Since the dawn of computing, the quest for a better understanding of Nature has been a driving force for technological development. Groundbreaking achievements by great scientists have paved the way from the abacus to the supercomputing power of today. When trying to replicate Nature in the computer's silicon test tube, there is need for precise and computable process descriptions. The scientific fields of Mathematics and Physics provide a powerful vehicle for such descriptions in terms of Partial Differential Equations (PDEs). Formulated as such equations, physical laws can become subject to computational and analytical studies. In the computational setting, the equations can be discreti ed for efficient solution on a computer, leading to valuable tools for simulation of natural and man-made processes. Numerical solution of PDE-based mathematical models has been an important research topic over centuries, and will remain so for centuries to come.

In the context of computer-based simulations, the quality of the computed results is directly connected to the model's complexity and the number of data points used for the computations. Therefore, computational scientists tend to fill even the largest and most powerful computers they can get access to, either by increasing the si e of the data sets, or by introducing new model terms that make the simulations more realistic, or a combination of both. Today, many important simulation problems can not be solved by one single computer, but calls for *parallel computing*. Whether being a dedicated multi-processor supercomputer or a loosely coupled cluster of office workstations, the concept of parallelism offers increased data storage and increased computing power. In theory, one gets access to the grand total of the resources offered by the individual units that make up the multi-processor environment. In practice, things are more complicated, and the need for data communication between the different computational units consumes parts of the theoretical gain of power.

Summing up the bits and pieces that go into a large-scale parallel computation, there are aspects of hardware, system software, communication protocols, memory management, and solution algorithms that have to be addressed. However, over time efficient ways of addressing these issues have emerged, better software tools have become available, and the cost of hardware has fallen considerably. Today, computational clusters made from commodity parts can be set up within the budget of a typical research department, either as a turn-key solution or as a do-it-yourself project. Supercomputing has become affordable and accessible.

About this book

This book addresses the major topics involved in numerical simulations on parallel computers, where the underlying mathematical models are formulated in terms of PDEs. Most of the chapters dealing with the technological components of parallel computing are written in a survey style and will provide a comprehensive, but still readable, introduction for students and researchers. Other chapters are more specialized, for instance focusing on a specific application that can demonstrate practical problems and solutions associated with parallel computations. As editors we are proud to put together a volume of high-quality and useful contributions, written by internationally acknowledged experts on high-performance computing.

The first part of the book addresses fundamental parts of parallel computing in terms of hardware and system software. These issues are vital to all types of parallel computing, not only in the context of numerical solution of PDEs. To start with, Ricky Kendall and co-authors discuss the programming models that are most commonly used for parallel applications, in environments ranging from a simple departmental cluster of workstations to some of the most powerful computers available today. Their discussion covers models for message passing and shared memory programming, as well as some future programming models. In a closely related chapter, Jim Teresco et al. look at how data should be partitioned between the processors in a parallel computing environment, such that the computational resources are utilized as efficient as possible. In a similar spirit, the contribution by Martin Rumpf and Robert Strzodka also aims at improved utilization of the available computational resources. However, their approach is somewhat unconventional, looking at ways to benefit from the considerable power available in graphics processors, not only for visualization purposes but also for numerical PDE solvers. Given the low cost and easy access of such commodity processors, one might imagine future cluster solutions with really impressive price-performance ratios.

Once the computational infrastructure is in place, one should concentrate on how the PDE problems can be solved in an efficient manner. This is the topic of the second part of the book, which is dedicated to parallel algorithms that are vital to numerical PDE solution. Luca Formaggia and co-authors present parallel domain decomposition methods. In particular, they give an overview of algebraic domain decomposition techniques, and introduce sophisticated preconditioners based on a multilevel approximative Schur complement system and a Schwarz-type decomposition, respectively. As Schwarz-type methods call for a coarse level correction, the paper also proposes a strategy for constructing coarse operators directly from the algebraic problem formulation, thereby handling unstructured meshes for which a coarse grid can be difficult to define. Complementing this multilevel approach, Frank Hülsemann et al. discuss how another important family of very efficient PDE solvers, geometric multigrid, can be implemented on parallel computers. Like domain decomposition methods, multigrid algorithms are potentially capable of being order-optimal such that the solution time scales linearly with the number of unknowns. However, this paper demonstrates that in order to maintain high computational performance the construction of a parallel multigrid solver is certainly problem-dependent. In the following chapter, Ulrike Meier Yang addresses parallel algebraic multigrid methods. In contrast to the geometric multigrid variants, these algorithms work only on the algebraic system arising from the discretization of the PDE, rather than on a multiresolution discretization of the computational domain. Ending the section on parallel algorithms, Nikos Chrisochoides surveys methods for parallel mesh generation. Meshing procedures are an important part of the discretization of a PDE, either used as a preprocessing step prior to the solution phase, or in case of a changing geometry, as repeated steps in course of the simulation. This contribution concludes that it is possible to develop parallel meshing software using off-the-shelf sequential codes as building blocks without sacrificing the quality of the constructed mesh.

Making advanced algorithms work in practice calls for development of sophisticated software. This is especially important in the context of parallel computing, as the complexity of the software development tends to be significantly higher than for its sequential counterparts. For this reason, it is desirable to have access to a wide range of software tools that can help make parallel computing accessible. One way of addressing this need is to supply high-quality software libraries that provide parallel computing power to the application developer, straight out of the box. The *hypre* library presented by Robert D. Falgout et al. does exactly this by offering parallel high-performance preconditioners. Their paper concentrates on the conceptual interfaces in this package, how these are implemented for parallel computers, and how they are used in applications. As an alternative, or complement, to the library approach, one might look for programming languages that tries to ease the process of parallel coding. In general, this is a quite open issue, but Xing Cai and Hans Petter Langtangen contribute to this discussion by considering whether the high-level language Python can be used to develop efficient parallel PDE solvers. They address this topic from two different angles, looking at the performance of parallel PDE solvers mainly based on Python code and native data structures, and through the use of Python to parallelize existing sequential PDE solvers written in a compiled language like FORTRAN, C or C++. The latter approach also opens for the possibility of combining different codes in order to address a multi-model or multiphysics problem. This is exactly the concern of Lois Curfman McInnes and her co-authors when they discuss the use of the Common Component Architecture (CCA) for parallel PDE-based simulations. Their paper gives an introduction to CCA and highlights several parallel applications for which this component technology is used, ranging from climate modeling to simulation of accidental fires and explosions.

To communicate experiences gained from work on some complete simulators, selected parallel applications are discussed in the latter part of the book. Xing Cai and Glenn Terje Lines present work on a full-scale parallel simulation of the electrophysiology of the human heart. This is a computationally challenging problem, which due to a multiscale nature requires a large amount of unknowns that have to be resolved for small time steps. It can be argued that full-scale simulations of this problem can not be done without parallel computers. Another challenging geodynamics problem, modeling the magma genesis in subduction zones, is discussed by Matthew G. Knepley et al. They have ported an existing geodynamics code to use PETSc, thereby making it parallel and extending its functionality. Simulations performed with the resulting application confirms physical observations of the thermal properties in subduction zones, which until recently were not predicted by computations. Finally, in the last chapter of the book, Carolin Körner et al. present parallel Lattice Boltzmann Methods (LBMs) that are applicable to problems in Computational Fluid Dynamics. Although not being a PDE-based model, the LBM approach can be an attractive alternative, especially in terms of computational efficiency. The power of the method is demonstrated through computation of 3D free surface flow, as in the interaction and growing of gas bubbles in a melt.

Acknowledgements

We wish to thank all the chapter authors, who have written very informative and thorough contributions that we think will serve the computational community well. Their enthusiasm has been crucial for the quality of the resulting book.

Moreover, we wish to express our gratitude to all reviewers, who have put time and energy into this project. Their expert advice on the individual papers has been useful to editors and contributors alike. We are also indebted to Dr. Martin Peters at Springer-Verlag for many interesting and useful discussions, and for encouraging the publication of this volume.

Fornebu
September, 2005

Are Magnus Bruaset
Aslak Tveito

Contents

Part I Parallel Computing

Part II Parallel Algorithms

Part III Parallel Software Tools

Part IV Parallel Applications

Part I

Parallel Computing

1

Parallel Programming Models Applicable to Cluster Computing and Beyond

Ricky A. Kendall[1], Masha Sosonkina[1], William D. Gropp[2], Robert W. Numrich[3], and Thomas Sterling[4]

[1] Scalable Computing Laboratory, Ames Laboratory, USDOE, Ames, IA 50011, USA
[rickyk,masha]@scl.ameslab.gov
[2] Mathematics and Computer Science Division, Argonne National Laboratory, Argonne, IL 60439, USA
gropp@mcs.anl.gov
[3] Supercomputing Institute, University of Minnesota, Minneapolis, MN 55455, USA
rwn@msi.umn.edu
[4] California Institute of Technology, Pasadena, CA 91125, USA
tron@cacr.caltech.edu

Summary. This chapter centers mainly on successful programming models that map algorithms and simulations to computational resources used in high-performance computing. These resources range from group-based or departmental clusters to high-end resources available at the handful of supercomputer centers around the world. Also covered are newer programming models that may change the way we program high-performance parallel computers.

1.1 Introduction

Solving a system of partial differential equations (PDEs) lies at the heart of many scientific applications that model physical phenomena. The solution of PDEs—often the most computationally intensive task of these applications—demands the full power of multiprocessor computer architectures combined with effective algorithms.

This synthesis is particularly critical for managing the computational complexity of the solution process when *nonlinear PDEs* are used to model a problem. In such a case, a mix of solution methods for large-scale nonlinear and linear systems of equations is used, in which a nonlinear solver acts as an "outer" solver. These methods may call for diverse implementations and programming models. Hence sophisticated software engineering techniques and a careful selection of parallel programming tools have a direct effect not only on the code reuse and ease of code handling but also on reaching the problem solution efficiently and reliably. In other words, these tools and techniques affect the numerical efficiency, robustness, and parallel performance of a solver.

For *linear PDEs*, the choice of a solution method may depend on the type of linear system of equations used. Many parallel direct and iterative solvers are designed to solve a particular system type, such as symmetric positive definite linear systems. Many of the iterative solvers are also specific to the application and data format. There exists only a limited selection of "general-purpose" distributed-memory iterative-solution implementations. Among the better-known packages that contain such implementations are PETSc [3, 46], *hypre* [11, 23], and pARMS [50]. One common feature of these packages is that they are all based on domain decomposition methods and include a wide range of parallel solution techniques, such as preconditioners and accelerators.

Domain decomposition methods simply divide the domain of the problem into smaller parts and describe how solutions (or approximations to the solution) on each part is combined to give a solution (or approximation) to the original problem. For hyperbolic PDEs, these methods take advantage of the finite signal speed property. For elliptic, parabolic, and mixed PDEs, these methods take advantage of the fact that the influence of distant parts of the problem, while nonzero, is often small (for a specific example, consider the Green's function for the solution to the Poisson problem). Domain decomposition methods have long been successful in solving PDEs on single processor computers (see, e.g, [72]), and lead to efficient implementations on massively parallel distributed-memory environments.[5] Domain decomposition methods are attractive for parallel computing mainly because of their "divide-and-conquer" approach, to which many parallel programming models may be readily applied. For example, all three of the cited packages use the message-passing interface MPI for communication. When the complexity of the solution methods increases, however, the need to mix different parallel programming models or to look for novel ones becomes important. Such a situation may arise, for example, when developing a nontrivial parallel incomplete LU factorization, a direct sparse linear system solver, or any algorithm where data storage and movement are coupled and complex. The programming model(s) that provide(s) the best portability, performance, and ease of development or expression of the algorithm should be used. A good overview of applications, hardware and their interactions with programming models and software technologies is [17].

1.1.1 Programming Models

What is a programming model? In a nutshell it is the way one thinks about the flow and execution of the data manipulation for an application. It is an algorithmic mapping to a perceived architectural moiety.

In choosing a programming model, the developer must consider many factors: performance, portability, target architectures, ease of maintenance, code revision mechanisms, and so forth. Often, tradeoffs must be made among these factors. Trading computation for storage (either in memory or on disk) or for communication of data is a common algorithmic manipulation. The complexity of the tradeoffs is compounded by the use of parallel algorithms and hardware. Indeed, a programmer may

[5] No memory is visible to all processors in a distributed-memory environment; each processor can only see their own local memory.

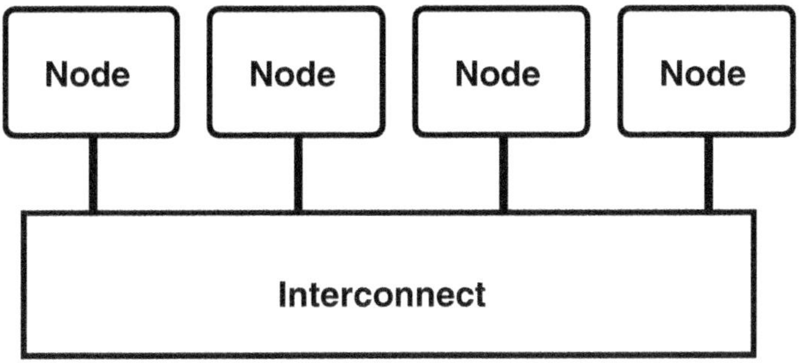

Fig. 1.1. Generic architecture for a cluster system.

have (as many libraries and applications do) multiple implementations of the same algorithm to allow for performance tuning on various architectures.

Today, many small and high-end high-performance computers are clusters with various communication interconnect technologies and with nodes[6] having more than one processor. For example, the Earth Simulator [20] is a cluster of very powerful nodes with multiple vector processors; and large IBM SP installations (e.g., the system at the National Energy Research Scientific Computing Center, http://hpcf.nersc.gov/computers/SP) have multiple nodes with 4, 8, 16, or 32 processors each. These systems are at an abstract level the same kind of system. The fundamental issue for parallel computation on such clusters is how to select a programming model that gets the data in the right place when computational resources are available. This problem becomes more difficult as the number of processors increases; the term *scalability* is used to indicate the performance of an algorithm, method, or code, relative to a single processor. The scalability of an application is primarily the result of the algorithms encapsulated in the programming model used in the application. No programming model can overcome the scalability limitations inherent in the algorithm. There is no free lunch.

A generic view of a cluster architecture is shown in Figure 1.1. In the early Beowulf clusters, like the distributed-memory supercomputer shown in Figure 1.2, each node was typically a single processor. Today, each node in a cluster is usually at least a dual-processor symmetric processing (SMP) system. A generic view of an SMP node or a general shared-memory system is shown in Figure 1.3. The number of processors per computational node varies from one installation to another. Often, each node is composed of identical hardware, with the same software infrastructure as well.

The "view" of the target system is important to programmers designing parallel algorithms. Mapping algorithms with the chosen programming model to the system architecture requires forethought, not only about how the data is moved, but also about what type of hardware transport layer is used: for example, is data moved over

[6] A node is typically defined as a set of processors and memory that have a single system image; one operating system and all resources are visible to each other in the "node" moiety.

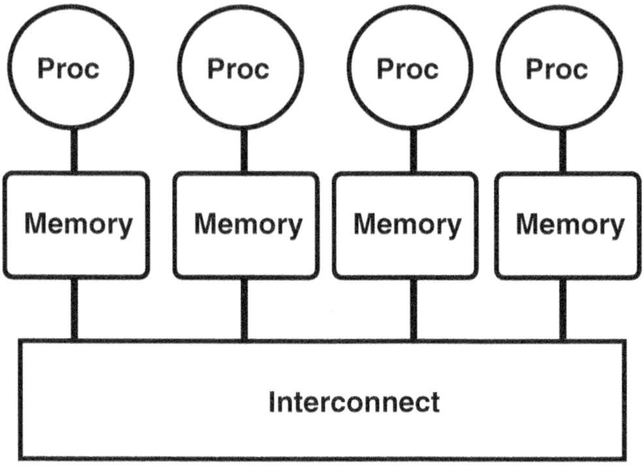

Fig. 1.2. Generic architecture for a distributed-memory cluster with a single processor.

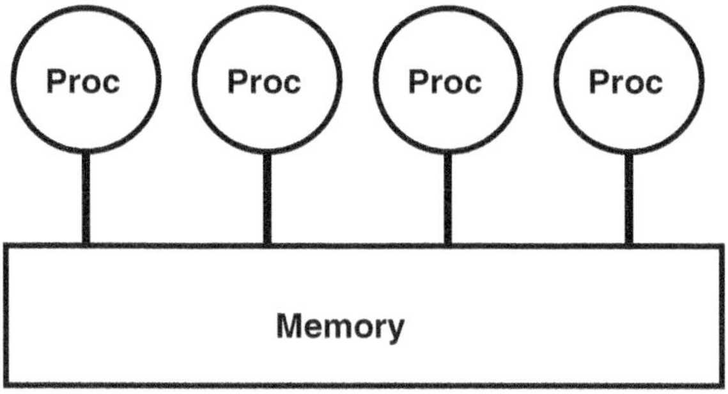

Fig. 1.3. Generic architecture for a shared-memory system.

a shared-memory bus between cooperating threads or over a fast Ethernet network between cooperating processes?

This chapter presents a brief overview of various programming models that work effectively on cluster computers and high-performance parallel supercomputers. We cannot cover all aspects of message-passing and shared-memory programming. Our goal is to give a taste of the programming models as well as the most important aspects of the models that one must consider in order to get an application parallelized. Each programming model takes a significant effort to master, and the learning experience is largely based on trial and error, with error usually being the better educational track. We also touch on newer techniques that are being used successfully and on a few specialty languages that are gaining support from the vendor community. We give numerous references so that one can delve more deeply into any area of interest.

1.1.2 Application Development Efforts

"Best practices" for software engineering are commonly applied in industry but have not been so widely adopted in high-performance computing. Dubois outlines ten such practices for scientific programming [18]. We focus here on three of these.

The first is the use of a revision control system that allows multiple developers easy access to a central repository of the software. Both commercial and open source revision control systems exist. Some commonly used, freely available systems include Concurrent Versions System (CVS), Subversion, and BitKeeper. The functionality in these systems includes

- branching release software from the main development source,
- comparing modifications between versions of various subunits,
- merging modifications of the same subunit from multiple users, and
- obtaining a version of the development or branch software at a particular date and time.

The ability to recover previous instances of subunits of software can make debugging and maintenance easier and can be useful for speculative development efforts.

The second software engineering practice is the use of automatic build procedures. Having such procedures across a variety of platforms is useful in finding bugs that creep into code and inhibit portability. Automated identification of the language idiosyncrasies of different compilers minimizes efforts of porting to a new platform and compiler system. This is essentially normalizing the interaction of compilers and your software.

The third software engineering practice of interest is the use of a robust and exhaustive test suite. This can be coupled to the build infrastructure or, at a minimum, with every software release. The test suite should be used to verify the functionality of the software and, hence, the viability of a given release; it also provides a mechanism to ensure that ports to new computational resources are valid.

The cost of these software engineering mechanisms is not trivial, but they do make the maintenance and distribution easier. Consider the task of making Linux software distribution agnostic. Each distribution must have different versions of particular software moieties in addition to the modifications that each distribution makes to that software. Proper application of these tasks is essentially making one's software operating system agnostic.

1.2 Message-Passing Interface

Parallel computing, with any programming model, involves two actions: *transferring data* among workers and *coordinating* the workers. A simple example is a room full of workers, each at a desk. The work can be described by written notes. Passing a note from one worker to another effects data transfer; receiving a note provides coordination (think of the note as requesting that the work described on the note be executed). This simple example is the background for the most common and most portable parallel computing model, known as *message passing*. In this section we briefly cover the message-passing model, focusing on the most common form of this model, the Message-Passing Interface (MPI).

1.2.1 The Message-Passing Interface

Message passing has a long history. Even before the invention of the modern digital computer, application scientists proposed halls full of skilled workers, each working on a small part of a larger problem and passing messages to their neighbors. This model of computation was formalized in computer science theory as communicating sequential processes (CSP) [36]. One of the earliest uses of message passing was for the Caltech Cosmic Cube, one of the first scalable parallel machines [71]. The success (perhaps more accurately, the potential success of highly parallel computing demonstrated by this machine) spawned many parallel machines, each with its own version of message passing.

In the early 1990s, the parallel computing market was divided among several companies, including Intel, IBM, Cray, Convex, Thinking Machines, and Meiko. No one system was dominant, and as a result the market for parallel software was splintered. To address the need for a single method for programming parallel computers, an informal group calling itself the MPI Forum and containing representatives from all stake-holders, including parallel computer vendors, applications developers, and parallel computing researchers, began meeting [33]. The result was a document describing a standard application programming interface (API) to the message-passing model, with bindings for the C and Fortran languages [52]. This standard quickly became a success. As is common in the development of standards, there were a few problems with the original MPI standard, and the MPI Forum released two updates, called MPI 1.1 and MPI 1.2. MPI 1.2 is the most widely available version today.

1.2.2 MPI 1.2

When MPI was standardized, most message-passing libraries at that time described communication between separate processes and contained three major components:

- Processing environment – information about the number of processes and other characteristics of the parallel environment.
- Point-to-point – messages from one process to another
- Collective – messages between a collection of processes (often all processes)

We will discuss each of these in turn. These components are the heart of the message passing programming model.

Processing Environment

In message passing, a parallel program comprises a number of separate processes that communicate by calling routines. The first task in an MPI program is to initialize the

```
#include "mpi.h"
#include <stdio.h>
int main( int argc, char *argv[] )
{
    int rank, size;
    MPI_Init( &argc, &argv );
    MPI_Comm_size( MPI_COMM_WORLD, &size );
    MPI_Comm_rank( MPI_COMM_WORLD, &rank );
    printf( "Hello World! I am %d of %d\n", rank, size );
    MPI_Finalize( );
    return 0;
}
```

Fig. 1.4. A simple MPI program.

MPI library; this is accomplished with `MPI_Init`. When a program is done with MPI (usually just before exiting), it must call `MPI_Finalize`. Two other routines are used in almost all MPI programs. The first, `MPI_Comm_size`, returns in the second argument the number of processes available in the parallel job. The second, `MPI_Comm_rank`, returns in the second argument a ranking of the calling process, with a value between zero and size-1. Figure 1.4 shows a simple MPI program that prints the number of processes and the rank of each process. `MPI_COMM_WORLD` represents all the cooperating processes.

While MPI did not specify a way to run MPI programs (much as neither C nor Fortran specifies how to run C or Fortran programs), most parallel computing systems require that parallel programs be run with a special program. For example, the program `mpiexec` might be used to run an MPI program. Similarly, an MPI environment may provide commands to simplify compiling and linking MPI programs. For example, for some popular MPI implementations, the following steps will run the program in Figure 1.4 with four processes, assuming that program is stored in the file `first.c`:

```
mpicc -o first first.c
mpiexec -n 4 first
```

The output may be

```
Hello World! I am 2 of 4
Hello World! I am 3 of 4
Hello World! I am 0 of 4
Hello World! I am 1 of 4
```

Note that the output of the process rank is not ordered from zero to three. MPI specifies that all routines that are *not* MPI routines behave independently, including I/O routines such as `printf`.

We emphasize that MPI describes communication between *processes*, not *processors*. For best performance, parallel programs are often designed to run with one process per processor (or, as we will see in the section on OpenMP, one thread per processor). MPI supports this model, but MPI also allows multiple processes to be run on a single-processor machine. Parallel programs are commonly developed on single-processor laptops, even with multiple processes. If there are more than a few processes per processor, however, the program may run very slowly because of contention among the processes for the resources of the processor.

Point-to-Point Communication

The program in Figure 1.4 is a very simple parallel program. The individual processes neither exchange data nor coordinate with each other. Point-to-point communication allows two processes to send data from one to another. Data is sent by using routines such as `MPI_Send` and is received by using routines such as `MPI_Recv` (we mention later several specialized forms for both sending and receiving).

We illustrate this type of communication in Figure 1.5 with a simple program that sums contributions from each process. In this program, each process first determines its rank and initializes the value that it will contribute to the sum. (In this case, the sum itself is easily computed analytically; this program is used for illustration only.) After receiving the contribution from the process with rank one higher, it adds the received value into its contribution and sends the new value to the process with rank one lower. The process with rank zero only receives data, and the process with the largest rank (equal to size-1) only sends data.

The program in Figure 1.5 introduces a number of new points. The most obvious are the two new MPI routines `MPI_Send` and `MPI_Recv`. These have similar arguments. Each routine uses the first three arguments to specify the data to be sent or received. The fourth argument specifies the destination (for `MPI_Send`) or source (for `MPI_Recv`) process, by rank. The fifth argument, called a *tag*, provides a way to include a single integer with the data; in this case the value is not needed, and a zero is used (the value used by the sender must match the value given by the receiver). The sixth argument specifies the collection of processes to which the value of rank is relative; we use `MPI_COMM_WORLD`, which is the collection of all processes in the parallel program (determined by the startup mechanism, such as `mpiexec` in the "Hello World" example). There is one additional argument to `MPI_Recv`: `status`. This value contains some information about the message that some applications may need. In this example, we do not need the value, but we must still provide the argument.

The three arguments describing the data to be sent or received are, in order, the address of the data, the number of items, and the type of the data. Each basic datatype in the language has a corresponding MPI datatype, as shown in Table 1.1.

MPI allows the user to define new datatypes that can represent noncontiguous memory, such as rows of a Fortran array or elements indexed by an integer array (also called scatter-gathers). Details are beyond the scope of this chapter, however.

This program also illustrates an important feature of message-passing programs: because these are separate, communicating processes, all variables, such as `rank` or `valOut`, are private to each process and may (and often will) contain different values. That is, each process has its own memory space, and all variables are private

```
#include "mpi.h"
#include <stdio.h>
int main( int argc, char *argv[] )
{
    int        size, rank, valIn, valOut;
    MPI_Status status;

    MPI_Init( &argc, &argv );

    MPI_Comm_size( MPI_COMM_WORLD, &size );
    MPI_Comm_rank( MPI_COMM_WORLD, &rank );

    /* Pick a simple value to add */
    valIn = rank;

    /* receive the partial sum from the right processes
       (this is the sum from i=rank+1 to size-1) */
    if (rank < size - 1) {
        MPI_Recv( &valOut, 1, MPI INT, rank + 1, 0,
          MPI_COMM_WORLD, &status );
        valIn += valOut;
    }
    /* Send the partial sum to the left (rank-1) process */
    if (rank > 0) {
        MPI_Send( &valIn, 1, MPI_INT, rank - 1, 0,
          MPI_COMM_WORLD );
    }
    else {
        printf( "The sum is %d\n", valOut );
    }

    MPI_Finalize( );
    return 0;
}
```

Fig. 1.5. A simple program to add values from each process.

Table 1.1. Some major predefined MPI datatypes.

C		Fortran	
int	MPI_INT	INTEGER	MPI_INTEGER
float	MPI_FLOAT	REAL	MPI_REAL
double	MPI_DOUBLE	DOUBLE PRECISION	MPI_DOUBLE_PRECISION
char	MPI_CHAR	CHARACTER	MPI_CHARACTER
short	MPI_SHORT		

to that process. The only way for one process to change or access data in another process is with the explicit use of MPI routines such as `MPI_Send` and `MPI_Recv`.

MPI provides a number of other ways in which to send and receive messages, including nonblocking (sometimes incorrectly called asynchronous) and synchronous routines. Other routines, such as `MPI_Iprobe`, can be used to determine whether a message is available for receipt. The nonblocking routines can be important in applications that have complex communication patterns and that send large messages. See [30, Chapter 4] for more details and examples.

Collective Communication and Computation

Any parallel algorithm can be expressed by using point-to-point communication. This flexibility comes at a cost, however. Unless carefully structured and documented, programs using point-to-point communication can be challenging to understand because the relationship between the part of the program that sends data and the part that receives the data may not be clear (note that well-written programs using point-to-point message passing strive to keep this relationship as plain and obvious as possible).

An alternative approach is to use communication that involves all processes (or all in a well-defined subset). MPI provides a wide variety of collective communication functions for this purpose. As an added benefit, these routines can be optimized for their particular operations (note, however, that these optimizations are often quite complex). As an example Figure 1.6 shows a program that performs the same computation as the program in Figure 1.5 but uses a single MPI routine. This routine, `MPI_Reduce`, performs a sum reduction (specified with `MPI_SUM`), leaving the result on the process with rank zero (the sixth argument).

Note that this program contains only a single branch (if) statement that is used to ensure that only one process writes the result. The program is easier to read than its predecessor. In addition, it is effectively parallel; most MPI implementations will perform a sum reduction in time that is proportional to the log of the number of processes. The program in Figure 1.5, despite being a parallel program, will take time that is proportional to the number of processes because each process must wait for its neighbor to finish before it receives the data it needs to form the partial sum.[7]

Not all programs can be conveniently and efficiently written by using only collective communications. For example, for most MPI implementations, operations on PDE meshes are best done by using point-to-point communication, because the data exchanges are between pairs of processes and this closely matches the point-to-point programming model.

---

[7]One might object that the program in Figure 1.6 doesn't do exactly what the program in Figure 1.5 does because, in the latter, all of the intermediate results are computed and available to those processes. We offer two responses. First, only the value on the rank-zero process is printed; the others don't matter. Second, MPI offers the collective routine `MPI_Scan` to provide the partial sum results if that is required.

```
#include "mpi.h"
#include <stdio.h>
int main( int argc, char *argv[] )
{
    int        rank, valIn, valOut;
    MPI_Status status;

    MPI_Init( &argc, &argv );

    MPI_Comm_rank( MPI_COMM_WORLD, &rank );

    /* Pick a simple value to add */
    valIn = rank;

    /* Reduce to process zero by summing the values */
    MPI_Reduce( &valIn, &valOut, 1, MPI_INT, MPI_SUM, 0,
                MPI_COMM_WORLD );
    if (rank == 0) {
        printf( "The sum is %d\n", valOut );
    }

    MPI_Finalize( );
    return 0;
}
```

Fig. 1.6. Using collective communication and computation in MPI.

Other Features

MPI contains over 120 functions. In addition to nonblocking versions of point-to-point communication, there are routines for defining groups of processes, user-defined data representations, and testing for the availability of messages. These are described in any comprehensive reference on MPI [73, 30].

An important part of the MPI design is its support for programming in the large. Many parallel libraries have been written that make use of MPI; in fact, many applications can be written that have no explicit MPI calls and instead use libraries that themselves use MPI to express parallelism. Before writing any MPI program (or any program, for that matter), one should check to see whether someone has already done the hard work. See [31, Chapter 12] for a summary of some numerical libraries for Beowulf clusters.

1.2.3 The MPI-2 Extensions

The success of MPI created a desire to tackle some of the features not in the original MPI (henceforth called MPI-1). The major features include parallel I/O, the creation of new processes in the parallel program, and one-sided (as opposed to point-to-point) communication. Other important features include bindings for Fortran 90 and

C++. The MPI-2 standard was officially released on July 18, 1997, and "MPI" now means the combined standard consisting of MPI-1.2 and MPI-2.0.

Parallel I/O

Perhaps the most requested feature for MPI-2 was parallel I/O. A major reason for using parallel I/O (as opposed to independent I/O) is performance. Experience with parallel programs using conventional file systems showed that many provided poor performance. Even worse, some of the most common file systems (such as NFS) are not designed to allow multiple processes to update the same file; in this case, data can be lost or corrupted. The goal for the MPI-2 interface to parallel I/O was to provide an interface that matched the needs of applications to create and access files in parallel, while preserving the flavor of MPI. This turned out to be easy. One can think of writing to a file as sending a message to the file system; reading a file is somewhat like receiving a message from the file system ("somewhat," because one must ask the file system to send the data). Thus, it makes sense to use the same approach for describing the data to be read or written as is used for message passing—a tuple of address, count, and MPI datatype. Because the I/O is parallel, we need to specify the group of processes; thus we also need a communicator. For performance reasons, we sometimes need a way to describe where the data is on the disk; fortunately, we can use MPI datatypes for this as well.

Figure 1.7 shows a simple program for reading a single integer value from a file. There are three steps, each similar to what one would use with non-parallel I/O:

1. Open the file. The `MPI_File_open` call takes a communicator (to specify the group of processes that will access the file), the file name, the access style (in this case, read-only), and another parameter used to pass additional data (usually empty, or `MPI_INFO_NULL`) and returns an `MPI_File` object that is used in MPI-IO calls.
2. Use all processes to read from the file. This simple call takes the file handle returned from `MPI_File_open`, the same buffer description (address, count, datatype) used in an `MPI_Recv` call, and (also like `MPI_Recv`) a status variable. In this case we use `MPI_STATUS_IGNORE` for simplicity.
3. Close the file.

Variations on this program, using other routines from MPI-IO, allow one to read different parts of the file to different processes and to specify from where in the file to read. As with message passing, there are also nonblocking versions of the I/O routines, with a special kind of nonblocking collective operation, called split-phase collective, available only for these I/O routines.

Writing files is similar to reading files. Figure 1.8 shows how each process can write the contents of the array `solution` with a single collective I/O call.

Figure 1.8 illustrates the use of collective I/O, combined with *file views*, to efficiently write data from many processes to a single file in a way that provides a natural ordering for the data. Each process writes `ARRAY_SIZE` double-precision values to the file, ordered by the MPI rank of the process. Once this file is written, another

```
    /* Declarations, including */
    MPI_File fh;
    int val;

    /* Start MPI */
    MPI_Init( &argc, &argv );

    /* Open the file for reading only */
    MPI_File_open( MPI_COMM_WORLD, "input.dat",
                   MPI_MODE_RDONLY, MPI_INFO_NULL, &fh );

    /* All processes access the file and read the same value
       into val */
    MPI_File_read_all( fh, &val, 1, MPI_INT,
                       MPI_STATUS_IGNORE );
    /* Close the file when no longer needed */
    MPI_File_close( &fh );
```

Fig. 1.7. A simple program to read a single integer from a file.

```
#define ARRAY_SIZE 1000
    /* Declarations, including */
    MPI_File fh;
    int rank;
    int solution[ARRAY_SIZE];

    /* Start MPI */
    MPI_Init( &argc, &argv );

    /* Open the file for reading only */
    MPI_File_open( MPI_COMM_WORLD, "output.dat",
                   MPI_MODE_WRONLY, MPI_INFO_NULL, &fh );

    /* Define where each process writes in the file */
    MPI_Comm_rank( MPI_COMM_WORLD, &rank );
    MPI_File_set_view( fh, rank * ARRAY_SIZE * sizeof(double),
                       MPI_DOUBLE, MPI_DOUBLE, "native",
       MPI_INFO_NULL );
    /* Perform the write */
    MPI_File_write_all( fh, solution, ARRAY_SIZE, MPI_DOUBLE,
                        MPI_STATUS_IGNORE );
    /* Close the file when no longer needed */
    MPI_File_close( &fh );
```

Fig. 1.8. A simple program to write a distributed array to a file in a standard order that is independent of the number of processes.

program, using a different number of processes, can read the data in this file. For example, a non-parallel program could read this file, accessing all of the data.

Several good libraries provide convenient parallel I/O for user applications. Parallel netCDF [49] and HDF-5 [24] can read and write data files in a standard format, making it easy to move files between platforms. These libraries also encourage the inclusion of *metadata* in the file that describes the contents, such as the source of the computation and the meaning and units of measurements of the data. Parallel netCDF in particular encourages a collective I/O style for input and output, which helps ensure that the parallel I/O is efficient. We recommend that an I/O library be used if possible.

Dynamic Processes

Another feature that was often requested for MPI-2 was the ability to create and use additional processes. This is particularly valuable for ad hoc collections of desktop systems. Since MPI is designed for use on all kinds of parallel computers, from collections of desktops to dedicated massively parallel computers, a scalable design was needed. MPI must also operate in a wide variety of environments, including ones where process creation is controlled by special process managers and schedulers.

In order to ensure scalability, process creation in MPI is collective, both over a group of processes that are creating new processes and over the group of processes created. The act of creating processes, or *spawning*, is accomplished with the routine `MPI_Comm_spawn`. This routine takes the name of the program to run, the command-line arguments for that program, the number of processes to create, the MPI communicator representing the group of processes that are spawning the new processes, a designated *root* (the rank of one process in the communicator that all members of that communicator agree to), and an `MPI_Info` object. The call returns a special kind of communicator, called an *intercommunicator*, that contains two groups of processes: the original group (from the input communicator) and the group of created processes. MPI point-to-point communication can then be used with this intercommunicator. The call also returns an array of error codes, one for each process.

Dynamic process creation is often used in master-worker programs, where the master process dynamically creates worker processes and then sends the workers tasks to perform. Such a program is sketched in Figure 1.9.

MPI also provides routines to spawn different programs on different processes with `MPI_Comm_spawn_multiple`. Special values used for the `MPI_Info` parameter allow one to specify special requirements about the processes, such as their working directory.

In some cases two parallel programs may need to connect to each other. A common example is a climate simulation, where separate programs perform the atmospheric and ocean modeling. However, these programs need to share data at the ocean-atmosphere boundary. MPI allows programs to connect to one another by using the routines `MPI_Comm_connect` and `MPI_Comm_accept`. See [32, Chapter 7] for more information.

```
MPI_Comm workerIntercomm;
int errcodes[10];
...
MPI_Init( &argc, &argv );
...
MPI_Comm_spawn( "./worker", MPI_ARGV_NULL, 10,
                MPI_INFO_NULL, 0, MPI_COMM_SELF,
&workerIntercomm, errcodes );

for (i=0; i<10; i++) {
    MPI_Send( &task, 1, MPI_INT, i, 0, workerIntercomm );
    ...
}
```

Fig. 1.9. Sketch of an MPI master program that creates 10 worker processes and sends them each a task, specified by a single integer.

One-Sided Communication

The message-passing programming model relies on the sender and receiver cooperating in moving data from one process to another. This model has many strengths but can be awkward, particularly when it is difficult to coordinate the sender and receiver. A different programming model relies on one-sided operations, where one process specifies both the source and the destination of the data moved between processes. Experience with BSP [35] and the Cray SHMEM [14] demonstrated the value of one-sided communication. The challenge for the MPI Forum was to design an interface for one-sided communication that retained the "look and feel" of MPI and could deliver good and reliable performance on a wide variety of platforms, including very fast computers without cache-coherent memory. The result was a compromise, but one that has been used effectively on one of the fastest machines in the world, the Earth Simulator.

In one-sided communication, a process may either *put* data into another process or *get* data from another process. The process performing the operation is called the *origin* process; the other process is the *target* process. The data movement happens without *explicit* cooperation between the origin and target processes. The origin process specifies both the source and destination of the data. A third operation, *accumulate*, allows the origin process to perform some basic operations, such as sum, with data at the target process. The one-sided model is sometimes called a put-get programming model.

Figure 1.10 sketches the use of `MPI_Put` for updating "ghost points" used in a one-dimensional finite difference grid. This has three parts:

1. One-sided operations may target only memory that has been marked as available for use by a particular memory window. The memory window is the one-sided analogue to the MPI communicator and ensures that only memory that the target process specifies may be updated by another process using MPI one-sided operations. The definition is made with the `MPI_Win_create` routine.

```
#     define ARRAYSIZE
      double x[ARRAYSIZE+2];
      MPI_Win win;
      int rank, size, leftNeighbor, rightNeighbor;

      MPI_Init( &argc, &argv );
      ...
      /* compute the neighbors. MPI_PROC_NULL means
         "no neighbor" */
      leftNeighbor = rightNeighbor = MPI_PROC_NULL;
      MPI_Comm_rank( MPI_COMM_WORLD, &rank );
      MPI_Comm_size( MPI_COMM_WORLD, &size );
      if (rank > 0) leftNeighbor = rank - 1;
      if (rank < size - 1) rightNeighbor = rank + 1;

      ...
      /* x[0] and x[ARRAYSIZE+1] are the ghost cells */
      MPI_Win_create( x, (ARRAYSIZE+2) * sizeof(double),
                      sizeof(double), MPI_INFO_NULL,
      MPI_COMM_WORLD, &win );
      MPI_Win_fence( 0, win );
      MPI_Put( &x[1], 1, MPI_DOUBLE,
               leftNeighbor, ARRAYSIZE+1, 1, MPI_DOUBLE, win );
      MPI_Put( &x[ARRAYSIZE], 1, MPI_DOUBLE,
               rightNeighbor, 0, 1, MPI_DOUBLE, win );
      MPI_Win_fence( 0, win );

      ...
      MPI_Win_free( &win );
```

Fig. 1.10. Sketch of a program that uses MPI one-sided operations to communicate ghost cell data to neighboring processes.

2. Data is moved by using the `MPI_Put` routine. The arguments to this routine are the data to put from the origin process (three arguments: address, count, and datatype), the rank of the target process, the destination of the data relative to the target window (three arguments: offset, count, and datatype), and the memory window object. Note that the destination is specified as an offset into the memory that the target process specified by using `MPI_Win_create`, not a memory address. This provides better modularity as well as working with heterogeneous collections of systems.
3. Because only the origin processes call `MPI_Put`, the target process needs some way to know when the data is available. This is accomplished with the `MPI_Win_fence` routine, which is collective over all the processes that created the memory window (in this example, all processes). In fact, in MPI the put, get, and accumulate calls are all nonblocking (for maximum performance), and

the `MPI_Win_fence` call ensures that these calls have completed at the origin processes.

While the MPI one-sided model is similar to other one-sided models, it has important differences. In particular, some models assume that the addresses of variables (particularly arrays) are the same on all processes. This assumption simplifies many features of the implementation and is true for many applications. MPI, however, does not assume that all programs are the same or that all runtime images are the same (e.g., running on heterogeneous platforms, which could be all IA32 processors but with different installed runtime libraries for C or Fortran). Thus, the address of `MyArray` in the program on one processor may not be the same as the address of the variable with the same name on another processor (some programming models, such as Co-Array Fortran, do make and require this assumption; see Section 1.5.2).

While we have touched on the issue of synchronization, this is a deep subject and is reflected in the MPI standard. Reading the standard can create the impression that the MPI model is very complex, and in some ways this is correct. However, the complexity is designed to allow implementors the greatest flexibility while delivering precisely defined behavior. A few simple rules will guarantee the kind of behavior that many users expect and use. The full rules are necessary only when trying to squeeze the last bits of performance from certain kinds of computing platforms, particularly machines without fully cache-coherent memory systems, such as certain vector machines that are among the world's fastest. In fact, rules of similar complexity apply to shared-memory programming and are related to the pragmatic issues of memory consistency and tradeoffs between performance and simplicity.

Other Features in MPI-2

Among the most important other features in MPI-2 are bindings for C++ and Fortran 90. The C++ binding provides a low-level interface that exploits the natural objects in MPI. The Fortran 90 binding includes an MPI module, providing some argument checking for Fortran programs. Other features include routines to specify levels of thread safety and to support tools that must work with MPI programs. More information may be found in [29].

1.2.4 State of the Art

MPI is now over twelve years old. Implementations of MPI-1 are widespread and mature; many tools and applications use MPI on machines ranging from laptops to the world's largest and fastest computers. See [55] for a sampling of papers on MPI applications and implementations. Improvements continue to be made in the areas of performance, robustness, and new hardware. In addition, the parallel I/O part of MPI-2 is widely available.

Shortly after the MPI-2 standard was released, Fujitsu had an implementation of all of MPI-2 except for `MPI_Comm_join` and a few special cases of the routine `MPI_Comm_spawn`. Other implementations, free or commercially supported, are now available for a wide variety of systems.

The MPI one-sided operations are less mature. Many implementations now support at least the "active target" model (these correspond to the BSP or put-get followed by barrier). In some cases, while the implementation of these operations is correct, the performance may not be as good as MPI's point-to-point operations. Other implementations have achieved good results, even on clusters with no special hardware to support one-sided operations [75]. Recent work exploiting the abilities of emerging network standards such as Infiniband shows how the MPI one-sided operations can provide excellent performance [42].

1.2.5 Summary

MPI provides a mature, capable, and efficient programming model for parallel computation. A large number of applications, libraries, and tools are available that make use of MPI. MPI applications can be developed on a laptop or desktop, tested on an ad hoc cluster of workstations or PCs, and then run in production on the world's largest parallel computers. Because MPI was designed to support "programming in the large," many libraries written with MPI are available, simplifying the task of building many parallel programs. MPI is also general and flexible; any parallel algorithm can be expressed in MPI. These and other reasons for the success of MPI are discussed in more detail in [28].

1.3 Shared-Memory Programming with OpenMP

Shared-memory programming on multiprocessor systems has been around for a long time. The typical generic architectural schematic for a shared-memory system or an individual SMP node in a distributed-memory system is shown in Figure 1.3. The memory of the system is directly accessible by all processors, but that access may be coupled by different bandwidth and latency mechanisms. The latter situation is often refered to as non-uniform memory access (NUMA). For optimal performance, parallel algorithms must take this into account.

The vendor community offers a huge number of shared-memory-based hardware systems, ranging from dual-processor systems to very large (e.g., 512-processor) systems. Many clusters are built from these shared-memory nodes, with two or four processors being common and a few now using 8-way systems. The relatively new AMD Opteron systems will be generally available in 8-way configurations within the coming year. More integrated parallel supercomputer systems such as the IBM SP have 16- or 32-way nodes.

Programming in shared memory can be done in a number of ways, some based on threads, others on processes. The main difference, by default, is that threads share the same process construct and memory, whereas multiple processes do not share memory. Message passing is a multiple process based programming model. Overall, thread-based models have some advantages. Creating an additional thread of execution is usually faster than creating another process, and synchronization and context switches among threads are faster than among processes. Shared-memory programming is in general incremental; a given section of code can be parallelized without modifying external data storage or data access mechanisms.

Many vendors have their own shared-memory programming models. Most offer System V interprocess communication (IPC) mechanisms, which include shared-memory segments and semaphores [77]. System V IPC usually shares memory segments among different processes. The Posix standard [41, 57] offers a specific threads model called Pthreads. It has a generic interface that makes it more suitable for systems-level programming than for high-performance computing applications. Only one compiler (as far as we know) supports the Fortran Pthreads standard; C/C++ support is commonplace in Unix; and there is a one-to-one mapping of the Pthreads API to the Windows threads API as well, so the latter is a common shared-memory programming model available to the development community. Java threads also provides a mechanism for shared-memory concurrent programming [40].

Many other thread-based programming libraries are available from the research community as well, for example, TreadMarks [44]. These libraries are supported across a wide variety of platforms principally by the library development teams. OpenMP, on the other hand, is a shared-memory, thread-based programming model or API supported by the vendor community. Most commercial compilers available for Linux provide OpenMP support.

Overall, thread-based models have some advantages. Creating an additional thread of execution is usually faster than creating another process. Synchronization and context switches among threads are faster than among processes.

In the remainder of this section, we focus on the OpenMP programing model.

1.3.1 OpenMP History

OpenMP [12, 15] was organized in 1997 by the OpenMP Architecture Review Board (ARB), which owns the copyright on the specifications and manages the standard development. The ARB is composed primarily of representatives from the vendor community; membership is open to corporate, research, or academic institutions, not to individuals [65]. The goal of the original effort was to provide a shared-memory programming standard that combined the best practices of the vendor community offerings and some specifications that were a part of previous standardization efforts of the Parallel Computing Forum [48, 26] and the ANSI X3H5 [25] committee.

The ARB keeps the standard relevant by expanding the standard to meet needs and requirements of the user and development communities. The ARB also works to increase the impact of OpenMP and interprets the standard for the community as questions arise. The currently available version 2 standards for C/C++ [64] and Fortran [63] can be downloaded from the OpenMP ARB Web site [65]. The ARB has combined these standards into one working specification (version 2.5) for all languages, clarifying previous inconsistencies and strengthening the overall standard. The merged draft was released in November, 2004.

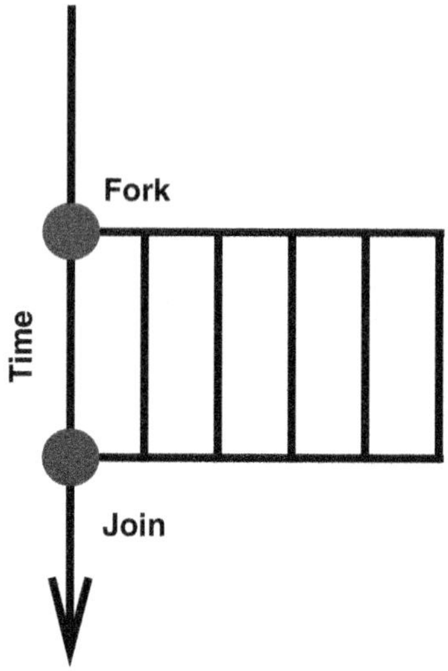

Fig. 1.11. Fork-and-join model of executing threads.

1.3.2 The OpenMP Model

OpenMP uses an execution model of fork and join (see Figure 1.11) in which the "master" thread executes sequentially until it reaches instructions that essentially ask the runtime system for additional threads to do concurrent work. Once the concurrent scope of execution has completed, these extra threads simply go away, and the master thread continues execution serially. The details of the underlying threads of execution are compiler dependent and system dependent. In fact, some OpenMP implementations are developed on top of Pthreads. OpenMP uses a set of compiler directives, environment variables, and library functions to construct parallel algorithms within an application code. OpenMP is relatively easy to use and affords the ability to do incremental parallelism within an existing software package.

OpenMP uses a variety of mechanisms to construct parallel algorithms within an application code. These are a set of compiler directives, environment variables, and library functions. OpenMP is essentially an implicit parallelization method that works with standard C/C++ or Fortran. Various mechanisms are available for dividing work among executing threads, ranging from automatic parallelism provided by some compiler infrastructures to the ability to explicitly schedule work based on the thread ID of the executing threads. Library calls provide mechanisms to determine the thread ID and number of participating threads in the current scope of execution. There are also mechanisms to execute code on a single thread atomically in order to protect execution of critical sections of code. The final application becomes a series of sequential and parallel regions, for instance connected segments of the single serial-parallel-serial segment as shown in Figure 1.12.

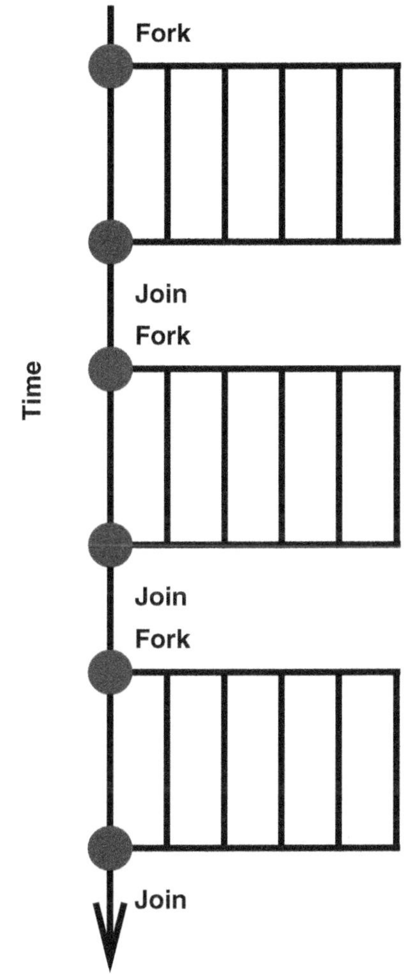

Fig. 1.12. An OpenMP application using the fork-and-join model of executing threads has multiple concurrent teams of threads.

Using OpenMP in essence involves three basic parallel constructs:

1. Expression of the algorithmic parallelism or controlling the flow of the code
2. Constructs for sharing data among threads or the specific communication mechanism involved
3. Synchronization constructs for coordinating the interactions among threads

These three basic constructs, in their functional scope, are similar to those used in MPI or any other parallel programming model.

OpenMP directives are used to define blocks of code that can be executed in parallel. The blocks of code are defined by the formal block structure in C/C++ and

C code

```
#include <stdio.h>
#include <omp.h>
int main(int argc, char *argv[])
{
  int tid;
#pragma omp parallel private(tid)
  {
    tid = omp_get_thread_num();
    printf("<%d>\n",tid);
  }
}
```

Fortran Code

```
      program hello
      implicit none
      integer tid
      integer omp_get_thread_num
      external omp_get_thread_num
!$omp parallel private(tid)
      tid = omp_get_thread_num()
      write(6,'(1x,a1,i4,a1)')
     &   '<',tid,'>'
!$omp end parallel
      end
```

Fig. 1.13. "Hello World" OpenMP code.

by comments in Fortran; both the beginning and end of the block of code must be identified. There are three kinds of OpenMP directives: parallel constructs, work-sharing constructs within a parallel construct, and combined parallel-work-sharing constructs.

Communication is done entirely in the shared-memory space of the process containing threads. Each thread has a unique stack pointer and program counter to control execution in that thread. By default, all variables are shared among threads in the scope of the process containing the threads. Variables in each thread are either shared or private. Special variables, such as reduction variables, have both a shared scope and a private scope that changes at the boundaries of a parallel region. Synchronization constructs include mutual exclusions that control access to shared variables or specific functionality (e.g., regions of code). There are also explicit and implied barriers, the latter being one of the subtleties of OpenMP. In parallel algorithms, there must be a communication of critical information among the concurrent execution entities (threads or processes). In OpenMP, nearly all of this communication is handled by the compiler. For example, a parallel algorithm has to know the number of entities participating in the concurrent execution and how to identify the appropriate portion of the entire computation for each entity. This maps directly to a process-count- and process-identifier-based algorithm in MPI.

A simple example is in order to whet the appetite for the details to come. In the code segments in Figure 1.13 we have a "Hello World"-like program that uses OpenMP. This generic program uses a simple parallel region that designates the block of code to be executed by all threads. The C code uses the language standard braces to identify the block; the Fortran code uses comments to identify the beginning and end of the parallel region. In both codes the OpenMP library function `omp_get_thread_num` returns the thread number, or ID, of the calling thread; the result is an integer value ranging from 0 to the number of threads minus 1. Note that type information for the OpenMP library function function does not follow the default variable type scoping in Fortran. To run this program, one would execute the binary like any other binary. To control the number of threads used, one would set the environment variable `OMP_NUM_THREADS` to the desired value. What output should be expected from this code? Table 1.2 shows the results of five runs with the number

Table 1.2. Multiple runs of the OpenMP "Hello World" program. Each column represents the output of a single run of the application on 3 threads.

Run 1	Run 2	Run 3	Run 4	Run 5
<0>	<2>	<1>	<0>	<0>
<1>	<1>	<0>	<1>	<1>
<2>	<0>	<2>	<2>	<2>

of threads set to 3. The output from this simple example illustrates an important point about thread-based parallel programs, in OpenMP or any other thread model: There is no control over which thread executes first within the context of a parallel region. This decision is determined by the runtime system. Any expectation or required ordering of the execution of threads must be explicitly coded. The simple concurrency afforded by OpenMP requires that each task, such as a single iteration of a loop, be an independent execution construct.

One of the advantages of OpenMP is *incremental parallelization*—the ability to parallelize loops at a time or even small segments of code at a time. By iteratively identifying the most time-consuming components of an application and then parallelizing those components, one eventually gets a fully parallelized application. Any programming model requires a significant amount of testing and code restructuring to get optimal performance.[8] Although the mechanisms of OpenMP are straightforward and easier than other parallel programming models, the cycle of restructuring and testing is still important. The programmer may introduce a bug by incorrectly parallelizing a code and introducing a dependency that goes undetected because the code was not then thoroughly tested. One should remember that the OpenMP user has no control on the order of thread execution; a few tests may detect a dependency—or may not. In other words the tests you run may just get "lucky" and give the correct results. We discuss dependency analysis further in Section 1.3.4.

1.3.3 OpenMP Directives

The mechanics of parallelization with OpenMP are relatively straightforward. The first step is to insert compiler directives into the source code identifying the code segments or loops to be parallelized. Table 1.3 shows the sentinel syntax of a general directive for OpenMP in the supported languages [64, 63]. The easiest way to learn how to develop OpenMP applications is through examples. We start with a simple algorithm, computing the norm of the difference of two vectors. This is a common way to compare vectors or matrices that are supposed to be the same. The serial code fragment in C and Fortran is shown in Figure 1.14. This simple example exposes some of the concepts needed to appropriately parallelize a loop with OpenMP. By thinking about executing each iteration of the loop independently, we can see several

[8]Some parallel software developers call parallelizing a code re-bugging a code, and this is often an apropos statement.

Table 1.3. General sentinel syntax of OpenMP directives.

Language	Syntax
Fortran 77	*$omp directive [options] C$omp directive [options] !$omp directive [options]
Fortran 90/95	!$omp directive [options]
Continuation Syntax	!$omp directive [options] !$omp+ directive [options]
C or C++	#pragma omp directive [options]
Continuation Syntax	#pragma omp directive [options] \\ directive [options]

```
C code fragment

norm = (double) 0.0;
for(i=0;i<len;i++) {
  diff = z[i]-zp[i];
  norm += diff*diff;
}
```

```
Fortran code fragment

norm = 0.0d00
do i = 1,len
   diff = z(i) - zp(i)
   norm = norm + diff*diff
enddo
```

Fig. 1.14. "Norm of vector difference" serial code.

issues with respect to reading from and writing to memory locations. First, we have to understand that each iteration of the loop essentially needs a separate `diff` memory location. Since `diff` for *each* iteration is *unique* and different iterations are being executed concurrently on multiple threads, `diff` cannot be shared. Second, with all threads writing to `norm`, we have to ensure that all values are appropriately added to the memory location. This process can be handled in two ways: We can protect the summation into `norm` by a critical section (an atomic operation), or we can use a reduction clause to sum a thread local version of `norm` into the final value of `norm` in the master thread. Third, all threads of execution have to read the values of the vectors involved and the length of the vectors.

Now that we understand the "data" movement in the loop, we can apply directives to make the movement appropriate. Figure 1.15 contains the parallelized code using OpenMP with a critical section. We have identified `i` as private so that only one thread will execute a given value of `i`; each iteration is executed only once. Also private is `diff` because each thread of execution must have a specific memory location to store the difference; if `diff` were not private, the overlapped execution of multiple threads would not guarantee the appropriate value when it is read in the norm summation step. The "atomic" directive allows only one thread at a time to

C code fragment

```
norm = (double) 0.0;
#pragma omp parallel for private(i,diff) shared(len,z,zp,norm)
 for(i=0;i<len;i++) {
   diff = z[i]-zp[i];
#pragma omp atomic
   norm += diff*diff;
}
```

Fortran code fragment

```
      norm = 0.0d00
!$OMP PARALLEL DO PRIVATE(i,diff) SHARED(len,z,zp,norm)
      do i = 1,len
        diff = z(i) - zp(i)
!$OMP ATOMIC
        norm = norm + diff*diff
      enddo
!$OMP END PARALLEL DO
```

Fig. 1.15. "Norm of vector difference" OpenMP code with a critical section.

do the summation of norm, thereby ensuring that the correct values are summed into the shared variable. This is important because summation involves the data load, register operations, and data store. If this were not protected, multiple threads could overlap these operations. For example, thread 1 could load a value of norm, thread 2 could store an updated value of norm, and then thread 1 would have the wrong value of norm for the summation.

Since all the threads have to execute the norm summation line atomically, there clearly will be contention for access to update the value of norm. This overhead, waiting in line to update the value, will severely limit the overall performance and scalability of the parallel loop.[9] A better approach would be to have each thread sum into a private variable and then use the partial sums in each thread to compute the total norm value. This is what is done with a reduction clause. The variable in a reduction clause is private during the execution of the concurrent threads, and the value in each thread is reduced over the given operation and returned to the master thread just as a shared variable operates. This dual nature provides a mechanism to parallelize the algorithm without the need for the atomic operation as in Figure 1.16. This eliminates the thread contention of the atomic operation.

The reduction mechanism is a useful technique, and another example of the use of the reduction clause is in order. In developing parallel algorithms, one often measures their performance by timing the event in each execution entity, either in each

[9] In fact, this simple example will not scale well regardless of the OpenMP mechanism used because the amount of work in each thread compared to the overhead of the parallelization is small.

thread or in each process. Knowing the minimum, maximum, and average time of concurrent tasks will give some indication of the level of load balance in the algorithm. If the minimum, maximum, and average times are all about the same, then the algorithm has good load-balance. If the minimum, maximum, or both are far away from the average then there is a load imbalance that has to be mitigated. This can be accomplished by some sort of regrouping of elements of each task or via some dynamic mechanism. As a specific example, we will show code fragment for a sparse matrix vector multiplication in Figure 1.17. The sparse matrix is stored in the compressed-row-storage (CRS) format, a standard format that many sparse codes use in their algorithms. See [68] for details of various sparse matrix formats.

To parallelize this loop using OpenMP, we have to determine the data flow in the algorithm. We will parallelize this code over the outer loop, `i`. Each iteration of that loop will be executed only once across all threads in the team. Each iteration is independent, so writing to `yvec(i)` is independent in each iteration. Therefore, we do not have to protect that write with an atomic directive as we did in the "norm" computation example. Hence, `yvec` needs to be shared because each thread will write to some part of the vector. The temporary summation variable `t` and the inner do loop variable `k` are different for each iteration of `i`. Thus, they must be private; that is, each thread must have a separate memory location. All other variables are only being read, so these variables are shared because all threads have to know all the values.

Figure 1.18 shows the parallelized code fragment. Timing mechanisms are inserted for the do loop and the reduction clause is inserted for each of the reduction variables, `timemin`, `timemax`, and `timeave`. The OpenMP library function `omp_get_wtime()` returns a double-precision clock tick based on some implementation dependent epoch. The library function `omp_get_num_threads()` returns the total number of threads in the team of the parallel region. The defaults are used for scheduling the iterations of the `i` loop across the threads. In other words, approximately $n/numthread$ iterations are assigned to each thread in the team. Thread 0 will have iterations `i` = 1, 2, ...,$n/numthread$, thread 1 will have `i` = $n/numthread$ + 1, ..., 2*$n/numthread$, and so on. Any remainder in $n/numthread$ is assigned to the team of threads via a mechanism determined by the OpenMP implementation.

Our example of a parallelized sparse matrix multiply where we determine the minimum, maximum, and average times of execution could show some measure of load-imbalance. Each row of the sparse matrix has a different number of elements. If the sparse matrix has a dense block banding a portion of the diagonal and mostly diagonal elements elsewhere there will be a larger "load" on the thread that computes the components from the dense block. Figure 1.19 shows the representation of such a matrix and how it would be split by using the default OpenMP scheduling mechanisms with three threads. With the "static" distribution of work among the team of three threads, a severe load imbalance will result. This problem can be mitigated in several ways. One way would be to apply a chunk size in the static distribution of work equal to the size of the dense block divided by the number of threads. This

```
                                                C code fragment

    norm = (double) 0.0;
#pragma omp parallel for private(i,diff) \
                     shared(len,z,zp,norm) reduction(+:norm)
    for(i=0;i<len;i++) {
       diff = z[i]-zp[i];
       norm += diff*diff;
    }
```

```
                                          Fortran code fragment

      norm = 0.0d00
!$OMP PARALLEL DO PRIVATE(i,diff) SHARED(len,z,zp,norm)
!$OMP+            REDUCTION(+:norm)
      do i = 1,len
         diff = z(i) - zp(i)
         norm = norm + diff*diff
      enddo
!$OMP END PARALLEL DO
```

Fig. 1.16. "Norm of vector difference" OpenMP code with a reduction.

```
!      compute yvec = Amat*xvec
!      Amat sparse matrix stored in CRS format
!           Flat linear storage of elements of A
!      row_ptr() points to the start and of each row of A
!                in the flat linear storage of A.  The last
!                element has the number of non-zero elements
!                of A + 1. Therefore each row has
!                row_ptr(i+1)-row_ptr(i)-1 elements
!      col_ind() provides the column index for each
!                element of A
!
      do i = 1,n
!
!     compute the inner product of row i with vector xvec
!
         t = 0.0d0
         do k=row_ptr(i), row_ptr(i+1)-1
            t = t + amat(k)*xvec(col_ind(k))
         enddo
!
!     store result in yvec(i)
!
         yvec(i) = t
      enddo
```

Fig. 1.17. Sequential sparse matrix multiply code fragment, in Fortran.

```
!       compute yvec = Amat*xvec
...
!$OMP PARALLEL REGION PRIVATE(i,t,k,timestart,timeend,numthread)
!$OMP+                SHARED(n,row_ptr,amat,xvec,col_ind,yvec)
!$OMP+                REDUCTION(MIN:timemin) REDUCTION(MAX:timemax)
!$OMP+                REDUCTION(+:timeave)
      timestart = omp_get_wtime()
!$OMP PARALLEL DO
      do i = 1,n
         t = 0.0d0  ! inner product of row i with vector xvec
         do k=row_ptr(i), row_ptr(i+1)-1
            t = t + amat(k)*xvec(col_ind(k))
         enddo
         yvec(i) = t !     store result in yvec(i)
      enddo
!$OMP END PARALLEL DO
      timeend = omp_get_wtime()
      numthread = omp_get_num_threads()
      timemin = timeend-timestart
      timemax = timeend-timestart
      timeave = (timeend-timestart)/numthread
!$OMP END PARALLEL REGION
```

Fig. 1.18. Parallel sparse matrix multiply code fragment, in Fortran, that times the operation and reduces the minimum, maximum, and average times.

would lead to the distribution of work shown in Figure 1.20. This can be accomplished by modifying the `PARALLEL DO` directive of Figure 1.18 to

```
!$OMP PARALLEL DO SCHEDULE(STATIC,(SIZE_OF_DENSE_BLOCK/numthreads))
```

where `SIZE_OF_DENSE_BLOCK` must be determined before the do loop construct in the parallel region. Determining this value is added overhead on the parallelization of the serial code.

At times, more explicit control may be necessary. The same kind of explicit control necessary in the equivalent message-passing implementation. The algorithm can be scheduled explicitly with similar constructs such as the number of threads and the thread identifier. This is another advantage of OpenMP; in addition to incremental parallelization, a programmer can take as much explicit control as is necessary for a given algorithm.

Figure 1.21 shows an explicit parallelization of the sparse matrix multiply. The OpenMP library function `omp_get_thread_num()` returns the thread identifier in the range from 0 ... the number of threads minus 1. Each thread starts with the iteration that matches a thread identifier, and the "parallel" loop now increments by the number of threads. There is no longer a need for the `PARALLEL DO` directive because of the explicit control! This interleaves each iteration in order to a different thread, so the issues of load balance are minimized.

The C/C++ version of the example in Figure 1.21 would be more complicated because the reduction clause operators available in C/C++ do not include `MIN` or `MAX` functionality. No intrinsic functions are available for use in the reduction clause.

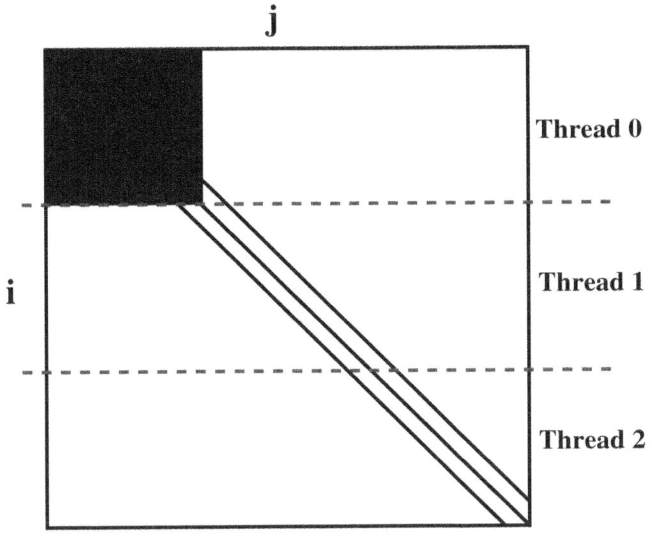

Fig. 1.19. A sparse matrix that is dense in one area. Using our sparse matrix vector algorithm on three threads, we would access the matrix as shown.

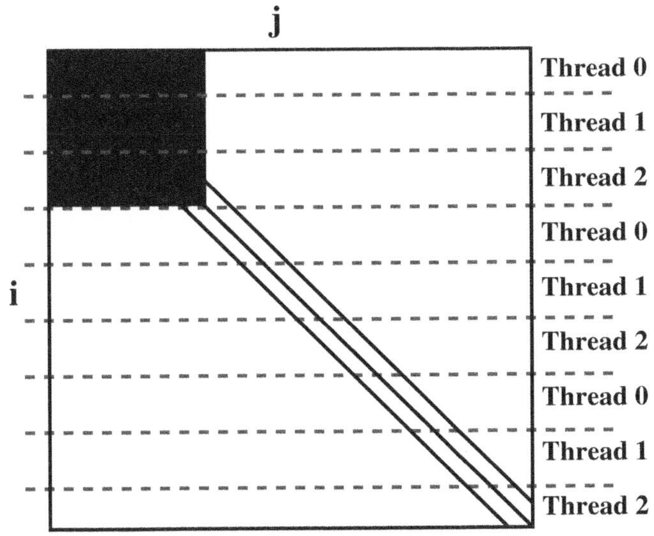

Fig. 1.20. A sparse matrix that is dense in one area. Using our sparse matrix vector algorithm with the appropriate chunk size on three threads we would access the matrix as shown. This is more load-balanced than the default distribution of iterations to the team of threads.

```
!       compute yvec = Amat*xvec
...
!$OMP PARALLEL REGION PRIVATE(i,t,k,timestart,timeend,numthread,tid)
!$OMP+                SHARED(n,row_ptr,amat,xvec,col_ind,yvec)
!$OMP+                REDUCTION(MIN:timemin) REDUCTION(MAX:timemax)
!$OMP+                REDUCTION(+:timeave)
      timestart = omp_get_wtime()
      tid = omp_get_thread_num() ! get the thread identifier
      numthread = omp_get_num_threads()

      do i = (tid+1),n,numthread
         t = 0.0d0    ! inner product of row i with vector xvec
         do k=row_ptr(i), row_ptr(i+1)-1
            t = t + amat(k)*xvec(col_ind(k))
         enddo
         yvec(i) = t    !     store result in yvec(i)
      enddo
      timeend = omp_get_wtime()
      timemin = timeend-timestart
      timemax = timeend-timestart
      timeave = (timeend-timestart)/numthread
!$OMP END PARALLEL REGION
```

Fig. 1.21. Parallel sparse matrix-multiply code fragment, in Fortran, that times the operation and reduces the minimum, maximum, and average times. The concurrency is explicitly controlled with the thread identifier and the number of threads.

1.3.4 Data Dependencies and False Sharing

In parallelizing algorithms, one has to ensure that every memory write operation is essentially independent of other memory operations from other threads in the team. If the programmer writes to a location in one thread and reads that same location in another thread of execution, a dependency exists. Since OpenMP provides no control over which thread executes, the programmer must deal with this dependency either by scoping the appropriate variables (private or shared) or introducing synchronization mechanisms to ensure that the dependency is met. Mitigating these data race conditions or dependencies is at the heart of shared-memory parallel programming, since data communication is through shared variables. Chandra et al. have a good, somewhat formal, discussion of the process of identifying and removing these dependencies [9].

The mechanisms for dealing with these data dependencies often require some restructuring of code. For example, it may be necessary to split a loop that computes multiple quantities. The "fissioned" loops can be run in parallel but the original construct cannot. In other situations new intermediate quantities may need to be introduced. These will add additional memory requirements and the overhead of generating those intermediates.

Code restructuring will certainly involve tradeoffs that may affect performance and thus force a specific way of parallelizing the algorithm. One such performance issue is that, although there is no formal data dependency, there is a performance degradation because of the nature of the memory locations being accessed by the threads in the team. If independent threads are writing to memory locations in the same cache line, there is no true data dependency because each thread is writing to separate memory locations. Unfortunately, since these locations are in the same

```
!      compute yvec = Amat*xvec
...
      blocksize = 5  ! the number of iterations each thread gets
      numblocks = n/blocksize  ! number of blocks of iterations
! a remainder means extra block
      if (mod(n,blocksize).ne.0) numblocks=numblocks+1

!$OMP PARALLEL REGION PRIVATE(ii,i,ilo,ihi,t,k)
!$OMP+                PRIVATE(timestart,timeend,numthread,tid)
!$OMP+                SHARED(n,row_ptr,amat,xvec,col_ind,yvec)
!$OMP+                SHARED(blocksize,numblocks)
!$OMP+                REDUCTION(MIN:timemin) REDUCTION(MAX:timemax)
!$OMP+                REDUCTION(+:timeave)
      timestart = omp_get_wtime()
      tid = omp_get_thread_num() ! get the thread identifier
      numthread = omp_get_num_threads()

      do ii = (tid+1),numblocks,numthread
         ilo = (ii-1)*blocksize + 1  ! start of each block
         ihi = min((ilo+blocksize-1),n)
         do i = ilo,ihi
            t = 0.0d0    ! inner product of row i with vector xvec
            do k=row_ptr(i), row_ptr(i+1)-1
               t = t + amat(k)*xvec(col_ind(k))
            enddo
            yvec(i) = t    !     store result in yvec(i)
         enddo
      enddo
      timeend = omp_get_wtime()
      timemin = timeend-timestart
      timemax = timeend-timestart
      timeave = (timeend-timestart)/numthread
!$OMP END PARALLEL REGION
```

Fig. 1.22. Parallel sparse matrix multiply code fragment, in Fortran, that times the operation and reduces the minimum, maximum, and average times. The concurrency is explicitly controlled with the thread identifier and the number of threads and appropriate blocking of the parallelized iterations to avoid false sharing.

cache line, performance is degraded because each write forces the data to be flushed from the other processor cache. This cache thrashing is called "false sharing."

Can "false sharing" really impact the performance of a parallel algorithm? Yes. In fact, the algorithm presented in Figure 1.21 will suffer from false sharing. The write to `yvec(i)` in the first iteration of each thread all have elements contiguous in memory; e.g., thread 0 and thread 1 will interact via the cache. As the algorithm proceeds the effect may decrease because of the varying size of the number of elements in each row of the matrix; each iteration will take a different time to execute. One way to mitigate this is to block the access to the iterations and thus the writes to `yvec(i)`. The block size simply has to be large enough to ensure that the writes to `yvec(i)` in each thread will not be in the same cache line. Figure 1.22 shows the blocked algorithm that will avoid false sharing using explicit control of the concurrency among the team of threads. This explicit blocking could be accomplished by modifying the `PARALLEL DO` directive of Figure 1.18 to

```
!$OMP PARALLEL DO SCHEDULE(STATIC,5)
```

Other mechanisms can be used to modify the way iterations are scheduled. They are explored in more detail in references [9, 63, 64].

```
        do j = 1, CCOLS
           do k = 1, BROWS
              Btmp = B(k,j)
              do i = 1, CROWS
                  C(i,j) = C(i,j) + A(i,k)*Btmp
              enddo
           enddo
        enddo
```

Fig. 1.23. Partially cache-optimized matrix-multiply: serial code.

```
!$OMP PARALLEL DO PRIVATE(i,j,k,Btmp)
        do j = 1, CCOLS
           do k = 1, BROWS
              Btmp = B(k,j)
              do i = 1, CROWS
                  C(i,j) = C(i,j) + A(i,k)*Btmp
              enddo
           enddo
        enddo
!$OMP END PARALLEL DO
```

Fig. 1.24. Partially cache-optimized matrix-multiply: parallel code.

As another example for analysis and parallelization, we examine the simple cache-optimized matrix multiply in Figure 1.23. Our examination of this code suggests that we should maximize the work in each thread with respect to the overhead of the OpenMP parallelization constructs. In particular, we should parallelize the outermost loop. A glance at the memory locations with write operations indicates that only `C(i,j)`, `Btmp`, `i`, `j`, and `k` are relevant. For effective parallelization, the loop index variables must be different for each thread, thus accessing only appropriate parts of the matrix. `A` and `B` have only read operations. Since all threads need to know the dimensions of the matrices, `CCOLS`, `CROWS`, and `BROWS` need to be shared among team members. Since we are parallelizing over the `j` loop, each thread has a unique set of `j` values; and since `Btmp` is a function of `j`, each thread should have a unique `Btmp` (i.e., `Btmp` should be private to each thread).

This loop structure can be parallelized in many ways. The most straightforward is to use the combined parallel work sharing DO constructs. The parallel code based on our analysis is shown in Figure 1.24. The data in Table 1.4 shows the performance on a four-processor SMP system. The scalability indicates some overhead. On four threads the efficiency ranges from 97.1% to 95.4% with increasing matrix sizes. The performance could be improved by further optimizing the cache with a blocking algorithm.

Table 1.4. Timings in seconds for multiple runs of the OpenMP parallelized matrix multiply code.

	Matrix Rank			
number of threads	500	1000	1500	2000
4	0.43	3.56	12.05	28.37
3	0.57	4.65	15.67	37.05
2	0.84	6.92	23.28	55.12
1	1.67	13.57	45.87	108.27

1.3.5 Future of OpenMP

We have described in this section a robust programming model for the development of applications using OpenMP on shared-memory systems. There are many ways to tackle a parallel algorithm, from the application of simple directives to essentially full control basing the execution on the thread identifiers available. At this point we have described both message passing with MPI and thread programming with OpenMP. Some applications use both, with mixed results [16, 43, 56]. Hybrid MPI/OpenMP applications are emerging in part due to the nature of how clusters are evolving with larger processor counts per node. Hybrid MPI/OpenMP software development presents several challenges. The programmer interested in this hybrid model should get a sound understanding of both programming models separately and then begin to merge them. The programmer interested in this should carefully understand the MPI-2 scope of thread policy set up in the initialization phase of MPI-2 codes. The real trick in merging these two programming models is getting the code to work in four different modes: serially, with just OpenMP, with just MPI, and with both MPI and OpenMP [38]. The hybrid code in any of these modes should generate correct results regardless of how many threads are used at the thread level or how data is distributed among multiple processes. Current hybrid applications have been developed with a subset of these four modes due to the complexity of the resultant application. Primarily MPI communications are done only in the master thread of execution. Hybrid applications is an advanced topic in programming models and more research is in progress addressing the issues involved.

Cluster-Based OpenMP

At a recent workshop, Intel described a new offering, Cluster OpenMP [37], that is in beta testing. The idea is to provide a runtime infrastructure that allows the OpenMP programming model to run on clusters. Intel's offering can serve as a reference implementation for this idea, but it is limited to Itanium clusters at the moment. Intel has added directives and library functions to make clear distinctions between private, shared, and "sharable" data (data that is among processes, i.e., on another cluster node). Cluster-based OpenMP is a current topic in the research community and should be monitored as the research efforts demonstrate the effectiveness [76, 39, 51].

The ultimate goal of these efforts is to have the runtime environment provide good performance on clusters for OpenMP; comparable performance to hybrid MPI and OpenMP is required. The programming syntax of the value-added standard would allow incremental parallelism that is often difficult with MPI code development. Many issues must be considered in this environment. Remote process invocation is an issue that will be of interest because the landscape of clusters and communication interconnects is vast.

Specifications 2.5 and 3.0

Currently the merged OpenMP 2.5 specification is completed and is available for public comment.[10] A major change in the OpenMP 2.5 specification is the merger of the Fortran and C/C++ specifications into a single document [5]. The ARB is also resolving inconsistencies in the specifications, expanding the glossary, improving the examples, and resolving some of the more difficult issues with respect to the flush semantics and the persistence of threadprivate data across multiple parallel regions.

The 3.0 specification is on hold until the 2.5 merger is done, but several topics are under discussion to expand the applicability of OpenMP. These include task parallelism to handle while loops and recursion, automatic scoping of variables, interaction with other thread models (e.g., POSIX threads), more control or definition of the memory model for NUMA-style shared-memory systems, and expanded schedule types and reusable schedules. As an example of the importance of the last issue, the guided schedule gives an exponential decay of the chunk size of iterations for a loop construct. The ability to control or change the decay rate is useful for improved performance of some algorithms. The 3.0 specification will also address many of the issues of nested parallelism that is in several implementations now. One major issue that needs to be considered is error reporting to the application. What happens if no more threads/resources are available? Currently, most implementations simply serialize the construct. A code developer may want to switch algorithms based on the runtime environment.

1.3.6 Availability of OpenMP

Most vendors provide OpenMP compilers, and several open source implementations are available. The OpenMP Web site [65] provides more information regarding their availability and function. There are also pointers for open-source implementations.

1.4 Distributed Shared-Memory Programming Models

Distributed shared-memory (DSM) programming models use a physically distributed memory architecture with some aspect of shared-memory technology. DSM models are not as popular as message-passing or direct shared-memory models but have many of the complications of both.

[10] See the `http://www.openmp.org` web site.

The goal for DSM technology is to facilitate the use of aggregate system memory, the most costly component of most high-end systems. Stated differently, most DSM programming models want to provide shared-memory-like programming models for distributed-memory systems. This aspect of shared memory can be effected in hardware or software. The hardware mechanisms are those with the highest performance and cost. Software mechanisms range from those that are transparent to the user to those coded explicitly by the user. Since obvious latencies exist in the software stacks of these implementations, performance still depends on the skill of the programmers using these technologies. DSM models are not as popular as message passing or direct shared-memory models but have many of the complications of both.

Software DSMs fall basically into three categories:

- Transparent operating system technology
- Language-supported infrastructure
- Variable/array/object-based libraries

DSMs that are transparent to the user often use a virtual-memory system with kernel modifications to allow for inter-node page accesses. This approach makes the programming straightforward in function, but getting good performance requires understanding the locality of the data and the way data movement happens. These systems include technologies such as ThreadMarks [44], InterWeave [10], Munin [7], and Cashmere [19].

Language-based infrastructure includes specialty languages such as High Performance Fortran (see section 1.4.1) and one that is now getting vendor support, Unified Parallel C (see Section 1.5.1).

Data-specific DSM libraries have been those most used by the high-performance computing community. They include the popular SHMEM programming model available on the Cray T3D and T3E systems [4]. These DSMs require that the programmer identify variables or objects that are shared, unlike OpenMP where everything is shared by default. Operations that separate shared and local variables require programmer control of the consistency appropriate for the algorithm. Data movement is neither automatic nor transparent; it must be coded explicitly or understood via implicit data movement from library interfaces.

1.4.1 High Performance Fortran

High Performance Fortran (HPF) is a distributed-memory version of Fortran 90 that, like OpenMP, relies on the use of directives to describe the features that support parallel programming. Because HPF uses directives, most HPF programs may be compiled by any Fortran 90 compiler and run on a single processor. HPF was developed by an informal group and published as a standard [34] in much the same way as MPI. In fact, the MPI Forum followed the same procedures used by the HPF Forum.

HPF is not as widely available as MPI and OpenMP but is still in use. A slight extension of HPF is in use on the Earth Simulator; an application using that version of HPF achieved a performance of 14.9 Teraflops and was awarded a Gordon Bell

```
      program matmult
      integer, parameter :: n=1000
      real a(n,n), b(n,n), c(n,n)
!HPF$ DISTRIBUTE(BLOCK,BLOCK)::C
!HPF$ ALIGN A(i,*) WITH C(i,*)
!HPF$ ALIGN B(*,j) WITH C(*,j)
!
      a = 1
      b = 2
      do i=1,n
          do j=1,n
              c(i,j) = dot_product(a(i,:),b(:,j))
          enddo
      enddo
      write (*,*) c
      end
```

Fig. 1.25. Simple HPF matrix multiply program.

prize in 2002 [69]. In this section, we will touch on a few of the features of HPF and give one example. More information and some examples may be found in [17]; the full HPF standard is also available [47].

One of the most important steps in implementing a parallel program is distributing the data across the processes. This step can often be burdensome and error prone. HPF provides several directives that allow the programmer to easily and efficiently describe many data distributions. The most important of these is the `distribute` directive. For example, to distribute an array across all processes in blocks, use

```
      real a(100)
!HPF$ DISTRIBUTE(BLOCK) a
```

Note that the HPF directive is a comment because it begins with an exclamation point and will be ignored by Fortran 90 compilers that do not support HPF. HPF supports several styles of data decomposition, including `BLOCK` (contiguous groups of elements across processes) and `CYCLIC` (round-robin assignment of elements across processes). One of the most attractive features of HPF is that the programmer may change the data distribution by changing only the `DISTRIBUTE` directive; the HPF compiler takes care of all of the changes to the code that are required by a different distribution.

The other important directive for data decomposition is the `ALIGN` directive. This tells the HPF compiler to align one distributed array with another. This lets the programmer provide information about the relationship between the use of elements of different distributed arrays to the compiler, which can be used by the compiler to produce more efficient code.

HPF provides additional directives; for example, there is a way to specify that a variable is involved in a reduction operation. Figure 1.25 shows a simple matrix-matrix multiply example. Note that, unlike the MPI case, program declares the sizes of the arrays, not just the part that is on a particular process. The HPF compiler handles all of the details of the data decomposition, including determining the sizes of the local versions of the arrays. This example does not include any of the code that would normally be used to implement cache and register blocking; such changes are necessary to achieve high performance.

1.4.2 SHMEM

SHMEM exists in implementations from various computer and interconnect vendors [4, 1, 70, 54]. In addition, a public-domain version—a generalized portable SHMEM, or GPSHMEM [67, 66]—has been augmented for use on clusters. The SHMEM model is an asynchronous one-sided message-passing or data-passing model. SHMEM assumes that computations are performed in separate address spaces and that data is explicitly passed. The asynchronous one-sided model assumes that a process can read ("get") data or write ("put") data from or to another process's address space without the active participation of the second process. These one-sided operations are now a component of MPI-2 [53], and we encourage programmers to use that functionality as opposed to SHMEM (see Section 1.2.3). It will, however, take time for the functionality to propagate through all the vendor-supported MPI implementations.

SHMEM relies on remotely accessible data objects that are symmetric. These are data objects that have a known relationship among the local and remote addresses, such as Fortran common blocks or variables with the SAVE attribute, data allocated with `shpalloc` in Fortran or `shmalloc` in C or C++. SHMEM has a robust set of collective routines based on a triplet of arguments: the starting processor, log of the stride, and the number of processors involved. This power-of-two stride was required for the hardware of the T3D and T3E systems, but it is not generally applicable to clusters. GPSHMEM augmented this behavior to include arbitrary stride counts. The collective routines operate on the same symmetric data objects in multiple processes; this a requirement is made to improve efficiency.

SHMEM can be thought of as a middle ground between message passing and a full DSM language. SHMEM supports other operations such as work-shared broadcast and reduction, barrier synchronization, and atomic memory operations. An atomic memory operation is an atomic read-and-update operation, such as a fetch-and-increment, on a remote or local data object. Full barriers, barriers on a subset of processes, and a locking mechanism are also provided. There are some problems with SHMEM in that there is a name-space explosion because the interface does not include the size of the object being passed. For example, five different broadcast calls are available in the T3E implementation: `shmem_broadcast`, `shmem_broadcast4`, `shmem_broadcast8`, `shmem_broadcast32`, and finally, `shmem_broadcast64`. Moreover, there is no standard for SHMEM, so other vendor or open-source implementations are free to augment the library as their needs arise. This augmentation is often via environment variables where the default values may or may not provide optimal performance.

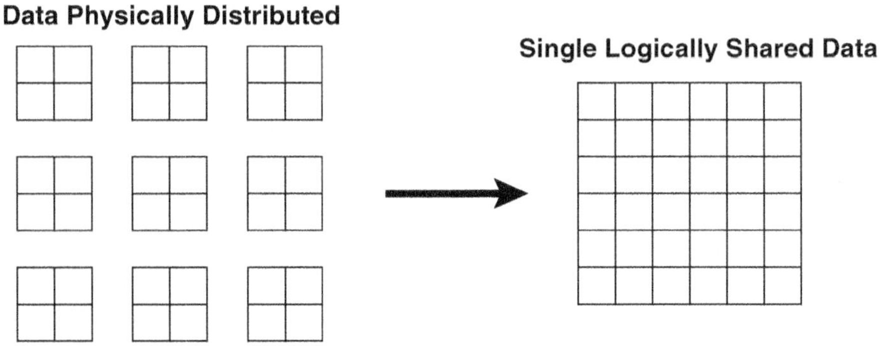

Fig. 1.26. View of data structures in Global Arrays.

1.4.3 Global Arrays

The Global Arrays (GA) Toolkit [59, 60, 58] was designed to offer the best functionality of both distributed-memory and shared-memory programming models. In fact, GA requires the use of a message-passing library so an application can use message-passing algorithms in addition to the GA algorithms. The data is divided into local data and logically shared data that can be accessed only through the user interface layer of the GA package. GA assumes that the data representation is arrays of multiple dimensions. This provides a NUMA view of the aggregate memory of the system. The data locality must be managed explicitly by the programmer, with the knowledge that remote data access is slower than local data access. Figure 1.26 represents the view of the data structures in an GA application. The cost of remote data access promotes data reuse and locality of reference.

The GA toolkit allows the user complete control over the data distribution to match any algorithmic needs. The user can have the library distribute the data automatically or can identify a specific dimension or block size for distribution. Complete irregular distributions are also possible. The locality information of data is also available. For example, a specific multidimensional patch of a GA that is required for an algorithmic computation may exist on one or more processes; the locality information is an array of the process identifiers.

Figure 1.27 shows the computational flow of a GA application. Data is extracted from "global" memory to "local" memory. The process then computes on that portion of the array copied to local memory. The results are copied or accumulated to global memory for further processing as the algorithm dictates.

Copy operations from the "global" data to "local" data and the reverse are the fundamental functionality of GA. In addition, the locality information provided allows direct access to data "owned" by a given process. This arrangement allows for virtually any needed data parallel operations. Several built-in data-parallel-like operations are provided, including zeroing an array, filling an array with an arbitrary value, printing an array, and scaling an array by a constant.

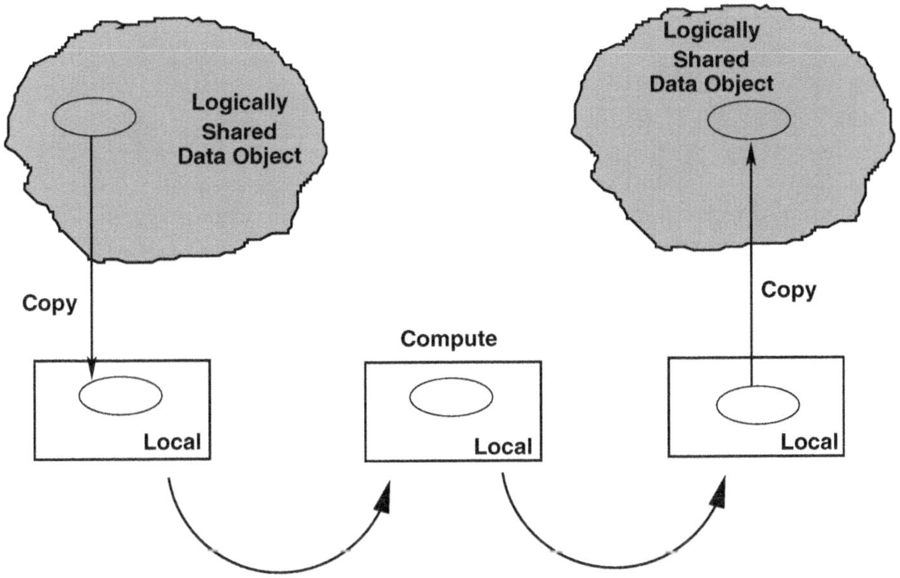

Fig. 1.27. Computational flow of a GA application.

GA has several language interfaces: C, C++, Fortran 77, Fortran 90/95, and Python. There is also a common component architecture (CCA) component version. In addition, the library provides language interoperability for mixed-language applications. Arrays created and used in Fortran can be accessed by using the other language interfaces. Internal storage is, by default, that of the Fortran language but can be made either row or column major.

The library has evolved from the initial development for NWChem, a computational chemistry suite [74, 45], to meet requirements of new application areas. Ghost cells and sparse data structures were added to provide functionality for halo-like simulations and Grid-based codes, respectively. The data movement engine was separated from the original implementation and now provides a portable one-sided communication tool, the Aggregate Remote Memory Copy Interface (ARMCI). ARMCI handles the actual data transfers, synchronization operations, and memory management. GA also has a secondary storage mechanism, disk resident arrays (DRAs). DRAs extend the memory hierarchy one additional level; they allow for out-of-core algorithm development as well as internal checkpointing of data. Furthermore, GA offers interfaces to third-party libraries such as ScaLAPACK.

GA provides portable performance; it runs on most major cluster interconnect technologies and high-end supercomputers. ARMCI, the data movement engine, is tuned to the fastest mechanisms available on various platforms. The developers have strong interactions with vendor software and hardware engineers to keep the infrastructure current and the performance at the highest level. GA will continue to expand to meet the requirements of the user community as the need arises.

To give a flavor of GA programming, we present a simple blocked matrix-multiply routine in Figure 1.28. The function stores the product of two matrices A and B in the resultant C matrix. The assumption is that the GAs for each matrix are created and A and B are filled prior to calling the routine and that all matrices are two dimensional.

GA provides a robust set of functionality. The toolkit is essentially the standard programming model for electronic structure computational chemistry codes, where most of the manipulations are contractions of multidimensional tensors of various orders into lower-order tensors. GA is also used in image processing, financial security forecasting, computational biology, fluid dynamics, and other areas. GA does not offer the full incremental parallelism of OpenMP, but the functional code is straightforward to generate and then tune for performance. Rapid prototyping is possible once the initial infrastructure is built. More information is provided on the Global Arrays home page [27].

1.5 Future Programming Models

We present here a few examples of what we delineate as future programming languages or models, not because they are new ideas, but because they are just now moving from the research community to the vendor community. Other programming languages should be considered if one is willing to live on the "bleeding edge" of technology—that is, with very robust features of the language and little support. Of particular note is Titanium [78], a high-performance Java dialect with extensions needed by scientific applications. We close this section with a view of what is next beyond near-term extrapolation of current technology and what is needed to really reach petaflops.

1.5.1 Unified Parallel C

Unified Parallel C [21] (UPC) is a parallel extension of the ANSI C standard. UPC, like Co-Array Fortran (see Section 1.5.2), has the advantage of extending a well-known and well-understood language for parallel computation. The development of the UPC language started with ANSI C and included experiences from various distributed parallel computing language efforts in the research community, with input from vendors, users, and academia.

UPC is a distributed shared-memory parallel programming language. The execution model assumes a number of threads working independently in a single-program multiple-data (SPMD) paradigm. The language provides synchronization when needed via barriers, the memory consistency model, and explicit locks. The memory in the language is logically split into private and shared memory with an affinity for a specific thread. Any thread can read from the globally shared address space, and the language extension includes identifying which data and pointers to data are "shared" among the threads. Figure 1.29 represents the memory layout in the UPC model.

```
      subroutine ga_simplematmul(g_c, g_a, g_b)
      implicit double precision (a-h,o-z)
! include files from the GA suite
#include "mafdecls.fh"
#include "global.fh"
c                  omitting declaration of variables
      parameter (blocksize = 32)  ! arbitrary block size
c
! get matrix dimensions
      ga_inquire(g_a,typea,rowsa,colsa)
      ga_inquire(g_b,typeb,rowsb,colsb)
      ga_inquire(g_c,typec,rowsc,colsc)
! check that types and dimensions match if not call ga_error
      call ga_zero(g_c)  ! zero the result
      blocksi = rowsc/blocksize + 1
      blocksj = colsc/blocksize + 1
      blocksk = colsa/blocksize + 1
! allocate local arrays loca[blocksize][blocksize],locb,locc
! get the number of processes
      nproc = ga_nnodes()
! atomically get the next task an ordered count 0, 1, ...
!                                across all processes
      mtask = nexttask(nproc)
      itask = -1
      do ib = 1,blocksi
         ilo = (ib-1)*blocksize + 1
         ihi = min((ilo+blocksize-1),rowsc)
         mdg = ihi-ilo + 1
         itask = itask + 1
! parallelize over i blocks (ib variable)
         if (itask.eq.mtask) then
            do kb = 1,blocksk
               klo = (kb-1)*blocksize + 1
               khi = min((klo+blocksize-1),colsa)
               kdg = khi - klo + 1
! get patch of global A copied into local array loca
               ga_get(g_a,ilo,ihi,klo,khi,loca,mdg)
               do jb = 1,blocksj
                  jlo = (jb-1)*blocksize + 1
                  jhi = min((jlo+blocksize-1),colsc)
                  ndg = jhi - jlo + 1
! get patch of global B copied into local array locb
                  ga_get(g_b,klo,khi,jlo,jhi,locb,kdg)
! use optimize BLAS locally to compute patch of in locc
                  call dgemm('n','n',mdg,ndg,kdg,1.0d00,
     +                  loca,mdg,locb,kdg,0.0d00,locc,mdg)
! accumulate into global array C from local locc
                  ga_acc(g_c,ilo,ihi,jlo,jhi,locc,mdg,1.0d00)
               enddo
            enddo
            mtask = nexttask(nproc)
         endif
      enddo
      end
```

Fig. 1.28. Simple GA matrix multiply routine.

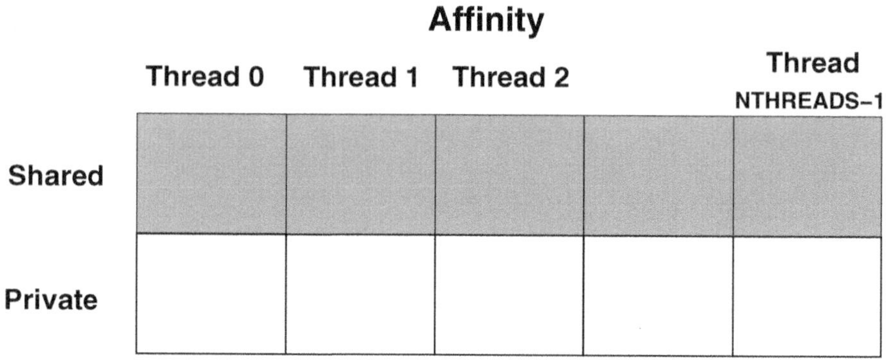

Fig. 1.29. UPC memory model with respect to the thread affinity.

The SPMD nature of the model allows for work distribution based on the thread identifier, MYTHREAD, and the number of threads involved, THREADS. MYTHREAD and THREADS are keywords in the UPC language. The actual translation depends on the underlying runtime infrastructure, but that is transparent to the user from a functional point of view and is the responsibility of the compiler.

Because threads share memory and because portions of shared memory have affinity to specific threads, access to that memory has a sequencing issue that depends on the underlying runtime environment. Developing a UPC application simply requires specifying either a "strict" or a "relaxed" memory consistency mode. This specification can be done for the entire program, for a defined block of code, or for a specific variable or array. The "relaxed" consistency mode allows memory accesses in each thread to follow normal ANSI C models, ignoring access to "local" shared-memory references from other threads. When using the relaxed mode, the programmer is ultimately responsible for handling any synchronization necessary. The "strict" mode follows normal ANSI C models while considering accesses from all threads.[11] Locks are provided to ensure atomic access to critical sections of code and the associated memory locations. Figure 1.30 shows a "Hello World" program similar to the one presented in the OpenMP discussion.

The sharing of data is explicitly coded in the use of the "shared" qualifier or in how memory is allocated dynamically with the UPC memory allocation routines. Since data being shared has affinity to threads, the user needs to control how data is laid out. Both the static and the dynamic memory modes allow for this. By default, elements of data arrays are distributed by element in a round-robin fashion to the shared-memory region of each thread. This can easily be blocked to distribute rows or columns of matrices to each thread.

In addition to the SPMD use of the thread identifier and the number of threads to share work among threads, there is a work-sharing construct `upc_forall`. The

[11] This is a simplification of the consistency model; consult the UPC specifications [21] for more details.

```
#include <stdio.h>
#include "upc_relaxed.h"
int main(int argc, char *argv[])
{
  int tid;
  {
    tid = MYTHREAD;
    printf("<%d> of %d Threads\n",tid,THREADS);
  }
}
```

Fig. 1.30. "Hello World" UPC code.

functional form of this construct is similar to the standard for loop construct but with an extra affinity parameter:

```
upc_forall ( init-expr ; cond-expr; incr-expr; affinity ),
```

where `init-expr`, `cond-expr`, and `incr-expr` are the ANSI C equivalent expressions. The affinity parameter can be either a variable or an address to a variable. The affinity expression controls which thread actually computes an iteration of the loop construct. For a variable the thread that executes the loop is `MYTHREAD == variable%THREADS`. For an address the thread that executes the loop is `MYTHREAD == upc_threadof(address)`. The UPC library function `upc_threadof` identifies the thread that has affinity for the address argument.

The UPC language has great potential for providing long-term portability and performance for a wide variety of applications. We have provided only a taste of the language. The UPC Web site[12] provides many more details, examples, and availability of compilers.

1.5.2 The Co-Array Fortran Extension to Fortran 95

Co-Array Fortran [62] is an alternative parallel programming language based on an extension to Fortran 95. It uses a simple syntax that is intuitively natural to a Fortran programmer. It adopts a purely local view of data and computation, but it allows the programmer to make local data globally visible by declaring some variables to be co-arrays. A co-array is a Fortran 95 object, whether an intrinsic object or a user-defined derived type, that is declared with a co-dimension. For example, the declaration,

```
real :: x[*]
```

defines a scalar co-array object that is replicated across program images. The asterisk notation `[*]` indicates that program images are *virtual* images, replicated copies of a program within the SPMD programming model.

The actual number of images is determined when the program starts execution. The runtime system assigns images to physical processors in a platform-specific

[12] See the `http://upc.gwu.edu` web site.

manner, for example, as processes or threads. The number of images is fixed; it may be the same as the number of physical processors, it may be greater, or it may be less. Each physical processor may be responsible for more than one image, for example, taking work from a task queue. Conversely, more than one physical processor may be responsible for the same image, for example, by spawning threads within a process to share the work. The programmer decides whether an image works only on its own local data or, using co-array syntax, works on data that it does not own, by making local copies of data owned by other images.

Co-dimensions may be multidimensional just like normal dimensions. Programmers can use them to represent a *logical* decomposition of virtual images that corresponds to a logical decomposition of a physical problem. For example, a two-dimensional field decomposed into blocks, as commonly used in weather, climate, and ocean codes, might be declared with two co-dimensions.

```
real :: field(m,n)[p,*]  .
```

In this case, each image holds a patch of the field of local size `(m x n)`. The asterisk notation indicates that the number of images is determined when the program starts execution, but the programmer wants to think of the images within a two-dimensional grid with `p` images in the first dimension.

For many applications that use finite difference operators to solve partial differential equations, for example, programmers often add halo cells around the local field data.

```
real :: field(0:m+1,0:n+1)[p,*]
```

The main communication requirement is the exchange of halo data, which, using Co-Array Fortran syntax, can be written with just a few lines of code [6, 61]. For example, the exchange in the east-west direction

```
field(1:m,0)    = field(1:m,n)[p,q-1]
field(1:m,n+1)  = field(1:m,1)[p,q+1]
```

can be written with two lines of code, where the programmer has adopted the convention that the first co-dimension represents the north-south direction and the second represents the east-west direction. The image corresponding to `[p,q]` fills its lower halo with data from its west neighbor `[p,q-1]` and its upper halo with data from its east neighbor `[p,q+1]`. Since co-array syntax allows an image to read or write data owned by any other image, it is the programmer's responsibility to provide appropriate synchronization.

Basing a parallel programming model on a simple extension to an existing language has a number of advantages. First, the programmer need not learn a new language. Co-array syntax is natural and familiar to the Fortran programmer. Second, the co-array extension can be implemented by using existing compiler technology. Co-dimensions behave, in most respects, like normal dimensions. Third, since the new parallel syntax becomes part of the language, the programmer can use it to write customized communication patterns that fit a particular problem, without being restricted solely to those patterns provided by a library. Fourth, the compiler can generate optimized code that takes advantage of specific features of specialized hardware on particular platforms. For example, in the halo exchange example, it can schedule communication to overlap with computation and to exercise multiple hardware channels simultaneously. Fifth, code written with Co-Array Fortran is portable [13]. Because the extension is part of the language, a compiler must implement it for all platforms it supports.

1.5.3 Beyond Future Programming Models

Programming language design follows system architecture development, at least in the domain of performance-critical computation, including high-performance computing. Language serves as the medium between a user's application and the underlying execution target platform. The challenge to programming is to extract the best possible performance from the target parallel computer system for a given application while retaining correctness. The degree of difficulty (length of programming time) strongly depends on the ease of performance tuning.

Historically, a healthy tension dominating language design has existed between language abstraction to hide system complexity from the programmer and low-level language constructs to expose the system mechanisms for direct and precise control to achieve the best performance. However, parallel programming methods have been heavily oriented toward constructs providing explicit control of low-level mechanisms because the principal target architectures, including massively parallel processors (MPPs) and commodity clusters, provide little or no support for automatic management of system-wide parallel computation—hence the popularity of models such as MPI (e.g., MPICH-2) that expose the underlying system architecture in detail and give the programmer complete control of how the application program is mapped to the system resources, as well as the synchronization of their cooperative operation.

Unfortunately, current-generation high-end systems not only are difficult to program but often exhibit significant inefficiencies in operation, negating much of the advantage of exploiting existing commodity components. Future system architectures for high-end capability (as opposed to capacity) computing in the trans-petaflops performance regime may be custom designed for the purpose of global parallel execution, unlike conventional MPPs. While this assertion is considered controversial today, important projects are under way to achieve this (e.g., the DARPA HPCS program).

If real parallel computing systems reemerge, replacing (at least in part) aggregated ensembles of commodity microprocessors in the arena of high-end computing, programming methodologies and languages that represent them will be devised to reflect their new underlying architectures. While we cannot know in absolute terms what future programming languages will look like in this new petaflops computing world, it is possible to identify key attributes of such languages based on reasonable assumptions about such future machines. Examples of such assumptions include the following:

- Global address space such that any part of the system state can be accessed efficiently from any other execution site within the distributed system

- Relaxed consistency methods for efficient copy semantics
- Hardware support for efficient parallel execution for coarse-, medium-, and fine-grained parallelism
- Rapid context-switching with multi-threaded execution
- Automatic hardware-supported latency hiding
- Efficient synchronization for many forms of coordination including message passing, producer-consumer, message-driven, and object-oriented
- Dynamic adaptive resource management and load balancing
- Streaming processing for high-temporal-locality computing
- In-memory processing for high-bandwidth, low-locality computing
- High-global-bandwidth, low-latency system-wide communication

Future programming languages for custom petaflops-scale system architectures incorporating some or all of these properties will differ from conventional programming practices by providing constructs that support a richer descriptive semantics of application parallelism and locality, rather than imperative specification of explicit mapping of data and code to hardware elements as is done today. Latency will be hidden in such future machines by a variety of automatic methods, and a much wider range of forms of parallelism will be efficiently supported. Thus, the key challenge to future programming is to make available to compilers, runtime systems, and hardware architecture descriptions of algorithmic/application parallelism and the synchronization relationships among coordinated computing actions.

A secondary feature of such future languages is the ability to represent locality relationships of data and tasks at various levels of granularity as a source of hints or heuristics for assisting and guiding the system in allocating and assigning physical resources. This is very different from the conventional practice of the programmer asserting the exact resource allocation mapping. Not only does this advanced approach simplify programming, but it also allows the system to exploit runtime information in conjunction with programmer and compile-time information to determine optimal placement of logical objects on the distributed physical resources.

While a rich set of semantics for parallelism representation and locality relationship description may constitute a major part of future programming languages for custom-scalable petaflops-scale system architectures, additional language capabilities will be incorporated to deal with practical aspects of very large systems. Three factors in particular will drive innovation in future language design:

1. *Performance monitoring* will become an integral part of the compiler and language, not just to show the programmer the bottlenecks, but to permit advanced compilation and runtime systems to make direct use of observed operation characteristics for automatic performance tuning, with some guidance by the programmer.
2. *Microcheckpointing* will be used to identify key locations in the by the programmer. Microcheckpointing identifies key locations in the execution trace where subsets of total program data may be temporarily archived until some follow on release point is correctly accomplished, at which point the snapshot of the partial state may be garbage collected. These minor fall-back points are employed

when an error is detected in subsequent execution without having to restart the entire program.

3. *Advanced input/output constructs* will be used for generating "information products." It will become increasingly impractical to attempt to store the full raw data from a simulation because the data sets will become prohibitively large. Also, the data itself, even if visualized, may not be useful in understanding the implications and consequences of the results. An output layer to process the raw data may be necessary to generate information products that can be many orders of magnitude smaller than the basic data values but far more meaningful to the scientist or engineer. Future languages will emphasize high-level information rather than raw data sets as the principal output content, and the I/O semantics of the language will reflect this new usage.

1.6 Final Thoughts

The information in this chapter touches only the tip of the iceberg with respect to the issues of writing parallel programs. Even long-term practitioners fall into the many pitfalls of developing parallel codes. Overall, writing parallel programs is best learned by "getting your hands dirty."

It is important to use the technology needed to get the job done, but it is also important to think about what changes might come in the future. The software development research community is producing new technologies rapidly and some of these technologies may be useful in high-performance application development. Although implementing object-oriented technology in Fortran 77 is impossible, some of the object-oriented concepts can build better-structured Fortran 77 codes. For example, abstraction and data-hiding are easily implemented with solid APIs for the functionality required.

What will come in the future? In this book, the chapter on common component architecture technology [2] discusses how the CCA framework has been used successfully to integrate functionality among multiple computational chemistry codes on parallel platforms. Also, the cross-cutting technologies of aspect-oriented programming [22] could change the way in which we construct software infrastructure for event logging, performance monitoring, or computational steering.

One additional comment is in order. Readers new to parallel computing on clusters might ask which is the best programming model with respect to performance and scalability. These are only two aspects of the interaction with a programming model and an application code with many different algorithms. The programming model also determines the ease of algorithmic development and thus application development and maintenance, Asking which is best is similar to asking which preconditioner, which Kyrlov subspace method, or which editor is the best to use. All programming models have strengths and weaknesses, and the choice is best made by those actually using the programming model for their particular purpose. MPI offers the greatest availability, portability, and scalability to large systems. OpenMP offers very good portability and availability with reasonable scalability on SMP systems.

The distributed shared-memory programming models are best when long-term availability is possible and there is an appropriate match to the algorithms or applications involved. Clearly, programming models and their associated execution models will have to evolve to be able to reach sustained petaflops levels of computing, which will in time move to computational resources known as clusters.

Finally, we will put together a series of examples of "working" code for many of the programming models discussed in this chapter. These will be designed around small computational kernels or simple applications in order to illustrate each model. The examples will be available at the Center for Programming Models for Scalable Parallel Computing website[13].

Acknowledgments

This work was supported by the Mathematical, Information, and Computational Sciences Division subprogram of the Office of Advanced Scientific Computing Research, Office of Science, U.S. Department of Energy, under Contract W-31-109-ENG-38 with Argonne National Laboratory and under Contract W-7405-ENG-82 at Ames Laboratory. The U.S. Government retains for itself, and others acting on its behalf, a paid-up, non-exclusive, irrevocable worldwide license in said article to reproduce, prepare derivative works, distribute copies to the public, and perform publicly and display publicly, by or on behalf of the Government. We thank the members of the Center for Programming Models for Scalable Parallel Computing [8] who have helped us better understand many of the issues of parallel software development and the associated programming models. We thank Brent Gorda, Angie Kendall, Gail W. Pieper, Douglas Fuller, and Professor Gary T. Leavens for reviewing the manuscript. We also thank the book series editors and referees for their many helpful comments.

References

1. Alphaserver SC user guide, 2000. Bristol, Quadrics Supercomputer World Ltd.
2. R. Armstrong, D. Gannon, A. Geist, K. Keahey, S. R. Kohn, L. McInnes, S. R. Parker, and B. A. Smolinski. Toward a common component architecture for high-performance scientific computing. In *Proceedings of the 8th High Performance Distributed Computing (HPDC'99)*, 1999. URL: `http://www.cca-forum.org`.
3. S. Balay, W. D. Gropp, L. C. McInnes, and B. F. Smith. PETSc users manual. Technical Report ANL-95/11 - Revision 2.1.0, Argonne National Laboratory, 2001.
4. R. Bariuso and A. Knies. SHMEM's User's Guide. SN-2515 Rev. 2.2, Cray Research, Inc., Eagan, MN, USA, 1994.
5. M. Bull. OpenMP 2.5 and 3.0. In *Proceedings of the Workshop on OpenMP Applications and Tools, WOMPAT 2004*, Houston, TX, May 17-18 2004. (Invited talk).
6. P. M. Burton, B. Carruthers, G. S. Fischer, B. H. Johnson, and R. W. Numrich. Converting the halo-update subroutine in the MET Office unified model to Co-Array Fortran. In

[13]At this URL: `http://www.pmodels.org/ppde`.

W. Zwieflhofer and N. Kreitz, editors, *Developments in Teracomputing: Proceedings of the Ninth ECMWF Workshop on the Use of High Performance Computing in Meteorology*, pp. 177–188. World Scientific Publishing, 2001.
7. J. B. Carter, J. K. Bennett, and W. Zwaenepoel. Implementation and performance of Munin. In *Proceedings of the 13th ACM Symp. on Operating Systems Principles (SOSP-13)*, pages 152–164, 1991.
8. Center for Programming Models for Scalable Parallel Computing. URL: `http://www.pmodels.org`.
9. R. Chandra, L. Dagum, D. Kohr, D. Maydan, J. McDonald, and R. Menon. *Parallel Programming in OpenMP*. Morgan Kaufmann Publishers, San Francisco, CA, 2001.
10. D. Chen, S. Dwarkadas, S. Parthasarathy, E. Pinheiro, and M. L. Scott. Interweave: A middleware system for distributed shared state. In *Languages, Compilers, and Run-Time Systems for Scalable Computers*, pages 207–220, 2000.
11. E. Chow, A. Cleary, and R. Falgout. HYPRE User's manual, version 1.6.0. Technical Report UCRL-MA-137155, Lawrence Livermore National Laboratory, Livermore, CA, 1998.
12. D. Clark. OpenMP: A parallel standard for the masses. *IEEE Concurrency*, 6(1):10–12, January–March 1998.
13. C. Coarfa, Y. Dotsenko, J. L. Eckhardt, and J. Mellor Crummey. Co-array Fortran performance and potential: An NPB experimental study. In *The 16th International Workshop on Languages and Compilers for Parallel Computing (LCPC 2003)*, College Station, Texas, October 2003.
14. Cray Research. *Application Programmer's Library Reference Manual*, 2nd edition, Nov. 1995. Publication SR-2165.
15. L. Dagum and R. Menon. OpenMP: An industry standard API for shared-memory programming. *IEEE Computational Science & Engineering*, 5(1):46–55, January–March 1998.
16. S. Dong and G. E. Karniadakis. Dual-level parallelism for deterministic and stochastic CFD problems. In *Proceedings of Supercomputing, SC02*, Baltimore, MD, 2002.
17. J. Dongarra, I. Foster, G. Fox, W. D. Gropp, K. Kennedy, L. Torczon, and A. White, editors. *Sourcebook of Parallel Computing*. Morgan Kaufmann, 2003.
18. P. F. Dubois. Ten Good Practices In Scientific Programming. *Computing in Science & Engineering*, 1(1), January-February 1999.
19. S. Dwarkadas, N. Hardavellas, L. Kontothanassis, R. Nikhil, and R. Stets. Cashmere-VLM: Remote memory paging for software distributed shared memory. In *Proceedings of the 13th International Parallel Processing Symposium and 10th Symposium on Parallel and Distributed Processing*, pages 153–159. IEEE Computer Society, Apr. 1999.
20. Earth Simulator home page, `http://www.es.jamstec.go.jp`.
21. T. A. El-Ghazawi, W. W. Carlson, and J. M. Draper. UPC Language Specifications Version 1.1.1, October 2003. URL: `http://www.gwu.edu/~upc/docs/upc_spec_1.1.1.pdf`.
22. T. Elrad, R. E. Filman, and A. Bader. Aspect-Oriented Programming. *Communications of the ACM*, 44(10):29–32, October 2001.
23. R. D. Falgout, J. E. Jones, and U. M. Yang. The design and implementation of *hypre*, a library of parallel high performance preconditioners. In A. M. Bruaset and A. Tveito, editors, *Numerical Solution of Partial Differential Equations on Parallel Computers*, volume 51 of *Lecture Notes in Computational Science and Engineering*, pages 267–294. Springer-Verlag, 2005.

24. M. Folk, A. Cheng, and K. Yates. HDF5: A file format and I/O library for high performance computing applications. In *Proceedings of Supercomputing'99 (CD-ROM)*. ACM SIGARCH and IEEE, Nov. 1999.
25. FORTRAN 77 Binding of X3H5 Model for Parallel Programming Constructs. Draft Version, ANSI X3H5, 1992.
26. P. C. Forum. PCF Parallel FORTRAN Extensions. *FORTRAN Forum*, 10(3), September 1991. (Special issue).
27. Global Array Project. URL: `http://www.emsl.pnl.gov/docs/global`.
28. W. D. Gropp. Learning from the success of MPI. In B. Monien, V. K. Prasanna, and S. Vajapeyam, editors, *High Performance Computing – HiPC 2001*, number 2228 in Lecture Notes in Computer Science, pages 81–92. Springer, Dec. 2001.
29. W. D. Gropp, S. Huss-Lederman, A. Lumsdaine, E. Lusk, B. Nitzberg, W. Saphir, and M. Snir. *MPI—The Complete Reference: Volume 2, The MPI-2 Extensions*. MIT Press, Cambridge, MA, 1998.
30. W. D. Gropp, E. Lusk, and A. Skjellum. *Using MPI: Portable Parallel Programming with the Message Passing Interface,* 2nd edition. MIT Press, Cambridge, MA, 1999.
31. W. D. Gropp, E. Lusk, and T. Sterling, editors. *Beowulf Cluster Computing with Linux*. MIT Press, 2nd edition, 2003.
32. W. D. Gropp, E. Lusk, and R. Thakur. *Using MPI-2: Advanced Features of the Message-Passing Interface*. MIT Press, Cambridge, MA, 1999.
33. R. Hempel and D. W. Walker. The emergence of the MPI message passing standard for parallel computing. *Computer Standards and Interfaces*, 21(1):51–62, 1999.
34. High Performance Fortran Forum. High Performance Fortran language specification. *Scientific Programming*, 2(1–2):1–170, 1993.
35. J. M. D. Hill, B. McColl, D. C. Stefanescu, M. W. Goudreau, K. Lang, S. B. Rao, T. Suel, T. Tsantilas, and R. H. Bisseling. BSPlib: The BSP programming library. *Parallel Computing*, 24(14):1947–1980, Dec. 1998.
36. C. A. R. Hoare. Communicating sequential processes. *Communications of the ACM*, 21(8):666–677, Aug. 1978.
37. J. Hoeflinger. Towards industry adoption of OpenMP. In *Proceedings of the Workshop on OpenMP Applications and Tools, WOMPAT 2004*, Houston, TX, May 17–18 2004. Invited Talk.
38. F. Hoffman. Writing hybrid MPI/OpenMP code. *Linux Magazine*, 6(4):44–48, April 2004. URL: `http://www.linux-mag.com/2004-04/extreme_01.html`.
39. Y. Hu, H. Lu, A. L. Cox, and W. Zwaenepoel. OpenMP for networks of SMPs. In *Proceedings of the 13th International Parallel Processing Symposium*, April 1999.
40. P. Hyde. *Java Thread Programming*. SAMS, 1999.
41. *IEEE Standard for Information Technology-Portable Operating System Interface (POSIX)*. IEEE Standard No.: 1003.1, 2004.
42. W. Jiang, J. Liu, H.-W. Jin, D. K. Panda, W. D. Gropp, and R. Thakur. High performance MPI-2 one-sided communication over InfiniBand. Technical Report ANL/MCS-P1119-0104, Mathematics and Computer Science Division, Argonne National Laboratory, 2004.
43. G. Jost, J. Labarta, and J. Gimenez. What multilevel parallel programs do when you are not watching: A performance analysis case study comparing MPI/OpenMP, MLP, and nested OpenMP. In *Proceedings of the Workshop on OpenMP Applications and Tools, WOMPAT 2004*, pages 29–40, Houston, TX, May 17-18 2004. (Invited talk).
44. P. Keleher, A. L. Cox, S. Dwarkadas, and W. Zwaenepoel. TreadMarks: Distributed shared memory on standard workstations and operating systems. In *Proceedings of the Winter 94 Usenix Conference*, pages 115–131, January 1994.

45. R. A. Kendall, E. Aprà, D. E. Bernholdt, E. J. Bylaska, M. Dupuis, G. I. Fann, R. J. Harrison, J. Ju, J. A. Nichols, J. Nieplocha, T. P. Straatsma, T. L. Windus, and A. T. Wong. High performance computational chemistry; an overview of NWChem a distributed parallel application. *Computer Physics Communications*, 128:260–283, 2002.
46. M. G. Knepley, R. F. Katz, and B. Smith. Developing a geodynamics simulator with petsc. In A. M. Bruaset and A. Tveito, editors, *Numerical Solution of Partial Differential Equations on Parallel Computers*, volume 51 of *Lecture Notes in Computational Science and Engineering*, pages 413–438. Springer-Verlag, 2005.
47. C. Koelbel, D. B. Loveman, R. S. Schreiber, G. L. Steele, and M. E. Zosel. *The High Performance Fortran Handbook*. MIT Press, 1994.
48. B. Leasure, editor. *PCF Fortran: Language Definitons, Version 3.1*. The Parallel Computing Forum, Champaign, IL, 1990.
49. J. Li, W. Liao, A. Choudhary, R. Ross, R. Thakur, W. D. Gropp, R. Latham, A. Siegel, B. Gallagher, and M. Zingale. Parallel netCDF: A high-performance scientific I/O interface. In *Proceedings of SC2003*, Nov. 2003.
50. Z. Li, Y. Saad, and M. Sosonkina. pARMS: A parallel version of the algebraic recursive multilevel solver. *Numerical Linear Algebra with Applications*, 10:485–509, 2003.
51. R. K. Lie Huang, Barbara Chapman. OpenMP on distributed memory via global arrays. In *Proceedings of Parallel Computing 2003 (ParCo2003)*, Dresden, Germany, September 2–5 2003.
52. Message Passing Interface Forum. MPI: A Message-Passing Interface standard. *International Journal of Supercomputer Applications*, 8(3/4):165–414, 1994.
53. Message Passing Interface Forum. MPI2: A Message Passing Interface standard. *International Journal of High Performance Computing Applications*, 12(1–2):1–299, 1998.
54. Message Passing Toolkit: MPI programmer's manual, document number : 007-3687-010, 2003. Mountain View, CA, Silicon Graphics Inc.
55. Mpi papers. URL: `http://www.mcs.anl.gov/mpi/papers`.
56. K. Nakajima and H. Okuda. Parallel Iterative Solvers for Unstructured Grids Using and OpenMP/MPI Hybrid Programming Model for GeoFEM Platfrom on SMP Cluster Architectures. *Lecture Notes in Computer Science*, 2327:437–448, 2002.
57. B. Nichols, D. Buttlar, and J. P. Farrel. *Pthreads Programming*. O'Reilly & Associates, Inc, 1996.
58. J. Nieplocha, R. Harrison, M. Krishnan, B. Palmer, , and V. Tipparaju. Combining shared and distributed memory models: Evolution and recent advancements of the Global Array Toolkit. In *Proceedings of POOHL'2002 workshop of ICS-2002*, New York, NY, 2002.
59. J. Nieplocha, R. J. Harrison, and R. J. Littlefield. Global Arrays: A portable "shared memory" programming model for distributed memory computers. In *Proceedings of Supercomputing 1994, SC94*, pages 340–349, 1994.
60. J. Nieplocha, R. J. Harrison, and R. J. Littlefield. Global Arrays: A nonuniform memory access programming model for high-performance computers. *The Journal of Supercomputing*, 10:197–220, 1996.
61. R. W. Numrich, J. Reid, and K. Kim. Writing a multigrid solver using Co-Array Fortran. In B. Kågström, J. Dongarra, E. Elmroth, and J. Waśniewski, editors, *Applied Parallel Computing: Large Scale Scientific and Industrial Problems*, volume 1541 of *Lecture Notes in Computer Science*, pages 390–399. Springer, 1998.
62. R. W. Numrich and J. K. Reid. Co-Array Fortran for parallel programming. *ACM Fortran Forum*, 17(2):1–31, 1998.
63. OpenMP Architecture Review Board. *OpenMP Fortran Application Program Interface, Version 2.0*. November 2000. URL: `http://www.openmp.org/drupal/mp-documents/fspec20.pdf`.

64. OpenMP Architecture Review Board. *OpenMP C and C++ Application Program Interface, Version 2.0.* March 2002. URL: `http://www.openmp.org/drupal/mp-documents/cspec20.pdf`.
65. OpenMP Architecture Review Board home page, `http://www.openmp.org`.
66. K. Parzyszek and R. A. Kendall. GPSHMEM: Application to kernel benchmarks. In *Proceedings of the Fourteenth IASTED International Conference on Parallel and Distributed Computing and Systems (PDCS 2002)*, pages 404–409. ACTA Press, Anaheim, CA, 2002.
67. K. Parzyszek, J. Nieplocha, and R. A. Kendall. A generalized portable SHMEM library for high performance computing. In M. Guizani and X. Shen, editors, *Proceedings of the IASTED Parallel and Distributed Computing and Systems 2000*, pages 401–406. IASTED, Calgary, 2000.
68. Y. Saad. SPARSKIT: A basic tool kit for sparse matrix computations. Technical Report 90-20, NASA Ames Research Center, Moffett Field, CA, 1990.
69. H. Sakagami, H. Murai, Y. Seo, and M. Yokokawa. 14.9 TFLOPS three-dimensional fluid simulation for fusion science with HPF on the Earth Simulator. In *Proceedings of Supercomputing*, 2002.
70. Scali Library User's Guide, 2002. Published by Scali, Oslo, Norway.
71. C. L. Seitz. The cosmic cube. *Communications of the ACM*, 28(1):22–33, Jan. 1985.
72. B. Smith, P. Bjørstad, and W. D. Gropp. *Domain Decomposition: Parallel Multilevel Methods for Elliptic Partial Differential Equations.* Cambridge University Press, New York, 1996.
73. M. Snir, S. W. Otto, S. Huss-Lederman, D. W. Walker, and J. Dongarra. *MPI: The Complete Reference.* MIT Press, Cambridge, MA, 1995.
74. T. Straatsma, E. Aprà, T. Windus, W. E. d. J. E. J. Bylaska, S. Hirata, M. Valiev, M. T. Hackler, L. L. Pollack, R. J. Harrison, M. Dupuis, D. Smith, J. Nieplocha, V. Tipparaju, M. Krishnan, A. A. Auer, E. Brown, G. Cisneros, G. I. Fann, H. Fruchtl, J. Garza, K. Hirao, R. A. Kendall, J. Nichols, K. Tsemekhman, K. Wolinski, J. Anchell, D. Bernholdt, P. Borowski, T. Clark, D. Clerc, H. Dachsel, M. Deegan, K. K. Dyall, D. Elwood, E. Glendening, M. Gutowski, A. Hess, J. Jaffe, B. Johnson, J. Ju, R. Kobayashi, R. Kutteh, Z. Lin, R. Littlefield, X. Long, B. Meng, T. Nakajima, S. Niu, M. Rosing, G. Sandrone, M. Stave, H. Taylor, G. Thomas, J. van Lenthe, A. Wong, and Z. Zhang. NWChem, A computational chemistry package for parallel computers, Version 4.6, 2004. Pacific Northwest National Laboratory, Richland, WA.
75. R. Thakur, W. D. Gropp, and B. Toonen. Minimizing synchronization overhead in the implementation of MPI one-sided communication. In D. Kranzlmüller, P. Kacsuk, and J. Dongarra, editors, *Recent Advances in Parallel Virtual Machine and Message Passing Interface*, Lecture Notes in Computer Science, pages 57–67. Springer Verlag, 2004. 11th European PVM/MPI User's Group Meeting, Budapest, Hungary.
76. The Cluster Enabled Omni OpenMP Compiler. URL: `http://phase.hpcc.jp/Omni/Omni-doc/omni-scash.html`.
77. The Open Group. *System Interfaces and Headers, Issue 4, Version 2.* 1992. URL: `http://www.opengroup.org/public/pubs/catalog/c435.htm`.
78. K. Yelick, L. Semenzato, G. Pike, C. Miyamoto, B. Liblit, A. Krishnamurthy, P. Hilfinger, S. Graham, D. Gay, P. Colella, and A. Aiken. Titanium: A high-performance Java dialect. *Concurrency: Practice And Experience*, 10(11–13):825–836, 1998.

2

Partitioning and Dynamic Load Balancing for the Numerical Solution of Partial Differential Equations

James D. Teresco[1], Karen D. Devine[2], and Joseph E. Flaherty[3]

[1] Department of Computer Science, Williams College, Williamstown, MA 01267, USA
terescoj@cs.williams.edu

[2] Discrete Algorithms and Mathematics Department, Sandia National Laboratories, Albuquerque, NM 87185, USA
kddevin@sandia.gov

[3] Department of Computer Science, Rensselaer Polytechnic Institute, Troy, NY 12180, USA
flaherje@cs.rpi.edu

Summary. In parallel simulations, partitioning and load-balancing algorithms compute the distribution of application data and work to processors. The effectiveness of this distribution greatly influences the performance of a parallel simulation. Decompositions that balance processor loads while keeping the application's communication costs low are preferred. Although a wide variety of partitioning and load-balancing algorithms have been developed, their effectiveness depends on the characteristics of the application using them. In this chapter, we review several partitioning algorithms, along with their strengths and weaknesses for various PDE applications. We also discuss current efforts toward improving partitioning algorithms for future applications and architectures.

The distribution of data among cooperating processes is a key factor in the efficiency of parallel solution procedures for partial differential equations (PDEs). This distribution requires a data-partitioning procedure and distributed data structures to realize and use the decomposition. In applications with constant workloads, a static partition (or static load balance), computed in a serial or parallel pre-processing step, can be used throughout the computation. Other applications, such as adaptive finite element methods, have workloads that are unpredictable or change during the computation, requiring dynamic load balancers that adjust the decomposition as the computation proceeds. Partitioning approaches attempt to distribute computational work equally, while minimizing interprocessor communication costs. Communication costs are governed by the amount of data to be shared by cooperating processes (communication volume) and the number of partitions sharing the data (number of messages). Dynamic load-balancing procedures should also operate in parallel on distributed data, execute quickly, and minimize data movement by making the new data distribution as similar as possible to the existing one. The partitioning problem is defined in more detail in Section 2.1.

Numerous partitioning strategies have been developed. The various strategies are distinguished by trade-offs between partition quality, amount of data movement, and partitioning speed. Characteristics of an application (e.g., computation-to-communication ratio, cost of data movement, and frequency of repartitioning) determine which strategies are most appropriate for it. For example, geometric algorithms like recursive bisection and space-filling curve partitioning provide high-speed, medium-quality decompositions that depend only on geometric information (e.g., particles' spatial coordinates, element centroids). Graph-based algorithms provide higher quality decompositions based on connectivity between application data, but at a higher cost. Several strategies, with their relative trade-offs, are described in detail in Section 2.2 and Section 2.3.

Many partitioning procedures have been implemented directly in applications, using application-specific data structures. While this approach can provide high execution efficiency, it usually limits the application to a single procedure and burdens the application programmer with partitioning concerns. A number of software libraries are available that provide high-quality implementations of partitioning procedures, provide flexibility to switch among available methods, and free the application programmer from those details. Some of these software packages are described in Section 2.4.

While existing methods have been very successful, research challenges remain. New models, such as hypergraphs, can more accurately model communication. Multi-criteria partitioning can improve efficiency when different phases of a computation have different costs. Resource-aware computation, achieved by adjusting the partitioning or other parts of the computation according to processing, memory and communication resources, is needed for efficient execution on modern hierarchical and heterogeneous computer architectures. Current research issues are explored further in Section 2.5.

2.1 The Partitioning and Dynamic Load Balancing Problems

The most common approach to parallelizing PDE solution procedures assigns portions of the computational domain to cooperating processes in a parallel computation. Typically, one process is assigned to each processor. Data are distributed among the processes, and each process computes the solution on its local data (its *subdomain*). Inter-process communication provides data that are needed by a process but "owned" by a different process. This model introduces complications including (i) assigning data to subdomains (i.e., *partitioning*, or when the data is already distributed, *dynamic load balancing*), (ii) constructing and maintaining distributed data structures that allow for efficient data migration and access to data assigned to other processes, and (iii) communicating the data as needed during the solution process. The focus of this chapter is on the first issue: data partitioning.

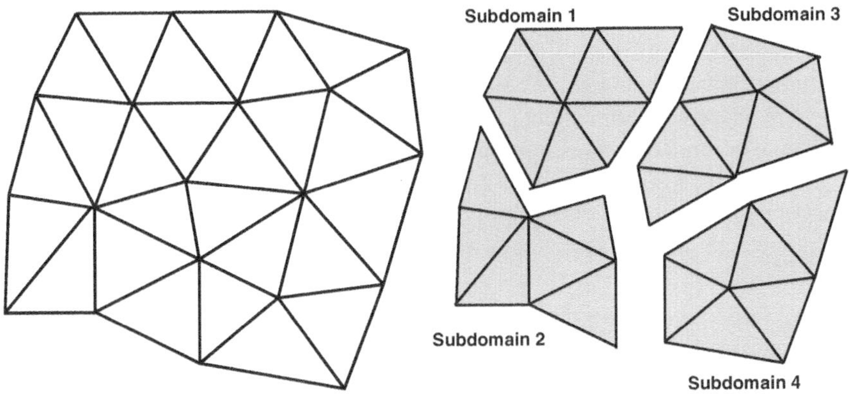

Fig. 2.1. An example of a two-dimensional mesh (left) and a decomposition of the mesh into four subdomains (right).

2.1.1 The Partitioning Problem

The computational work of PDE simulation is often associated with certain "objects" in the computation. For particle simulations, computation is associated with the individual particles; adjusting the distribution of particles among processors changes the processor load balance. For mesh-based applications, work is associated with the entities of the mesh — elements, surfaces, nodes — and decompositions can be computed with respect to any of these entities or to a combination of entities (e.g., nodes and elements). The partitioning problem, then, is the division of objects into groups or subdomains that are assigned to cooperating processes in a parallel computation.

At its simplest, a partitioning algorithm attempts to assign equal numbers of objects to partitions while minimizing communication costs between partitions. A partition's subdomain, then, consists of the data uniquely assigned to the partition; the union of subdomains is equal to the entire problem domain. For example, Figure 2.1 shows a two-dimensional mesh whose elements are divided into four subdomains. Often communication between partitions consists of exchanges of solution data for adjacent objects that are assigned to different partitions. For example, in finite element simulations, "ghost elements" representing element data needed by but not assigned to a subdomain are updated via communication with neighboring subdomains. While this data distribution is the most commonly used one for parallelization of PDE applications (and, indeed, will be assumed without loss of generality in the rest of this chapter), other data layouts are possible. In Mitchell's full-domain partition (FuDoP) [77], for example, each process is assigned a disjoint subdomain of a refined mesh. Then within each process, a much coarser mesh is generated for the rest of the problem domain, giving each process a view of the entire domain. This layout reduces the amount of communication needed to update subdomain boundary values during adaptive multigrid, at the cost of extra degrees of freedom and computation. A similar idea has been applied to parallel solution procedures by Bank and Holst to reduce communication costs for elliptic problems [3].

Objects may have weights proportional to the computational costs of the objects. These nonuniform costs may result from, e.g., variances in computation time due to different physics being solved on different objects, more degrees of freedom per element in adaptive p-refinement [1, 105], or more small time steps taken on smaller elements to enforce timestep contraints in local mesh-refinement methods [42]. Similarly, nonuniform communication costs may be modeled by assigning weights to connections between objects. Partitioning then has the goal of assigning equal total object weight to each subdomain while minimizing the weighted communication cost.

2.1.2 Dynamic Repartitioning and Load Balancing Problem

Workloads in dynamic computations evolve in time, so a partitioning approach that works well for a static problem or for a slowly-changing problem may not be efficient in a highly dynamic computation. For example, in finite element methods with adaptive mesh refinement, process workloads can vary dramatically as elements are added and/or removed from the mesh. Dynamic repartitioning of mesh data, often called dynamic load balancing, becomes necessary.

Dynamic repartitioning is also needed to maintain geometric locality in applications like crash simulations and particle methods. In crash simulations, for example, high parallel efficiency is obtained when subdomains are constructed of geometrically close elements [96]. Similarly, in particle methods, particles are influenced by physically near particles more than by distant ones; assigning particles to processes based on their geometric proximity to other particles reduces the amount of communication needed to compute particle interactions.

Dynamic load balancing has the same goals as partitioning, but with the additional constraints that procedures (i) must operate in parallel on already distributed data, (ii) must execute quickly, as dynamic load balancing may be performed frequently, and (iii) should be incremental (i.e., small changes in workloads produce only small changes in the decomposition) as the cost of redistribution of mesh data is often the most significant part of a dynamic load-balancing step. While a more expensive procedure may produce a higher-quality result, it is sometimes better to use a faster procedure to obtain a lower-quality decomposition, if the workloads are likely to change again after a short time.

2.1.3 Partition Quality Assessment

The goal of partitioning is to minimize time to solution for the corresponding PDE solver. A number of statistics may be computed about a decomposition that can indicate its suitability for use in an application.

The most obvious measure of partition quality is computational load balance. Assigning the same amount of work to each processor is necessary to avoid idle time on some processors. The most accurate way to measure imbalance is by instrumenting software to determine processor idle times. However, imbalance is often reported

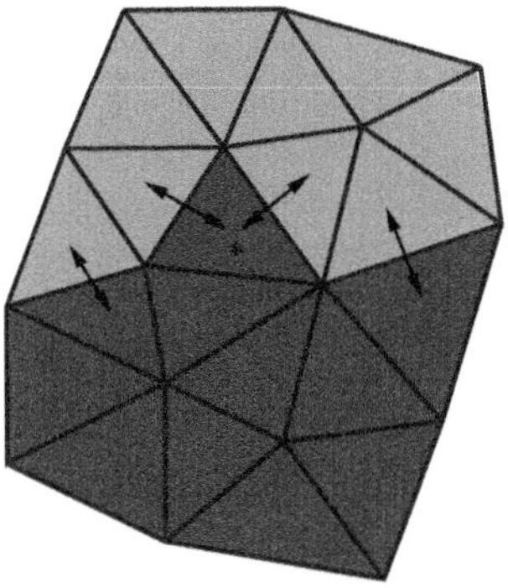

Fig. 2.2. Example where the number of elements on the subdomain boundary is not an accurate measure of communication costs. The shading indicates subdomain assignments. The element indicated by "*" needs to send its value to two neighbors in the other subdomain, but the value need only be communicated once.

with respect to the number of objects assigned to each subdomain (or the sum of object weights, in the case of non-uniform object computation costs).

Computational load balance alone does not ensure efficient parallel computation. Communication costs must also be considered. This task often corresponds to minimizing the number of objects on sharing data across subdomain boundaries, since the number of adjacencies on the bounding surface of each subdomain approximates the amount of local data that must be communicated to perform a computation. For example, in element decompositions of mesh-based applications, this communication cost is often approximated by the number of element faces on boundaries between two or more subdomains. (In graph partitioning, this metric is referred to as "edge cuts"; see Section 2.2.2.) A similar metric is a subdomain's *surface index*, the percentage of all element faces within a subdomain that lie on the subdomain boundary. Two variations on the surface index can be used to estimate the cost of interprocess communication. The *maximum local surface index* is the largest surface index over all subdomains, and the *global surface index* measures the percentage of all element faces that are on subdomain boundaries [14]. In three dimensions, the surface indices can be thought of as surface-to-volume ratios if the concepts of surface and volume are expanded beyond conventional notions; i.e., the "volume" is the whole of a subdomain, and the elements on subdomain boundaries are considered the "surface." The global surface index approximates the total communication volume, while the maximum local surface index approximates the maximum communication needed by any one subdomain.

A number of people [14, 50, 111] have pointed out flaws in minimizing only the edge cut or global surface index statistics. First, the number of faces shared by subdomains is not necessarily equal to the communication volume between the subdomains [50]; an element could easily share two or more faces with elements in a neighboring subdomain, but the element's data would be communicated only once to the neighbor (Figure 2.2). Second, interconnection network latency is often a significant component of communication cost; therefore, interprocess connectivity (the number of processes with which each process must exchange information during the solution phase) can be as significant a factor in performance [14] as the total volume of communication. Third, communication should be balanced, not necessarily minimized [95]. A balanced communication load often corresponds to a small maximum local surface index.

Another measure of partition quality is the internal connectivity of the subdomains. Having multiple disjoint connected components within a subdomain (also known as *subdomain splitting* [57]) can be undesirable. Domain decomposition methods for the solution of the linear systems will converge slowly for partitions with this property [25, 38]. Additionally, if a relatively small disjoint part of one subdomain can be merged into a neighboring subdomain, the boundary size will decrease, thereby improving the surface indices.

Subdomain aspect ratio has also been reported as an important factor in partition quality [32, 38], particularly when iterative methods such as Conjugate Gradient (CG) or Multigrid are used to solve the linear systems. Diekmann, et al. [32] provide several definitions of subdomain aspect ratio, the most useful being the ratio of the square of the radius of smallest circle that contains the entire subdomain to the subdomain's area. They show that the number of iterations needed for a preconditioned CG procedure grows with the subdomain aspect ratio. Furthermore, large aspect ratios are likely to lead to larger boundary sizes.

Geometric locality of elements is an important indicator of partition effectiveness for some applications. While mesh connectivity provides a reasonable approximation to geometric locality in some simulations, it does not represent geometric locality in all simulations. (In a simulation of an automobile crash, for example, the windshield and bumper are far apart in the mesh, but can be quite close together geometrically.) Geometric locality is also important in particle methods, where a natural representation of connectivity is not often available. Quality metrics based on connectivity are not appropriate for these types of simulations.

2.2 Partitioning and Dynamic Load Balancing Taxonomy

A variety of partitioning and dynamic load balancing procedures have been developed. Since no single procedure is ideal in all situations, many of these alternatives are commonly used. This section describes many of the approaches, grouping them into geometric methods, global graph-based methods, and local graph-based methods. Geometric methods examine only coordinates of the objects to be partitioned. Graph-based methods use the topological connections among the objects. Most geometric or graph-based methods operate as global partitioners or repartitioners. Local graph-based methods, however, operate among neighborhoods of processes in an existing decomposition to improve load balance. This section describes the methods; their relative merits are discussed in Section 2.3.

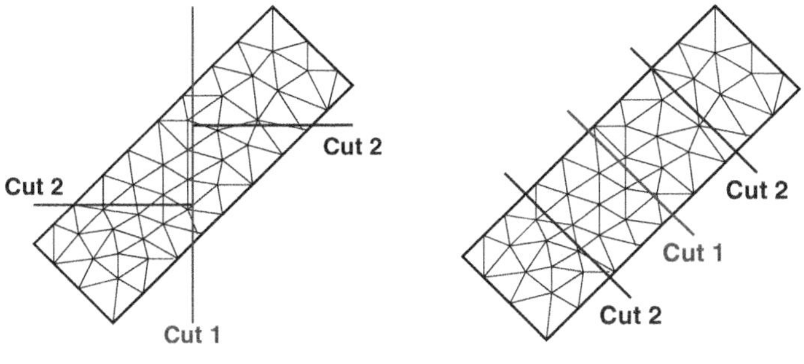

Fig. 2.3. Example of RCB cuts along coordinate axes (left) and RIB cuts along the principal axis of inertia (right).

2.2.1 Geometric Methods

Geometric methods are partitioners that use only objects' spatial coordinates and objects' computational weights in computing a decomposition. For example, in mesh partitioning, any mesh entities' coordinates (e.g., nodal coordinates, element centroids, surface element centroids) can be used. Geometric methods assign objects that are physically close to each other to the same partition in a way that balances the total weight of objects assigned to each partition. This goal is particularly effective for applications in which objects interact only if they are geometrically close to each other, as in particle methods and crash simulations.

Recursive Bisection

Methods using recursive bisection divide the simulation's objects into two equally weighted sets; the bisection algorithm is then applied to each set until the number of sets is equal to the number of desired partitions. (This description implies that the number of partitions must be a power of two; however, only minor changes in the algorithm are needed to allow an arbitrary number of partitions.)

Perhaps the most well-known geometric bisection method is Recursive Coordinate Bisection (RCB), developed by Berger and Bokhari [9]. In RCB, two sets are computed by cutting the problem geometry with a plane orthogonal to a coordinate axis (see Figure 2.3, left). The plane's direction is selected to be orthogonal to the longest direction of the geometry; its position is computed so that half of the object weight is on each side of the plane. In a twist on RCB, Jones' and Plassmann's Unbalanced Recursive Bisection (URB) algorithm [61] halves the problem geometry (instead of the set of objects) and then assigns processes to each half proportionally to the total object weight within the half.

Like RCB, Recursive Inertial Bisection (RIB) [107, 113] uses cutting planes to bisect the geometry. In RIB, however, the direction of the plane is computed to be orthogonal to long directions in the actual geometry, rather than to a coordinate axis

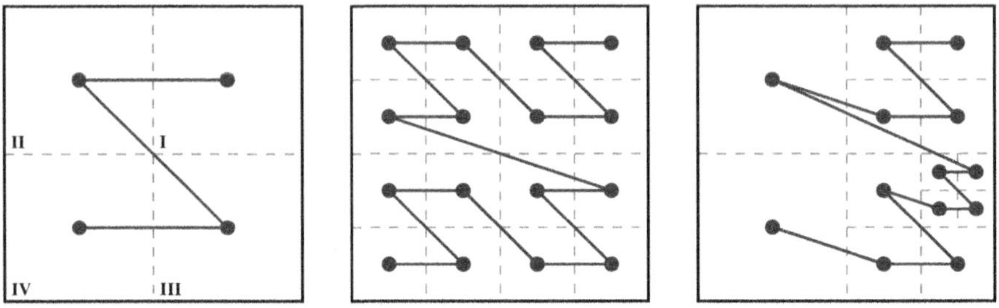

Fig. 2.4. Template curve for the Morton ordering (left), its first level of refinement (center), and an adaptive refinement (right).

(see Figure 2.3, right). Treating objects as point masses, the direction of principle inertia in the geometry is found by computing eigenvectors of a 3×3 matrix assembled from the point masses.

Space-Filling Curves

A second class of geometric partitioners utilizes a one-dimensional "traversal" or linearization to order objects or groups of objects. After determining a one-dimensional ordering, subdomains are formed from contiguous segments of the linearization. This technique produces well-formed subdomains if the ordering preserves *locality*, i.e., if objects that are close in the linearization are also close in the original coordinate space.

The linearization is often achieved using space-filling curves (SFCs). SFCs provide continuous mappings from one-dimensional to d-dimensional space [99]. They have been used to linearize spatially-distributed data for partitioning [2, 17, 33, 87, 89, 94], storage and memory management [22, 79], and computational geometry [7].

SFCs are typically constructed recursively from a single stencil. Each level of refinement replaces segments of the SFC with a new copy of the curve's stencil, subject to spatial rotations and reflections. The SFC can come arbitrarily close to any point in space. Most importantly for partitioning, some SFCs preserve locality, which Edwards [33] defines formally. Several orderings with different degrees of complexity and locality are possible; only the commonly-used Morton and Hilbert orderings are included here.

The Morton (Z-code or Peano) ordering [80, 84] is a simple SFC that traverses a quadrant's children in a "Z"-like pattern (in the order I, II, III, IV in Figure 2.4). The pattern at each refinement is identical to that used by its ancestors; no rotations or reflections are performed. However, there are large "jumps" in its linearization, particularly as the curve transitions from quadrant II to quadrant III, so Morton ordering does not always preserve locality. The jumps are even more apparent in three dimensions. Nevertheless, because of its simplicity, Morton ordering is viable in some circumstances, and provides a base ordering for all SFCs [17, 60].

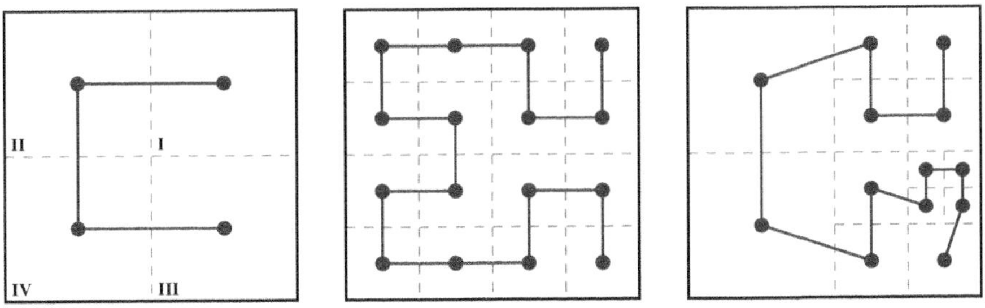

Fig. 2.5. Template curve for the Hilbert ordering (left), its first level of refinement (center), and an adaptive refinement (right).

The Hilbert ordering uses the Peano-Hilbert SFC [11, 90, 91] to order quadrants. It uses a bracket-like template with rotations and inversions to keep quadrants closer to their neighbors. (Figure 2.5). Hilbert ordering is locality preserving, and tends to be the most useful for partitioning.

SFC orderings can be applied directly to objects given only the objects' spatial coordinates [2]. Each object is assigned a unique "key" representing the object's position along the SFC. This key is a number in the range $[0, 1]$ that specifies the point on the SFC that passes closest to the object. The object are then ordered by their keys; this ordering can be done via global sorting, binning [8, 27, 29], or traversing an octree representing the SFC [17, 42, 44, 71, 75]. The one-dimensional ordering is then partitioned into appropriately sized pieces; all objects within a piece are assigned to one subdomain.

SFC partitioning was first used by Warren and Salmon [127] in particle-based gravitational simulations. They used a Morton ordering, but acknowledged that Hilbert ordering would improve locality. Patra and Oden [81, 89], Parashar and Browne [87], and Edwards [33] used Hilbert SFC ordering for finite element meshes. Patra and Oden choose cuts along the SFC to balance computational work in their hp-adaptive computation. Pilkington and Baden [94] apply SFCs for dynamic load balancing with a uniform mesh where computational workloads vary. Steensland, et al. [111] looked at SFCs for partitioning structured grids which undergo adaptive refinement. Octree partitioning [42, 71, 75] implements SFC partitioning using octree data structures commonly used in mesh generation. Mitchell's Refinement Tree partitioning [76, 78] uses nodal connectivity in adaptively refined meshes (instead of coordinate values) to generate a SFC through mesh elements; while this approach is not strictly a geometric method, the resulting decompositions are qualitatively identical to SFC-produced decompositions.

2.2.2 Global Graph-Based Partitioning

A popular and powerful class of partitioning procedures make use of connectivity information rather than spatial coordinates. These methods use the fact that the partitioning problem in Section 2.1.1 can be viewed as the partitioning of an induced

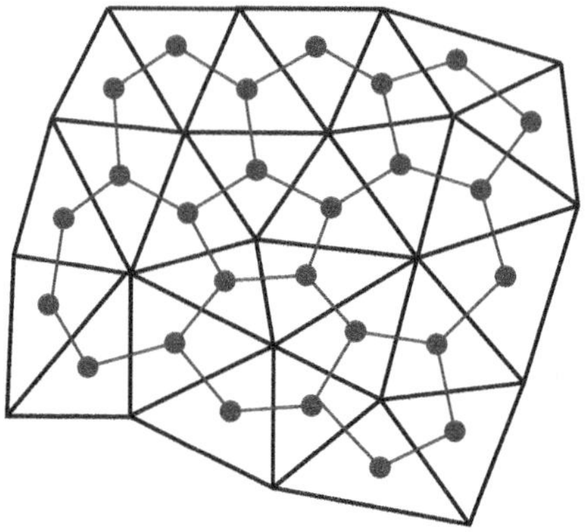

Fig. 2.6. Example two-dimensional mesh from Figure 2.1 (left) with its induced graph. (For the color version, see Figure A.1 on page 467).

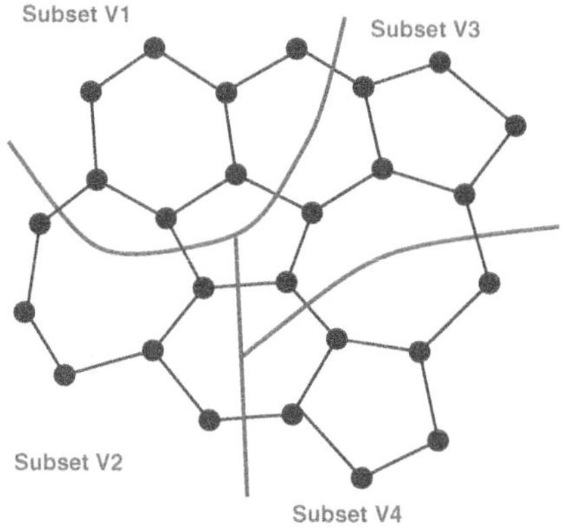

Fig. 2.7. Four-way partitioning of the graph from Figure 2.6. (For the color version, see Figure A.2 on page 467).

graph $G = (V, E)$, where objects serve as the graph vertices (V) and connections between objects are the graph edges (E). For example, Figure 2.6 shows an induced graph for the mesh in Figure 2.1; here, elements are the objects to be partitioned and, thus, serve as vertices in the graph, while shared element faces define graph edges.

A k-way partition of the graph G is obtained by dividing the vertices into subsets $V_1, ..., V_k$, where $V = V_1 \cup ... \cup V_k$, and $V_i \cap V_j = \oslash$ for $i \neq j$. Figure 2.7 shows one possible decomposition of the graph induced by the mesh in Figure 2.6. Vertices and edges may have weights associated with them representing computation and communication costs, respectively. The goal of graph partitioning, then, is to create subsets V_k with equal vertex weights while minimizing the weight of edges "cut" by subset boundaries. An edge e_{ij} between vertices v_i and v_j is cut when v_i belongs to one subset and v_j belongs to a different one. In Figure 2.7, eight edges are cut. Algorithms to provide an optimal partitioning are NP-complete [46, 47],

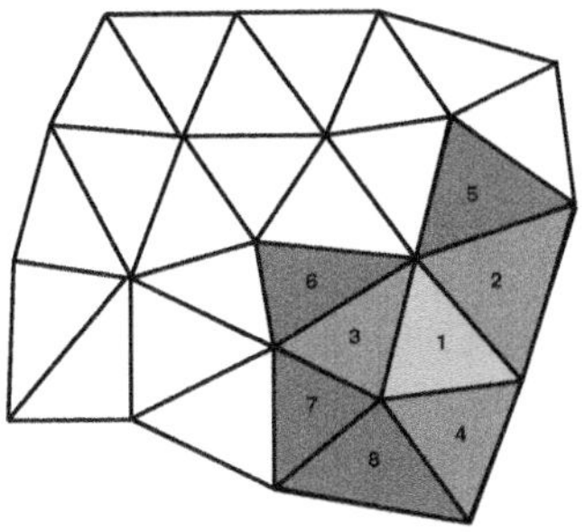

Fig. 2.8. Example greedy partitioning of a small mesh. Numbers indicate the order in which elements are added to the subdomain being constructed.

so heuristic algorithms are generally used. The graph partitioning is related back to the mesh partitioning problem by creating subdomains of the mesh corresponding to each subset V_i. Figure 2.1 (right) shows the partitioning of the mesh based on the graph partitioning of Figure 2.7.

A number of algorithms have been developed to partition graphs. Many of these were developed as static partitioners, intended for use as a preprocessing step rather than as a dynamic load balancing procedure. Some of the multilevel procedures do operate in parallel and can be used for dynamic load balancing.

Greedy Partitioning

Farhat [36] applied graph partitioning to the mesh partitioning problem. The graph is partitioned by a *greedy algorithm* (GR) that builds each subdomain by starting with a vertex and adding adjacent vertices until the subdomain's target size has been reached, see Figure 2.8. The procedure then chooses another unassigned vertex and builds the next subdomain. Farhat [38] reports success using these procedures. Such greedy procedures can also be components of the more commonly used multilevel partitioners described below.

Spectral Partitioning

A very well known static graph partitioning method is Recursive Spectral Bisection (RSB) [97, 107]. In RSB, the Laplacian matrix L of a graph is constructed. Each diagonal entry l_{ii} is the degree of vertex i; non-diagonal entries l_{ij} are -1 if edge e_{ij} exists in the graph, and 0 otherwise. The eigenvector x associated with the smallest non-zero eigenvalue of L is then used to divide the vertices into two sets. The median value of x is found. Then, for each x_i, if x_i is less than the median, vertex i is assigned to the first set; otherwise, it is assigned to the second set. This bisection procedure is repeated on the subgraphs until the number of sets is equal to the number of desired partitions.

RSB generally produces high quality partitions. The eigenvector calculation, however, is very expensive and, thus, RSB is used primarily for static partitioning.

Strategies using additional eigenvectors to compute four or eight partitions in each stage have proven to be effective while reducing the cost to partition [54].

Multilevel Partitioning

By far, the most successful global graph-based algorithms for static partitioning are multilevel graph partitioners [15, 55, 66], as evidenced by the number of static graph partitioning packages available [53, 64, 92, 98]. Multilevel methods' operation is much like the V-cycle used in multigrid solvers, in that an initial solution is computed on a coarse representation of the graph and used to obtain better solutions on finer representations.

Multilevel graph partitioning involves three major phases: (i) *coarsening*, the construction of a sequence of smaller graphs that approximate the original, (ii) *partitioning* of the coarsest graph, and (iii) *uncoarsening*, the projection of the partitioning of the coarsest graph onto the finer graphs, with a local optimization applied to improve the partitioning at each step. A simple example of this procedure for a small graph with two levels of coarsening is shown in Figure 2.9.

Coarsening procedures typically use a vertex matching algorithm that identifies vertices that can be combined to create coarse vertices. The set of edges from the coarse vertex is taken as the union of the edges for the combined vertices. The sum of the combined vertices' weights is used as the coarse vertex's weight. In this way, the structure and workloads of the input graph are preserved in the coarse representations. Matching at each level can be done by randomly selecting unmatched vertices [15, 55, 66, 125] or using heuristics [6, 23, 45, 48, 49, 66]. For example, heavy-edge matching combines the two vertices sharing the edge with the heaviest edge weight [66], suggesting that vertices with the strongest affinity toward each other should be combined.

The coarsest graph is then partitioned. Since this graph is small, a spectral method [55, 66] can be used efficiently. Faster greedy methods [66] can also be used; while they produce lower quality coarse partitions, local optimizations during the uncoarsening phase improve partition quality. A local optimization may also be used at this point to attempt to encourage incrementality [103]. A geometric procedure such as an SFC may also be used for this coarse partitioning [68].

The coarse decomposition is projected to the finer graphs, with refinements of the partitions made at each graph level. Typically a local optimization technique reduces a communication metric while maintaining and improving load balance. Most of the local optimization approaches are based on the Kernighan-Lin (KL) graph bisection algorithm [69] or its linear-time implementation by Fiduccia and Maytheses (FM) [39]. These techniques make a series of vertex moves from one partition to another, measuring the "gain" or improvement in the metric for each move; moves with high gain are accepted. Karypis and Kumar [66] perform only a few iterations of their KL-like procedure, noting that most of the gain is usually achieved by the first iteration. Hendrickson and Leland [55] continue their KL-like procedure to allow for the discovery of sequences of moves that, while individually making the decomposition worse, may lead to a net improvement. This allows the procedure to escape

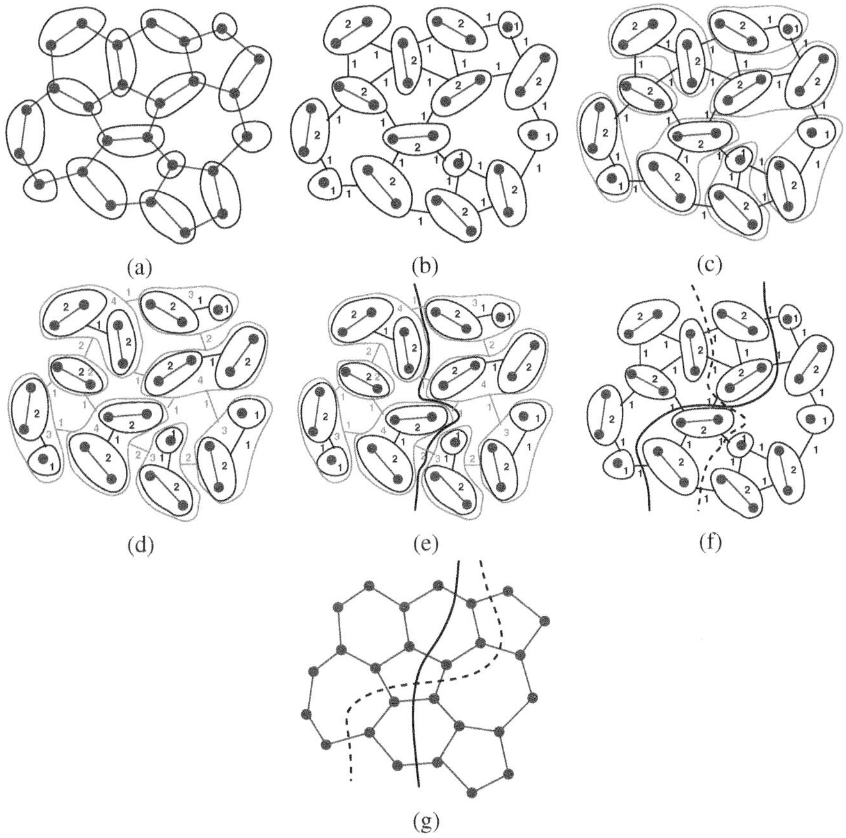

Fig. 2.9. Multilevel partitioning of the induced graph of Figure 2.6. Vertex matching in (a) leads to the coarse graph in (b). A second round of vertex matching in (c) produces the coarse graph in (d). This coarsest graph is partitioned in (e). The graph is uncoarsened one level and the partitioning is optimized in (f). The second level of uncoarsening, and another round of local optimization on this partitioning produces the final two-way partitioning shown in (g). (For the color version, see Figure A.3 on page 468).

from local minima. Walshaw, et al. [125] define a relative gain value for each vertex, intended to avoid collisions (i.e., vertices on opposite sides of a boundary each being selected to move).

Parallel implementation of multilevel graph partitioners has allowed them to be used for dynamic load balancing [67, 125]. These methods produce very high quality partitionings, but at a higher cost than geometric methods. Graph partitioners are not inherently incremental, but modifications such as the local methods described below can make them more effective for dynamic repartitioning.

2.2.3 Local Graph-based Methods

In an adaptive computation, dynamic load balancing may be required frequently. Applying globalpartitioning strategies after each adaptive step can be costly relative to solution time. Thus, a number of dynamic load balancing techniques that are intended to be fast and incrementally migrate data from heavily to lightly loaded processes have been developed. These are often referred to as local methods.

Unlike global partitioning methods, local methods work with only a limited view of the application workloads. They consider workloads within small, overlapping sets of processors to improve balance within each set. Heavily loaded processors within a set transfer objects to less heavily loaded processors in the same set. Sets can be defined by the parallel architecture's processor connectivity [70] or by the connectivity of the application data [58, 130]. Sets overlap, allowing objects to move between sets through several iterations of the local method. Thus, when only small changes in application workloads occur through, say, adaptive refinement, a few iterations of a local method can correct imbalances while keeping the amount of data migrated low. For dramatic changes in application workloads, however, many iterations of a local method are needed to correct load imbalances; in such cases, invocation of a global partitioning method may result in a better, more cost-effective decomposition.

Local methods typically consist of two steps: (i) computing a map of how much work (nodal weight) must be shifted from heavily loaded to lightly loaded processors, and (ii) selecting objects (nodes) that should be moved to satisfy that map. Many different strategies can be used for each step.

Most strategies for computing a map of the amount of data to be shifted among processes are based on the diffusive algorithm of Cybenko [24]. Using processor connectivity or application communication patterns to describe a computational "mesh," an equation representing the workflow is solved using a first-order finite-difference scheme. Since the stencil of the scheme is compact (using information only from neighboring processes), the method is local.

Several variations of this strategy have been developed to reduce data movement or improve convergence. Hu and Blake [58] take a more global view of load distributions, computing a diffusion solution while minimizing work flow over edges of a graph of the processes. Their method is used in several parallel graph-partitioning libraries [100, 125]. Such diffusion methods have been coupled with multilevel graph partitioners (see Section 2.2.2) to further improve their effectiveness [56, 100, 103, 124, 125].

Other techniques for accelerating the convergence of diffusion schemes include use of higher-order finite difference schemes and dimensional exchange. Watts, et al. [128, 129] use a second-order implicit finite difference scheme to solve the diffusion equation; this scheme converges to global balance in fewer iterations, but requires more work and communication per iteration. In dimensional exchange [24, 31, 132, 134], a hypercube architecture is assumed. (The algorithm can be used on other architectures by logically mapping the architecture to a hypercube.)

Processes exchange work with neighbors along each dimension of the hypercube; after looping over all dimensions, the workloads are balanced.

Demand-driven models are also common [26, 34, 70, 86, 130, 131, 132]. These models operate in either of two ways: (i) underloaded processes request work from their overloaded neighboring processes, or (ii) overloaded processes send work to their underloaded neighbors. The result is similar to diffusion algorithms, except that nodes are transferred to only a subset of neighbors rather than distributed to all neighbors. Version (i) of this model has shown to be more effective than (ii), as the majority of load-balancing work is performed by the underloaded process and overloading of the receiving process is avoided [132]. As in the diffusion algorithm, neighbors can be defined by following the physical processor network [70] or the logical data connections [131]. Ozturan's iterative tree-balancing procedure [26, 86] groups processes into trees based upon their work requests, moving work among processes within trees. This more global view accelerates the convergence of the diffusion, but also increases the average number of neighboring processes per process in the application's communication graph.

The second step of a local method is deciding which objects (graph nodes) to move to satisfy the workload transfers computed in the first step. Typically, variants of the KL [69] or FM [39] local optimization algorithms (used for refinement of multilevel partitions) are used. For each object, the gain toward a specific goal achieved by transferring the object to another process is computed. Many options for the gain function have been used (e.g., [28, 55, 125, 131, 134]). Most commonly, the weight of graph edges cut by subdomain boundaries is minimized. However, other goals might include minimizing the amount of data migrated [28, 100], minimizing the number of process neighbors, optimizing subdomain shape [30, 118], or some combination of these goals. The set of objects producing the highest gain is selected for migration. Selection continues until the actual workload transferred is roughly equal to the desired workload transfers.

2.3 Algorithm Comparisons

A number of theoretical and empirical comparisons of various partitioning strategies have been performed [14, 27, 37, 38, 41, 51, 57, 111, 112, 115]. Selection of the method that is most effective for an application depends on trade-offs between incrementality, speed and quality that can be tolerated by the application. A PDE solver which uses a single decomposition throughout the computation should consider strategies that produce high-quality partitions, with less concern for execution speed of the partitioner. A solver which uses frequent adaptivity will want to consider strategies that execute quickly and are incremental, with less emphasis on partition quality. A procedure which does not readily provide adjacency information will be restricted to geometric methods. This section summarizes and cites key results.

- RCB, URB
  - Geometric method: only coordinate information needed.

  - Incremental and suitable for dynamic load balancing [51].
  - Executes very quickly [115].
  - Moderate quality decompositions. Cutting planes help keep the number of objects on subdomain boundaries small for well-shaped meshes [18]. Unfortunate cuts through highly refined regions [115] or complex domain geometry [75, 73] can lead to poor decompositions. URB produces more uniform subdomain aspect ratios than RCB when there is a large variation in object density [61].
  - Conceptually simple; straightforward to implement in parallel [29].
  - Maintains geometric locality [51, 123].
  - Simple to determine intersections of objects with subdomains, e.g., for parallel contact detection and smoothed particle hydrodynamics simulations [96]; subdomains are described by simple parallelepipeds.
- RIB
  - Geometric method: only coordinate information needed.
  - Not incremental; may be unsuitable for dynamic load balancing [42].
  - Executes almost as quickly as RCB [115].
  - Slightly higher quality decompositions than RCB; lower quality than spectral and multilevel graph partitioning [38, 115]. Unfortunate cuts through highly refined regions can cause poor decompositions [115].
  - Conceptually simple; straightforward to implement in parallel [29, 106].
  - Maintains geometric locality.
  - Simple to determine intersections of objects with subdomains, e.g., for parallel contact detection and smoothed particle hydrodynamics simulations [96].
- SFC
  - Geometric method: only coordinate information needed.
  - Incremental and suitable for dynamic load balancing [42, 51].
  - Executes very quickly [42, 94, 112].
  - Slightly lower quality decompositions than geometric bisection methods [89].
  - Conceptually simple; straightforward to implement in parallel [94].
  - Choice of SFC used depends on locality requirements; Hilbert is usually best [17].
  - The global ordering induced by sorting SFC keys can be exploited to order data to improve cache performance during computation, and can provide automated translations between global and per-process numbering schemes [33, 88].
  - Possible to determine intersections of objects with subdomains, e.g., for parallel contact detection and smoothed particle hydrodynamics simulations [27].
- Greedy partitioning
  - Graph-based method: connectivity information is required.
  - Not incremental; may be unsuitable for dynamic load balancing.
  - Executes quickly [38, 118, 123, 124].

  - Medium-quality decompositions [123], better than RIB [38], and good with respect to subdomain aspect ratio [38]. Tends to leave non-connected or strip-wise subdomains in the last few partitions computed [57].
  - Difficult to implement in parallel.
  - Does not maintain geometric locality [123].
- Spectral graph partitioning
  - Graph-based method: connectivity information is required.
  - Not incremental; may be unsuitable for dynamic load balancing [112]. Van Driessche and Roose [117] developed modifications to include incrementality.
  - Executes very slowly [124].
  - Very high quality decompositions [124].
  - More difficult to parallelize than geometric methods [5, 109].
  - Does not maintain geometric locality [123].
  - Suitable primarily for static partitioning.
- Multilevel graph partitioning
  - Graph-based method: connectivity information is required.
  - Not incremental; may be unsuitable for dynamic load balancing [123]. Metrics may include migration cost to improve incrementality [103].
  - Executes slowly [115, 124].
  - Very high quality decompositions [67, 125].
  - Difficult to implement in parallel [67].
  - Does not maintain geometric locality.
- Local graph-based methods
  - Graph-based method: connectivity information is required.
  - Incremental; suitable for dynamic load balancing [26, 86].
  - Usually execute quickly, but several iterations may be needed for global balance. Also, more sophisticated techniques can be more expensive [112].
  - High quality decompositions, given a good starting decomposition.
  - Straightforward to implement in parallel [26, 86]; can be incorporated into multilevel strategies [55, 67, 125].
  - Useful as a post-processing step for other methods to improve partition quality [42, 75].

2.4 Software

Many software packages are available to provide static and dynamic load balancing to applications. Using these packages, application developers can access a variety of high-quality implementations of partitioning algorithms. Many packages also include supporting functionality (e.g., data migration tools and unstructured communication tools) commonly needed by applications using load balancing. Use of these packages saves application developers the effort of learning and implementing partitioning algorithms themselves, while allowing them to compare partitioning strategies within their applications. Moreover, many of the packages are available as open-source software; see bibliography entries for the packages cited for distribution details.

Static partitioning software is typically used as a pre-processor to the application. It can be used in two ways: as a stand-alone tool or as a function call from the application. In stand-alone mode, input files describe the problem domain to be partitioned; the format of these files is determined by the partitioning software. The computed decomposition is also written to files. The application must then read the decomposition files to distribute data appropriately. Function-call interfaces to static partitioners allow them to be called directly by applications during pre-processing phases of the application.

Several graph partitioning packages have been developed for static load balancing; they include Chaco [53], Metis [64], Jostle [120], Party [98] and Scotch [93]. These tools run in serial and have both stand-alone and function interfaces. For the stand-alone mode, users provide input files describing the problem domain in terms of a graph, listing vertices (objects), edges between vertices, vertex and edge weights, and possibly coordinates. The function-call interfaces accept a graph description of the problem through arrays using compressed sparse row (CSR) format. In both modes, applications have to convert their application data into the appropriate graph format.

By necessity, dynamic load-balancing software uses function call interfaces, as file-based interfaces would be unacceptable for balancing during a computation. Similarly, dynamic load-balancing software is executed in parallel, assuming an existing distribution of data; parallel execution is required to maintain scalability of the application. Two types of dynamic partitioning software are available: algorithm-specific libraries and toolkits of partitioning utilities.

ParMETIS [68] and PJostle [120] are two widely used algorithm-specific libraries. Both provide multi-level and diffusive graph partitioning. Like their serial counterparts, they accept input in CSR format, with extensions describing the existing partition assignment of the vertices; the arrays describing the application data as a graph in this compressed format must be built by the application. ParMETIS includes support for multiple weights per vertex [65] and edge [101], enabling multi-constraint and multi-objective partitioning (see Section 2.5.2). PJostle allows multiple vertex weights for multiphase applications [126] (see Section 2.5.2) and a network description [122] to allow partitioning for heterogeneous computing systems (see Section 2.5.3).

Load-balancing toolkits such as Zoltan [29] and DRAMA [72] incorporate suites of load-balancing algorithms with additional functionality commonly needed by dynamic applications. Both Zoltan and DRAMA include geometric partitioners (through implementations in the toolkits) and graph-based partitioners (through interfaces to ParMETIS and PJostle). They enable comparisons of various methods by providing a common interface for all partitioners and allowing applications to select a method via a single parameter. They also provide support for moving data between processors to establish a new decomposition.

The Zoltan toolkit [29] provides parallel dynamic load balancing and data management services to a wide range of applications, including particle simulations, mesh-based simulations, circuit simulations, and linear solvers. It includes geometric bisections methods (RCB, RIB), space-filling curve methods (HSFC, Octree, Refinement Tree), and graph-based partitioning (through ParMETIS and PJostle). Unlike the graph-partitioning libraries, Zoltan's design is "data-structure neutral"; i.e., Zoltan does not require the application to use or build particular data structures for Zoltan. Instead, a callback-function interface provides a simple, general way for applications to provide data to Zoltan. Applications provide simple functions returning, e.g., lists of objects to be partitioned, coordinates for the objects, and relationships between objects. Zoltan calls these functions to obtain application information needed to build its data structures. Once an application implements these callback functions, switching between load-balancing methods requires changing only one parameter.

Zoltan also includes a number of utilities that simplify development of dynamic applications. Its data migration tools assist in the movement of data among processors as they move from an old decomposition to a new one. Because Zoltan does not have information about application data structures, it cannot update them during migration. But given callback functions that pack and unpack data from communication buffers, its migration tools perform all communication needed for data migration. Zoltan's unstructured communication package provides a simple mechanism for complex communication among processors, freeing application developers from the details of individual message sends and receives. Its distributed data directory provides an efficient, scalable utility for locating data in the memory space of other processes. Key kernels of contact detection simulations—finding the partitions owning points and regions in space—are included for Zoltan's geometric and HSFC methods.

The DRAMA (Dynamic Re-Allocation of Meshes for parallel finite element Applications) toolkit [72] provides parallel dynamic load balancing and support services to mesh-based applications. DRAMA assumes a basic data structure of a mesh and enables partitioning of the mesh nodes, elements or both. The mesh is input to DRAMA through array-based arguments. Like Zoltan, DRAMA provides a number of partitioning strategies, including recursive bisection methods and graph partitioning through interfaces to ParMETIS and PJostle.

DRAMA includes a robust cost-model for use in partitioning. This model accounts for both computation and communication costs in determining effective decompositions. Because it assumes a mesh data structure, DRAMA includes more sophisticated support for data migration than Zoltan. It migrates its input mesh to its new location; this migrated mesh can then serve as a starting point for the application data migration. DRAMA provides support for heterogeneous computing architectures through PJostle's network description [121] (see Section 2.5.3). It also includes extensive support for contact detection and crash simulations.

Load-balancing tools that are tied more closely to specific applications also exist. For example, the PLUM system [82] provides dynamic load balancing for applications using adaptively refined meshes. Its goal is to minimize load-balancing overhead during adaptive computations. To do so, it balances with respect to a coarse mesh in the adaptive simulation using element weights proportional to the number of elements into which each coarse element has been refined. It uses an external partitioning library (e.g., ParMETIS) to compute a decomposition, and then uses a similarity matrix to remap partitions in a way that minimizes data movement between the old and new decompositions. Another example, the VAMPIRE library [110], produces decompositions for structured adaptive mesh refinement applications. Assuming the refined mesh is represented as a tree of uniform grids, it uses a SFC algorithm to distribute the grids to processors to evenly distribute work while attempting to minimize communication between the grids. Load-balancing systems are also included as parts of larger parallel run-time systems; see, e.g., CHARM++ [62] and PREMA [4].

2.5 Current Challenges

As parallel simulations and environments become more sophisticated, partitioning algorithms must address new issues and application requirements. Software design that allows algorithms to be compared and reused is an important first step; carefully designed libraries that support many applications benefit application developers while serving as test-beds for algorithmic research. Existing partitioners need additional functionality to support new applications. Partitioning models must more accurately represent a broader range of applications, including those with non-symmetric, non-square, and/or highly-connected relationships. And partitioning algorithms need to be sensitive to state-of-the-art, heterogeneous computer architectures, adjusting work assignments relative to processing, memory and communication resources.

2.5.1 Hypergraph Partitioning

Development of robust partitioning models is important in load-balancing research. While graph models (see Section 2.2.2) are often considered the most effective models for mesh-based PDE simulations, they have limitations for larger classes of problems (e.g., electrical systems, computational biology, linear programming). These new problems are often more highly connected, more heterogeneous, and less symmetric than mesh-based PDE problems.

As an alternative to graphs, hypergraphs can be used to model application data [19, 20]. A hypergraph $HG = (V, HE)$ consists of a set of vertices V representing the data objects to be partitioned and a set of hyperedges HE connecting two *or more* vertices of V. By allowing larger sets of vertices to be associated through edges, the hypergraph model overcomes many of the limitations of the graph model.

A key limitation of the graph model is that its edge-cut metric only approximates communication volume induced by a decomposition (see Section 2.1.3). While this approximation is adequate for traditional finite-element, finite-volume, and finite-difference simulations, it is not sufficient for more highly connected and unstructured data. In the hypergraph model, however, the number of hyperedge cuts is equal to the communication volume, providing a more effective partitioning metric.

Catalyurek and Aykanat [20] also describe the greater expressiveness of hypergraph models over graph models. Because edges in the graph model are non-directional, they imply symmetry in all relationships, making them appropriate only for problems represented by square, structurally symmetric matrices. Systems A with non-symmetric structure must be represented by a symmetrized model $A + A^T$, adding new edges to the graph and further skewing the communication metric. While a directed graph model could be adopted, it would not improve the accuracy of the communication metric. Likewise, graph models can not represent rectangular matrices, such as those arising in linear programming. Kolda and Hendrickson [52] propose using bipartite graphs. For an $m \times n$ matrix A, vertices $m_i, i = 1, \ldots, m$ represent rows, and vertices $n_j, j = 1, \ldots, n$ represent columns. Edges e_{ij} connecting m_i and n_j exist for non-zero matrix entries a_{ij}. But as in other graph models, the number of edge cuts only approximates communication volume.

Hypergraph models, on the other hand, do not imply symmetry in relationships, allowing both structurally non-symmetric and rectangular matrices to be represented. For example, the rows of a rectangular matrix could be represented by the vertices of a hypergraph. Each matrix column would be represented by a hyperedge connecting all non-zero rows in the column [20].

The improved communication metric and expressiveness of hypergraph models lead to impressive results. Using hypergraph partitioning, Catalyurek and Aykanat [20] report reductions in communication volume of 12-15% compared to graph partitioning for matrices from traditional finite difference applications. But for a broader range of matrices, including examples from linear programming, circuit simulations and stochastic programming, hypergraph partitioning produced reductions of 30-38% on average. Time to compute the hypergraph decomposition was 34-130% greater than that required to compute a graph decomposition.

Hypergraph partitioning's effectiveness has been demonstrated in many areas, including VLSI layout [16], sparse matrix decompositions [20, 119], and database storage and data mining [21, 85]. Serial hypergraph partitioners are available (e.g., hMETIS [63], PaToH [20, 19], Mondriaan [119]). Research into parallel hypergraph partitioning includes a disk-based implementation used for partitioning Markov matrices [116] and a distributed memory implementation in Zoltan [13]. Parallel implementation is needed for hypergraph partitioning to be viable for very large simulations. Additionally, incremental hypergraph algorithms (analogous to diffusive graph algorithms [24]) will be needed for dynamic applications.

2.5.2 Multi-criteria Partitioning

Most load-balancing research has focused on cases having a single load to be balanced. Multi-phase simulations, however, might have different work loads in each phase of a simulation. For example, a multiphysics simulation might include both fluid flow and solid mechanics phases. Crash simulations typically have a finite-element solve phase and a contact-detection phase. Even within a finite element simulation, the matrix assembly and matrix solve phases may have significantly different load characteristics depending on the physics of the problem.

One approach to balancing multi-phase simulations is to use separate decompositions for each phase, mapping data between decompositions when needed. This approach has been used with great success in crash simulations, where static graph-based decompositions were used for the finite element phase and dynamic geometric decompositions were used for contact detection [96]; data were transferred between the decompositions as needed between phases.

Still, the idea of having a single decomposition that is balanced with respect to multiple loads is attractive. With such a decomposition, no mapping of data is needed between phases, reducing application communication costs. Each object to be balanced would have a vector v of weights associated with it; the jth component of v would represent the object's workload in phase j. A single decomposition would then be generated that balances each vector component.

Walshaw, et al. [126] developed a multiphase graph partitioner in Jostle [120]. Assuming components of weight vector v represent a vertex's participation in a phase, they say the "type" of the vertex is the first phase j in which the vertex participates, i.e., for which $v[j] > 0$. They then balance each type of vertex separately, maintaining partition information from lower types as "stationary" vertices in the partitioning of higher types. That is, in computing a partition for vertices of type k, $k > j$, all vertices of type j within a partition are represented by a single "super-vertex" whose partition assignment is fixed to a particular partition; edges between these stationary vertices and vertices of type k are maintained to represent data dependencies between the phases. A standard graph partitioner is used to partition each type of vertices; in attempting to minimize cut edges, the graph partitioner is likely to assign type k vertices to the same partition as type j vertices to which they are connected, keeping inter-phase communication costs low.

The multi-constraint graph-partitioning model of Karypis, et al. [65, 104] in METIS [64] and ParMETIS [67] uses vertex weight vectors to create multiple load-balancing constraints. Using this model, they can compute both multiphase decompositions and decompositions with respect to multiple criteria, e.g., workloads and memory usage. Their approach is built on the multi-level framework commonly used in graph partitioning (see Section 2.2.2), with modifications made in the coarsening, coarse-partitioning, and refinement steps to accommodate multiple vertex weights. During coarsening, the same heavy-edge metric used in single-constraint partitioning is used to select vertices to be combined; this metric combines a vertex with the neighboring vertex sharing the heaviest edge weight. In multi-constraint partitioning, ties between combinations with the same metric value are broken by a "balanced-edge" metric that attempts to make all weights of the combined vertex as close to the same value as possible, as more uniform weights are easier to balance in the coarse-partitioning and refinement steps. A greedy recursive graph bisection algorithm is used to compute the coarse partition; at each level of recursion, two subdomains A and B are created by removing vertices from A (which initially contains the entire domain) and adding them to B (which initially is empty). In the multi-constraint case, vertices are selected based on their ability to reduce the heaviest weight of A the most. In refinement, KL [69] or FM [39] procedures are used. For multi-constraint partitioning, queues of vertices that can be moved are maintained for each weight and neighboring partition; vertices are again selected based on their ability to reduce the maximum imbalance over all weights while reducing the number of edges cut. To enforce the balance constraints in multi-constraint partitioning, an additional shifting of vertices among processors without regard to increases in the edge cut weight is sometimes needed before refinement.

Because geometric partitioners are preferred for many applications, Boman, et al. pursued multi-criteria partitioning for geometric partitioners, specifically RCB [12]. Their implementation is included in Zoltan [29]. RCB consists of a series of one-dimensional partitioning problems; objects i are ordered linearly by their coordinate values corresponding to the direction of the cut. Like other approaches, objects i have vector weights v_i representing the load-balance criteria. Instead of imposing multiple constraints, however, Boman, et al. formulate each one-dimensional problem as an optimization problem where the objective is to find a cut s such that

$$\min_s \max(g(\sum_{i \leq s} v_i), g(\sum_{i > s} v_i)),$$

where g is a monotonically non-decreasing function in each component of the input vector (typically $g(x) = \sum_j x_j^p$ with $p = 1$ or $p = 2$, or $g(x) = \|x\|$ for some norm). This objective function is unimodal with respect to s. In other words, starting with $s = 1$ and increasing s, the objective decreases, until at some point the objective starts increasing; that point defines the optimal bisection value s. (Note that the objective may be locally flat (constant), so there is not always a unique minimizer.) An optimal cut is computed in each coordinate direction; the cut producing the best balance is accepted.

In general, computing multi-criteria decompositions becomes more difficult as the number of criteria and/or number of partitions increases. As a result, partition quality can degrade. Likewise, multi-criteria partitions are more expensive to compute than single-criterion partitions; the extra cost, however, may be justified by the improved load balance and reduction of data transfer.

2.5.3 Resource-Aware Balancing

Cluster and grid computing have made hierarchical and heterogeneous computing systems increasingly common as target environments for large-scale scientific computation. Heterogeneity may exist in processor computing power, network speed, and memory capacity. Clusters may consist of networks of multiprocessors with varying computing and memory capabilities. Grid computations may involve communication across slow interfaces between vastly different architectures. Modern supercomputers are often large clusters with hierarchical network structures. Moreover, the characteristics of an environment can change during a computation due to increased multitasking and network traffic. For maximum efficiency, software must adapt dynamically to the computing environment and, in particular, data must be distributed in a manner that accounts for non-homogeneous, changing computing and networking resources. Several projects have begun to address resource-aware load balancing in such heterogeneous, hierarchical, and dynamic computing environments.

Minyard and Kallinderis [74] use octree structures to conduct partitioning in dynamic execution environments. To account for the dynamic nature of the execution environment, they collect run-time measurements based on the "wait" times of the processors involved in the computation. These "wait" times measure how long each CPU remains idle while all other processors finish the same task. The objects are assigned load factors that are proportional to the "wait" times of their respective owning processes. Each octant load is subsequently computed as the sum of load factors of the objects contained within the octant. The octree algorithm then balances the load factors based on the weight factors of the octants, rather than the number of objects contained within each octant.

Walshaw and Cross [122] conduct multilevel mesh partitioning for heterogeneous communication networks. They modify a multilevel algorithm in PJostle [120] seeking to minimize a cost function based on a model of the heterogeneous communication network. The model gives a static quantification of the network heterogeneity as supplied by the user in a Network Cost Matrix (NCM). The NCM implements a complete graph representing processor interconnections. Each graph edge is weighted as a function of the length of the path between its corresponding processors.

Sinha and Parashar [108] present a framework for adaptive system-sensitive partitioning and load balancing on heterogeneous and dynamic clusters. They use the Network Weather Service (NWS) [133] to gather information about the state and capabilities of available resources; then they compute the load capacity of each node as a weighted sum of processing, memory, and communications capabilities. Reported experimental results show that system-sensitive partitioning resulted in significant decrease of application execution time.

Faik, et al. [35] present the Dynamic Resource Utilization Model (DRUM) for aggregating information about the network and computing resources of an execution environment. Through minimally instrusive monitoring, DRUM collects dynamic information about computing and networking capabilities and usage; this information determines computing and communication "powers" that can be used as the percentage of total work to be assigned to processes. DRUM uses a tree structure to represent the underlying interconection of hierarchical network topologies (e.g., clusters of clusters, or clusters of multiprocessors). Using DRUM's dynamic monitoring and power computations, they achieved 90% of optimal load distribution for heterogeneous clusters [35].

Teresco [114] has implemented hierarchical partitioning procedures within the software package Zoltan. These procedures can be used alone, or can be guided by DRUM [35]. Hierarchical partitioning allows any combination of Zoltan's load-balancing procedures to be used on different levels and subtrees of hierarchical machine models. Tradeoffs in execution time, imbalance, and partition quality (*e.g.*, surface indices, interprocess connectivity) can hold greater importance in heterogeneous environments [115], making different methods more appropriate in certain types of environments. For example, consider the cluster of SMPs connected by Ethernet shown in Figure 2.10. A more costly graph partitioning can be done to partition into two subdomains assigned to the SMPs, to minimize communication across

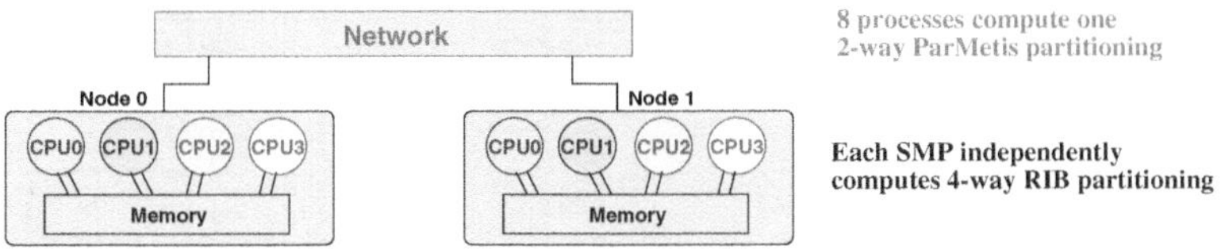

Fig. 2.10. Hierarchical balancing algorithm selection for two 4-way SMP nodes connected by a network. (For the color version, see Figure A.4 on page 468).

the slow network interface, possibly at the expense of some computational imbalance. Then, a fast geometric algorithm can be used to partition independently within each SMP. Teresco [114] reports that while multilevel graph partitioning alone often achieves the fastest computation times, there is some benefit to using this hierarchical load balancing, particularly in maintaining strict load balance within the SMPs.

2.5.4 Migration Minimization

The costs of dynamic load balancing include (i) preparation of the input to the partitioner, (ii) execution of the partitioning algorithm, and (iii) migration of application data to achieve the new decomposition. The migration step is often the most expensive, leading to efforts to reduce this cost. As described in Section 2.3, selection of appropriate load-balancing procedures contributes to reduced migration costs. Incremental procedures (e.g., RCB, SFC, Octree, diffusive graph partitioning) are preferred when data migration costs must be controlled. The unified partitioning strategy in ParMETIS computes both a multilevel graph decomposition ("scratch-remap") and a diffusive decomposition [102, 103]; it then selects the better decomposition in terms of load balance and migration costs.

Clever techniques can be used within an application to reduce data migration costs. For example, the most straightforward way to use partitioning and dynamic load balancing in a parallel adaptive computation is shown on the left in Figure 2.11. Here, an initial mesh is partitioned, and the computation proceeds, checking periodically to determine whether the solution resolution is sufficient. If not, the mesh is enriched adaptively, the load is rebalanced, and the computation continues. Alternatively, the rebalancing can be done *before* the mesh is actually enriched, if the error indicators used to predict refinement can also predict appropriate weights for the mesh before enrichment [43, 83] (Figure 2.11, right). This "predictive balancing" approach can improve computational balance during the refinement phase, and leads to less data migration, as redistribution occurs on the smaller mesh. Moreover, without predictive balancing, individual processors may have nearly all of their elements scheduled for refinement, leading to a memory overflow on those processors, when in fact the total amount of memory available across all processors is sufficient for the computation to proceed following refinement [40]. If the error indicators predict the resulting refinement with sufficient accuracy, the predictive balancing step also achieves a balanced partitioning of the refined mesh. In some cases, a corrective load

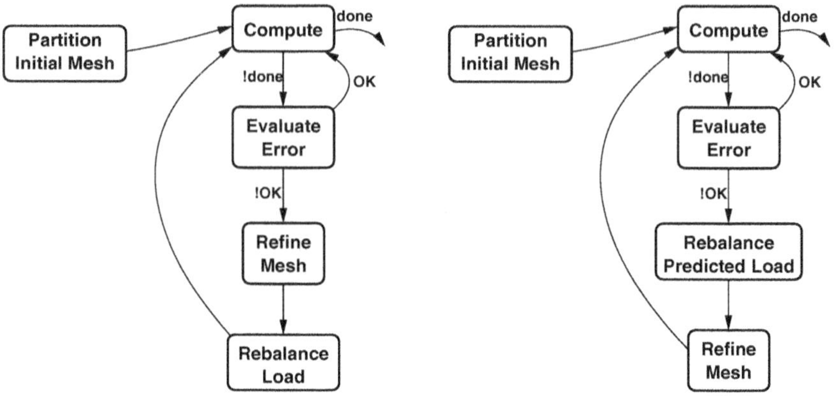

Fig. 2.11. Non-predictive (left) and predictive (right) program flows for a typical parallel adaptive computation.

balancing step, e.g., with one of the local methods outlined in Section 2.2.3, may be beneficial.

Techniques within load-balancing procedures can also reduce migration costs. The similarity matrix in PLUM [82] represents a maximal matching between an old decomposition and a new one. Old and new partitions are represented by the nodes of a bipartite graph, with edges between old and new partitions representing the amount of data they share. A maximal matching, then, numbers the new partitions to provide the greatest overlap between old and new decompositions and, thus, the least data movement. Similar strategies have been adopted by ParMETIS [68] and Zoltan [29].

Load-balancing objectives can also be adjusted to reduce data migration. Heuristics used in local refinement (see Section 2.2.3) can select objects for movement that have the lowest data movement costs. They can also select a few heavily weighted objects to satisfy balance criteria rather than many lightly weighted objects. Hu and Blake compute diffusive decompositions to achieve load balance subject to a minimization of data movement [59]. Berzins extends their idea by allowing greater load imbalance when data movement costs are high [10]; he minimizes a metric combining load imbalance and data migration to reduce actual time-to-solution (rather than load imbalance) on homogeneous and heterogeneous networks.

Acknowledgments

The authors thank the following people for their collaborations and discussions: Andrew Bauer, Diane Bennett, Rob Bisseling, Erik Boman, Paul Campbell, Laura Effinger-Dean, Jamal Faik, Luis Gervasio, Robert Heaphy, Bruce Hendrickson, Steve Plimpton, Robert Preis, Arjun Sharma, Lida Ungar, and Courtenay Vaughan.

References

1. S. Adjerid, J. E. Flaherty, P. Moore, and Y. Wang. High-order adaptive methods for parabolic systems. *Physica-D*, 60:94–111, 1992.
2. S. Aluru and F. Sevilgen. Parallel domain decomposition and load balancing using space-filling curves. In *Proc. International Conference on High-Performance Computing*, pages 230–235, 1997.
3. R. E. Bank and M. J. Holst. A new paradigm for parallel adaptive meshing algorithms. *SIAM J. Scien. Comput.*, 22:1411–1443, 2000.
4. K. J. Barker and N. P. Chrisochoides. An evaluation of a framework for the dynamic load balancing of highly adaptive and irregular parallel applications. In *Proc. Supercomputing 2003*, Phoenix, 2003.
5. S. T. Barnard. PMRSB: parallel multilevel recursive spectral bisection. In F. Baker and J. Wehmer, editors, *Proc. Supercomputing '95*, San Diego, December 1995.
6. S. T. Barnard and H. D. Simon. Fast multilevel implementation of recursive spectral bisection for partitioning unstructured problems. *Concurrency: Practice and Experience*, 6(2):101–117, 1994.
7. J. J. Bartholdi and L. K. Platzman. An $O(n \log n)$ travelling salesman heuristic based on spacefilling curves. *Operation Research Letters*, 1(4):121–125, September 1982.
8. A. C. Bauer. *Efficient Solution Procedures for Adaptive Finite Element Methods – Applications to Elliptic Problems*. PhD thesis, State University of New York at Buffalo, 2002.
9. M. J. Berger and S. H. Bokhari. A partitioning strategy for nonuniform problems on multiprocessors. *IEEE Trans. Computers*, 36:570–580, 1987.
10. M. Berzins. A new metric for dynamic load balancing. *Appl. Math. Modelling*, 25:141–151, 2000.
11. T. Bially. Space-filling curves: their generation and their application to band reduction. *IEEE Trans. Inform. Theory*, IT-15:658–664, Nov. 1969.
12. E. Boman, K. Devine, R. Heaphy, B. Hendrickson, M. Heroux, and R. Preis. LDRD report: Parallel repartitioning for optimal solver performance. Technical Report SAND2004–0365, Sandia National Laboratories, Albuquerque, NM, February 2004.
13. E. Boman, K. Devine, R. Heaphy, B. Hendrickson, W. F. Mitchell, M. S. John, and C. Vaughan. Zoltan: Data-management services for parallel applications. URL: `http://www.cs.sandia.gov/Zoltan`.
14. C. L. Bottasso, J. E. Flaherty, C. Özturan, M. S. Shephard, B. K. Szymanski, J. D. Teresco, and L. H. Ziantz. The quality of partitions produced by an iterative load balancer. In B. K. Szymanski and B. Sinharoy, editors, *Proc. Third Workshop on Languages, Compilers, and Runtime Systems*, pages 265–277, Troy, 1996.
15. T. Bui and C. Jones. A heuristic for reducing fill in sparse matrix factorization". In *Proc. 6th SIAM Conf. Parallel Processing for Scientific Computing*, pages 445–452. SIAM, 1993.
16. A. Caldwell, A. Kahng, and J. Markov. Design and implementation of move-based heuristics for VLSI partitioning. *ACM J. Experimental Algs.*, 5, 2000.
17. P. M. Campbell, K. D. Devine, J. E. Flaherty, L. G. Gervasio, and J. D. Teresco. Dynamic octree load balancing using space-filling curves. Technical Report CS-03-01, Williams College Department of Computer Science, 2003.
18. F. Cao, J. R. Gilbert, and S.-H. Teng. Partitioning meshes with lines and planes. Technical Report CSL–96–01, Xerox PARC, 1996. `ftp://parcftp.xerox.com/pub/gilbert/index.html`.

19. U. Catalyurek and C. Aykanat. Decomposing irregularly sparse matrices for parallel matrix-vector multiplications. *Lecture Notes in Computer Science*, 1117:75–86, 1996.
20. U. Catalyurek and C. Aykanat. Hypergraph-partitioning based decomposition for parallel sparse-matrix vector multiplication. *IEEE Trans. Parallel Dist. Systems*, 10(7):673–693, 1999.
21. C. Chang, T. Kurc, A. Sussman, U. Catalyurek, and J. Saltz. A hypergraph-based workload partitioning strategy for parallel data aggregation. In *Proc. of 11th SIAM Conf. Parallel Processing for Scientific Computing*. SIAM, March 2001.
22. S. Chatterjee, A. R. Lebeck, P. K. Patnala, and M. Thottethodi. Recursive array layouts and fast parallel matrix multiplication. In *ACM Symposium on Parallel Algorithms and Architectures*, pages 222–231, 1999.
23. C.-K. Cheng and Y.-C. A. Wei. An improved two-way partitioning algorithm with stable performance. *IEEE Trans. Computer Aided Design*, 10(12):1502–1511, 1991.
24. G. Cybenko. Dynamic load balancing for distributed memory multiprocessors. *J. Parallel Distrib. Comput.*, 7:279–301, 1989.
25. L. Dagum. Automatic partitioning of unstructured grids into connected components. In *Proc. Supercomputing Conference 1993*, pages 94–101, Los Alamitos, 1993. IEEE, Computer Society Press.
26. H. L. de Cougny, K. D. Devine, J. E. Flaherty, R. M. Loy, C. Özturan, and M. S. Shephard. Load balancing for the parallel adaptive solution of partial differential equations. *Appl. Numer. Math.*, 16:157–182, 1994.
27. K. D. Devine, E. G. Boman, R. T. Heaphy, B. A. Hendrickson, J. D. Teresco, J. Faik, J. E. Flaherty, and L. G. Gervasio. New challenges in dynamic load balancing. *Appl. Numer. Math.*, 52(2–3):133–152, 2005.
28. K. D. Devine and J. E. Flaherty. Parallel adaptive *hp*-refinement techniques for conservation laws. *Appl. Numer. Math.*, 20:367–386, 1996.
29. K. D. Devine, B. A. Hendrickson, E. Boman, M. St. John, and C. Vaughan. *Zoltan: A Dynamic Load Balancing Library for Parallel Applications; User's Guide*. Sandia National Laboratories, Albuquerque, NM, 1999. Tech. Report SAND99-1377. Open-source software distributed at `http://www.cs.sandia.gov/Zoltan`.
30. R. Diekmann, D. Meyer, and B. Monien. Parallel decomposition of unstructured fem-meshes. In *Proc. Parallel Algorithms for Irregularly Structured Problems*, pages 199–216. Springer LNCS 980, 1995.
31. R. Diekmann, B. Monien, and R. Preis. Load balancing strategies for distributed memory machines. In B. Topping, editor, *Parallel and Distributed Processing for Computational Mechanics: Systems and Tools*, pages 124–157, Edinburgh, 1999. Saxe-Coburg.
32. R. Diekmann, R. Preis, F. Schlimbach, and C. Walshaw. Shape-optimized mesh partitioning and load balancing for parallel adaptive fem. *Parallel Comput.*, 26(12):1555–1581, 2000.
33. H. C. Edwards. *A Parallel Infrastructure for Scalable Adaptive Finite Element Methods and its Application to Least Squares C^∞ Collocation*. PhD thesis, The University of Texas at Austin, May 1997.
34. R. Enbody, R. Purdy, and C. Severance. Dynamic load balancing. In *Proc. 7th SIAM Conference on Parallel Processing for Scientific Computing*, pages 645–646. SIAM, February 1995.
35. J. Faik, L. G. Gervasio, J. E. Flaherty, J. Chang, J. D. Teresco, E. G. Boman, and K. D. Devine. A model for resource-aware load balancing on heterogeneous clusters. Technical Report CS-04-03, Williams College Department of Computer Science, 2004. Presented at Cluster '04.

36. C. Farhat. A simple and efficient automatic FEM domain decomposer. *Computers and Structures*, 28(5):579–602, 1988.
37. C. Farhat, S. Lanteri, and H. D. Simon. TOP/DOMDEC: a software tool for mesh partitioning and parallel processing. *Comp. Sys. Engng.*, 6(1):13–26, 1995.
38. C. Farhat and M. Lesoinne. Automatic partitioning of unstructured meshes for the parallel solution of problems in computational mechanics. *Int. J. Numer. Meth. Engng.*, 36:745–764, 1993.
39. C. M. Fiduccia and R. M. Mattheyses. A linear time heuristic for improving network partitions. In *Proc. 19th IEEE Design Automation Conference*, pages 175–181. IEEE, 1982.
40. J. E. Flaherty, M. Dindar, R. M. Loy, M. S. Shephard, B. K. Szymanski, J. D. Teresco, and L. H. Ziantz. An adaptive and parallel framework for partial differential equations. In D. F. Griffiths, D. J. Higham, and G. A. Watson, editors, *Numerical Analysis 1997 (Proc. 17th Dundee Biennial Conf.)*, number 380 in Pitman Research Notes in Mathematics Series, pages 74–90. Addison Wesley Longman, 1998.
41. J. E. Flaherty, R. M. Loy, C. Özturan, M. S. Shephard, B. K. Szymanski, J. D. Teresco, and L. H. Ziantz. Parallel structures and dynamic load balancing for adaptive finite element computation. *Appl. Numer. Math.*, 26:241–263, 1998.
42. J. E. Flaherty, R. M. Loy, M. S. Shephard, B. K. Szymanski, J. D. Teresco, and L. H. Ziantz. Adaptive local refinement with octree load-balancing for the parallel solution of three-dimensional conservation laws. *J. Parallel Distrib. Comput.*, 47:139–152, 1997.
43. J. E. Flaherty, R. M. Loy, M. S. Shephard, B. K. Szymanski, J. D. Teresco, and L. H. Ziantz. Predictive load balancing for parallel adaptive finite element computation. In H. R. Arabnia, editor, *Proc. PDPTA '97*, volume I, pages 460–469, 1997.
44. J. E. Flaherty, R. M. Loy, M. S. Shephard, and J. D. Teresco. Software for the parallel adaptive solution of conservation laws by discontinuous Galerkin methods. In B. Cockburn, G. Karniadakis, and S.-W. Shu, editors, *Discontinous Galerkin Methods Theory, Computation and Applications*, volume 11 of *Lecture Notes in Compuational Science and Engineering*, pages 113–124. Springer, 2000.
45. J. Garbers, H. J. Promel, and A. Steger. Finding clusters in VLSI circuits. In *Proc. IEEE Intl. Conf. on Computer Aided Design*, pages 520–523, 1990.
46. M. Garey, D. Johnson, and L. Stockmeyer. Some simplified NP-complete graph problems. *Theoretical Computer Science*, 1(3):237–267, 1976.
47. M. R. Garey and D. S. Johnson. *Computers and Intractability: A Guide to the Theory of NP-Completeness.* W. H. Freeman and Company, 1979.
48. L. Hagen and A. Kahng. Fast spectral methofs for ratio cut partitioning and clustering. In *Proc. IEEE Intl. Conf. on Computer Aided Design*, pages 10–13, 1991.
49. L. Hagen and A. Kahng. A new approach to effective circuit clustering. In *Proc. IEEE Intl. Conf. on Computer Aided Design*, pages 422–427, 1992.
50. B. Hendrickson. Graph partitioning and parallel solvers: Has the emperor no clothes? In *Proc. Irregular'98*, volume 1457 of *Lecture Notes in Computer Science*, pages 218–225. Springer-Verlag, 1998.
51. B. Hendrickson and K. Devine. Dynamic load balancing in computational mechanics. *Comput. Methods Appl. Mech. Engrg.*, 184(2–4):485–500, 2000.
52. B. Hendrickson and T. G. Kolda. Graph partitioning models for parallel computing. *Parallel Comput.*, 26:1519–1534, 2000.
53. B. Hendrickson and R. Leland. The Chaco user's guide, version 2.0. Technical Report SAND94–2692, Sandia National Laboratories, Albuquerque, 1994. Open-source software distributed at `http://www.cs.sandia.gov/~bahendr/chaco.html`.

54. B. Hendrickson and R. Leland. An improved spectral graph partitioning algorithm for mapping parallel computations. *SIAM J. Scien. Comput.*, 16(2):452–469, 1995.
55. B. Hendrickson and R. Leland. A multilevel algorithm for partitioning graphs. In *Proc. Supercomputing '95*, 1995.
56. G. Horton. A multi-level diffusion method for dynamic load balancing. *Parallel Comput.*, 19:209–218, 1993.
57. S.-H. Hsieh, G. H. Paulino, and J. F. Abel. Evaluation of automatic domain partitioning algorithms for parallel finite element analysis. Structural Engineering Report 94-2, School of Civil and Environmental Engineering, Cornell University, Ithaca, 1994.
58. Y. F. Hu and R. J. Blake. An optimal dynamic load balancing algorithm. Preprint DL-P-95-011, Daresbury Laboratory, Warrington, WA4 4AD, UK, 1995.
59. Y. F. Hu, R. J. Blake, and D. R. Emerson. An optimal migration algorithm for dynamic load balancing. *Concurrency: Practice and Experience*, 10:467 – 483, 1998.
60. H. V. Jagadish. Linear clustering of objects with multiple attributes. In *Proc. ACM SIGMOD*, pages 332–342, 1990.
61. M. T. Jones and P. E. Plassmann. Computational results for parallel unstructured mesh computations. *Comp. Sys. Engng.*, 5(4–6):297–309, 1994.
62. L. V. Kale and S. Krishnan. CHARM++: A portable concurrent object oriented system based on C++. *ACM SIGPLAN notices*, 28(10):91–128, 1993.
63. G. Karypis, R. Aggarwal, V. Kumar, and S. Shekhar. Multilevel hypergraph partitioning: application in VLSI domain. In *Proc. 34th conf. Design automation*, pages 526 – 529. ACM, 1997.
64. G. Karypis and V. Kumar. Metis: Unstructured graph partitioning and sparse matrix ordering system. Tech. Report, University of Minnesota, Department of Computer Science, Minneapolis, MN, 1995. Open-source software distributed at `http://www-users.cs.umn.edu/~karypis/metis`.
65. G. Karypis and V. Kumar. Multilevel algorithms for multiconstraint graph paritioning. Technical Report 98-019, Department of Computer Science, University of Minnesota, 1998.
66. G. Karypis and V. Kumar. A fast and high quality multilevel scheme for partitioning irregular graphs. *SIAM J. Scien. Comput.*, 20(1), 1999.
67. G. Karypis and V. Kumar. Parallel multivelel k-way partitioning scheme for irregular graphs. *SIAM Review*, 41(2):278–300, 1999.
68. G. Karypis, K. Schloegel, and V. Kumar. *ParMetis Parallel Graph Partitioning and Sparse Matrix Ordering Library, Version 3.1.* University of Minnesota Department of Computer Science and Engineering, and Army HPC Research Center, Minneapolis, 2003. Open-source software distributed at `http://www-users.cs.umn.edu/~karypis/metis`.
69. B. Kernighan and S. Lin. An efficient heuristic procedure for partitioning graphs. *Bell System Technical Journal*, 29:291–307, 1970.
70. E. Leiss and H. Reddy. Distributed load balancing: design and performance analysis. *W. M. Kuck Research Computation Laboratory*, 5:205–270, 1989.
71. R. M. Loy. *Adaptive Local Refinement with Octree Load-Balancing for the Parallel Solution of Three-Dimensional Conservation Laws.* PhD thesis, Computer Science Dept., Rensselaer Polytechnic Institute, Troy, 1998.
72. B. Maerten, D. Roose, A. Basermann, J. Fingberg, and G. Lonsdale. DRAMA: A library for parallel dynamic load balancing of finite element applications. In *Proc. Ninth SIAM Conference on Parallel Processing for Scientific Computing*, San Antonio, 1999. Library distributed under license agreement from `http://www.ccrl-nece.de/~drama/drama.html`.

73. T. Minyard and Y. Kallinderis. Octree partitioning of hybrid grids for parallel adaptive viscous flow simulations. *Int. J. Numer. Meth. Fluids*, 26:57–78, 1998.
74. T. Minyard and Y. Kallinderis. Parallel load balancing for dynamic execution environments. *Comput. Methods Appl. Mech. Engrg.*, 189(4):1295–1309, 2000.
75. T. Minyard, Y. Kallinderis, and K. Schulz. Parallel load balancing for dynamic execution environments. In *Proc. 34th Aerospace Sciences Meeting and Exhibit*, number 96-0295, Reno, 1996.
76. W. F. Mitchell. Refinement tree based partitioning for adaptive grids. In *Proc. Seventh SIAM Conf. on Parallel Processing for Scientific Computing*, pages 587–592. SIAM, 1995.
77. W. F. Mitchell. The full domain partition approach to distributing adaptive grids. *Appl. Numer. Math.*, 26:265–275, 1998.
78. W. F. Mitchell. The refinement-tree partition for parallel solution of partial differential equations. *NIST Journal of Research*, 103(4):405–414, 1998.
79. B. Moon, H. V. Jagadish, C. Faloutsos, and J. H. Saltz. Analysis of the clustering properties of the Hilbert space-filling curve. *IEEE Trans. Knowledge and Data Engng.*, 13(1):124–141, January/February 2001.
80. G. M. Morton. A computer oriented geodetic data base and a new technique in file sequencing. Technical report, IBM Ltd., March 1966.
81. J. T. Oden, A. Patra, and Y. Feng. Domain decomposition for adaptive *hp* finite element methods. In *Proc. Seventh Intl. Conf. Domain Decomposition Methods*, State College, Pennsylvania, October 1993.
82. L. Oliker and R. Biswas. PLUM: Parallel load balancing for adaptive unstructured meshes. *J. Parallel Distrib. Comput.*, 51(2):150–177, 1998.
83. L. Oliker, R. Biswas, and R. C. Strawn. Parallel implementaion of an adaptive scheme for 3D unstructured grids on the SP2. In *Proc. 3rd International Workshop on Parallel Algorithms for Irregularly Structured Problems*, Santa Barbara, 1996.
84. J. A. Orenstein. Spatial query processing in an object-oriented database system. In *Proc. ACM SIGMOD*, pages 326–336, May 1986.
85. M. Ozdal and C. Aykanat. Hypergraph models and algorithms for data-pattern based clustering. *Data Mining and Knowledge Discovery*, 9:29–57, 2004.
86. C. Özturan. *Distributed Environment and Load Balancing for Adaptive Unstructured Meshes*. PhD thesis, Computer Science Dept., Rensselaer Polytechnic Institute, Troy, 1995.
87. M. Parashar and J. C. Browne. On partitioning dynamic adaptive grid hierarchies. In *Proc. 29th Annual Hawaii International Conference on System Sciences*, volume 1, pages 604–613, Jan. 1996.
88. M. Parashar, J. C. Browne, C. Edwards, and K. Klimkowski. A common data management infrastructure for adaptive algorithms for PDE solutions. In *Proc. SC97*, San Jose, CA, 1997.
89. A. Patra and J. T. Oden. Problem decomposition for adaptive *hp* finite element methods. *Comp. Sys. Engng.*, 6(2):97–109, 1995.
90. E. A. Patrick, D. R. Anderson, and F. K. Brechtel. Mapping multidimensional space to one dimension for computer output display. *IEEE Trans. Computers*, C-17(10):949–953, October 1968.
91. G. Peano. Sur une courbe, qui remplit toute une aire plane. *Mathematische Annalen*, 36:157–160, 1890.
92. F. Pellegrini. SCOTCH 3.1 User's guide. Technical Report 1137-96, LaBRI, Université Bordeaux I, August 1996. Library available at `http://www.labri.fr/Perso/~pelegrin/scotch/`.

93. F. Pellegrini and J. Roman. Experimental analysis of the dual recursive bipartitioning algorithm for static mapping. Technical Report 1038-96, Université Bordeaux I, 1996.
94. J. R. Pilkington and S. B. Baden. Dynamic partitioning of non-uniform structured workloads with spacefilling curves. *IEEE Trans. on Parallel and Distributed Systems*, 7(3):288–300, 1996.
95. A. Pınar and B. Hendrickson. Graph partitioning for complex objectives. In *Proc. 15th Int'l Parallel and Distributed Processing Symp. (I PDPS)*, San Francisco, CA, April 2001.
96. S. Plimpton, S. Attaway, B. Hendrickson, J. Swegle, C. Vaughan, and D. Gardner. Transient dynamics simulations: Parallel algorithms for contact detection and smoothed particle hydrodynamics. *J. Parallel Distrib. Comput.*, 50:104–122, 1998.
97. A. Pothen, H. Simon, and K.-P. Liou. Partitioning sparse matrices with eigenvectors of graphs. *SIAM J. Mat. Anal. Appl.*, 11(3):430–452, 1990.
98. R. Preis and R. Diekmann. PARTY – a software library for graph partitioning. In B. Topping, editor, *Advances in Computational Mechanics with Parallel and Distributed Processing*, pages 63–71. CIVIL-COMP PRESS, 1997. Library distributed under free research and academic license at `http://wwwcs.upb.de/fachbereich/AG/monien/RESEARCH/PART/party.html`.
99. H. Sagan. *Space-Filling Curves*. Springer-Verlag, 1994.
100. K. Schloegel, G. Karypis, and V. Kumar. Multilevel diffusion schemes for repartitioning of adaptive meshes. *J. Parallel Distrib. Comput.*, 47(2):109–124, 1997.
101. K. Schloegel, G. Karypis, and V. Kumar. A new algorithm for multi-objective graph partitioning. Tech. Report 99-003, University of Minnesota, Department of Computer Science and Army HPC Center, Minneapolis, 1999.
102. K. Schloegel, G. Karypis, and V. Kumar. A unified algorithm for load-balancing adaptive scientific simulations. In *Proc. Supercomputing*, Dallas, 2000.
103. K. Schloegel, G. Karypis, and V. Kumar. Wavefront diffusion and LMSR: Algorithms for dynamic repartitioning of adaptive meshes. *IEEE Trans. Parallel Distrib. Syst.*, 12(5):451–466, 2001.
104. K. Schloegel, G. Karypis, and V. Kumar. Parallel static and dynamic multiconstraint graph partitioning. *Concurrency and Computation – Practice and Experience*, 14(3):219–240, 2002.
105. M. S. Shephard, S. Dey, and J. E. Flaherty. A straightforward structure to construct shape functions for variable p-order meshes. *Comp. Meth. in Appl. Mech. and Engng.*, 147:209–233, 1997.
106. M. S. Shephard, J. E. Flaherty, H. L. de Cougny, C. Özturan, C. L. Bottasso, and M. W. Beall. Parallel automated adaptive procedures for unstructured meshes. In *Parallel Computing in CFD*, number R-807, pages 6.1–6.49. Agard, Neuilly-Sur-Seine, 1995.
107. H. D. Simon. Partitioning of unstructured problems for parallel processing. *Comp. Sys. Engng.*, 2:135–148, 1991.
108. S. Sinha and M. Parashar. Adaptive system partitioning of AMR applications on heterogeneous clusters. *Cluster Computing*, 5(4):343–352, October 2002.
109. A. Sohn and H. Simon. S-HARP: A scalable parallel dynamic partitioner for adaptive mesh-based computations. In *Proc. Supercomputing '98*, Orlando, 1998.
110. J. Steensland. Vampire homepage. `http://user.it.uu.se/~johans/research/vampire/vampire1.html`, 2000. Open-source software distributed at `http://user.it.uu.se/~johans/research/vampire/download.html`.

111. J. Steensland, S. Chandra, and M. Parashar. An application-centric characterization of domain-based SFC partitioners for parallel SAMR. *IEEE Trans. Parallel and Distrib. Syst.*, 13(12):1275–1289, 2002.
112. J. Steensland, S. Söderberg, and M. Thuné. A comparison of partitioning schemes for blockwise parallel SAMR algorithms. In *Proc. 5th International Workshop on Applied Parallel Computing, New Paradigms for HPC in Industry and Academia*, volume 1947 of *Lecture Notes in Computer Science*, pages 160–169, London, 2000. Springer-Verlag.
113. V. E. Taylor and B. Nour-Omid. A study of the factorization fill-in for a parallel implementation of the finite element method. *Int. J. Numer. Meth. Engng.*, 37:3809–3823, 1994.
114. J. D. Teresco, J. Faik, and J. E. Flaherty. Hierarchical partitioning and dynamic load balancing for scientific computation. Technical Report CS-04-04, Williams College Department of Computer Science, 2004. To appear in the Proceedings of PARA'04.
115. J. D. Teresco and L. P. Ungar. A comparison of Zoltan dynamic load balancers for adaptive computation. Technical Report CS-03-02, Williams College Department of Computer Science, 2003. Presented at COMPLAS '03.
116. A. Trifunovic and W. J. Knottenbelt. Towards a parallel disk-based algorithm for multilevel k-way hypergraph partitioning. In *Proc. 18th International Parallel and Distributed Processing Symposium (IPDPS'04)*, page 236b, Santa Fe, 2004.
117. R. Van Driessche and D. Roose. An improved spectral bisection algorithm and its application to dynamic load balancing. *Parallel Comput.*, 21:29–48, 1995.
118. D. Vanderstraeten, C. Farhat, P. Chen, R. Keunings, and O. Ozone. A retrofit based methodology for the fast generation and optimization of large-scale mesh partitions: beyond the minimum interface size criterion. *Comput. Methods Appl. Mech. Engrg.*, 133:25–45, 1996.
119. B. Vastenhouw and R. H. Bisseling. A two-dimensional data distribution method for parallel sparse matrix-vector multiplication. Preprint 1238, Dept. of Mathematics, Utrecht University, May 2002.
120. C. Walshaw. *The Parallel JOSTLE Library User's Guide, Version 3.0*. University of Greenwich, London, UK, 2002. Library distributed under free research and academic license at `http://staffweb.cms.gre.ac.uk/~c.walshaw/jostle/`.
121. C. Walshaw and M. Cross. Multilevel Mesh Partitioning for Heterogeneous Communication Networks. Tech. Rep. 00/IM/57, Comp. Math. Sci., Univ. Greenwich, London SE10 9LS, UK, March 2000.
122. C. Walshaw and M. Cross. Multilevel Mesh Partitioning for Heterogeneous Communication Networks. *Future Generation Comput. Syst.*, 17(5):601–623, 2001. (Originally published as Univ. Greenwich Tech. Rep. 00/IM/57).
123. C. Walshaw and M. Cross. Dynamic mesh partitioning and load-balancing for parallel computational mechanics codes. In B. H. V. Topping, editor, *Computational Mechanics Using High Performance Computing*, pages 79–94. Saxe-Coburg Publications, Stirling, 2002. (Invited Chapter, Proc. Parallel & Distributed Computing for Computational Mechanics, Weimar, Germany, 1999).
124. C. Walshaw, M. Cross, and M. Everett. A localized algorithm for optimizing unstructured mesh partitions. *Intl. J. of Supercomputer Applications*, 9(4):280–295, 1995.
125. C. Walshaw, M. Cross, and M. Everett. Parallel dynamic graph-partitioning for unstructured meshes. *J. Parallel Distrib. Comput.*, 47(2):102–108, 1997.
126. C. Walshaw, M. Cross, and K. McManus. Multiphase mesh partitioning. *Appl. Math. Modelling*, 25(2):123–140, 2000. (Originally published as Univ. Greenwich Tech. Rep. 99/IM/51).

127. M. S. Warren and J. K. Salmon. A parallel hashed oct-tree n-body algorithm. In *Proc. Supercomputing '93*, pages 12–21. IEEE Computer Society, 1993.
128. J. Watts. A practical approach to dynamic load balancing. Master's Thesis, October 1995.
129. J. Watts, M. Rieffel, and S. Taylor. A load balancing technique for multiphase computations. In *Proc. High Performance Computing '97*, pages 15–20. Society for Computer Simulation, 1997.
130. S. Wheat. *A Fine Grained Data Migration Approach to Application Load Balancing on MP MIMD Machines*. PhD thesis, University of New Mexico, Department of Computer Science, Albuquerque, 1992.
131. S. Wheat, K. Devine, and A. MacCabe. Experience with automatic, dynamic load balancing and adaptive finite element computation. In H. El-Rewini and B. Shriver, editors, *Proc. 27th Hawaii International Conference on System Sciences*, pages 463–472, Kihei, 1994.
132. M. Willebeek-LeMair and A. Reeves. Strategies for dynamic load balancing on highly parallel computers. *IEEE Parallel and Distrib. Sys.*, 4(9):979–993, 1993.
133. R. Wolski, N. T. Spring, and J. Hayes. The Network Weather Service: A distributed resource performance forecasting service for metacomputing. *Future Generation Comput. Syst.*, 15(5-6):757–768, October 1999.
134. C. Xu, F. Lau, and R. Diekmann. Decentralized remapping of data parallel applications in distributed memory multiprocessors. Tech. Rep. tr-rsfb-96-021, Dept. of Computer Science, University of Paderborn, Paderborn, Germany, Sept. 1996.

3

Graphics Processor Units: New Prospects for Parallel Computing

Martin Rumpf[1] and Robert Strzodka[2]

[1] University of Bonn, Institute for Numerical Simulation, Wegelerstr. 6, 53115 Bonn, Germany
`martin.rumpf@ins.uni-bonn.de`

[2] caesar research center, Ludwig-Erhard-Allee 2, 53044 Bonn, Germany
`strzodka@caesar.de`

Summary. This chapter provides an introduction to the use of Graphics Processor Units (GPUs) as parallel computing devices. It describes the architecture, the available functionality and the programming model. Simple examples and references to freely available tools and resources motivate the reader to explore these new possibilities. An overview of the different applications of GPUs demonstrates their wide applicability, yet also highlights limitations of their use. Finally, a glimpse into the future of GPUs sketches the growing prospects of these inexpensive parallel computing devices.

3.1 Introduction

This introductory section motivates the use of Graphics Processor Units (GPUs) as parallel computing devices and explains the different computing and programming models. Section 3.1.3 reports on hands-on experience with this kind of processing. A comparison with shared memory machines clarifies the similarities and differences.

The rest of the chapter is organized as follows. Section 3.2 presents the most important aspects of graphics hardware related to scientific computing. For a wider context and specific GPU topics, the appendix (Section 3.5) is referenced in various places. Building on the experience from Section 3.1.3, Section 3.3 explains how to construct efficient linear equation solvers and presents partial differential equation (PDE) applications. It also contains a section with links to examples of code and other resources. In Section 3.4 we conclude with an outlook on future functionality and the use of multiple GPUs.

3.1.1 Motivation

Over the last decade, GPUs have developed rapidly from being primitive drawing devices to being major computing resources. The newest GPUs have as many as 220 million transistors, approximately twice as many as a typical Central Processor

Unit (CPU) in a PC. Moreover, the L2 cache consumes most of the transistors in a CPU, while GPUs use only small caches and devote the majority of transistors to computation. This large number of parallel processing elements (PEs) converts the GPU into a parallel computing system. Current devices have up to 16 parallel pipelines with one or two PEs each. A single processing element (PE) is capable of performing an addition or multiplication of four component vectors (4-vectors) of single-precision floating-point numbers in one clock cycle. This amounts to a total of 128 floating point operations per clock cycle. With a clock frequency of up to 500 MHz, peak performance of a single unit approaches 64 GFLOPS, and with the introduction of the PCI Express (PCIe) bus a motherboard will be able to accommodate several graphics cards.

We will concentrate on the exploitation of the high internal *parallelism* of a single GPU. Section 3.1.4 explains how the parallel PEs of a single GPU can be viewed in light of the familiar shared memory computing model, although all PEs reside on the same chip. The development of GPU clusters, where several graphics cards work in parallel, has recently been initiated and Section 3.4.2 provides some information on this quickly growing area of research.

High performance and many successful implementations of PDE solvers on GPUs have already caught the attention of the scientific computing community. Implemented PDE types include Navier-Stokes equations, Lattice Boltzmann equations, reaction-diffusion systems, non-linear diffusion processes, level-set equations, and Euler equations for variational functional minimization. The web site [10] offers an overview. In 2D, problem sizes go up to 4096^2 nodes, in 3D up to 256^3 nodes. The limiting factor is the size of the *video memory* on the graphics cards (current maximum 512Mb). If one is willing to accept a slower rate of data exchange by using the main memory, the problem size is not limited by the graphics card. Reported speedup factors, as compared to a modern single CPU solver, are often in the range 5-20.

Manufacturers of GPUs are now considering the potential of their devices for parallel computing, although the driving force of the development is still the computer game market. This influences the balance in some of the common antonyms:

- Performance - Accuracy
  For optimal performance GPUs offer different floating point formats of 16, 24 and 32 bit, but native support for a double precision format is unlikely in the near future. A hardware-assisted emulation could be feasible.
- Processing - Communication
  GPUs process large data sets quickly with many parallel processing elements (PEs), but direct communication between them does not exist.
- Generality - Specialty
  Fairly general high-level languages for GPU programming exist, but the setup of the execution environment for the programs and the data handling still requires some graphics-specific knowledge.

Luckily, more and more physical simulation is being used in computer games, which increases the demand for general computing capabilities. The fast development cycle of GPUs reacts with high flexibility to the changing requirements and tends towards a general parallel computing device. At some stage, the growing demand for more scientifically -orientated GPUs could even make a separate production line worthwhile, including, for example, double precision arithmetic. Simultaneously, the recently emerging support for the utilization of multiple GPUs will increase. Current GPUs are not yet able to replace CPU-based parallel systems in scientific computations on a large scale. However, we want to familiarize the reader with the looming possibilities and demonstrate that many algorithms can already benefit from their being executed on GPUs.

3.1.2 Data-Stream-Based Architectures

Peak performance of computer systems is often in excess of actual application performance, due to the *memory gap* problem [32], the mismatch of memory and processor performance. In data-intensive applications, the processing elements (PEs) often spend most of the time waiting for data. GPUs have traditionally been optimized for high data throughput, with wide data buses (256 bit) and the latest memory technology (GDDR3). In contrast to instruction-stream-based (ISB) CPUs, they also subscribe to the data-stream-based (DSB) computing paradigm [13]. In DSB computing one exploits the situation in which the same operation is applied to many data items. Thus, the processing is not based on an instruction stream, but, rather, on a data stream. The PEs are first configured for the execution of the desired operation. Then, the data streams through the so configured pipeline of PEs undergoing the configured operations. The stream of data stops only when a new configuration must be applied. So, for the performance of DSB architectures, it is crucial that the configuration does not change frequently, but rather remains constant for a large data stream, e.g. for all components of a large vector.

The DSB model separates the two tasks of configuring the PEs and controlling the data-flow to and from the PEs. By contrast, an instruction prescribes both the operation to be executed and the required data. The separation of tasks deals much better with the memory gap problem, because the individual elements of the data streams can be assembled from memory before the actual processing. This allows the optimization of the memory access patterns, minimizing latencies and maximizing the sustained *bandwidth*. In ISB architectures only a limited prefetch of the input data can occur, as jumps are expected in the instruction stream. By contrast, it is inherent in the DSB model that no such jumps will occur for a long time. Thus, the resources can be concentrated on efficient data retrieval and parallel processing rather than jump predictions and speculative execution. Clearly, the advantage applies only to algorithms that exhibit this kind of regular behavior. Therefore, for some irregular algorithms, it is advantageous to increase the operation count in favor of more regular behavior, and thus faster execution, on DSB hardware.

The DSB architectures comprise reconfigurable logic, reconfigurable computing, processor-in-memory and stream architectures. GPUs may be seen as a restricted form of a stream processor. They are not the most powerful or efficient architecture, but offer an unrivaled price-performance ratio, which makes this advantageous

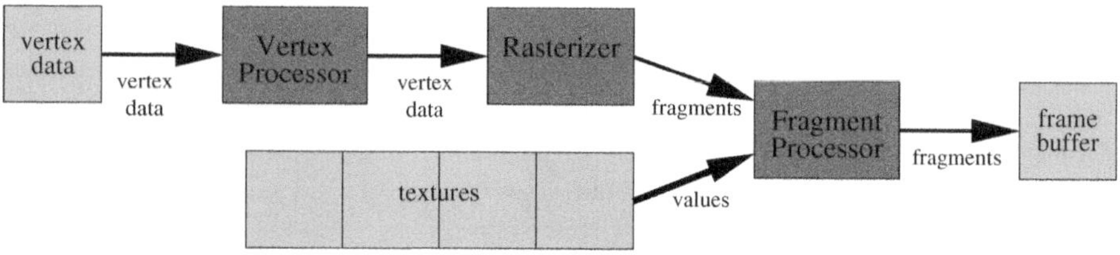

Fig. 3.1. A simplified diagram of the Direct X 9 graphics pipeline. Light gray represents data containers, dark gray processing units. The data containers are becoming fully interchangeable, i.e. a 2D data array can serve as an array of vertex data, a texture or a destination buffer within the frame-buffer. See Figure 3.5 for a more detailed diagram.

processing concept easily available on any PC, and not only on specially configured hardware systems. Moreover, GPUs have the great advantage that there exist widespread platform (Direct X) and operating system (OpenGL) independent Application Programming Interfaces (APIs) for access to their functionality, whereas other architectures require a proprietary environment. The API guarantees that despite the different hardware components of GPUs from different companies, the programmer can access a common set of operations through the same software interface, namely the API. Similarly to the situation with CPUs, the programming model for GPUs has evolved from assembly to high level languages, which now allow a clear and modular configuration of the graphics pipeline.

3.1.3 GPU Programming Model

Graphics Processor Units (GPUs) are, as the name suggests, designed to process graphics. Put simply, GPUs render geometric primitives such as points, lines, triangles or quads into a discrete representation of the $[-1, 1] \times [-1, 1]$ domain, called the *frame-buffer*. The geometric primitives are defined by *vertex* coordinates. The discrete elements in the frame-buffer are pixels. Because the primitives are continuous objects and the frame-buffer is a discrete representation, GPUs contain a so-called *rasterizer* that decomposes a primitive into fragments, which correspond to the set of affected pixels (see below why fragments and pixels are not the same). In the case of 2D primitives, one can choose whether to rasterize the contour or the interior, and we will always assume the latter.

The rasterizer divides the *graphics pipeline* into two parts where manipulation of data can take place. Prior to rasterization we have the Vertex Processor (VP), which operates on data associated with a vertex. Following rasterization we have the Fragment Processor (FP), which operates on data associated with a fragment (see Figure 3.1). Logically, the VP processes one vertex and the FP one fragment at a time, without any interaction with other vertices or fragments. Physically, there are several independent parallel pipelines, more for the FP than for the VP. See Section 3.5.2 for more details of the graphics pipeline and Section 3.2.3 for a discussion of the available parallelism.

Which data is associated with a vertex or fragment? In addition to the vertex coordinates, a vertex can also carry colors, a normal, and so-called texture coordinates (and a few more parameters). The VP can change all this data, including the vertex coordinates. The rasterizer interpolates the data between the vertices of a primitive when the fragments are generated. Therefore, each fragment has its own set of the above parameters. The FP combines the parameters to a final value, which is then assigned to the corresponding pixel in the frame-buffer. Currently, the frame-buffer position of a fragment generated by the rasterizer cannot be changed in the FP. Hence, there is a one-to-one correspondence between a fragment and the pixel to which the result of the FP will be written (unless it is discarded altogether). However, a fragment carries much more information than a pixel. For example, the texture coordinates associated with a fragment are typically used to retrieve values from textures, i.e. previously defined or computed 1D to 4D (typically 2D) data arrays (Figure 3.1). So, the FP reduces the information of a fragment to the single color value of a pixel. To be precise, a pixel may carry several values (see Section 3.5.2).

Both the VP and FP support a rich set of arithmetic, exponential and trigonometric functions on floating point numbers. They can be programmed by C-like high-level languages, which, to a certain extent extent, also support flow control instructions such as conditional statements, loops and function calls. The VP, and above all the FP, is decisive for accuracy and performance of computations on GPUs. Section 3.2.2 specifies the exact number formats and operations.

How can this setting be used for scientific computing? If we think of a square grid of dimension $N_x \times N_y$, then the node values form a vector of length $N_x \cdot N_y$ and can be represented by a 2D array, naturally preserving the neighbor relations. In GPUs we use a 2D texture for such a 2D array. Let us first describe the execution of a single operation; namely, the addition of two nodal vectors $\bar{A}$ and $\bar{B}$. Once the graphics environment is set up for this simple operation, it will be easy to add more functionality. For the addition of $\bar{A}$ and $\bar{B}$ on the GPU, we need to add the texels (elements) of the corresponding textures. For this to happen we must configure the graphics pipeline appropriately and then define the data streams to be processed. First we need a configuration, a so called *shader*, for the FP that executes an addition for a pair of vector components:

```
// shader FP_ADD2
float add2(float2 texCoord       : TEXCOORD0, // texture coords
     uniform sampler2D Tex_A : texunit0,  // texture A
     uniform sampler2D Tex_B : texunit1)  // texture B
  : COLOR                                 // color as output
{
  float valA= f1tex2D(Tex_A, texCoord);   // texel from A
  float valB= f1tex2D(Tex_B, texCoord);   // texel from B
  return valA+valB;                       // addition
}
```

The configuration of the VP and FP are nowadays usually written in a high-level graphics language. This is a listing in the language C for graphics (Cg). We will list all shaders in Cg, but the graphics languages are actually very similar and the reader may prefer a different one (see Section 3.5.4). In addition to C we have the colon notation, which specifies the semantics of the input and output. Here, we use one set of coordinates (`texCoord`), two textures (`Tex_A, Tex_B`) and a float color value as output. The function call `f1tex2D(.,.)` reads one float from the given texture and coordinate. The actual addition happens in the last but one line. As noted earlier, *logically* the shader operates on one fragment at a time in a sequence, but *physically*, the parallel FP pipelines run this configuration simultaneously. The loop over the texture elements is implicit.

To start the processing of the loop, we must first configure the pipeline with our shader and bind the textures $\bar{A}$ and $\bar{B}$ as input sources:

```
cgGLBindProgram(fpProg[FP_ADD2]);                // bind shader

glActiveTexture(GL_TEXTURE0);                    // texunit0
glBindTexture(GL_TEXTURE_2D, texID[TEX_A]);     // bind TEX_A

glActiveTexture(GL_TEXTURE1);                    // texunit1
glBindTexture(GL_TEXTURE_2D, texID[TEX_B]);     // bind TEX_B
```

This code is a part of a normal C/C++ file. The functions are defined by the OpenGL and Cg API. We assume that `fpProg[FP_ADD2]` is a handle to our shader configuration from above, and `texID` is a vector that contains the OpenGL IDs of our textures. Now everything is configured and we only need to specify the geometry to be rendered. The above C/C++ code continues with the appropriate calls of OpenGL functions:

```
// function drawTex2D()
glBegin(GL_QUADS);                          // render quad
  glMultiTexCoord2f(GL_TEXTURE0, 0,0);// texture bottom left
  glVertex2f(-1,-1);                        // vertex  bottom left
  glMultiTexCoord2f(GL_TEXTURE0, 0,1);
  glVertex2f(-1,1);
  glMultiTexCoord2f(GL_TEXTURE0, 1,1);
  glVertex2f(1,1);
  glMultiTexCoord2f(GL_TEXTURE0, 1,0);// texture bottom right
  glVertex2f(1,-1);                         // vertex  bottom right
glEnd();
```

With the function call `glEnd()` the processing of the data streams starts and we obtain the result $\bar{C} = \bar{A} + \bar{B}$ at the end of the pipeline in the *frame-buffer* (see Figure 3.1).

The reader may have noticed that the dimensions of our textures $N_x \times N_y$ do not show up anywhere. This is because graphics APIs work mostly with normalized coordinates. From the code above, we see that a *texture* is accessed via coordinates from $[0, 1]^2$ and the *frame-buffer* with vertex coordinates from $[-1, 1]^2$. Hence, the values N_x, N_y are only used in the definition of the textures and the viewport of the frame-buffer, and not in the rendering. As the VP can change all parameters, including the texture and vertex coordinates, we could also address the textures and the frame-buffer by other number ranges $T \subset \mathbb{R}^2$ and $F \subset \mathbb{R}^2$, if we were to bind appropriate constant matrices to the VP that performs the mappings $T \to [0,1]^2$ and $F \to [-1,1]^2$. In this way, it is possible to use the integer indices of the texels in the textures and pixels in the frame-buffer for the addressing, but this requires that the constant matrices be changed and re-bound each time the texture size or viewport size of the frame-buffer changes.

So far, we have described the execution of a single operation; namely addition, on the vector components. Now, we could easily add many more operations to our FP shader. The entire data from up to 32 different textures can be involved in the computations. In particular, we can read different elements of the textures to compute discrete gradients or, in general, any filters. However, usually we cannot map the entire problem into one shader (see Section 3.3.4). So the question arises, how can we use the result from the frame-buffer in a subsequent computation? Logically, the most elegant way is to define a texture $\bar{C}$ before the operation, and then render the result directly into that texture, using it as a destination buffer. After the operation we would bind a different texture as destination, say $\bar{D}$, and $\bar{C}$ could be bound as a source. Section 3.2.1 explains the details and also other possibilities.

As it becomes apparent that there are even more issues, not discussed above, that must be addressed, it will also become apparent to the reader that the handling of this kind of processing is very demanding. In fact, the low-level setup of the graphics pipeline can sometimes frustrating, even to experts. Luckily, though, there exist several libraries that will do most of the tedious work and let the programmer concentrate on the algorithm. Section 3.3 presents examples at this higher level of abstraction. The end of Section 3.5.4 discusses the development of even more abstract data-stream-based (DSB) programming approaches. For someone new to GPUs this will still feel unfamiliar at first, but parallel programming with the Message Passing Interface (MPI) [16, Section 2.2] or OpenMP [16, Section 2.3] also needs some practice before it can be performed comfortably.

3.1.4 Comparison with Shared Memory Model

Because the pipelines of GPU operate independently of each other on a common memory, the graphics card is similar to a shared memory parallel computer. (See Figure 1.3 in [16].) This section describes the GPU architecture from this perspective.

On graphics cards, not entire processors, but relatively few PEs constitute a pipeline. By a graphics processing element (PE) we mean an element that can perform a general multiplication or addition on a 4-vector or a special function on a scalar (e.g. $\sqrt{x}, 1/x$) for the largest available number format in one clock cycle. By viewing each pipeline as an individual node of a parallel computer, the GPU can be regarded as a restricted form of a shared memory machine. The following restrictions apply:

- All pipelines read from the same memory.
  This functionality is very similar to that of a general shared memory machine. The pipelines interact with each other by reading common memory. There is no

direct communication between them. In detail, a FP pipeline cannot specify a general memory address for reading directly, but it can read from any position within the bound textures. In other words, it can *gather* data from textures without restriction. Sufficiently large textures will cover all of the available *video memory* space on the graphics card. Hence, practically all problem data can be accessed, but it must be grouped in textures. This is fairly natural, because data arrays also provide some kind of logical memory grouping to a CPU. The memory access behavior is also similar, with local data reads being cheaper than random accesses. Currently, up to 32 textures can be bound during the execution of a FP shader and the choice must be made before the processing. Both restrictions are expected to disappear almost completely with Windows Graphics Foundation (WGF), the next generation of graphics API. (See Section 3.5.3.)

- All pipelines operate on the same stream of data.

  This is different from most shared memory machines, which utilize the Multiple Instruction Multiple Data (MIMD) model with no restriction on the sources from of the multiple data. Older GPUs use the Single Instruction Multiple Data (SIMD) model exclusively. This is efficient in terms of transistor count, but if the execution requires different branches, performance loss ensues. For small branches, the solution is to use *predication*, in which both branches are evaluated and thereafter the appropriate changes to the registers are written. Long branches are usually inefficient in pure SIMD architecture.

  In the latest graphics hardware supporting VS3 Vertex Shader and PS3 Pixel Shader models (see Section 3.5.3) two different solution paths have been taken. The VP pipelines are fully MIMD capable and thus need dynamic load balancing, but since the pipelines work on the same data stream this can be done automatically with little overhead. The FP is basically still SIMD but can evaluate different branches consecutively by keeping track of the current state and invalidating the results of individual pipelines. This results in some loss of performance, but such loss is acceptable if the executed branch changes very infrequently in the data stream. In the future, the FP will probably become fully MIMD too, although there is an ongoing debate as to whether this is really desirable, because the additional logic could also be used for more SIMD *parallelism*, which benefits the common cases.

- All pipelines write to the same destination arrays (frame-buffer).

  Shared memory machines usually do not have this restriction, although it avoids synchronization problems. A GPU pipeline cannot decide to output its data to an arbitrary position in memory. The destination memory must be defined beforehand and it cannot be read during the processing (in the general case). Therefore, there are no write-read collisions and no problems occur with cache coherency. For the FP, the restrictions go even further. Currently, the FP pipeline cannot change the destination address of a fragment at all. In other words, it cannot *scatter* data. This avoids completely any write collisions and allows parallel out-of-order processing. However, because the VP can scatter within the frame-buffer, fragments are roughly sorted by the primitives from which they were created.

Future FPs are likely to allow scatter at some stage, but the bound on chosen memory regions as destinations seems reasonable to avoid the general synchronization problems.

So, the two main restrictions are the lack of scattering in the FP and the poor ability to handle branching. With respect to lack of scattering, it is possible to turn to the gathers for help. The gathers are almost fully general and often exploited to alleviate other problems. In particular, scatters can be reformulated as gathers. Concerning branching, it is usual to try to move the branch condition away from the FP into the VP or even higher into the main program where one decides about the geometry to be rendered. A common practice is to divide the domain into tiles. A classification step determines which tiles need to take which branch of the code. Then each tile is streamed through a shader that contains the appropriate branch. This assumes that the branch condition will most likely evaluate to the same value for all pixels within one tile. Tiles that contain pixels scheduled for different branches must be processed twice (if both branches are non-empty), which is basically equivalent to predication. See [20, 30, 4] for application-specific implementations of this technique.

Since a GPU is so similar in certain respects to a shared memory machine, the reader may wonder why the programming model is so different (Section 3.1.3). A particular difference is that while OpenMP allows an incremental parallelization of an existing code [16, Section 2.3], the GPU forces us from the beginning into a new design with a distinction between shaders for the configuration of the VP and FP and the specification of the dataflow in the form of geometry to be rendered. Remember that this distinction is innate to DSB architectures (Section 3.1.2), which assume implicitly that changing data and non-changing instructions dominate the work load. This requires different data processing in the hardware and a different programming model. It also brings new opportunities and new restrictions. The massively parallel performance depends heavily on some of these restrictions and therefore a general incremental way to replace serial code with GPU parallelism is not feasible. The required generality would destroy the envisioned advantage. Future GPU programming will look more and more like CPU programming, and in the long run they might even use the same code basis. However, such code will have to respect the hardware characteristics of the GPUs, which often is not the case for current software. For efficient *parallelism* the programming model must support the hardware.

3.2 Theory

In Section 3.3 we extend the example from Section 3.1.3 to a linear equation system solver. For an exact derivation, more background is required on the data containers, control of global data-flow, the available operations and parallelism. However, the reader may choose to continue directly with Section 3.3 and look up the necessary information as needed.

3.2.1 Dataflow

The general dataflow in a GPU is prescribed by the *graphics pipeline* (Figure 3.1). The standard data path from the main memory and the textures to the *frame-buffer* always has been fast, but in iterative PDE solvers we need more than one pass and intermediate results must be reused for subsequent computations. This means that the content of the frame-buffer must be resent through the graphics pipeline repeatedly. The efficiency of this general data handling has improved significantly over the years, but a fully flexible solution is still in development. There are several possibilities for further processing of the results from the frame-buffer:

- Read-back (`glReadPixels`).
  We can read the selected content of the frame-buffer back to the main memory. This is a slow operation, because data transfer has always been optimized in the direction from main memory to the graphics card. With the PCI Express bus with a symmetric bandwidth in both directions, this has finally changed this year. However, even then, the available bandwidth on the card is much higher than over the bus, so transferring data to the main memory and back onto the card is inefficient. Data should be read back only if it requires analysis by the CPU.
- Copy-to-texture (`glCopyTexSubImage1D/2D/3D`).
  The frame-buffer content can be used to redefine parts of an existing texture, or to create a new one. This also requires copying data, but the high data bandwidth on the card makes this operation much faster than read-back.
- Copy-to-frame-buffer (`glCopyPixels`).
  It is possible to copy data from the frame-buffer onto itself. The per-fragment operations can be applied to the copied data, but not the programmable FP programs (see Section 3.5.2). Hence, this operation is mainly useful for copying data between different buffers in the frame-buffer and possibly combining the content of the source and destination with simple operations.
- Render-to-texture (`WGL_ARB_pbuffer, WGL_ARB_render_texture`).
  This is the current state of the art, but currently supported only under Windows. It is possible to allocate a *pbuffer*, i.e. a non-visible target buffer that serves as the destination for the output data stream. As soon as the pbuffer is not a render target any more, it can be used as a texture. Ultimately, this means that it is possible to render directly to a texture. Hence, we can continue to talk about textures, which now can be rendered to. The only problem with pbuffers is that they carry a lot of static information, which causes a performance penalty when binding a new pbuffer as the destination for the output data stream. The switch between the use of a texture as a data source and data destination is fast only for special configurations; see below.
- Architectural Review Board (ARB) superbuffers.
  Current graphics driver development addresses the problem of slow pbuffer switches by introducing a new, light-weight mechanism for using raw data arrays as source or destination at various points in the graphics pipeline. The idea

is to define a memory array together with properties that describe the intended usage. The graphics driver then decides where to allocate the memory (cacheable, Accelerated Graphics Port (AGP) or video memory), depending on these properties. To some extent, the functionality is already available with the Vertex Buffer Object (VBO) and the Pixel Buffer Object (PBO) extensions, but the OpenGL ARB superbuffer group works on a more general and fully flexible solution.

Apart from the incurred switch delay, pbuffers serve the purpose of flexible data handling on GPUs well. In actual code, the mechanism for binding the pbuffer as the source or destination is encapsulated in a class. When the superbuffers appear, a reimplementation of this class immediately yields the additional benefits without any further changes to the applications themselves. A great problem for the job sharing between the CPU and the GPU is the limited bus width between the chipset and the graphics card. Even the PCI Express bus, which promises a theoretical 4GB/s data transfer rate in each direction, cannot approach the excess of 30GB/s of onboard bandwidth. Systems with multiple GPUs must also respect this discrepancy; see Section 3.4.2.

The *frame-buffer* and pbuffers are actually collections (typically 1-6) of 2D data arrays (surfaces) of equal dimensions (Section 3.5.2). Current GPUs support Multiple Render Targets (MRTs), i.e. the shaders can output results to several of these surfaces simultaneously. No *scatter* is allowed here, i.e. exactly the same position in all surfaces is written to, but more than four float results can be output at once. This technique is compatible with the render-to-texture mechanism, i.e. each surface is a different texture and all of them can be written to in one pass. However, each write goes to exactly the same position in each of the textures.

Multi-surface pbuffers also help to avoid general pbuffer switches. Those surfaces that are not the destinations of the current render pass can be used as sources (textures). Swapping the roles of destination and source on the surfaces is far less expensive than a general pbuffer switch. Thus, iterations are usually performed on a multi-surface pbuffer in a ping-pong manner, i.e. for iteration 0 we have surface 0 as source and surface 1 as destination; for iteration 1 we have surface 1 as source and surface 0 as destination, etc. In addition, more surfaces and other textures can be sources and the MRT technique even allows the use of several of the surfaces as destinations simultaneously. During the ping-pong procedure the same pbuffer is read from and written to, but the source and destination memory is disjoint, so that no write-read collisions can occur. In comparison to the superbuffers, however, multi-surface pbuffers are still restricted, because GPUs offer only a few surfaces (up to 6) and they must have the same size and format. Moreover, pbuffers in general do not support all of the available texture formats.

3.2.2 Operations

First let us examine the available floating point number formats. Three different formats have been introduced with the development of GPUs that support Direct X 9 (Table 3.1). Soon, the standard IEEE s23e8 format (without denormalized numbers)

Table 3.1. Precision of floating point formats supported in graphics hardware. These formats were introduced with Direct X 9, which required the graphics hardware to have a format with at least the fp32 precision in the VP and fp24 in the FP. The unit roundoff, i.e. the upper bound on the relative error in approximating a real number with the corresponding format, is half the machine epsilon ε.

format	fp16	fp24	fp32
GPUs with FP precision	Wildcat Realizm, GeForceFX 5800/5900/6800	DeltaChrome S4/S8, Volari V8, Radeon 9700/9800/X800	Wildcat Realizm, GeForceFX 5800/5900/6800
GPUs with VP precision	-	-	all Direct X 9 chips, Wildcat Realizm (fp36)
setup	s10e5	s16e7	s23e8
ε	$9.8 \cdot 10^{-4}$	$1.5 \cdot 10^{-5}$	$1.2 \cdot 10^{-7}$

will be a common standard, because chips that support the PS3 model are required to have a corresponding PEs throughout the pipeline. Hence, the half-precision format will be mainly useful to save memory and bandwidth, and possibly for fragment blending, which to date has no full floating point support. The implementation of a double float format is unlikely in the near future, though a hardware emulation could be feasible.

Both the VP and FP support a rich set of operations. There is a difference between the functionality offered by the high-level languages and the assembly languages, as the latter more closely express which functional units really reside in the hardware. However, since the languages intentionally include more primitive functions with the expectation that they will receive hardware support in future GPU, we want to present the functionality at this language level. Unfortunately, there is, as yet, no unified shader model for the VP and the FP. The FP imposes some additional restrictions, although this is not caused by a lack of language constructs, but rather by their use. In the following we will use the Cg syntax, but Direct X High-Level Shading Language (HLSL) is almost identical and OpenGL Shading Language (GLSL) very similar (see Section 3.5.4).

- Floating-point types: `half, float, half2, float4, float4x4`.
  The `half` is a s10e5 and the `float` a s23e8 (or s16e7) floating-point format (see Table 3.1). For both scalar types there exist native vector types of up to 4 components and all matrix types up to 4×4. Components of the native vectors can be arbitrarily swizzled, i.e. they can be duplicated and their order can be changed, e.g.:

```
float4 a(0, 1, 2, 3);
float4 b= a.xyzw;   // b==float4(0, 1, 2, 3)
float4 c= a.wyxz;   // c==float4(3, 1, 0, 2)
float3 d= a.ywy;    // d==float3(1, 3, 1)
```

Most graphics PEs operate internally on 4-vectors, so using the native vector types can greatly reduce the number of required operations.

- Data types: `float[5], float[6][3], struct`.

  General vectors and arrays can be defined, but there is only a limited number of temporary registers (up to 32 float 4-vectors), so for practical purposes, the size is extremely limited. There are more constant registers (up to 256 float 4-vectors). Arrays are first-class types, i.e. they are copied when assigned, since there are no pointers or associated functionality. In the VP constant vectors/arrays can be indexed with variables. Only the newest PS3 model for the FP supports such indexing for the texture coordinates.
- Mathematical functions.

Arithmetic	`+, -, *, /, fmod`
Sign, Comparison	`abs, sign, min, max, clamp`
Integers	`ceil, floor, round, frac`
Exponential	`sqrt, exp, exp2, log, log2, pow`
Trigonometric	`sin, cos, tan, asin, ..., sinh, ...`
Interpolation	`step, smoothstep, lerp`
Vector	`dot, cross, length, normalize, distance`
Matrix	`mul, transpose, determinant`

  Almost all scalar functions can also operate component-wise on the native floating-point vector types.
- Data access: `tex1D, tex2D, tex3D, texRECT`.

  In the FP, one to three dimensional textures (data arrays) can be read from arbitrary positions, e.g.:

```
float4 coord= IN.texCoord; // current texture coordinates
float4 a= tex1D(Tex_A, coord.x);
float4 b= tex2D(Tex_B, coord.xy);
float4 c= tex3D(Tex_C, coord.xyz);
```

  Currently, normalized coordinates from $[0,1]^2$ are used for texture access and only special rectangular 2D textures (`texRECT`) are accessed by coordinates from $[0,w]\times[0,h]$, which depend on the width (w) and height (h) of the texture. The texture samplers `Tex_A, Tex_B, Tex_C` cannot be chosen dynamically. This is expected to change in the future. In the newest VS3 model the VP can also access textures.
- Conditions: `bool, bool4`, $\&\&, ||, !, <, >, ==, !=, ?:$.

  Conditions must evaluate to a Boolean type. The operations can work component-wise on Boolean vectors. In case of the operator ?: this allows an individual decision for each vector component, e.g.

```
bool4 cond(true, false, false, true);
float4 a= cond? float4(0,2,4,6) : float4(1,3,5,7);
// Now a==float4(0,3,5,6)
```

- Control flow: `int, int4, if/else, while, for`.

  The conditions must be scalar Booleans. In the VP dynamic branches are fully supported, so there are no further restrictions on the constructs. In the newest PS3 model, there is restricted support for dynamic branching in the FP (see Section 3.1.4). Otherwise `if/else` is resolved with *predication*, i.e. both branches are evaluated and conditional writes update the registers with the correct results. Without PS3 loops are unrolled, which must be possible. The integer types are currently not supported in hardware and are internally represented as floats. They are basically supposed to be used as loop counters and in case of unrolled loops, for example, they do not show up at all in the end.

- Abstraction: `struct, typedef`, functions, function overloading, interfaces. The high level languages offer increasingly more of the abstract constructs known from C/C++ or Java, although some restrictions apply. As the abstraction is not dependent on the available processing elements (PEs) in the hardware, it is likely to evolve further.

Since the PS2 model (Direct X 9 GPUs) and the introduction of floating-point number formats, the desire for arithmetic functionality has been basically fulfilled. The limits are now set by the control flow and variable indexing of vectors/arrays. For some configurations, the number of available temporary registers may also be a restriction. Future GPUs will relax these constraints further.

3.2.3 Parallelism

Figure 3.1 visualizes the stream processor nature of GPUs. We see two types of *parallelism* there: (i) the parallelism in depth innate to the pipeline concept, and (ii) the parallelism in breadth given by the breadth (4-vectors) and number of parallel vertex (up to 6) and fragment pipelines (up to 16). Because there is no configurable routing in GPUs, unlike FPGAs for example, these numbers are fixed, which has several consequences.

The deep pipeline makes frequent invocations or reconfigurations inefficient, i.e. for each rendering call the same operations should be applied to at least several thousand data items; the more the better. This does not mean that we cannot treat primitives smaller than 32×32 efficiently, but small regions undergoing the same operations should store their geometry in a common VBO. Then, one invocation suffices to execute the configured operations on all defined regions. Unfortunately, in the case of points, even then performance is highly reduced, because GPUs are optimized for processing 2D regions. Therefore, it is currently difficult to implement algorithms that require updates of singular, spatially unrelated nodes.

Up to 128 floating point operations can be executed per clock cycle in the FP, but the 256 bit wide Double Data Rate (DDR) memory interface delivers only 64 bytes. This means that to avoid a memory *bandwidth* problem the *computational intensity* should be, on average, above 8, i.e. eight or more operations should be performed in the FP on each floating point value read from the memory (assuming four bytes per float). Because the memory on graphics cards clocks higher than the GPU, and because of small internal caches, in practice the computational intensity may be a bit lower, but the general rule remains. The significant overbalance of processing power against bandwidth has arisen only with the recent generation of graphics hardware. This trend is likely to continue, because computer games now also use programs with higher computational intensity and the integration of additional PEs into the GPUs is cheaper than the corresponding bandwidth increases. Note that, despite less internal parallelism, the high clock frequencies of the CPUs, and less bandwidth from the main memory system require a similarly high or even higher computational intensity for the CPUs. However, the bandwidth efficient programming methodologies for CPUs that exploit the large and fast on-chip caches cannot be directly applied to GPUs, which have only small caches. GPUs reduce the bandwidth requirements best in the case of strong data locality, e.g. node neighbors in a grid. See Section 3.3.3 for a discussion of efficient matrix vector products.

3.3 Practice

Section 3.1.3 offers a glimpse of the programming of GPUs. Now, after getting to know the dataflow and processing functionality in more detail, we want to demonstrate how to build up an efficient solver for a linear equation system on a GPU. Then we will present some of the existing PDE applications and list links to resources for GPU programming.

3.3.1 Setup

So far, we have talked about rendering to a frame-buffer. However, what we see on the screen are individual windows controlled by a window manager. Window management, the allocation of pbuffers and initialization of extensions depend on the operating system. Luckily, there exist libraries that abstract dependencies in a common interface. We will use the GLUT library for the Graphics User Interface (GUI), the GLEW library for the extension initialization and the RenderTexture utility class for the handling of pbuffers. Links to all resources used in the code samples are given in Section 3.3.6.

With the libraries, the main function for the addition of two vectors $\bar{A}$ and $\bar{B}$ as discussed in Section 3.1.3 needs only few lines of code:

```
#include <GL/glew.h>           // extension initializer GLEW
#include <GL/glut.h>           // window manager GLUT
#include "WinGL.h"             // my GUI
#include "AppVecAdd.h"         // my GPU application

int main(int argc, char *argv[]) {
  glutInit(&argc, argv);       // init GLUT window manager
  glutInitDisplayMode(GLUT_DOUBLE | GLUT_RGB);

  // simple example: addition C= A+B
  WinGL winAdd;                // my GUI based on GLUT
  glewInit();                  // init extensions with GLEW
```

```
    AppVecAdd add;                  // my GPU application
    winAdd.attachApp(&add);         // attach App to GUI

    glutMainLoop();                 // start main loop of GLUT
    return 0;
  }
```

The first lines in main initialize the GLUT manager and set the default display mode. Then we create a GUI. The GUI manages a window, keyboard strokes and a menu with GLUT. Via the resource section (Section 3.3.6) the reader may find many tutorials that demonstrate the ease with which a GUI may be created with GLUT. The GLEW library call initializes all available OpenGL extensions. The extensions are necessary to use pbuffers and the programmable pipeline, for example. Most current GPUs support them. Then, we create our application class and attach it to the GUI such that the user can call the application functions. Finally, we start the event loop of GLUT and wait for the invocation of these functions by user interaction.

It is in the application class that the more interesting things happen. The constructor uses the Cg API to compile the application-specific shaders and load them to the graphics card. For the vector addition example above, we need a trivial VP shader that passes the vertex data through unchanged, and the `fpProg[FP_ADD2]` shader for the addition from Section 3.1.3. However, for other application classes, more shaders are loaded and can be later accessed by the vectors `vpProg[]`, `fpProg[]`. The constructor also uses the RenderTexture utility class to allocate the textures and store them in a vector `texP[]`:

```
  // enum EnumTex { TEX_A, TEX_B, TEX_C, TEX_NUM };
  for(i= 0; i<TEX_NUM; i++) {              // allocate textures
    RenderTexture* tp= new RenderTexture("r=32f tex2D rtt");
    tp->Initialize(256, 256);              // texture size
    texP.push_back(tp);                    // store in a vector
  }
```

The mode-string requests a 32 bit float 2D texture suitable for the render-to-texture (`rtt`) mechanism (see Section 3.2.1). Currently, only Windows supports the render-to-texture mechanism, so on other systems copy-to-texture (`ctt`) should be used instead. The RenderTexture class has a simple interface for using the textures as either a destination or source of a data stream, possibly emulating render-to-texture by copy-to-texture internally. To set the initial data in a texture we simply define the values in a float array (`floatData`) and then render it:

```
texP[TEX_A]->BeginCapture();                // TEX_A is destination
glDrawPixels(texP[TEX_A]->GetWidth(),
             texP[TEX_A]->GetHeight(),
             GL_RED, GL_FLOAT, floatData);
texP[TEX_A]->EndCapture();                  // TEX_A is source
```

In this way, the application class contains vectors with the shaders `vpProg[]`, `fpProg[]` and a vector with the initialized textures `texP[]`. These are the main steps during the initialization of a GPU application and are independent of which operations will be performed later.

After the above initialization, the entire function for the addition of the vectors $\bar{A}$ and $\bar{B}$, which gets called via the GUI, reads as follows:

```
void AppVecAdd::exec() {
  CGprogram curVp= vpProg[VP_IDENTITY];// vertex shader
  CGprogram curFp= fpProg[FP_ADD2];    // fragment shader

  texP[TEX_C]->BeginCapture();        // TEX_C is destination

  cgGLEnableProfile(cgGetProgramProfile(curVp)); // enable
  cgGLEnableProfile(cgGetProgramProfile(curFp)); // profiles

  cgGLBindProgram(curVpProg);          // bind
  cgGLBindProgram(curFpProg);          // shaders

  glActiveTexture(GL_TEXTURE0);        // texunit0
  texP[TEX_A] >Bind();                 // bind TEX_A
  glActiveTexture(GL_TEXTURE1);        // texunit1
  texP[TEX_B]->Bind();                 // bind TEX_B

  drawTex2D();                         // render-to-texture

  cgGLDisableProfile(cgGetProgramProfile(curVp)); // disable
  cgGLDisableProfile(cgGetProgramProfile(curFp)); // profiles

  texP[TEX_C]->EndCapture();           // TEX_C is source
}
```

The shader `fpProg[FP_ADD2]` and the function `drawTex2D()` are listed in Section 3.1.3. All other calls fall within the functionality of the Cg API or the RenderTexture class. Because we have two different program sources, namely this C++ code and the Cg shaders, the passing of arguments to the shaders relies on the numbering of the texture units: `GL_TEXTURE0` corresponds to `texunit0` in the shader. The numbering corresponds to the numbering of arguments passed to the multidimensional function realized by the shader. Above, we do not see the OpenGL texture IDs explicitly, as in Section 3.1.3, because they are taken care of automatically by the RenderTexture utility class. Alternatively to the numbering, Cg also allows the association of the OpenGL texture IDs with the sampler names of the textures in the shader.

Even without the preparatory work of the initialization, the `exec()` function above seems a lot of code for the simple addition of two vectors $\bar{C} = \bar{A} + \bar{B}$. In practice, such operations are always encapsulated into a single function call. One option is to derive a class from RenderTexture and add functionality to it, such that the destination texture manages the operation:

```
texP[TEX_C]->execOp(fpProg[FP_ADD2],texP[TEX_A],texP[TEX_B]);
```

For convenience, one could even define an `operator+` function in this way, but this would be of little use since practical shaders do not represent elementary functions. Another option is to have a class for each shader and then simply write something like

```
fpProg[FP_ADD2].exec(texP[TEX_C], texP[TEX_A], texP[TEX_B]);
```

For a particularly short notation it is possible to have a function within our application class that only takes the indices:

```
execOp(TEX_C, FP_ADD2, TEX_A, TEX_B);   // C= A+B
```

We will use this notation in what follows. Clearly, the graphics setup described here is an example of how to get started fairly quickly. For large projects, more abstraction is recommended.

3.3.2 Vector Operations

Once the graphics specific-parts are abstracted, it is easy to realize further operations on vectors. We simply need to write a new FP shader, e.g. `FP_ATAN2` and then call it:

```
execOp(TEX_C, FP_ATAN2, TEX_A, TEX_B);     // C= atan(A/B)
```

Remember that the high-level languages support most of the standard mathematical functions directly (Section 3.2.2). So to write the `FP_ATAN2` we only need to exchange the `return` value in `FP_ADD2` (Cg listing in Section 3.1.3) for

```
return atan2(valA,valB);
```

For the vectors $\bar{A}, \bar{B}$ and $\bar{C}$ represented by the textures `TEX_A, TEX_B, TEX_C` this would correspond to

$$\bar{C}_\alpha = \text{atan}(\bar{A}_\alpha / \bar{B}_\alpha) \,.$$

With a standard set of such shaders it is easy to evaluate formulae, e.g. linear interpolation

```
execOp(TEX_C, FP_SUB2, TEX_B, TEX_A);     // C= B-A
execOp(TEX_D, FP_MUL2, TEX_C, TEX_M);     // D= C*M
execOp(TEX_R, FP_ADD2, TEX_A, TEX_D);     // R= A+D= A+M(B-A)
```

However, it is much more efficient to have one shader that does exactly the same in one pass. This avoids the costly pbuffer switches (Section 3.2.1) and increases the computational intensity (Section 3.2.3). So, in general, the shaders of an application should execute as much as possible in one go. The application class `AppVecAdd` that contains the addition as the only shader is, in this respect, an unrealistic example.

Unfortunately, the instruction to pack everything into one shader whenever possible easily results in the generation of a multitude of shaders. Consider, for example, the task of applying $h(f_i(\bar{V}_\alpha), g_j(\bar{V}_\alpha)), 1 \leq i, j \leq N$ to the components of a vector $\bar{V}$, where the choice of i, j depends on some computed entity. For optimal performance we would have to write N^2 shaders S_{ij}. Duplication of code could be minimized by using include files that implement the $2N + 1$ functions $h(.,.), f_i(.), g_i(.), 1 \leq i \leq N$, but even so, N^2 short, different files would have to be generated. The remedy in such cases is to generate the few changing lines of code at run time and use run time compilation. If only a few run time compilations are necessary the performance does not degrade, especially if the GPU does not have to wait for the compiler but can be occupied with some other task during the software compilation process.

3.3.3 Matrix Vector Product

Matrix vector products are ubiquitous in scientific computing. The application of a matrix can be seen as a series of *gather* operations on the vector components. The matrix rows $\bar{A}_{\alpha,.}$ define what to gather and the weights. The gathers are inner products between the rows and the vector:

$$AV = \left(\bar{A}_{\alpha,.} \cdot \bar{V}\right)_{\alpha}, \qquad \bar{A}_{\alpha,.} := (A_{\alpha,\beta})_{\beta} .$$

For a 2D problem we store the nodal vector $\bar{V}$ of the corresponding 2D grid as a 2D texture. Then α and β must be seen as 2-dimensional multi-indices as above. This means that without renumbering of indices, a full matrix for a 2D problem is a 4D structure. Due to the fact that GPUs restrict each dimension of a texture to 4096 or less, we usually cannot store a full matrix in a 1D or 2D texture. The best option is to use a 3D texture (4D textures are rarely supported), where in each 2D slice of the texture we pack several 2D row vectors $\bar{A}_{\alpha,.}$. Depending on the current pixel position, the FP shader retrieves the correct weights and performs the multiplications and additions. The result is obtained in one pass.

Because of the packing, a certain amount of address translation must be performed to retrieve the correct values. The Vertex Processor (VP) is usually not the bottleneck in scientific computing and is, therefore, used for the task of precomputing the offsets to the packed rows. The offset texture coordinates are passed to the FP. Another way of packing is to define a 4-valued texture and thus quadruple the number of values that are stored per texel. The point of optimizing operations for an execution on 4-vectors is discussed at the end of this section. From the point of view of memory, the packing of floats into 4-vectors is a disadvantage, because the large amount of data that has to be retrieved with a single command can lead to a local *memory gap* problem. Reading the four floats individually gives the compiler more freedom to place some computation between the reads and thus hide memory *latency* and insufficient *bandwidth*.

In general, full matrices can be handled only for small dimensions. A full-float matrix for a 128x128 grid requires 1Gb of memory, which exceeds the present *video memory* of graphics cards. Future GPU will offer hardware virtualization, such that a texture can also reside (partly) in main memory. However, the necessary data transfer to the graphics card is a strong bound on the performance in this case. Luckily, in practice most matrices are sparse. Typical examples are band matrices. Each band can be stored in a texture of the same dimension as the grid. The bands can be packed together in one texture, which will reduce the lines of code necessary for the texture binding. This also avoids the problem that a maximum of 32 different textures can be bound to a FP shader. The VP can perform the offset computation, and the matrix vector product can be obtained in one pass again.

In the case of Finite Element codes and the discretizations of local operators, it is also possible to store the matrix in the form of elemental matrices (see (3.2)). Then, for each element, the components of the elemental matrices must be stored in separate textures or a packed arrangement. This is a special case of the general idea of partial matrix assembly that is presented below. It is particularly advantageous if the elemental matrices possess some common structure, such as symmetry or parameterization by few variables, since this greatly reduces the storage requirements. In the case of parameterization, one would only store the few values from which the elemental matrices can be built up (see (3.3)), and thus favorably increase the computational intensity of the matrix vector product. One problem arises, however, for the output of the local application of an elemental matrix. The GPU cannot scatter data, which means that only one node within the element can be updated from the result of the FP. The updating of the other nodes in the same pass would require the recomputation of the matrix vector product on the element for each node. One remedy for this is to output the complete result vector on each element, and from this gather the information onto the nodes in a second pass. This is a typical strategy on GPUs for reformulating a regular *scatter* operation in terms of a *gather*.

For irregular sparse matrices, two strategies are available: (i) cover the non-zero entries efficiently with some regular structures, i.e. rows, columns, sub-diagonals, and encode this structure statically into the shader, or one (ii) use a level of indirection in the processing, such that, e.g. the matrix entries contain not only the value but also the address (or offset) needed to access the corresponding component in the vector. The result can be computed in one pass. However, the irregularity of the entries can result in a serious performance problem if the number of entries per row differs significantly. Therefore, different vector components may require very different numbers of multiplications and additions. The PS2 model for the FP cannot stop the computation dynamically, i.e. all gather operations take the same time within the same shader. In the worst case one full row in the matrix suffices to make the matrix vector product as expensive as one with a full matrix. The newer PS3 model can make the distinction, but in terms of performance it is only beneficial if all spatially coherent vector components require approximately the same number of operations (see Section 3.1.4). Otherwise, the longest case dominates the execution time again.

Recapitulating, we can say that within the noted restrictions, matrix vector products can be computed for arbitrary matrices. Usually the matrices are not constant and have to be assembled first (see (3.1)). In many cases it is best not to assemble matrices explicitely, or at least not fully. Recall from Section 3.2.3 that current GPUs require a *computational intensity* of approximately 8 to avoid bandwidth shortage (in the case of floats). However, in a matrix vector product, we read both the matrix entry and the vector component and perform just one assembly operation: a multiply and add (MAD). In other words we exploit only 6.25% of the available processing power. Consider three flavors of the matrix vector product for improvement:

- On-the-fly product: compute entries of A for each $A\bar{V}$ application.
  At first, it may seem foolish to compute the entries of A over and over again. However, this can still be faster than simply reading the precomputed data, because the comparably slow reading will stall the computation. Clearly, the advantage can only be gained for simple entries that can be computed quickly from little data. This technique has the lowest memory requirement and thus may also be applied in cases when the entire A would not fit into memory.
- Partial *assembly*: apply A on-the-fly with some precomputed results.
  This is a flexible technique which allows the computation and bandwidth resources to be balanced. On-the-fly products are infeasible if many computations are required to build up the matrix entries. In this case, a few intermediate results that already encompass most of the required operations should be generated. Then, during the matrix vector product, these few intermediate results are retrieved and the matrix finishes the entry computation on-the-fly. This reduces the bandwidth requirement and targets an optimal computational intensity. The few intermediate results have also the advantage of modest memory consumption.
- Full *assembly*: precompute all entries of A, use these in $A\bar{V}$.
  This makes sense if additional operations of high computational intensity hide the bandwidth problem of the pure matrix vector product. To achieve this, it even makes sense to execute operations unrelated to the current matrix vector product in the same shader. The Multiple Render Target (MRT) technique (Section 3.2.1) allows the unrelated results to be output into separate textures. If the bandwidth shortage can be hidden (though this is hard to achieve), full assembly is the fastest option for the matrix vector product, but also the one with the highest memory requirements.

The above discussion is not specific to GPUs, because the same considerations apply to CPUs. Yet there is a relevant and important difference between GPUs and CPUs. While block matrix techniques exploit the large caches on typical CPUs, this is not possible in the case of GPUs, because they have only small caches and rely strongly on the right balance of operations and *bandwidth* capacity. This is a crucial factor and should be checked carefully in the case of poor performance on the GPU. Pure matrix matrix multiplications, for example, are not faster on GPUs than on current CPUs [6].

Following the considerations about bandwidth, it is also possible to opt for low-level optimizations of the matrix vector product [2], related to the fact that the processing elements (PEs) operate, in general, on 4-vectors. However, newer GPUs can split the 4-component PEs into a 3:1 or even 2:2 processing mode, evaluating two different commands on smaller vectors simultaneously. The high-level language compilers optimize for this feature by automatically reordering commands whenever possible. The resource section offers links to tools that analyze the efficiency of shaders for a given GPU.

3.3.4 Solvers of Linear Equation Systems

We have repeatedly encouraged the reader to put as many operations as possible into one shader. Large shaders avoid pbuffer switches (Section 3.2.1) and help to hide bandwidth shortage (Section 3.2.3, Section 3.3.3). Why, then, should the entire problem not be solved in one shader? If this can be done without unnecessary additional operations or data access, it is the right choice. However, the implementation of a separable filter in one pass is a waste of resources. The same applies to the iterative solution of a linear equation system $A\bar{X} = \bar{R}$,

$$\bar{X}^0 = \text{initial guess}, \qquad \bar{X}^{l+1} = F(\bar{X}^l),$$

where $F(.)$ is the update method, e.g. conjugate gradient. The implementation of several iterations in one shader is unwise, because it multiplies the number of operations and, in particular, data accesses.

Which solvers are suitable for GPUs? They must allow parallel independent processing of vector components, and do so without direct write-read cycles. The first is important to exploit the parallel pipelines, while the second is a restriction of the FP which, in general, has no access to the destination buffer during the processing. An alternative is to process the vector in blocks with several passes, such that during the processing of a block the previously computed blocks can be accessed.

The conjugate gradient solver and its variants (preconditioned, asymmetric) rely on operations of the following forms:

$$F(\bar{X}^l) = \bar{X}^l + \frac{\bar{r}^l \cdot \bar{p}^l}{A\bar{p}^l \cdot \bar{p}^l}\bar{p}^l, \qquad \bar{p}^l = \bar{r}^l + \frac{\bar{r}^l \cdot \bar{r}^l}{\bar{r}^{l-1} \cdot \bar{r}^{l-1}}\bar{p}^{l-1}, \qquad \bar{r}^l = \bar{R} - A\bar{X}^l .$$

The main ingredients are the matrix vector product, which was discussed in the previous section, and the inner product, which is a *reduction* operation.

Reductions are not directly supported in hardware on GPUs. An easy solution is to write a shader with a loop that runs over all texels and performs the summation. By rendering a single pixel with this shader, the result is obtained; but this does not utilize the parallel pipelines. At least 16 pixels with subtotals should be rendered before a final summation is performed. A second possibility is to perform consecutive additions of neighboring texels and thus reduce the dimensions of the texture by 2 in each pass. In the end, the result is also a 1x1 texture. Which option is better depends strongly on how data is arranged in textures: linearly or hierarchically. Traditionally, GPUs are optimized for the second option of fast access to neighbor texels. With the first, there may be general problems with the maximal instruction number in the FP (see Section 3.5.3). The result of the reduction can be either read back to the CPU or left in a 1×1 texture for further use. In an interactive approximation, the read-back is necessary at some stage to retrieve the norm of the residual and decide whether the iterations should be stopped. However, the asynchronous read-back mechanism does not stop the computation.

We see that all ingredients necessary for a solver of a linear equation system can be implemented on a GPU. The initialization of the graphics pipeline requires some effort (Section 3.3.1), but once the work is done or prepared by some library, it is possible to concentrate on the algorithm. Concerning the matrix vector product (Section 3.3.3), attention should be paid to the high ratio of processing elements (PEs) against the bandwidth in GPUs. The use of a fully-assembled matrix consumes a lot of bandwidth and is only appropriate if this disadvantage can be hidden with accompanying computations. Finally, solvers of linear equation systems must allow for parallel processing of the vector components. Reduction operations are not native to GPUs, but can be resolved efficiently. Several researchers have solved PDE problems along these lines. The next section discusses some applications.

3.3.5 PDE Applications

We consider the discretization of partial differential equations on GPUs. In the field of continuum mechanics, various physical processes have been simulated in graphics hardware [18, 17, 21, 19, 12]. Beyond physical simulation, GPU-accelerated PDE methods are also very popular in geometric modeling and image processing [20, 3, 11, 29, 14]. The GPU Gems book series also contains an increasing number of GPU-accelerated PDE solvers [7, 25] and the site [10] offers an extensive overview of GPU-based computations. The processing of triangular grids, shading and texturing of highly resolved meshes, and the processing of images (usually regarded as surface textures), are the original applications for which graphics cards have been designed and optimized. Before we provide an overview of a number of applications in this field, we outline the basic setup for the treatment of PDEs on GPUs.

Consider a general differential operator A that acts on functions u defined on a domain Ω and ask for a solution of the differential equation

$$A[u] = f$$

for a given right-hand side f. In addition, require certain boundary condition to be fulfilled on $\partial\Omega$. In the case of variational problems, we ask for minimizers of energies E over functions u, such that a differential equation appears as the Euler Lagrange equation, with $A[u] = \operatorname{grad} E[u]$ and $f = 0$. If we take into account some time-dependent propagation, relaxation or diffusion process, we frequently obtain a differential equation of the form

$$\partial_t u + A[u] = f\,.$$

Now, we ask for solutions u that depend on the time t and the position x on Ω. In the case of a second-order diffusion we usually deal with $A[u] = -\mathrm{div}(a[u]\nabla u)$, where $a[u]$ is a diffusion coefficient or tensor that possibly depends on the unknown solution u. In the case of Hamilton Jacobi equations that describe, for instance, the propagation of interfaces, we deal with $A[u] = H(\nabla u)$. E. g. $H(\nabla u) = v(t,x)\,\|\nabla u(t,x)\|$ corresponds to the propagation of the level-sets of the function u at time t and position x with a speed $v(t,x)$ in the direction of the normaly. In many cases, v itself depends non-linearly on u.

Now consider the discretization of these differential equations based on Finite Elements. Obviously, other ways to discretize PDEs such as Finite Volume or Finite Difference approaches lead to fairly similar computational requirements. We consider a simplicial or rectangular mesh $\mathcal{M}_h$ on Ω with grid size h and a Finite Element space V_h with a $N = \#I$-dimensional basis $\{\Phi_\alpha\}_{\alpha\in I}$ consisting of basis functions Φ_α with local support on the domain. Now, we ask for a discrete solution

$$U(x) = \sum_{\alpha\in I} \bar{U}_\alpha \, \Phi_\alpha(x)$$

of the stationary problem, such that U approximates the continuous solution u, or we compute space and time discrete solutions $U^k(x) = \sum_{\alpha\in I} \bar{U}^k_\alpha \, \Phi_\alpha(x)$, with $u(t_k, x) \approx U^k(x)$, for $t_k = k\,\tau$ and some time-step τ.

Usually, Finite Element algorithms consists of two main ingredients; namely, the *assembly* of certain discrete vectors in $\mathbb{R}^N$ or matrices in $\mathbb{R}^{N^2}$ and the *discrete solution update*, with an iterative linear equation system solver, an explicit update scheme in time, or a combination of both in case of an iterative scheme with an inner linear system of equations to be solved:

- Assembly.
  In an explicit gradient descend algorithm, we usually compute the discrete gradient

  $$\left(\mathrm{grad}_{V_h} E[U]\right)_\alpha = \langle E'[U], \Phi_\alpha\rangle$$

  via a traversal over the grid $\mathcal{M}_h$. Locally on each element we collect contributions to the integral $\langle E'[U], \Phi_\alpha\rangle$ for all Φ_α such that its support intersects the current element. Similarly, the assembly of a Finite Element matrix, e. g. the stiffness matrix in the above-mentioned diffusion process

  $$L_{\alpha,\beta} = \int_\Omega a[U]\nabla\Phi_\alpha \cdot \nabla\Phi_\beta \,\mathrm{d}x \tag{3.1}$$

  starts by initializing $L = 0$, followed by a traversal of all elements. On each element E a corresponding local *elemental matrix*

  $$l_{\alpha,\beta}(E) = \int_E a[U]\nabla\Phi_\alpha \cdot \nabla\Phi_\beta \,\mathrm{d}x \tag{3.2}$$

  is computed first, corresponding to all pairings of local basis functions relevant on this element. Then, we can either store the collection of elemental matrices or assemble them into the global matrix L (see Section 3.3.3).
  All these operations match the data-stream-based (DSB) computing paradigm perfectly. The instruction set is always the same. Only the data to be processed changes and this data can be gathered by simple texture reads. In the case of a linear Finite Element space, the relation between the texels in the textures and the degrees of freedom is particularly simple. For example, if we process an image,

the values at the image pixels correspond directly to the degrees of freedom in the Finite Element space, and thus a coordinate vector in the Finite Element space is again an image. Similarly, we can treat each row in an irregular sparse matrix as a floating point texture and the corresponding index field, which contains the global position of the entries, as an integer texture [3]. The indirect access is less efficient because it partly breaks the paradigm of reading all data in streams. However, GPUs have also internal mechanisms to reduce the incurred performance penalty in such cases. The same problem cannot appear for the output because GPUs do not support the scattering of data.

For vector-valued functions u, e.g. positions in space, deformations, gradients or 2D Jacobians, the data can be kept in 4-valued textures. Then, it is also easy to take advantage of the processing elements (PEs) operating on 4-vectors (see Section 3.2.2). However, for larger texels (4 floats = 16B) it is more difficult to hide the incurred memory latency, so storage of the individual components is often the better choice (see Section 3.3.3). After realization of the correct functionality, the optimal option can be determined by a profiling tool.

- Discrete solution update.

In the case of a simple update scheme, it is preferable to interleave the assembly with the update. That is, for a time-step of a discrete gradient descent

$$U^{k+1} = U^k + \tau \text{grad}_{V_h} E[U^k]\,,$$

we immediately add the element-wise components of the update to the old discrete solution. When an iterative solver for a linear equation system is involved, i.e. the matrix is required in more than one matrix vector product, there are three possibilities: on-the-fly products, a partial or a full *assembly*. These possibilities were discussed in Section 3.3.3.

For a regular grid, standard linear stiffness matrices or mass matrices can be applied efficiently on-the-fly, because the matrix entries are locally the same for all elements, and can be stored in constants in the shaders. This changes if we consider non-linear stiffness matrices as defined above for the diffusion problem. For example, if $a[u]$ is a diffusion tensor and we use the midpoint integration rule for $a[u]$ in (3.2), we precompute the midpoint values $a[u]_E^{i,j}$ in a texture and store the constants $C_{\alpha,\beta}^{i,j} = \int_E \partial_i \Phi_\alpha \cdot \partial_j \Phi_\beta \, dx$ in the shader. For isometric elements, the constants are few because they depend only on the difference $\alpha - \beta$. Then, the elemental matrices are parameterized by $a[u]_E^{i,j}$ and can be quickly reconstructed on-the-fly:

$$l_{\alpha,\beta}(E) = \sum_{i,j} a[u]_E^{i,j} C_{\alpha,\beta}^{i,j} \,. \tag{3.3}$$

The advantages are higher *computational intensity* in the matrix vector product and reduced memory requirements. Recall that for very large triangular meshes or images, the full assembly of a matrix still conflicts with the limited video memory size of graphics cards.

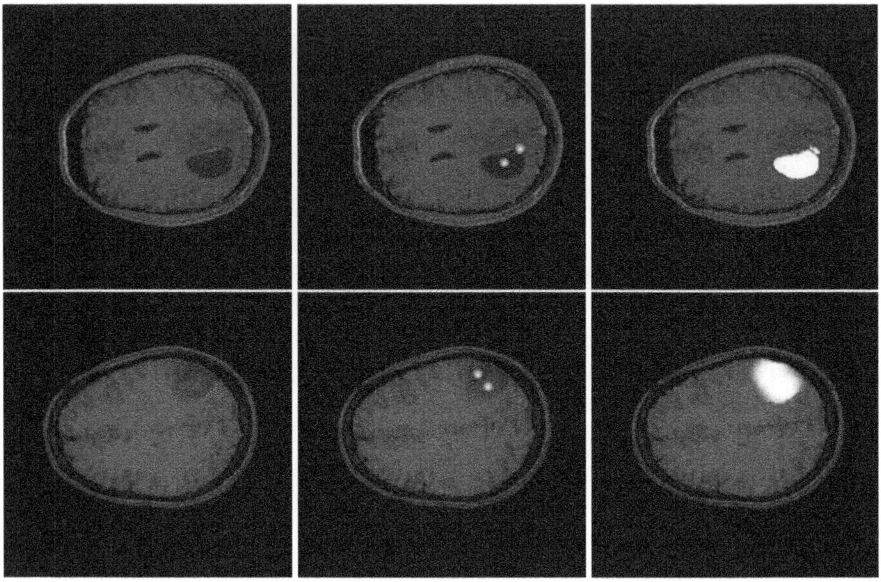

Fig. 3.2. Segmentation of tumors computed in Direct X 7 graphics hardware. The expansion of the level-set depends on the image values, its gradient and the placement of the seeds.

Now consider the processing of images as a concrete application field. Classical tasks in image processing are

- segmentation,
- feature-preserving image denoising,
- image registration.

Images are perfectly matched to rectangular textures and can be regarded as functions in a piecewise bilinear Finite Element space. Furthermore, if it comes to real-time processing and visualization, the results of our PDE algorithm reside already on the graphics boards, where they are needed for display. This underlines the conceptual benefits of PDE-based image processing directly on the GPU. In what follows we provide a brief sketch of some methods:

- Segmentation.
  Consider a region-growing algorithm for the segmentation of image regions whose boundaries are indicated by steep gradients. Therefore, a segment domain is represented by a level-set of a function u and sets $v(t,x) = \eta(\|\nabla I\|)$, where I is the image. Here, $\eta(\cdot)$ is some non-negative edge-indicating function, which is zero for $\|\nabla I\|$ larger than a certain threshold. Now, we ask for a family of level-set functions and corresponding evolving segment domains, such that

$$\partial_t u + v(t,x)\,\|\nabla u\| = 0\,.$$

Fig. 3.3. Anisotropic diffusion for image denoising computed in Direct X 8 graphics hardware. The anisotropy allows the smoothing of noisy edges without blurring them completely.

The initial data $u(0,\cdot)$ is supposed to represent a user-defined mark on the image [26] (see Figure 3.2).

- Feature-preserving image denoising.

  Multiscale methods in image denoising are fairly common nowadays. The desired result is a family of images that exhibit different levels of detail, fine scale to coarse scale, and which are successively coarser representations of the initial fine-scale image. The aim of denoising is to filter out fine-scale noise on coarser scales while preserving important edges in the image. Such a scale of images can be generated solving a non-linear diffusion problem of the type

$$\partial_t u - \operatorname{div}(a[u]\nabla u) = 0\,,$$
$$a[u] = g(\|\nabla(G^\sigma * u)\|)\,.$$

  The diffusivity $g(s) = (1+\frac{s^2}{\lambda^2})^{-1}$ is large away from the edges and small in the vicinity of the edges, as indicated by large image gradients. To ensure robustness a prefiltering of the image by some Gaussian filter G^σ of filter width σ is invoked here. One can further improve the results, allowing, in addition, for a smoothing along the tangential direction on the edge [27] (see Figure 3.3).

- Image registration.

  Matching of a template image T with a reference image R via a non-rigid deformation ϕ - often called registration - can be formulated naturally as a variational problem. The aim is to achieve a good correlation of the template image T and the deformed reference image R:

$$T \circ \phi \approx R\,.$$

  In the simplest case of unimodal registration we can ask for a deformation ϕ given on image domain Ω, such that the energy

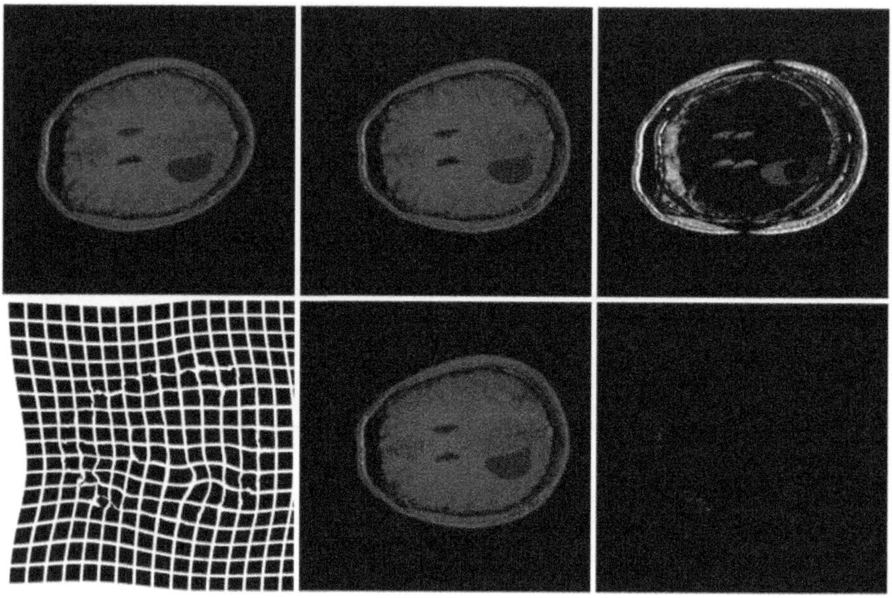

Fig. 3.4. Registration of medical images with a possible acquisition artefact computed in Direct X 9 graphics hardware. The six tiles are arranged in the following way: on the upper left we see the template that should be deformed to fit the reference image to the right of it; on the lower left we see the computed deformation applied to a uniform grid and to the right the registration result, i.e. the template after the deformation. The rightmost column shows the scaled quadratic difference between the template and the reference image before (upper row) and after (lower row) the registration.

$$E[\phi] = \int_{\Omega} |T \circ \phi - R|^2 \, \mathrm{d}x$$

is minimal in a class of suitable deformations. This problem turns out to be ill-posed and requires a regularization, by, for example, adding an elastic energy $\int_{\Omega} W(D\phi) \, \mathrm{d}x$ that measures the quality of the deformation itself and not only the quality of the match. Alternatively, a regularized gradient flow, which ensures smoothness of the resulting deformation, can be applied. After a discretization, the result is a global, highly non-linear optimization problem. Thus, the procedure is to consider a scale of matching problems ranging from coarse to fine. First, match on the coarse scale is found and then successively finer scales are treated [29] (see Figure 3.4).

The next section lists websites that point to many other PDE applications realized on GPUs including demos and code examples.

3.3.6 Resources

Up-to-date links to the sites below and the code samples discussed in this chapter are available online at the Springer site associated with this book.

The low-level programming of GPUs can be very tedious. Therefore, one usually uses libraries that facilitate the programming and abstract the details. The code examples in this chapter are based on the following resources:

- Graphics API: OpenGL
  `www.opengl.org`
- Shader language and API: Cg
  `developer.nvidia.com/page/cg_main.html`
- Window manager: GLUT
  `www.opengl.org/resources/libraries/glut.html`
- Extension initializer: GLEW
  `glew.sourceforge.net`
- Pbuffer handler: RenderTexture
  `gpgpu.sourceforge.net`

The choices are fairly common, apart from the last one where many still use self-made *pbuffer* handlers. However, we encourage the reader to explore the links below and discover other possibilities that might suit them better. To all of the above there are good alternatives and the authors themselves have used different tools, depending on the project requirements. The different combinations of graphics APIs and shader languages are discussed in more detail in Section 3.5.4. The rest of this section is a collection of useful links related to GPU programming.

- Scientific Computing on GPUs
  - GPGPU - General Purpose Computation on GPUs
    `www.gpgpu.org`
    This site addresses specifically general purpose computations, while other resources have usually a stronger focus on graphics applications. Related news, papers, code and links to resources are given and a forum for discussion is maintained. The site also features two full-day tutorials from the SIGGRAPH 2004 and Visualization 2004 conferences on scientific use of GPUs.
  - ShaderTech - Real-Time Shaders
    `www.shadertech.com`
    Here, shaders in general are discussed, and scientific examples are included. The site features news, articles, forums, source code and links to tools and other resources.
- Major Development Sites
  From time to time, one encounters technical GPU problems that have been solved already. The following development sites contain a huge store of examples, libraries, white papers, presentations, demonstrations, and documentation for GPUs. In particular, they offer well-assembled Software Development Kits (SDKs) that demonstrate various graphics techniques.

  - OpenGL Resources
    `www.opengl.org`
  - DirectX Resources
    `msdn.microsoft.com/directx`
  - ATI Developer Site
    `www.ati.com/developer`
  - NVIDIA Developer Site
    `developer.nvidia.com`
- Developer Tools
  These sites are good starting points for exploring the numerous freely available tools for GPUs. They include advanced Integrated Development Environments (IDEs) for shader development, debugging and performance analysis.
  - ShaderTech Tool Archive
    `www.shadertech.com/tools`
  - OpenGL Coding Resources
    `www.opengl.org/resources/index.html`
  - Microsoft Direct X Tools
    `www.msdn.microsoft.com/library/default.asp?url=/library/en-us/directx9_c/directx/graphics/Tools/Tools.asp`
  - ATI Tools
    `www.ati.com/developer/tools.html`
  - NVIDIA Tools
    `www.developer.nvidia.com/page/tools.html`
  - Babelshader - Pixel to Fragment Shader Translator (D. Horn)
    `www.graphics.stanford.edu/~danielrh/babelshader.html`
  - Imdebug - The Image Debugger (B. Baxter)
    `www.cs.unc.edu/~baxter/projects/imdebug/`
  - Shadesmith - Shader Debugger (T. Purcell, P. Sen)
    `www.graphics.stanford.edu/projects/shadesmith`

3.4 Prospects

The development of GPUs is rapid. Performance doubles approximately every nine months. Many new features are introduced with each generation and they are quickly picked up by implementations. This fast pace is likely to continue for at least several more years. What should we expect in the future?

3.4.1 Future GPUs

Throughout the chapter we have pointed to expected developments of GPUs. The information is based mainly on the features of the Windows Graphics Foundation

(WGF) announced by Microsoft for the new Windows generation (Longhorn [22]). Let us summarize the points.

- Parallelism/Bandwidth
  The parallelism will continue to grow rapidly, as well as the bandwidth. However, since increasing the first is cheaper than the second, programming will have to focus on *computational intensity* even more strongly than at present.
- Shaders
  Unlimited instruction counts and a unified shader model will be introduced. Much of the fixed pipeline functionality will be replaced by the use of programmable shaders. Without the fixed functionality, GPUs will basically be a collection of parallel PEs. This will introduce scheduling tasks that are likely to be hidden from the programmer. Memory will be virtualized to operate on data that would otherwise not fit the *video memory*. Further improvements will include indexing of textures and greater temporary storage for intermediate results in shaders.
- Dataflow
  We have emphasized the view that textures, pbuffers, *vertex* data and the *frame-buffer* can all be seen as interchangeable collections of 2D data arrays, although full flexibility is not yet available. Future GPUs will fully incorporate this view and shaders will decide on their own how they want to interpret the data. The *graphics pipeline* will also offer several exit points for the data streams and not only the one at the end of the pipeline. As a result it will, for example, be possible to manipulate a mesh iteratively with the VP.
- Precision
  The latest VS3, PS3 model prescribes 32 bit float precision throughout the pipeline. Several GPUs offer this already and many more will soon follow. The support for double floats is unlikely in the near future, although there are, in principle, no barriers. The problem of development lies, rather, in the difficulties of creating a demand: a strong demand for double precision GPUs would make production feasible, yet at present, such demand is unlikely from the scientific community, because GPUs receive little attention from that quarter precisely because they do not have double precision. Further, a demand is unlikely to come from the graphics or computer game community where GPU vendors earn their money.
- Dynamic branching/MIMD in the FP
  Currently, GPUs with PS3 support are only efficient at infrequent dynamic branching. The problems with MIMD are additional transistors and scheduling problems, but the development of the processing elements (PEs) points clearly towards MIMD in the near future. The unification of the shader model does not necessarily mean that the PEs become the same, however, common PEs would allow better resource utilization in view of changing loads on vertices and fragments.

- Scatter
  The read-only, write-only regions avoid many synchronization problems. One-sided communication models are also known for their efficiency from CPU-based parallel computers; see [16, Section 2.2.3, page 17] Writing to arbitrary memory addresses would destroy too many of these advantages. However, scattering within a specified region does not have a negative effect on synchronization. Current GPUs can already scatter data by rendering it as a set of points. WGF will allow the generation of new primitives so that data duplication of individual items will be possible too. However, current GPU are not efficient at point processing; and this, it will be difficult to change.

Many of the expected features are already available, to some extent, through different extensions (Section 3.5.1). Respecting the current limitations on resources and performance, this already allows current development to be directed towards the new hardware. In other words, it is worth exploring the GPU as a general parallel processor, even if some restrictions still apply.

3.4.2 GPU Cluster

A single GPU already offers a lot of parallelism, but similar to CPUs, demand for higher performance suggests the use of multiple GPUs to work on a common task. The integration hierarchy is developing similarly to that of CPU-based parallel computers. One node, represented by a mainboard with several PCI Express slots, can accommodate several graphics cards. Clusters of multiple nodes are connected with the usual fast interconnects. However, both developments are in their infancy. NVIDIA offers a technology to couple two of their newest GPUs [23], ATI is expected to present a similar technology for their products, and Alienware announced a solution for all PCI Express graphics cards [1]. The solutions claim full transparency, so that the programmer only has to consider a number of general rules that will minimize the implicit synchronization between the cards. In addition, extensions to more than two boards seem feasible.

Initial academic work on the utilization of GPU clusters for parallel visualization [28, 15, 9] and computing [8, 5] also exists. Clearly, these approaches carry with them the same complexity as do CPU clusters. In particular, the considerations on partitioning and dynamic load balancing in [31] apply. The communication is even more complicated, because the cluster interconnects transport the data to the main memory and there is another stage of indirection in exchanging this data with the video memory of the graphics cards. In addition, once we are willing to pay the price of the comparably slow data transport between the graphics card and the main memory, it makes sense to involve the CPU in the processing too. We see the cluster as eventually being a pool of heterogenous processors with different computing paradigms and interconnects between them. While future graphics APIs will address the topics of job sharing and multiple GPUs and research on heterogeneous computer systems in general is ongoing, the efficient utilization of all available resources in GPU clusters is likely to remain a challenge for a long time.

3.5 Appendix: GPUs In-Depth

Graphics hardware has undergone a rapid development over the last 10 years. Starting as a primitive drawing device, it is now a major computing resource. We here outline the technological development, the logic layout of the graphics pipeline, a rough classification of the different hardware generations, and the high-level programming languages.

3.5.1 Development

Up to the early 1990s, standard graphics cards were fairly unimpressive devices from a computational point of view, although having 16 colors in a 640x350 display (EGA) as opposed to four colors in a 320x200 display (CGA) did make a big difference. Initially, the cards were only responsible for the display of a pixel array prepared by the CPU. The first available effects included the fast changing of color tables, which enabled color animations and the apparent blending of images. Then the cards started to be able to process 2D drawing commands and some offered additional features, such as video frame grabbing or multi-display support.

The revolutionary performance increase of graphics cards started in the mid 1990s, with the availability of graphics accelerators for 3D geometry processing. The already well-established game market welcomed this additional processing power with open arms and soon no graphics card would sell without 3D acceleration features. Since then, the GPU has taken over more and more computational tasks from the CPU. The performance of GPUs has grown much faster than that of CPUs, doubling performance approximately every nine months, which is the equivalent of a 'Moore's Law squared'.

During the late 1990s the number of GPU manufacturers decreased radically, at least for PC graphics cards. Although other companies are trying to gain or regain ground in the market, NVIDIA and ATI have clearly been dominant, both in performance and market shares, for several years now. Hence, the following discussions we cite primarily their products. Concerning the market, we should mention that actually Intel is the largest producer of graphics chips, in the form of integrated chip-sets. However, these are inexpensive products and rank low on the performance scale, so we will deal only with stand-alone GPUs on graphics cards.

Together with the reduction of GPU designers, the number of different APIs to access their functionality has also decreased. The OpenGL API and the Direct X API are the survivors. The API guarantees that despite the different hardware internals of GPUs from different companies, the programmer can access a common set of operations through the same software interface, namely the API. The proprietary graphics driver is responsible for translating the API calls into the proprietary commands understood by the specific GPU. In this respect, the API is similar to an operating system, which also abstracts the underlying hardware for the programmer and offers standardized access to its functionality, although an operating system does more than that.

If the hardware offers new features and is downward compatible, an old API still functions, but it lacks the new functionality. However, the use of new features in a new API results in an incompatibility with older hardware. Therefore, programmers are reluctant to use new features as long as they expect a significant demand for their applications on older hardware. The hardware vendor can promote the use of the new API by emulating the new hardware features in software on older systems, but this may turn out very demanding or impractical if the software emulation is too slow. So, in practice, programmers opt to assume very low requirements for the hardware and ignore incompatibility issues. Only the time-critical parts of the code are sometimes implemented for each hardware standard separately and chosen dynamically upon identification of the hardware. The above applies both to programs for different versions of an operating system and programs (mainly games) for different versions of graphics APIs. However, graphics hardware has evolved much quicker and game performance is often a critical factor, such that the changes of API versions and the lowest common requirements are moving faster than in the CPU market.

OpenGL and Direct X have been incorporating the quickly evolving feature set of GPUs differently. OpenGL uses a very flexible extension system. Each vendor can expose the whole functionality of its hardware product by proprietary extensions to the API. The OpenGL ARB [24], which includes the main players in the graphics field, helps in the standardization of these extensions to prevent the undermining of the common interface idea through too many incompatible proprietary extensions. In practice, the proprietary extensions appear first and then the standard access points evolve over time. The different versions of Direct X on the other hand, are prescribed by Microsoft and thus simply define a fixed set of requirements. Naturally, these requirements are discussed with the GPU designers beforehand. If the hardware supersedes them quantitatively, then Direct X often allows the use of these additional resources, but qualitatively new features have to wait for the next generation of APIs. So, we may say that the Direct X API changes more or less step in step with the new graphics hardware generations, while OpenGL evolves continuously, first on proprietary and subsequently on ARB paths. Currently, OpenGL is undergoing its first major revision since 1992, from the 1.x versions to version 2.0 [24] in an attempt to include many of the already well-established and new extensions into the core and prepare the API for future developments.

3.5.2 Graphics Pipeline

The Graphics Processor Unit (GPU), the central computational chip on a graphics card, may be seen as a restricted form of a stream processor (see Section 3.1.2). Via a set of commands, a particular state of the *graphics pipeline* in the GPU is configured and then data streams are sent through that pipeline. The output stream is visualized on the screen or resent through the pipeline after a possible reconfiguration. Although graphics cards have not, in the past, been seen in this context, current developments show a clear tendency towards the production of a general parallel computing device.

A schematic view of the graphics pipeline is presented in Figure 3.5. The abstraction omits some details but offers a clear perspective on the available functionality.

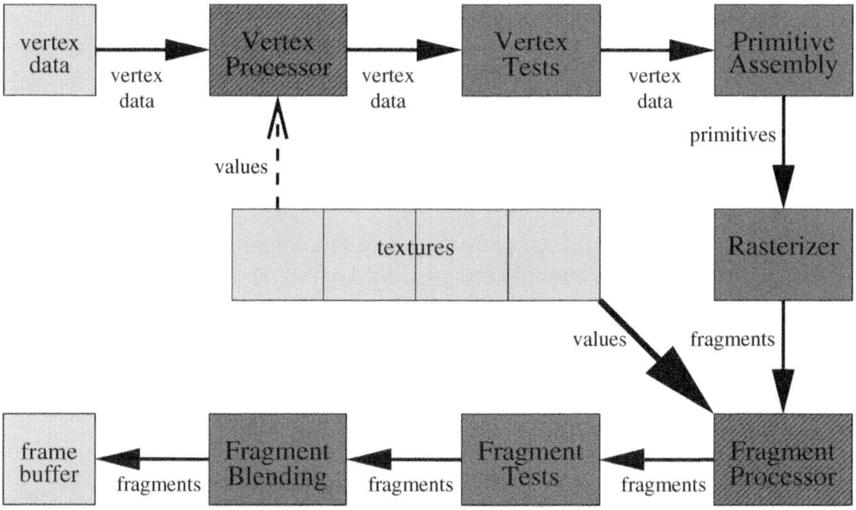

Fig. 3.5. A diagram of the graphics pipeline. Light gray represents data containers, dark gray processing units. The emphasized VP and FP are the units that evolved most in the graphics pipeline over the years, up to the stage where they accept freely programmable shader programs as configurations. Actually, the names VP and FP refer only to the new programmable pipeline stages, but the older functionality was located in the same place.
The thick arrow from the textures to the FP represents the largest data streams in the pipeline. Accordingly, the FP consumes the majority of resources in a GPU. The access to textures from the VP is a recent feature, as is the upcoming full interchangeability of the data containers in the pipeline, which allows a 2D data array to serve as an array of vertex data, a texture, or a destination buffer within the frame-buffer.

The logical pipeline has remained basically the same during the evolution of graphics hardware and changes can be identified by the increased flexibility and functionality of the individual components. Let us describe the operational tasks of the individual components:

- *Vertex* data
  We need an array that defines the geometry of the objects to be rendered. Beside the vertex coordinates, the vertex data may also contain color, normal and texture coordinate information (and a few more parameters). Although the data may be specified with one to four components, both coordinates (XYZW) and colors (RGBA) are internally always processed as 4-vectors. During the evolution of graphics hardware, it was principally the choices for the places where the vertex data can be stored (cacheable, AGP or video memory) and the efficiency of handling that data that increased. Modern VBOs allow us to specify the intended use and let the graphics driver decide which type of memory is ideally suited for the given purpose.
- Vertex Processor (VP)
  The VP manipulates the data associated with each vertex individually. Over the

years, the number of possible operations has increased dramatically. In the beginning, only multiplications with predefined matrices could be performed. Nowadays, the VP runs shader programs on the vertex data and the new generation has a restricted texture access from the VP. However, each vertex is still processed individually without any implicit knowledge about the preceding or succeeding vertices.

- Vertex tests

  Vertex tests determine the further processing of geometric primitives on the vertex level. They include mainly back-face culling, which eliminates polygons facing backwards (if the object is opaque one cannot see its back) and clipping, which determines the visible 3D space with an intersection of several 3D half spaces, defined by clipping planes. The vertex tests are still controlled by parameters and there have been only quantitative improvements in the number of clipping planes over time.
- Primitive assembly, *rasterizer*

  The geometric primitives that can be rendered are points, line segments, triangles, quads and polygons. Each vertex is processed individually and the clipping of primitives may introduce new vertices such that primitives have to be reassembled before rasterization. In addition, for simplicity, the rasterizer in many graphics architectures operates exclusively on triangles, so other primitives must be converted into a set of triangles before processing. Given a triangle and the vertex data associated with each of its vertices, the rasterizer interpolates the data for all the pixels inside the triangle. The resulting data associated with a pixel position is called a *fragment*. The rasterization could be controlled with parameters, for example defining patterns for lines or the interior of objects.
- Textures

  Textures are user-defined 1D to 4D (typically 2D) data arrangements stored in the *video memory* of the graphics card. Their elements, which can have up to four components (RGBA), are called texels. In general, the dimensions of all textures had to be powers of 2, but now there exists a general extension for textures with other dimensions.

  Input images of a problem are usually represented as textures on the graphics card and their values are processed by the FP and fragment blending. Over the years, quantitative improvements of textures have included their maximal number, their maximal size and the precision of the used fixed-point number format. Qualitative improvements are the support of various dimensionalities, the different access modes, the floating-point number format, and flexibility in the creation and reuse of texture data in different contexts. From the modern point of view, textures represent just a special use of data arrays that can serve as input to the FP (texture mode), as the destination for the output stream of the graphics pipeline (output mode), or even as an array defining vertex data (vertex mode).
- Fragment Processor (FP)

  The FP manipulates the individual fragments. Similarly to the way in which vertices are processed, each fragment is processed independently of the others in the same data stream. With the interpolated texture coordinates, the FP can access additional data from textures. The functionality of the FP has improved enormously over the years. In a qualitative sense, the range of available access modes of texture data and operations on these values in the FP has grown rapidly, culminating in a FP controlled by assembly or high-level code with access to arbitrary texture positions and a rich set of mathematical and control operations. In a quantitative sense, the number of accessible textures and the number of admissible fragment operations has increased significantly.

- *Frame-buffer*
  The frame-buffer is the 2D destination of the output data stream. It contains different buffers of the same dimensions for the color, depth and stencil (and accumulation) values. Not all buffers need to be present at once. In addition, while each buffer allows certain data formats, some combinations may not be available. There exists at least one color buffer, but typically there is a front buffer, which contains the scene displayed on the screen, and a back buffer, where the scene is built up. Over the years, it has mainly been the maximal size, the number and the precision of the buffers that has increased. A recent development, already sketched in the discussion of textures, regards the frame-buffer as an abstract frame for a collection of equally-sized 2D data arrays. After rendering, the same 2D data arrays may be used as textures or vertex data.
- Fragment tests
  Equivalent to the vertex tests for vertices, the fragment tests determine whether the current fragment should be processed further or discarded. However, the fragment tests are more numerous and powerful than the vertex tests and some of them allow a comparison against the values stored at the associated pixel position of the fragment in the depth or stencil buffer, and also a restricted manipulation of these values, depending on the outcome of the tests. Because they access the frame-buffer directly their functionality cannot be realized in one pass, even in the newest FP.
- Fragment blending
  Before the FP became a powerful computational resource, computations were mainly performed by different blending modes. The blending operation combines the color value of the fragment with the color value in the color buffer, controlled by weighting factors and the blending mode. For instance, the blending operation can be a convex combination of the values using a certain weight. Blending has become less popular in recent years, because on most GPUs it has not supported the higher precision number formats, while the much more powerful FP does. However, currently, support for higher precision blending is increasing again. The advantage of blending is the direct access to the destination value in the frame-buffer, which is not supported by the FP on most GPUs.
  The blending modes are continuous functions of the input values. In addition, logical operations can be performed at the end of the pipeline, but these are

seldom used, because they have received no hardware support from the manufacturers of GPU.

As outlined in Section 3.1.3, for general purpose computations the FP is the most relevant part of the pipeline. The VP can be often used to reduce the workload of the FP by precomputing data that depends bilinearly on the vertex data across the domain, e.g. positions of node neighbors in a regular grid. The vertex and fragment tests are useful for masking out certain regions of the computational domain for special treatment and fragment blending can be used for a fast and simple combination of the output value with the destination value, e.g. accumulation.

3.5.3 Classification

Because of the almost synchronous evolution of the Direct X API and the generations of graphics hardware in recent years, it is easiest to classify GPUs according to the highest version of Direc tX that they support. In fact, it is only the Direct3D API that concerns us, but Microsoft releases the different APIs in a bundle, so it is usually the version of the whole release that is referred to. From Direct X 8 on, it is possible to differentiate the versions further by the functionality of the Vertex Shaders (VSs), which configure the VP, and the Pixel Shaders (PSs), which configure the FP. This Direct X (DX), VS, PS classification is useful, even if the OpenGL API is used for the implementation, because in contrast to Direct X, OpenGL evolves continuously with the introduction of individual extensions. In what follows, we provide an overview of the recent graphics hardware generations and list some typical representatives. The paragraphs point out the main functionality associated with the VS1 to VS3 and PS1 to PS3 shader models.

- Direct X 8 (VS1, PS1) GPUs, 2001-2002,

  e.g. 3DLabs Wildcat VP, Matrox Parhelia 512 (VS2, PS1), NVIDIA GeForce 3/4, ATI Radeon 8500.

  These GPUs introduced programmability to the graphics pipeline, i.e. assembly programs for the VP and highly restricted programs for the FP. However, the number formats were still restricted to low-precision fixed-point number systems.
- Direct X 9 (VS2, PS2) GPUs, 2002-2004,

  e.g. S3 DeltaChrome S8, XGI Volari Duo V8, NVIDIA GeForceFX 5800/5900, ATI Radeon 9700/9800.

  Direct X 9 is the current standard. With these GPUs, floating-point number formats appear. The programmability of the VP gains function calls, dynamic branching and looping. The PS2 model finally allows freely programmable code for the FP. High-level languages (HLSL, GLSL, Cg) facilitate the programming of the VP and FP.
- Direct X 9+ (VS2-VS3, PS2-PS3) GPUs, 2004,

  e.g. 3DLabs Wildcat Realizm (VS2, PS3), NVIDIA GeForce 6800 (VS3, PS3), ATI Radeon X800 (VS2, PS2).

  In the VS3 model, the VP gains additional functionality in the form of restricted

Table 3.2. The number of supported instructions in the VP and FP for the different *shader* models.

VS1	VS2+loops	VS3+loops	PS1	PS2	PS3	WGF
128	256	512-32768	8-14	96-512	512-32768	unlimited

texture access and more functionality for register indexing. The PS3 FP now also supports the features of function calls and restricted forms of dynamic branching, looping and variable indexing of texture coordinates.

- WGF 2, 2006?
  The next Windows generation (Longhorn [22]) will contain a new Windows specific graphics interface labeled WGF. The main expected features are a unified shader model, resource virtualization, better handling of state changes, and a general IO model for data streams. Future GPU generations will probably support all these features in hardware. See Section 3.4.1 for a more detailed discussion.

The number of supported instructions in the VP and FP for the different *shader* models is given in Table 3.2.
This classification shows a clear tendency of GPUs to be developing in the direction of a general parallel computing device. Clearly, the WGF functionality will offer more flexibility than the current APIs, but this should not deter the reader from working with the current standard Direct X 9 (VS2, PS2), because it is expected to be the baseline functionality for many years to come.

3.5.4 Programming Languages

With the advent of a fully programmable pipeline in Direct X 9, three high-level languages for the programming of shaders, i.e. VP and FP configurations, appeared. The differences between them are fairly small and stem from the underlying graphics Application Programming Interface (API).

- Direct X - HLSL
  `msdn.microsoft.com/library/default.asp?url=/library/en-us/directx9_c/directx/graphics/ProgrammingGuide/ProgrammablePipeline/HLSL/ProgrammableHLSLShaders.asp`
  The HLSL is used to define shaders for the VP and FP under Direct X. Usually, the shaders are configured directly with the high-level code, but the compiler can also be instructed to output the generated assembly code as vertex or pixel shaders. If desired, the assembly code can be changed or written from scratch, but this option will probably disappear in the future.
- OpenGL - GLSL
  `www.opengl.org/documentation/oglsl.html`
  In GLSL, shaders are defined for the VP and FP under OpenGL. Different extensions also allow the use of assembly code for the configuration. There exist ARB

extensions, which cover the common set of functionality among the different GPUs, and proprietary extensions, which expose additional features. However, the direct use of assembly code has become uncommon, because the GLSL compiler embedded in the graphics driver offers automatic optimization towards the specific GPU in use.

- Direct X, OpenGL - Cg
  `developer.nvidia.com/page/cg_main.html`
  Cg follows the same idea as HLSL and GLSL; namely, to allow high-level configuration of the VP and FP. However, Cg as such is independent of the particular graphics API used. The compiler can generate code for different hardware profiles. The profiles comprise different versions of the vertex and pixel shaders under Direct X and different versions of the vertex and fragment shaders under OpenGL. So, plugging the generated assembly code into the appropriate API slots establishes the desired configuration. To hide this intermediate layer, the Cg Toolkit also provides an API that accepts Cg code directly. To the programmer, it looks as though Direct X and OpenGL have a native Cg interface, just as they have a HLSL or GLSL one, respectively.

As the languages are very similar, the reader may wonder why there is any difference between them at all. They differ because the languages were not released at the same time and, more importantly, a smoother integration into the existing APIs was desired. Clearly, a common interface would have been nicer, since even slightly different syntax disturbs the work flow. It is to be hoped that there will be more standardization in the future. In the meantime, we encourage the reader to be pragmatic about the choice of languages and other resources (see Section 3.3.6).

In Section 3.1.3 we saw that the coding of the shaders is only one part of the work. In addition, the pipeline with the shaders and textures must be configured, and the geometry to be rendered must be defined. Hence, more information is needed to obtain the same result from the same shader. The Direct X FX and the Cg FX for Direct X and OpenGL formats allow this additional information to be stored. Appropriate API calls set up the entire environment to implement the desired functionality. These formats can also include alternative implementations for the same operation, e.g. to account for different hardware functionality. The only problem with these convenient tools is that in a foolproof solution, unnecessarily many state calls may be provided, even though the required changes from one operation to another are minimal.

A more general approach to GPU programming is to use stream languages that are not targeted directly at the VP and FP but, rather, at the underlying data-stream-based (DSB) processing concept. A compiler generates machine-independent intermediate code from the stream program. Then, back-ends for different hardware platforms map this code to the available functionality and interfaces. This generality is very attractive, but it does not mean that the stream program can be written without any consideration of the chosen hardware platform. Some language features might be difficult to realize on certain hardware and would lead to a significant performance loss. In these cases, less optimal solutions that avoid these features must be chosen.

Using code that is more hardware-specific clearly delivers better performance, but nobody opts for coding everything on the lowest level. Hence, offering a trade-off between performance and abstraction to the programmer makes sense. We sketch two prominent stream languages with a focus on GPUs.

- Sh - University of Waterloo
  `libsh.org`
  Sh uses the C++ language for the meta-programming of stream code. This has the advantage that the C++ language features are immediately available and the compiler does the necessary analysis and code processing. In addition, it addresses the above-mentioned problem of the specification of an appropriate accompanying graphics environment to the shaders. Sh allows a direct interaction of shader definition with texture and state configuration. With a fast compilation process, dynamic manipulation of the stream code is feasible. Another advantage of working within the familiar C++ environment is the potential for incremental introduction of GPU usage into suitable existing software. We say suitable, because the processing of many of the general methods of organizing data, such as trees, lists or even stacks, are difficult to accelerate on GPUs. Current back-ends support GPUs under OpenGL and different CPUs.
- Brook - Stanford University
  `www.graphics.stanford.edu/projects/brookgpu`
  The Brook language is based on the concepts of streams and kernels. It is an abstraction of the data streams and shaders of GPUs. This abstraction frees us from the entire consideration of texture handling and geometry processing. In particular, it breaks the emphasis on rendering passes. The focus is on the actual data and its processing. Hardware virtualization also overcomes the limits that the graphics API places on the number of bound textures, their sizes and types of manipulation. However, for the generation of efficient code, programmers must be aware of which features map well to the hardware and which require costly workarounds. The richer feature set is well-suited for the development and simulation of programs that assume additional hardware functionality in future GPUs. Current back-ends support GPUs under Direct X and OpenGL, and different CPUs.

The code examples in this chapter use the OpenGL API and the Cg language. For someone new to graphics programming, the use of the stream languages, which already include a lot of abstraction, would make the examples look simpler. We have chosen a medium level of abstraction to illustrate how efficient programming depends on the hardware characteristics of GPUs. This understanding is equally important for the more abstract approaches to GPU programming, because the abstraction does not free the programmer from considering the hardware characteristics during the implementation. Similarly, the abstraction offered by MPI for parallel computers presumes implicit knowledge about the architecture and its functionality. Nevertheless, the higher abstraction is very attractive, because the stream languages preserve the main performance characteristics of GPUs by construction. In practice, the type of problem still determines whether it really is possible to obtain an efficient implementation on the high level. However, the stream languages are under active development and are extending their 'domain of efficiency' continuously. We recommend that the reader follow the links provided above for the download of the language libraries and detailed documentation.

Acronyms

AGP Accelerated Graphics Port
API Application Programming Interface
ARB Architectural Review Board
Cg C for graphics (high-level language)
CPU Central Processor Unit
DDR Double Data Rate (memory)
DSB data-stream-based
DX Direct X
FP Fragment Processor
GLSL OpenGL Shading Language
GPU Graphics Processor Unit
GUI Graphics User Interface
HLSL Direct X High-Level Shading Language
IDE Integrated Development Environment
ISB instruction-stream-based
MIMD Multiple Instruction Multiple Data
MPI Message Passing Interface
MRT Multiple Render Target
PBO Pixel Buffer Object
PCI Peripheral Component Interconnect
PCIe PCI Express
PDE partial differential equation
PE processing element
PS Pixel Shader
SDK Software Development Kit
SIMD Single Instruction Multiple Data
VBO Vertex Buffer Object
VP Vertex Processor
VS Vertex Shader
WGF Windows Graphics Foundation

References

1. Alienware. Alienware's Video Array.
   `http://www.alienware.com/alx_pages/main_content.aspx`.
2. C. Bajaj, I. Ihm, J. Min, and J. Oh. SIMD optimization of linear expressions for programmable graphics hardware. *Computer Graphics Forum*, 23(4), Dec 2004.
3. J. Bolz, I. Farmer, E. Grinspun, and P. Schröder. Sparse matrix solvers on the GPU: Conjugate gradients and multigrid. In *Proceedings of SIGGRAPH 2003*, 2003.
4. G. Coombe, M. J. Harris, and A. Lastra. Radiosity on graphics hardware. In *Proceedings Graphics Interface 2004*, 2004.
5. Z. Fan, F. Qiu, A. Kaufman, and S. Yoakum-Stover. GPU cluster for high performance computing. In *Proceedings of the ACM/IEEE SuperComputing 2004 (SC'04)*, Nov 2004.
6. K. Fatahalian, J. Sugerman, and P. Hanrahan. Understanding the efficiency of GPU algorithms for matrix-matrix multiplication. In *Graphics Hardware 2004*, 2004.
7. R. Fernando, editor. *GPU Gems: Programming Techniques, Tips, and Tricks for Real-Time Graphics*. Addison-Wesley Professional, 2004.
8. J. Fung and S. Mann. Using multiple graphics cards as a general purpose parallel computer : Applications to computer vision. In *Proceedings of the 17th International Conference on Pattern Recognition (ICPR 2004)*, volume 1, pages 805–808, 2004.
9. N. K. Govindaraju, A. Sud, S.-E. Yoon, and D. Manocha. Interactive visibility culling in complex environments using occlusion-switches. In *ACM SIGGRAPH Symposium on Interactive 3D Graphics*, 2003.
10. GPGPU - general purpose computation using graphics hardware.
    `http://www.gpgpu.org/`.
11. M. Harris. *Real-Time Cloud Simulation and Rendering*. PhD thesis, UNC Chapel Hill, Sep. 2003.
12. M. J. Harris, G. Coombe, T. Scheuermann, and A. Lastra. Physically-based visual simulation on graphics hardware. In *Proceedings of Graphics Hardware 2002*, pages 109–118, 2002.
13. R. Hartenstein. Data-stream-based computing: Models and architectural resources. In *International Conference on Microelectronics, Devices and Materials (MIDEM 2003)*, Ptuj, Slovenia, Oct. 2003.
14. R. Hill, J. Fung, and S. Mann. Reality window manager: A user interface for mediated reality. In *Proceedings of the 2004 IEEE International Conference on Image Processing (ICIP 2004)*, 2004.
15. G. Humphreys, M. Houston, R. Ng, R. Frank, S. Ahern, P. D. Kirchner, and J. T. Klosowski. Chromium: a stream-processing framework for interactive rendering on clusters. In *SIGGRAPH'02*, pages 693–702, 2002.
16. R. A. Kendall, M. Sosonkina, W. D. Gropp, R. W. Numrich, and T. Sterling. Parallel programming models applicable to cluster computing and beyond. In A. M. Bruaset and A. Tveito, editors, *Numerical Solution of Partial Differential Equations on Parallel Computers*, volume 51 of *Lecture Notes in Computational Science and Engineering*, pages 3–54. Springer-Verlag, 2005.
17. T. Kim and M. Lin. Visual simulation of ice crystal growth. In *Proc. ACM SIGGRAPH / Eurographics Symposium on Computer Animation*, 2003.
18. P. Kipfer, M. Segal, and R. Westermann. UberFlow: A GPU-based particle engine. In *Graphics Hardware 2004*, 2004.
19. J. Krueger and R. Westermann. Linear algebra operators for GPU implementation of numerical algorithms. *ACM Transactions on Graphics (TOG)*, 22(3):908–916, 2003.

20. A. Lefohn, J. Kniss, C. Handen, and R. Whitaker. Interactive visualization and deformation of level set surfaces using graphics hardware. In *Proc. Visualization*, pages 73–82. IEEE CS Press, 2003.
21. W. Li, X. Wei, and A. Kaufman. Implementing Lattice Boltzmann computation on graphics hardware. *The Visual Computer*, 2003.
22. Microsoft. Longhorn Developer Center. `http://msdn.microsoft.com/longhorn`.
23. NVIDIA. NVIDIA scalable link interface (SLI). `http://www.nvidia.com/page/sli.html`.
24. OpenGL Architectural Review Board (ARB). *OpenGL: graphics application programming interface*. `http://www.opengl.org/`.
25. M. Pharr and R. Fernando, editors. *GPU Gems 2 : Programming Techniques for High-Performance Graphics and General-Purpose Computation*. Addison-Wesley Professional, 2005.
26. M. Rumpf and R. Strzodka. Level set segmentation in graphics hardware. In *Proceedings ICIP'01*, volume 3, pages 1103–1106, 2001.
27. M. Rumpf and R. Strzodka. Using graphics cards for quantized FEM computations. In *Proceedings VIIP'01*, pages 193–202, 2001.
28. R. Samanta, T. Funkhouser, K. Li, and J. P. Singh. Hybrid sort-first and sort-last parallel rendering with a cluster of PCs. In *Proceedings of SIGGRAPH/Eurographics Workshop on Graphics Hardware 2000*, pages 97–108, 2000.
29. R. Strzodka, M. Droske, and M. Rumpf. Image registration by a regularized gradient flow - a streaming implementation in DX9 graphics hardware. *Computing*, 2004. to appear.
30. R. Strzodka and A. Telea. Generalized distance transforms and skeletons in graphics hardware. In *Proceedings of EG/IEEE TCVG Symposium on Visualization VisSym '04*, 2004.
31. J. D. Teresco, K. D. Devine, and J. E. Flaherty. Partitioning and dynamic load balancing for the numerical solution of partial differential equations. In A. M. Bruaset and A. Tveito, editors, *Numerical Solution of Partial Differential Equations on Parallel Computers*, volume 51 of *Lecture Notes in Computational Science and Engineering*, pages 55–88. Springer-Verlag, 2005.
32. M. Wilkes. The memory gap (keynote). In *Solving the Memory Wall Problem Workshop*, 2000. `http://www.ece.neu.edu/conf/wall2k/wilkes1.pdf`.

Part II

Parallel Algorithms

4

Domain Decomposition Techniques

Luca Formaggia[1], Marzio Sala[2], and Fausto Saleri[1]

[1] MOX, Dipartimento di Matematica "F. Brioschi", Politecnico di Milano, Italy
[luca.formaggia,fausto.saleri]@polimi.it

[2] Sandia National Laboratories, Albuquerque, USA
msala@sandia.gov

Summary. We introduce some parallel domain decomposition preconditioners for iterative solution of sparse linear systems like those arising from the approximation of partial differential equations by finite elements or finite volumes. We first give an overview of algebraic domain decomposition techniques. We then introduce a preconditioner based on a multilevel approximate Schur complement system. Then we present a Schwarz-based preconditioner augmented by an algebraic coarse correction operator. Being the definition of a coarse grid a difficult task on unstructured meshes, we propose a general framework to build a coarse operator by using an agglomeration procedure that operates directly on the matrix entries. Numerical results are presented aimed at assessing and comparing the effectiveness of the two methodologies. The main application will concern computational fluid dynamics (CFD), and in particular the simulation of compressible flow around aeronautical configurations.

4.1 Introduction

Modern supercomputers are often organized as a distributed environment and an efficient solver for partial differential equations (PDEs) should exploit this architectural framework. Domain decomposition (DD) techniques are a natural setting to implement existing single processor algorithm in a parallel context.

The basic idea, as the name goes, is to decompose the original computational domain Ω into subdomains $\Omega_i, i = 1, \ldots .M$, which may or may not overlap, and then rewrite the global problem as a "sum" of contributions coming from each subdomain, which may be computed in parallel. Parallel computing is achieved by distributing the subdomain to the available processors; often, the number of subdomains equals the number of processors, even if this is not, in general, a requirement. Clearly one cannot achieve a perfect (ideal) parallelism, since interface conditions between subdomains are necessary to recover the original problem, which introduce the need of inter-processor communications. The concept of parallel efficiency is clearly stated for the case of homogeneous systems (see [13]). An important concept is that of scalability: an algorithm is scalable if its performance is proportional to the number of processor employed. For the definition to make sense we should keep the processor workload approximately constant, so the problem size has to grow proportionally to the number of processor.

This definition is only qualitative and indeed there is not a quantitative definition of scalability which is universally accepted, and a number of scalability models proposed in the last years [14]. They are typically based on the selection of a measure which is used to characterize the performance of the algorithm, see for instance [22, 23, 11]. We may consider the algorithm *scalable* if the ratio between the performance measure and the number of processors is sub-linear. In fact, the ideal value of this ratio would be 1. Yet, since this ideal value cannot be reached in practice, a certain degradation should be tolerated.

Typical quantities that have been proposed to measure system performance include CPU time, latency time, memory, etc. From the user point of view, global execution time is probably the most relevant measure. A possible definition is the following. If $E(s, N)$ indicates the execution time for a problem of size s when using N processor on a given algorithm, then the scalability from N to $M > N$ processors is given by

$$S_{M,N} = \frac{E(M\gamma, M)}{E(N\gamma, N)},$$

where γ is the size of the problem on a single processor.

A few factors may determine the loss of scalability of a parallel code: the cost of inter-processor communication; the portion of code that has to be performed in a scalar fashion, may be replicated on each processor (for instance i/o if your hardware does not support parallel i/o). A third factor is related to a possible degradation of the parallel algorithm as the number of subdomain increases. In this work we will address exclusively the latter aspect, the analysis of the other two being highly dependent on the hardware. In particular, since we will be concerned with the solution of linear systems by parallel iterative schemes, the condition number of the parallel solver [17] is the most important measure related to the algorithm scalability properties. In this context, the algorithm is scalable if the condition number remains (approximately) constant as the ratio between problem size and number of subdomains is kept constant. In a domain decomposition method applied to the solution of PDE's (by finite volumes or finite elements) in $\mathbb{R}^d$ this ratio is proportional to $(H/h)^d$, being H and h the subdomain and mesh linear dimension, respectively. We are assuming a partition with subdomains of (approximately) the same size and a quasi-uniform mesh.

Domain decomposition methods may be classified into two main groups [17, 21]. The first includes methods that operate on the differential problem, we will call them *differential domain decomposition methods*. Here, a differential problem equivalent to the single domain one is written on the decomposed domain. Conditions at the interface between subdomains are recast as boundary conditions for local differential problems on each Ω_i. Then, the discretisation process is carried out on each subdomain independently (even by using different discretisation methods, if this is considered appropriate).

The second group includes the DD techniques that operates at the algebraic level. In this case the discretisation is performed (at least formally) on the original, single domain, problem and the decomposition process is applied on the resulting algebraic system. The latter technique has the advantage of being "problem independent" and may often be interpreted as a preconditioner of the global solver. The former, however, may better exploit the characteristics of the differential problem to hand and allows to treat problems of heterogeneous type more naturally. We refer to the relevant chapter in [17] for more insights on heterogeneous DD techniques.

In this chapter, we deal with DD schemes of algebraic type. In particular, we address methods suited to general problems, capable of operating on unstructured mesh based discretisations. For a more general overview of the DD method the reader may refer to the already cited literature, the review paper [5] and the recent monograph [25].

We will focus on domain decomposition techniques that can be applied to the numerical solution of PDEs on complex, possibly three dimensional, domains. We will also consider discretisations by finite element or finite volume techniques, on unstructured meshes. The final result of the discretisation procedure is eventually a large, sparse linear system of the type

$$A\mathbf{u} = \mathbf{f}\,, \tag{4.1}$$

where $A \in \mathbb{R}^{n \times n}$ is a sparse and often non-symmetric and ill conditioned real matrix. Indeed, also non-linear problems are usually treated by an iterative procedure (e.g. a Newton iteration) that leads to the solution of a linear system at each iteration. This is the case, for instance, of implicit time-advancing schemes for computational fluid dynamics (CFD) problems.

The decomposition of the domain will induce a corresponding block decomposition of the matrix A and of the vector $\mathbf{f}$. This decomposition may be exploited to derive special parallel solution procedures, or parallel preconditioners for iterative schemes for the solution of system (4.1). Perhaps the simplest preconditioner is obtained using a block-Jacobi procedure, where each block is allocated to a processor and is possibly approximated by an incomplete factorization [18] (since usually an exact factorization is too expensive). This approach may work well for simple problems, yet its performance degrades rapidly as the size of the matrix increases, leading to poor scalability properties. Other popular techniques are the Schwarz methods with a coarse grid correction [2] and the preconditioners based on the Schur complement system, like the balancing Neumann/Neumann [24], the FETI [10, 25] and the wire-basket method [1, 21]. In this work we will address preconditioners based either on an approximate Schur complement (SC) system or on Schwarz techniques, because of their generality and relatively simple implementation.

Schwarz iterations is surely one of the DD based parallel preconditioner with the simplest structure. In its basic form, it is equivalent to a block-Jacobi preconditioner, where each block is identified by the set of unknowns contained in each subdomain. In order to improve the performance of Schwarz iterations, the partitions of the original domain are extended, so that they overlap and the overlapping region acts as means of communication among the subdomains. In practice, the domain subdivision is commonly carried out at discrete level, that is by partitioning the computational mesh. In the minimal overlap version of the Schwarz method the overlap between subdomains is reduced to a single layer of elements. Although a bigger overlap may improve convergence, a minimal overlap allows to use the same data structure normally used for the parallel matrix-vector multiplication, thus saving memory. However, the scalability is scarce and a possible cure consists in augmenting the preconditioner by a coarse operator, either in an additive or in a multiplicative fashion [21] . The coarse operator may be formed by discretising the problem to hand on a much coarser mesh. The conditions by which a coarse mesh is admissible and is able to provide an appropriate coarse operator have been investigated in [3]. Another possible way to construct the coarse operator is to form a reduced matrix by resorting to a purely algebraic procedure [28, 20]. We will here describe a rather general setting for the construction of the coarse operator.

The set up of the Schur complement system in a parallel setting is only slightly more involved. The major issue here is the need of preconditioning the Schur complement system in order to avoid the degradation of the condition number as the number of subdomain increases. We will here present a technique to build preconditioners for the Schur complement system starting from a preconditioner of the original problem.

The chapter is organized as follows. Schur complement methods are introduced in Sections 4.2 and 4.3. Schwarz methods are detailed in Section 4.4. Numerical results for a model problem and for the solution of the compressible Euler equations are presented in Section 4.5. Section 4.6 gives some further remarks on the techniques that have been presented.

4.2 The Schur Complement System

Let us consider again (4.1) which we suppose has been derived by a finite element discretisation of a differential problem posed on a domain $\Omega \subset \mathbb{R}^d, d = 2, 3$. Indeed, the considerations of this Section may be extended to other type of discretisations as well, for instance finite volumes, yet we will here focus on a finite element setting. More precisely, we can think of (4.1) as being the algebraic counterpart of a variational boundary value problem which reads: find $u_h \in V_h$ such that

$$a\left(u_h, v_h\right) = \left(f, v_h\right) \qquad \forall v_h \in V_h\,, \tag{4.2}$$

where V_h is a finite element space, a the bilinear form associated to the differential problem to hand and $(f, v) = \int_\Omega f v d\Omega$ is the standard L^2 product.

We consider a decomposition of the domain Ω made in the following way. We first triangulate Ω and indicate by $\mathcal{T}_h^{(\Omega)}$ the corresponding mesh. For the sake of simplicity we assume that the boundary of Ω coincides with the boundary of the triangulation and we consider the case where the degrees of freedom of the discrete problem are located at mesh vertices, like in linear finite elements. In particular, a partition into two subdomains is carried out by splitting $\mathcal{T}_h^{(\Omega)}$ into 3 parts, namely $\mathcal{T}_h^{(1)}$, $\mathcal{T}_h^{(2)}$ and $\Gamma^{(1,2)}$ such that $\mathcal{T}_h^{(1)} \cup \mathcal{T}_h^{(2)} \cup \Gamma^{(1,2)} = \mathcal{T}_h^{(\Omega)}$. We may associate to $\mathcal{T}_h^{(1)}$ and $\mathcal{T}_h^{(2)}$ the two disjoint subdomains $\Omega^{(1)}$ and $\Omega^{(2)}$ formed by the interior

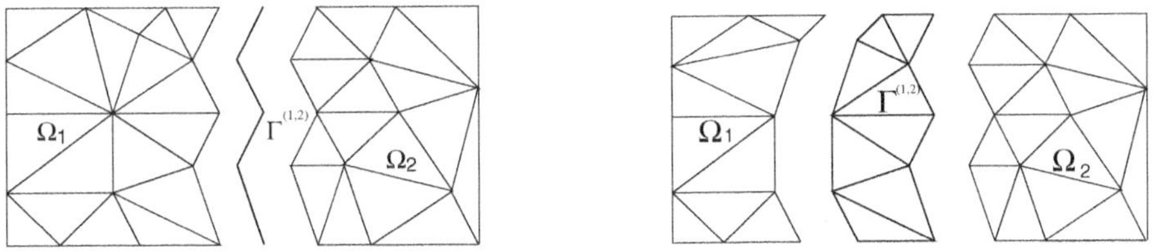

Fig. 4.1. Example of element-oriented (left) and vertex-oriented (right) decomposition.

of the union of the elements of $\mathcal{T}_h^{(1)}$ and $\mathcal{T}_h^{(2)}$ respectively, while $\Gamma^{(1,2)} \equiv \Gamma^{(2,1)}$ is clearly equal to $\Omega \setminus (\Omega^{(1)} \cup \Omega^{(2)})$.

Two notable cases are normally faced, namely

- $\Gamma^{(1,2)}$ reduces to a finite number of disjoint measurable $d-1$ manifolds. An example of this situation is illustrated in the left drawing of Figure 4.1, where $\Gamma^{(1,2)} = \overline{\Omega}^{(1)} \cap \overline{\Omega}^{(2)}$, i.e. $\Gamma^{(1,2)}$ is the common part of the boundary of $\Omega^{(1)}$ and $\Omega^{(2)}$. This type of decomposition is called *element oriented* (EO) decomposition, because each element of $\mathcal{T}_h$ belongs exclusively to one of the subdomains $\overline{\Omega}^{(i)}$, while the vertices laying on $\Gamma^{(1,2)}$ are shared between the subdomains triangulations.
- $\Gamma^{(1,2)} \subset \mathbb{R}^d, d = 2, 3$ is formed by one layer of elements of the original mesh laying between $\Omega^{(1)}$ and $\Omega^{(2)}$. In Figure 4.1, right, we show an example of such a decomposition, which is called *vertex oriented* (VO), because each vertex of the original mesh belongs to just one of the two subdomains $\overline{\Omega}^{(i)}$. We may also recognize two extended, overlapping, sub-domains: $\widetilde{\Omega}^{(1)} = \Omega^{(1)} \cup \Gamma^{(1,2)}$ and $\widetilde{\Omega}^{(2)} = \Omega^{(2)} \cup \Gamma^{(1,2)}$. We have here the minimal overlap possible. Thicker overlaps may be obtained by adding more layers of elements to $\Gamma^{(1,2)}$.

Both decompositions may be readily extended to any number of subdomains, as shown in Figure 4.2.

The choice of a VO or EO decomposition largely affects the data structures used by the parallel code. Sometimes, the VO approach is preferred since the transition region $\Gamma^{(1,2)}$ may be replicated on the processors which holds $\Omega^{(1)}$ and $\Omega^{(2)}$ respectively and provides a natural means of data communication among processor which also allow to implement a parallel matrix-vector product. Furthermore, the local matrices may be derived directly from the global matrix A, with no work needed at the level of the (problem dependent) assembly process. For this reason the VO technique is also the matter of choice of many parallel linear algebra packages. We just note that for the sake of simplicity, in this work we are assuming that the graph of the matrix A coincides with the computational mesh (which is the case of a linear finite element approximation of a scalar problem). This means that if the element a_{ij} of the matrix is different from zero, then the vertices $\mathbf{v}_i$ and $\mathbf{v}_j$ of the computational mesh are connected by an edge. However, the techniques here proposed may be generalized quite easily to more general situations.

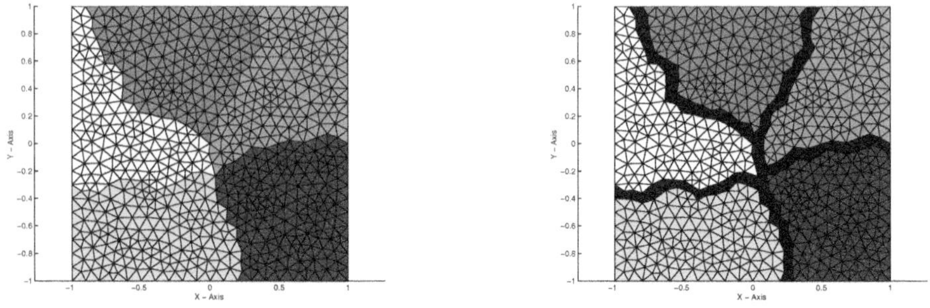

Fig. 4.2. Example of element-oriented (left) and vertex-oriented (right) decomposition in the case of a partition of Ω into several subdomains. (For the color version, see Figure A.5 on page 469).

We now introduce some additional notations. The nodes at the intersection between subdomains and $\Gamma^{(1,2)}$ are called *border* nodes. More precisely, those in $\mathcal{T}_h^{(i)} \cap \Gamma^{(1,2)}$ are the border nodes of the domain $\Omega^{(i)}$. A node of $\mathcal{T}_h^{(i)}$ which is not a border node is said to be *internal* to $\Omega^{(i)}, i = 1, 2$. We will consistently use the subscripts I and B to indicate internal and border nodes, respectively, while the superscript (i) will denote the subdomain we are referring to. Thus, $\mathbf{u}_I^{(i)}$ will indicate the vector of unknowns associated to nodes internal to $\Omega^{(i)}$, while $\mathbf{u}_B^{(i)}$ is associated to the border nodes. For ease of notation, in the following we will often avoid to make a distinction between a domain Ω and its triangulation $\mathcal{T}_h$ whenever it does not introduce any ambiguity.

4.2.1 Schur Complement System for EO Domain Decompositions

Let us consider again the left picture of Figure 4.1. For the sake of simplicity, we assume that the vector $\mathbf{u}$ is such that the unknowns associated to the points $\mathbf{u}_I^{(1)}$ internal to $\Omega^{(1)}$ are numbered first, followed by those internal to $\Omega^{(2)}$ ($\mathbf{u}_I^{(2)}$), and finally by those on $\Gamma^{(1,2)}$ ($\mathbf{u}_B = \mathbf{u}_B^{(1)} = \mathbf{u}_B^{(2)}$) (obviously this situation can always be obtained by an appropriate numbering of the nodes). Consequently, equation (4.1) can be written in the following block form

$$\begin{pmatrix} A_{II}^{(1)} & 0 & A_{IB}^{(1)} \\ 0 & A_{II}^{(2)} & A_{IB}^{(2)} \\ A_{BI}^{(1)} & A_{BI}^{(2)} & A_{BB}^{(1)} + A_{BB}^{(2)} \end{pmatrix} \begin{pmatrix} \mathbf{u}_I^{(1)} \\ \mathbf{u}_I^{(2)} \\ \mathbf{u}_B \end{pmatrix} = \begin{pmatrix} \mathbf{f}_I^{(1)} \\ \mathbf{f}_I^{(2)} \\ \mathbf{f}_B \end{pmatrix}. \tag{4.3}$$

Here, $A_{II}^{(i)}$ contains the elements of A involving the nodes internal to subdomain $\Omega^{(i)}$, while the elements in $A_{IB}^{(i)}$ are formed by the contribution of the boundary nodes to the rows associated to the internal ones. Conversely, in $A_{BI}^{(i)}$ we have the terms that link border nodes with internal ones (for a symmetric matrix $A_{BI}^{(i)} =$

$A_{IB}^{(i)^T}$). Finally, $A_{BB} = A_{BB}^{(1)} + A_{BB}^{(2)}$ is the block that involves only border nodes, which can be split into two parts, each built from the contribution coming from the corresponding subdomain. For instance, in a finite element procedure we will have

$$[A_{BB}^{(i)}]_{kj} = a(\phi_k|_{\Omega^{(i)}}, \phi_j|_{\Omega^{(i)}}), \tag{4.4}$$

where ϕ_k, ϕ_j are the finite element shape functions associated to border nodes k and j, respectively, restricted to the subdomain $\Omega^{(i)}$ and a is the bilinear form associated to the differential problem under consideration. An analogous splitting involves the right hand side $\mathbf{f}_B = \mathbf{f}_B^{(1)} + \mathbf{f}_B^{(2)}$.

A formal LU factorization of (4.3) leads to

$$\begin{pmatrix} A_{II}^{(1)} & 0 & 0 \\ 0 & A_{II}^{(2)} & 0 \\ A_{BI}^{(1)} & A_{BI}^{(2)} & I \end{pmatrix} \begin{pmatrix} I & 0 & A_{II}^{(1)^{-1}} A_{IB}^{(1)} \\ 0 & I & A_{II}^{(2)^{-1}} A_{IB}^{(2)} \\ 0 & 0 & S_h \end{pmatrix} \begin{pmatrix} \mathbf{u}_I^{(1)} \\ \mathbf{u}_I^{(2)} \\ \mathbf{u}_B \end{pmatrix} = \begin{pmatrix} \mathbf{f}_I^{(1)} \\ \mathbf{f}_I^{(2)} \\ \mathbf{f}_B \end{pmatrix},$$

where S_h is the **Schur complement** (SC) matrix, given by $S_h = S_h^{(1)} + S_h^{(2)}$ where

$$S_h^{(i)} = A_{BB}^{(i)} - A_{BI}^{(i)} A_{II}^{(i)^{-1}} A_{IB}^{(i)} \tag{4.5}$$

is the contributions associated to the subdomain $\Omega^{(i)}$, for $i = 1, 2$. We may note that we can solve the system (at least formally) using the following procedure. We first compute the border values $\mathbf{u}_B$ by solving

$$S_h \mathbf{u}_B = \mathbf{g}, \tag{4.6}$$

where $\mathbf{g} = \mathbf{g}^{(1)} + \mathbf{g}^{(2)}$, with

$$\mathbf{g}^{(i)} = \mathbf{f}_B^{(i)} - A_{BI}^{(i)} A_{II}^{(i)^{-1}} \mathbf{f}_I^{(i)} \quad , \; i = 1, 2 .$$

Then, we build the internal solutions $\mathbf{u}_I^{(i)}$, for $i = 1, 2$, by solving the two completely independent linear systems

$$A_{II}^{(i)} \mathbf{u}_I^{(i)} = \mathbf{f}_I^{(i)} - A_{IB}^{(i)} \mathbf{u}_B \quad , \; i = 1, 2 . \tag{4.7}$$

The second step is perfectly parallel. Furthermore, thanks to the splitting of S_h and g, a parallel iterative scheme for the solution of (4.6) can also be devised. However, some communications among subdomains is here required. The construction of the matrices $A_{BB}^{(i)}$ in (4.4) requires to operate at the level of the matrix assembly by the finite element code. In general, there is no way to recover them from the assembled matrix A. Therefore, this technique is less suited for "black box" parallel linear algebra packages. More details on the parallel implementation of the Schur complement system are given in Section 4.2.3.

4.2.2 Schur Complement System for VO Domain Decompositions

Let us consider again problem (4.1) where we now adopt a VO partition into two subdomains like the one on the right of Figure 4.1. The matrix A can be written again in a block form, where this time we have

$$A\mathbf{u} = \begin{pmatrix} A_{II}^{(1)} & A_{IB}^{(1)} & 0 & 0 \\ A_{BI}^{(1)} & A_{BB}^{(1)} & 0 & E^{(1,2)} \\ 0 & 0 & A_{II}^{(2)} & A_{IB}^{(2)} \\ 0 & E^{(2,1)} & A_{BI}^{(2)} & A_{BB}^{(2)} \end{pmatrix} \begin{pmatrix} \mathbf{u}_I^{(1)} \\ \mathbf{u}_B^{(1)} \\ \mathbf{u}_I^{(2)} \\ \mathbf{u}_B^{(2)} \end{pmatrix} = \begin{pmatrix} \mathbf{f}_I^{(1)} \\ \mathbf{f}_B^{(1)} \\ \mathbf{f}_I^{(2)} \\ \mathbf{f}_B^{(2)} \end{pmatrix}. \tag{4.8}$$

Here, the border nodes have been subdivided in two sets: the set $B^{(1)}$ of nodes of $\Gamma^{(1,2)}$ which lay on the boundary of $\Omega^{(1)}$ (the *border nodes* of $\Omega^{(1)}$) and the analogous set $B^{(2)}$ of the border nodes of $\Omega^{(2)}$. Correspondingly, we have the blocks $\mathbf{u}_B^{(1)}$ and $\mathbf{u}_B^{(2)}$ in the vector of unknowns and $\mathbf{f}_B^{(1)}$ and $\mathbf{f}_B^{(2)}$ in the right hand side. The entries in $E^{(i,j)}$ are the contribution to the equation associated to nodes in $B^{(i)}$ coming from the nodes in $B^{(j)}$. We call the nodes in $B^{(j)}$ contributing to $E^{(i,j)}$ *external nodes* of domain $\Omega^{(i)}$.

The nodes internal to $\Omega^{(i)}$ are the nodes of the triangulation $\mathcal{T}_h^{(i)}$ whose "neighbors" all belong to $\Omega^{(i)}$. In a matrix-vector product, values associated to internal nodes may be updated without communication with the adjacent subdomains. The update of the border nodes requires instead the knowledge of the values at the corresponding external nodes (which are in fact border nodes of neighboring subdomains). This duplication of information lends itself to efficient implementation of inter-processor communications.

Analogously to the previous section we can construct a Schur complement system operating on the border nodes, obtaining

$$S_h\mathbf{u}_B = \begin{pmatrix} S_h^{(1)} & E^{(1,2)} \\ E^{(2,1)} & S_h^{(2)} \end{pmatrix} \begin{pmatrix} \mathbf{u}_B^{(1)} \\ \mathbf{u}_B^{(2)} \end{pmatrix} = \begin{pmatrix} \mathbf{g}^{(1)} \\ \mathbf{g}^{(2)} \end{pmatrix},$$

where $S_h^{(1)}$ and $S_h^{(2)}$ are defined as in (4.5). Note, however, that now the entries in $A_{BB}^{(i)}$, $i = 1, 2$, are equal to the corresponding entries in the original matrix A. Thus they can be built directly from A as soon as the topology of the domain decomposition is known.

Once we have computed the border values $\mathbf{u}_B$, the internal solutions $\mathbf{u}_I^{(i)}, i = 1, 2$, are obtained by solving the following independent linear systems,

$$A_{II}^{(i)}\mathbf{u}_I^{(i)} = \mathbf{f}_I^{(i)} - A_{IB}^{(i)}\mathbf{u}_B^{(i)}, \quad i = 1, 2.$$

In Figure 4.3 we report the sparsity pattern of S_h in the case of a decomposition with 2 subdomains.

This procedure, like the previous one, can be generalized for an arbitrary number of subdomains. If we have M subdomains the decomposition of system (4.8) may be written in a compact way as

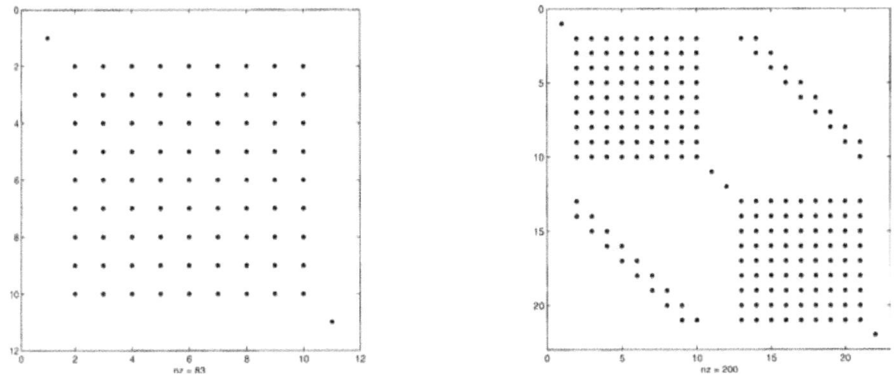

Fig. 4.3. Sparsity pattern for SC matrix derived from an EO decomposition (left) and a VO one (right).

$$\begin{pmatrix} A_{II} & A_{IB} \\ A_{BI} & A_{BB} \end{pmatrix} \begin{pmatrix} \mathbf{u}_I \\ \mathbf{u}_B \end{pmatrix} = \begin{pmatrix} \mathbf{f}_I \\ \mathbf{f}_B \end{pmatrix}, \tag{4.9}$$

where

$$A_{II} = \begin{pmatrix} A_{II}^{(1)} & & 0 \\ & \ddots & \\ 0 & & A_{II}^{(M)} \end{pmatrix}, \qquad A_{BB} = \begin{pmatrix} A_{BB}^{(1)} & E^{(1,2)} & \dots & E^{(1,M)} \\ E^{(2,1)} & & & \\ \vdots & & \ddots & \vdots \\ E^{(M,1)} & & & A_{BB}^{(M)} \end{pmatrix},$$

$$A_{IB} = \begin{pmatrix} A_{IB}^{(1)} \; A_{IB}^{(2)} \; \dots \; A_{IB}^{(M)} \end{pmatrix}, \qquad A_{BI} = \begin{pmatrix} A_{BI}^{(1)} \; A_{BI}^{(2)} \; \dots \; A_{BI}^{(M)} \end{pmatrix}^T$$

and

$$\mathbf{u}_I = \begin{pmatrix} \mathbf{u}_I^{(1)} \dots \mathbf{u}_I^{(M)} \end{pmatrix}^T, \qquad \mathbf{u}_B = \begin{pmatrix} \mathbf{u}_B^{(1)} \dots \mathbf{u}_B^{(M)} \end{pmatrix}^T,$$
$$\mathbf{f}_I = \begin{pmatrix} \mathbf{f}_I^{(1)} \dots \mathbf{f}_I^{(M)} \end{pmatrix}^T, \qquad \mathbf{f}_B = \begin{pmatrix} \mathbf{f}_B^{(1)} \dots \mathbf{f}_B^{(M)} \end{pmatrix}^T.$$

For the sake of space, we have transposed some matrices. Note however that here the transpose operator acts on the block matrix/vector, not on the blocks themselves, i.e. $(\mathbf{a}\ \mathbf{b})^T$ equals to $\begin{pmatrix} \mathbf{a} \\ \mathbf{b} \end{pmatrix}$ and not $\begin{pmatrix} \mathbf{a}^T \\ \mathbf{b}^T \end{pmatrix}$.

The Schur complement system of problem (4.1) can now be written as $S_h \mathbf{u}_B = \mathbf{g}$, where

$$S_h = A_{BB} - A_{BI} A_{II}^{-1} A_{IB}, \quad \text{and } \mathbf{g} = \mathbf{f}_B - A_{BI} A_{II}^{-1} \mathbf{f}_I .$$

To conclude this Section, we wish to note that an EO arrangement is often the direct result of a domain decomposition carried out at a differential level. In this case, the Schur complement matrix may be identified as the discrete counterpart of a particular differential operator acting on the interface Γ (the Steklov-Poincarè operator [17]). Instead, a VO decomposition is normally the result of a purely algebraic manipulation and in general is lacking an immediate interpretation at differential level. Finally, the VO arrangement produces a larger number of degrees of freedom in the resulting Schur complement system.

4.2.3 Parallel Solution of the Schur Complement System

Schur complement matrices are usually full and in large scale problems there is no convenience in building them in an explicit way. Thus, the SC system is normally solved by an iterative method, such as a Krylov acceleration method [26], which requires only the multiplication of the matrix with a vector. To compute $\mathbf{w}_B = S_h\,\mathbf{v}_B$, one may proceed as indicated in the following algorithm, where R_i is the restriction operator from the global border values $\mathbf{v}_B$ to those associated to subdomain $\Omega^{(i)}$.

ALGORITHM 1: COMPUTATION OF $\mathbf{w}_B = S_h\,\mathbf{v}_B$

1. Restrict $\mathbf{v}_B$ to each subdomain boundary,
$$\mathbf{v}_B^{(i)} = R_i \mathbf{v}_B, \quad i = 1, \dots, M\,.$$
2. For every $\Omega^{(i)}, \quad i = 1, \dots M$ solve
$$A_{II}^{(i)} \mathbf{u}_I^{(i)} = -A_{IB}^{(i)} \mathbf{v}_B^{(i)}\,,$$
then compute
$$\mathbf{w}_B^{(i)} = \sum_{j=1}^{M} E^{(ij)} v_B^j + A_{BB}^{(i)} \mathbf{v}^{(i)}{}_B - A_{BI}^{(i)} \mathbf{u}_I^{(i)}\,.$$
3. Apply the prolongation operators to get $\mathbf{w}_B = \sum_{i=1}^{M} R_i^T \mathbf{w}_B^{(i)}$.

In general, Steps 1 and 3 are just formal if we operate in a parallel environment since each processor already handles the restricted vectors $\mathbf{u}_B{}^{(i)}$ instead of the whole vector. Note that the linear system in Step 2 must be solved with high accuracy in order to make the Schur complement system equivalent to the original linear system (4.1).

Algorithm 1 requires four matrix-vector products and the solution of a linear system for each subdomain. Even if carried out in parallel, the latter operation can be rather expensive and is one of the drawbacks of a SC based method. The local problems have to be computed with high accuracy if we want to recover an accurate global solution.

Although (at least in the case of symmetric and positive-definite matrices) the condition number of the Schur matrix is no larger than that of the original matrix

Table 4.1. Convergence rate for different preconditioner of the Schur complement system with respect to the discretisation size h and the subdomain size H, for an elliptic problem. The constants C (which are different for each method) are independent from h and H, yet they may depend on the coefficients of the differential operator. The $\delta \in (0, 1]$ in the Vertex Space preconditioner is the overlap fraction, see the cited reference for details.

Preconditioner	*Estimation of the condition number of the preconditioned Schur complement operator*
P_h^J	$K((P_h^J)^{-1}S_h) \leq CH^{-2}(1 + \log(H/h))^2$
P_h^{BPS}	$K((P_h^{BPS})^{-1}S_h) \leq C(1 + \log(H/h))^2$
P_h^{VS}	$K((P_h^{VS})^{-1}S_h) \leq C(1 + \log \delta^{-1})^2$
P_h^{WB}	$K((P_h^{WB})^{-1}S_h) \leq C(1 + \log(H/h))^2$
$P_h^{NN,b}$	$K((P_h^{NN,b})^{-1}S_h) \leq C(1 + \log(H/h))^2$

A [21, 17], nevertheless it increases when h decreases (for a fixed number of subdomains), but also when H decreases (for a fixed h, i.e. for a fixed problem size). This is a cause of loss of scalability. A larger number of subdomains imply a smaller value of H, the consequent increase of the condition number causes in turn a degradation of the convergence of the iterative linear solver. The problem may be alleviated by adopting an outer preconditioner for S_h, we will give a fuller description in the next Paragraph.

We want to note that if we solve also Step 2 with an iterative solver, a good preconditioner must be provided also for the local problems in order to achieve a good scalability (for more details, see for instance [21, 24]).

Preconditioners for the Schur Complement System

Many preconditioners have been proposed in the literature for the Schur complement system with the aim to obtain good scalability properties. Among them, we briefly recall the Jacobi preconditioner P_h^J, the Dirichlet-Neumann P_h^{ND}, the balancing Neumann-Neumann $P_h^{NN,b}$ [24], the Bramble-Pasciak P_h^{BPS} [1], the Vertex-Space P_h^{VS} [8] and the wire-basket P_h^{WB} preconditioner. We refer [17, 25] and to the cited references for more details. Following [17], we summarize in Table 4.1 their preconditioning properties with respect to the geometric parameters h and H, for an elliptic problem (their extension to non-symmetric indefinite systems is, in general, not straightforward). We may note that with these preconditioners the dependence on H of the condition number becomes weaker, a part from the Jacobi preconditioner which is rather inefficient. The most effective preconditioners are also the ones more difficult to implement, particularly on arbitrary meshes.

Alternative and rather general ways to build a preconditioner for the SC system exploits the identity

$$S_h^{-1} = (0\ I)\, A^{-1} \begin{pmatrix} 0 \\ I \end{pmatrix}, \tag{4.10}$$

where I is the $n_B \times n_B$ identity matrix, being n_B the size of S_h. We can construct a preconditioner P_{Schur} for S_h from *any* preconditioner P_A^{-1} of the original matrix A by writing

$$P_{Schur} = \begin{pmatrix} 0 & I \end{pmatrix} P_A^{-1} \begin{pmatrix} 0 \\ I \end{pmatrix}.$$

If we indicate with $\mathbf{v}_B$ a vector of size n_B and with R_B the restriction operator on the interface variables, we can compute the operation $P_{Schur}^{-1}\mathbf{v}_B$ (which is indeed the one requested by an iterative solver) by an application of P_A^{-1}, as follows

$$P_{Schur}^{-1}\mathbf{v}_B = R_B P_A^{-1} \begin{pmatrix} 0 \\ \mathbf{v}_B \end{pmatrix} = R_B P_A^{-1} R_B^T \mathbf{v}_B . \tag{4.11}$$

In a parallel setting, we will of course opt for a parallel P_A^{-1}, like a Schwarz-based preconditioner of the type outlined in Section 4.4. This is indeed the choice we have adopted to precondition the SC matrix in many examples shown in this work.

4.3 The Schur Complement System Used as a Preconditioner

Although the Schur complement matrix is better conditioned than A, its multiplication with a vector is in general expensive. Indeed Step 2 of Algorithm 1 requires the solution of M linear systems, which should be carried out to machine precision, otherwise the iterative scheme converges slowly or may even diverge.

An alternative is to adopt a standard (parallel) iterative scheme for the global system (4.1) and use the SC system as a preconditioner. This will permit us to operate some modifications on the SC system in order to make it more computationally efficient. Precisely, we may replace S_h with a suitable approximation $\widetilde{S}$ that is cheaper to compute. The preconditioning matrix can then be derived as follows. We consider again the block decomposition (4.9) and we write A as a product of two block-triangular matrices,

$$A = \begin{pmatrix} A_{II} & 0 \\ A_{BI} & I \end{pmatrix} \begin{pmatrix} I & A_{II}^{-1} A_{IB} \\ 0 & S_h \end{pmatrix}.$$

Let us assume that we have good, yet cheaper, approximations of A_{II} and S_h, which we indicate as $\widetilde{A}_{II}$ and $\widetilde{S}$, respectively, a possible preconditioner for A is then

$$P_{\mathrm{ASC}} = \begin{pmatrix} \tilde{A}_{II} & 0 \\ A_{BI} & I \end{pmatrix} \begin{pmatrix} I & \tilde{A}_{II}^{-1} A_{IB} \\ 0 & \tilde{S} \end{pmatrix},$$

where ASC stands for *Approximate Schur Complement*. Indeed the approximation $\widetilde{A}_{II}$ may be used also to build $\widetilde{S}$ by posing $\widetilde{S} = A_{BB} - A_{BI}\widetilde{A}_{II}^{-1}A_{IB}$. Note that P_{ASC} operates on the whole system while $\widetilde{S}$ on the interface variables only, and that P_{ASC} does not need to be explicitly built, as we will show later on. A possible approximation for A_{II} is an incomplete LU decomposition [18], i.e. $\widetilde{A}_{II} = \tilde{L}\tilde{U}$, where

$\tilde{L}$ and $\tilde{U}$ are obtained from an incomplete factorization of A_{II}. Another possibility is to approximate the action of the inverse of A_{II} by carrying out a few cycles of an iterative solver or carrying out a multigrid cycle [12].

The solution of the preconditioned problem $P_{\text{ASC}} = \mathbf{zr}$, where $\mathbf{r} = (\mathbf{r}_I, \mathbf{r}_B)^T$ and $\mathbf{z} = (\mathbf{z}_I, \mathbf{z}_B)$ may be effectively carried out by the following Algorithm.

ALGORITHM 2: APPLICATION OF THE ASC PRECONDITIONER

1. Apply the lower triangular part of P_{ASC}. That is solve $\widetilde{A}_{II}\mathbf{y}_I = \mathbf{r}_I$ and compute $\mathbf{y}_B = \mathbf{r}_B - A_{BI}\mathbf{y}_I$.
2. Apply the upper triangular part of P_{ASC}. That is solve

$$\widetilde{S}\mathbf{z}_B = \mathbf{y}_B, \tag{4.12}$$

with $\widetilde{S} = A_{BB} - A_{BI}\widetilde{A}_{II}^{-1}A_{IB}$, and compute $\mathbf{z}_I = \mathbf{y}_I - \tilde{A}_{II}^{-1}A_{IB}\mathbf{z}_B$. The solution of (4.12) may be accomplished by an iterative scheme exploiting Algorithm 1.

Notice that the Step 1 and the computation of $\mathbf{z}_I$ in Step 2 are perfectly parallel. On the contrary (4.12) is a *global* operation, which, however, may be split into several parallel steps preceded and followed by scatter and gather operations involving communication among subdomains. Note that the matrix $\widetilde{S}$ may itself be preconditioned by using the technique outlined in Section 4.2.3, in particular by using a Schwarz-type preconditioner, as illustrated in Section 4.3.

As already said, $\widetilde{A}_{II}$ may be chosen as the ILU(f) incomplete factorization of A_{II}, where f is the fill-in factor. Furthermore, (4.12) may be solved inexactly, using a fixed number L of iterations of a Krylov solver. We will denote such preconditioner as **ASC-L-iluf**. Alternatively, one may avoid to factorize A_{II} and build its approximation implicitly by performing a fixed number of iteration when computing the local problems in Step (2) of Algorithm 1.

In both cases the action the ASC preconditioner corresponds to that of a matrix which changes at each iteration of the outer iterative solver. Consequently, one needs to make a suitable choice of the Krylov subspace accelerator for the the solution of (4.1) like, for instance, GMRESR [27] or FGMRES [18]. The former is the one we have used for the numerical results shown in this work.

We mention that the ASC preconditioner lends itself to a multilevel implementation, where a family of increasingly coarser approximations of the Schur complement matrix is used to build the preconditioner. The idea is that the coarsest approximation should be small enough to be solved directly. The drawback is the need of assembling and storing a number of matrices equal to the number of levels.

4.4 The Schwarz Preconditioner

The Schwarz iteration is a rather well known parallel technique based on an overlapping domain decomposition strategy. In a VO framework, it is normally built as follows (we refer again to Figure 4.1).

Each subdomain $\Omega^{(i)}$ is extended to $\widetilde{\Omega}^{(i)}$ by adding the strip $\Gamma^{(1,2)}$, i.e. $\widetilde{\Omega}^{(i)} = \Omega^{(i)} \cup \Gamma^{(1,2)}$, $i = 1, 2$. A parallel solution of the original system is then obtained by an iterative procedure involving local problems in each $\widetilde{\Omega}^{(i)}$, where on $\partial\widetilde{\Omega}^i \cap \Omega^{(j)}$ we apply Dirichlet conditions by getting the data from the neighboring subdomains.

What we have described here is the implementation with *minimum overlap*. A larger overlap may be obtained by adding further layers of elements. The procedure may be readily extended to an arbitrary number of subdomains. More details on the algorithm with an analysis of its main properties may be found, for instance, in [17].

A *multiplicative* version of the procedure is obtained by ordering the subdomains and solving the local problems sequentially using the latest available interface values. This is indeed the original Schwarz method and is sequential. Parallelism can be obtained by using the *additive* variant where all subdomain are advanced together, by taking the interface values at the previous iteration.

From an algebraic point of view, multiplicative methods can be reformulated as a block Gauss-Seidel procedure, while additive methods as block Jacobi procedure [18].

If used as a stand-alone solver, the Schwarz iteration algorithm is usually rather inefficient in terms of iterations necessary to converge. Besides, a damping parameter has to be added, see [17], in order to ensure that the algorithm converges. Instead, the method is a quite popular parallel preconditioner for Krylov accelerators. In particular its minimum overlap variant, which may exploit the same data structure normally used for the parallel implementation of the matrix-vector product, allowing a saving in memory requirement.

Let $B^{(i)}$ be the local matrix associated to the discretisation on the extended subdomain $\widetilde{\Omega}^{(i)}$, $R^{(i)}$ a restriction operator from the nodes in Ω to those in $\widetilde{\Omega}^{(i)}$, and $P^{(i)}$ a prolongation operator (usually, $P^{(i)} = (R^{(i)})^T$). Using this notation, the Schwarz preconditioner can be written as

$$P_{AS}^{-1} = \sum_{i=1}^{M} P^{(i)} B^{(i)} R^{(i)}, \tag{4.13}$$

being M the number of subdomains.

The matrices $B^{(i)}$ can be extracted directly from the set of rows of the global matrix A corresponding to the local nodes, discarding all coefficients whose indexes are associated to nodes exterior to the subdomain. The application of $R^{(i)}$ is trivial, since it returns the locally hosted components a the vector; the prolongation operator $P^{(i)}$ just does the opposite operation.

Although simple to implement, the scalability of the Schwarz preconditioner is hindered by the weak coupling between far away subdomains. We may recover a good scalability by the addition of a coarse operator [9, 21]. If the linear system arises from the discretisation of a PDE system a possible technique to build the coarse operator matrix A_H consists in discretising the original problem on a (very) coarse mesh, see for instance [4]. However the construction of a coarse grid and of the associated restriction and prolongation operators may become a rather complicated task when dealing with three dimensional problems and complicated geometries. An alternative is to resort to algebraic procedures, such as the aggregation or agglomeration technique [21, 1, 4, 2, 28, 20], which are akin to procedures developed in the context of algebraic multigrid methods, see also [29]. Here, we will focus on the latter, and in particular we will propose an agglomeration technique.

4.4.1 The Agglomeration Coarse Operator

To fix the ideas let us consider a finite element formulation (4.2). Thanks to a VO partitioning we can split the finite element function space V_h as

$$V_h = \bigcup_{i=1}^{M} V_h^{(i)},$$

where M is the number of subdomains, and $V_h^{(i)}$ is set of finite element functions associated to the triangulation of $\widetilde{\Omega}^{(i)}$ with zero trace on $\partial\widetilde{\Omega}^{(i)} \setminus \partial\Omega$. We suppose to operate in the case of minimal overlap among subdomains as described in the previous section and we indicate with $n^{(i)}$, the dimension of the space $V_h^{(i)}$. By construction, $n = \sum_{i=1}^{M} n^{(i)}$.

We can build a *coarse space* considering for each $\Omega^{(i)}$ a set of vectors $\{\boldsymbol{\beta}_s^{(i)} \in \mathbb{R}^{n^{(i)}}, s = 1, \ldots, l^{(i)}\}$ of nodal weights $\boldsymbol{\beta}_s^{(i)} = \left(\beta_{s,1}^{(i)}, \ldots, \beta_{s,n^{(i)}}^{(i)}\right)$ that are linearly independent, with $\beta_{s,n^{(i)}}^{(i)} \neq 0$. The value $l^{(i)}$, $i = 1, \ldots,$ will be the (local) dimension of the coarse operator on the corresponding subdomain. Clearly, we must have $l^{(i)} \leq n^{(i)}$ and, in general, $l^{(i)} << n^{(i)}$. We indicate with l the global dimension of the coarse space, i.e.

$$l = \sum_{i=1}^{M} l^{(i)}.$$

With the help of the vectors $\boldsymbol{\beta}_s^{(i)}$, we can define a set of local coarse space functions as linear combination of basis functions, i.e.

$$\mathcal{V}_H^{(i)} = \left\{ \Phi_s^{(i)} : \Omega \to \mathbb{R} \mid \Phi_s^{(i)} = \sum_{k=1}^{n^{(i)}} \beta_{s,k}^{(i)} \phi_k^{(i)},\ s = 1, \ldots, l^{(i)} \right\}.$$

It is easy to verify that the functions in $\mathcal{V}_H^{(i)}$ are linearly independent. Finally, the set $\mathcal{V}_H = \bigcup_{i=1}^{M} \mathcal{V}_H^{(i)}$ is the base of the global coarse grid space V_H, i.e. we take $V_H = \mathrm{span}\{\mathcal{V}_H\}$. By construction, $\dim(V_H) = \mathrm{card}(\mathcal{V}_H) = l$.

We note that $V_H \subset V_h$, as it is built by linear combinations of function in V_h, and any function $u_H \in V_H$ may be written as

$$u_H(\mathbf{x}) = \sum_{i=1}^{M} \sum_{s=1}^{l^{(i)}} U_s^{(i)} \Phi_s^{(i)}(\mathbf{x}), \tag{4.14}$$

where the $U_s^{(i)}$ are the "coarse" degrees of freedom. Finally, the coarse problem is built as:
Find $u_H \in V_H$ so

$$a(u_H, w_H) = f(w_H), \quad \forall w_H \in V_H .$$

From an algebraic point of view, we have

$$\sum_{i=1}^{M}\sum_{s=1}^{l^{(i)}} U_s^{(i)} \sum_{k=1}^{n^{(i)}}\sum_{t=1}^{n^{(m)}} \beta_{s,k}^{(i)}\beta_{q,t}^{(m)} a(\phi_k^{(i)}, \phi_t^{(m)}) = \sum_{t=1}^{n^{(m)}} \beta_{q,t}^{(m)}(f, \phi_t^{(m)})$$
$$q = 1, \dots, l^{(m)}, \; m = 1, \dots, M .$$

To complete the procedure we need to define a restriction operator $R_H : V_h \to V_H$ which maps a generic finite element function to a coarse grid function. Since $u \in V_h$ may be written as

$$u_h = \sum_{i=1}^{M}\sum_{k=1}^{n^{(i)}} \phi_k^{(i)} u_k^{(i)},$$

where the $u_k^{(i)}$ are the degrees of freedom associated to the triangulation of $\Omega^{(i)}$, a restriction operator may be defined by computing $u_H = R_h u$ as

$$u_H = \sum_{i=1}^{M}\sum_{s=1}^{l^{(i)}} U_s^{(i)} \Phi_s^{(i)},$$

where

$$U_s^{(i)} = \sum_{k=1}^{n^{(i)}} \beta_{s,k}^{(i)} u_k^{(i)}, \; s = 1, \dots l^{(i)}, \quad i = 1 \dots, M.$$

At the algebraic level, we build a global vector $\mathbf{U}_H$ by assembling the $U_s^{(i)}$ on a subdomain basis, i.e.

$$\mathbf{U}_H = \left(U_1^{(1)} \dots U_{l^{(1)}}^{(1)} \dots U_{l^{(M)}}^{(M)}\right)^T ,$$

and we arrange similarly the vector $\mathbf{u}_h$ of the nodal values of u_h, i.e.

$$\mathbf{u}_h = \left(u_1^{(1)} \dots u_{n^{(1)}}^{(1)} \dots u_{n^{(M)}}^{(M)}\right)^T .$$

The prolongation matrix $R_H^T \in \mathbb{R}^{n \times l}$ will then have the following block structure,

$$R_H^T = \begin{pmatrix} \boldsymbol{\beta}_1^{(1)^T} & \boldsymbol{\beta}_2^{(1)^T} & \dots & \boldsymbol{\beta}_{l(1)}^{(1)^T} & 0 & 0 & \dots & 0 & 0 & 0 & \dots & 0 \\ 0 & 0 & \dots & 0 & \boldsymbol{\beta}_1^{(2)^T} & \boldsymbol{\beta}_2^{(2)^T} & \dots & \boldsymbol{\beta}_{l(2)}^{(2)^T} & 0 & 0 & \dots & 0 \\ \vdots & \vdots & \ddots & \vdots & \vdots & \vdots & \ddots & \vdots & \vdots & \vdots & \ddots & \vdots \\ 0 & 0 & \dots & 0 & 0 & 0 & \dots & 0 & \boldsymbol{\beta}_1^{(M)^T} & \boldsymbol{\beta}_2^{(M)^T} & \dots & \boldsymbol{\beta}_{l(M)}^{(M)^T} \end{pmatrix},$$

and the coarse matrix A_H and right-hand side of the coarse system can be written as

$$A_H = R_H A R_H^T, \quad \mathbf{f}_H = R_H \mathbf{f},$$

respectively.

The conditions imposed on the $\boldsymbol{\beta}_s^{(i)}$ vectors guarantees that R_H has full rank. Moreover, if A is symmetric and positive definite, then A_H will share the same property. The application of the agglomeration coarse grid operator does not require to build R_H explicitly. Even the construction of the vectors $\boldsymbol{\beta}$ can be avoided if they have a simple structure. In fact the construction of A_H involves just a weighted sum of the element of A. Concerning the parallel implementation, the overhead of the coarse problem depends in general on the number of local coarse degrees of freedom. In general $l^{(i)} << n^{(i)}$ (in the limit we may even take $l^{(i)} = 1$ for all i!), and consequently the matrix A_H is rather small compared to A. This a major difference from the use of this technique in an algebraic multigrid setting, where many levels of coarse operator are considered.

The build up of the coarse linear system can be carried out as follows. Each processor computes the contribution to A_H and $\mathbf{f}_H$ corresponding to the associated subdomains, then it broadcasts the results to the other processors. Being the coarse system small, the cost of the broadcast operation is limited. Furthermore, it is usually carried out only once.

The domain decomposition technique may be used also for the set up of the $\boldsymbol{\beta}_s^{(i)}$ vectors. Indeed, after having set up the basic Schwarz preconditioner by assigning each extended subdomain $\widetilde{\Omega}_i$ to a different processor, at a second stage, each subdomain $\widetilde{\Omega}^{(i)}$ can be further partitioned into $l^{(i)}$ connected parts with minimum overlap. We will indicate this second level partition as $\omega_s^{(i)}$ with $s = 1, \ldots, l^{(i)}$. Then we may take

$$\beta_{s,k} = \begin{cases} 1 & \text{if node } k \text{ belongs to } \omega_s^{(i)} \backslash \partial\omega_s^{(i)}, \\ 0 & \text{otherwise.} \end{cases}$$

As already explained, the coarse grid operator is used to improve the scalability of a Schwarz-type parallel preconditioner P_S. We will indicate with P_{ACM} a preconditioner augmented by the application of the coarse operator (ACM stands for *agglomeration coarse matrix*) and we illustrate two possible strategies for its construction.

A one-step preconditioner, $P_{\mathrm{ACM},1}$, may be formally written as

$$P_{\mathrm{ACM},1}^{-1} = P_S^{-1} + R_H^T A_H^{-1} R_H, \tag{4.15}$$

and it corresponds to an additive application of the coarse operator.

An alternative formulation adopts the following two-steps Richardson method

$$\begin{aligned} \mathbf{u}^{n+1/2} &= \mathbf{u}^n + P_S^{-1}\mathbf{r}^n, \\ \mathbf{u}^{n+1} &= \mathbf{u}^{n+1/2} + R_H^T A_H^{-1} R_H \mathbf{r}^{n+1/2}, \end{aligned} \tag{4.16}$$

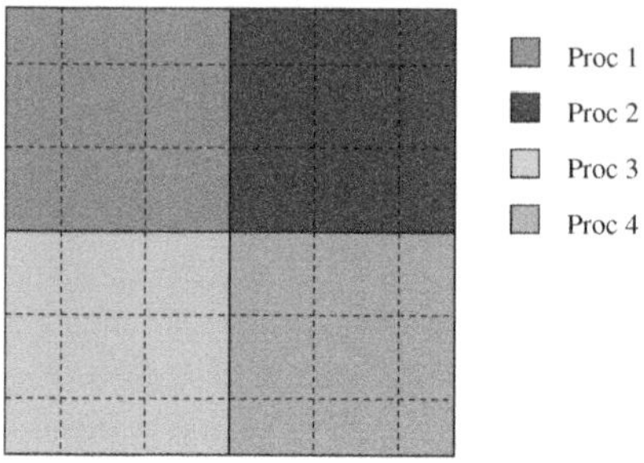

Fig. 4.4. Example of two-level decomposition. First level in continuous line, and second level in dashed line. Typically, each first-level decomposition subdomain is given to a different processor. (For the color version, see Figure A.6 on page 469).

where $\mathbf{r}^n$ and $\mathbf{u}^n$ are respectively the residual and the approximate solution at the $n-th$ iteration of the outer iterative solver. The corresponding preconditioning matrix can be formally written as

$$P_{\mathrm{ACM},2}^{-1} = P_S^{-1} + R_H^T A_{\mathrm{ACM}}^{-1} R_H - P_S^{-1} A R_H^T A_{\mathrm{ACM}}^{-1} R_H \,. \tag{4.17}$$

The implementation of (4.15) and (4.17) requires the parallel solution of the coarse problem

$$A_H \mathbf{u}_H = \mathbf{r}_H, \tag{4.18}$$

where $\mathbf{r}_H = R_H \mathbf{r}$. If one has only a limited number of processors at disposal, perhaps the best approach is to gather the entire coarse matrix A_H on one processor (say, processor 0), and perform each solution phase of (4.18) using an appropriate sequential linear solver. Each processor computes its contribution to $\mathbf{r}_H$ and sends it to processor 0 (gathering phase). The solution computed by processor 0 will then be scattered to the other processors, for example by a broadcast operation and the final prolongation operation will be carried out independently on each processor. As the coarse problem is likely to be small, its sequential solution causes little overhead and the cost of broadcast communication is also likely to be negligible as well.

Instead, if hundreds or thousands of processors are used, the coarse problem is likely to have a non-negligible size, and furthermore, the cost of the gathering-scattering communication phases may be substantial. Therefore, in this case a parallel direct solver is to be preferred. Generic interfaces to serial and parallel direct solvers are available, see [19]. We also mention that the presented two-level algebraic preconditioner (see Figure 4.4) can be extended in a multilevel fashion; see for instance [16].

4.5 Applications

In this Section we show some applications of the techniques here illustrated. We will report some academic tests used mainly to assess the basic properties of the scheme

Table 4.2. 2D Poisson problem. Comparison of different preconditioners. Number of iterations needed to reduce the initial residual by a factor of 10^6, M is the number of subdomains. The starting mesh has 180×180 squares, each of them has been divided into 2 triangles.

	solver	$M = 4$	$M = 9$	$M = 16$	$M = 25$
P_S	GMRES	-	57	70	76
P_C	GMRES	-	42	40	39
$P_{\mathrm{ACM},1}$	GMRES	-	56	69	70
$P_{\mathrm{ACM},2}$	GMRES	-	51	49	46
ASP-2-ilu0	GMRESR	99	97	97	99
ASP-4-ilu0	GMRESR	82	78	75	71
ASP-2-ilu1	GMRESR	68	68	70	69
ASP-2-ilu2	GMRESR	52	53	56	52

and more realistic applications. The latter include the solution of compressible flow around aircrafts and free surface hydrodynamics problems. For the decomposition of the domain, we have adopted the software package Metis [15], which operates on the finite element mesh, and can produce both EO and VO decompositions. For the overlapping Schwarz preconditioner, wider levels of overlap are recursively created by adding internal nodes that are linked with a side to the current overlap region. If we use higher order finite elements, we may still make the basic partitioning using just the mesh information. Then the new nodes corresponding to the additional degrees of freedom are added and their nature (interior, exterior or border) can be immediately identified by the geometrical entity that is associated to the node. For instance, additional in a VO framework, additional nodes on a side linking to nodes internal to subdomain Ω_i are internal to Ω_i etc.

4.5.1 2D Poisson Problem

We have considered the following Poisson problem

$$\begin{cases} -\Delta u = f & \text{in } \Omega, \\ u = 0 & \text{on } \partial\Omega, \end{cases}$$

where $\Omega = (0,1) \times (0,1)$. For the discretisation we have used P1 finite elements on a regular mesh. The linear system has been solved using GMRES(60), in the case of the Schwarz preconditioner, or GMRESR with the ASC preconditioner, up to a tolerance of 10^{-6}. The right-hand side is made up of random numbers between 0 and 1.

Being the mesh structured, we have divided the computational domain into M square subdomains, and we have used these subdomains to build a "classical" coarse correction for a 2-level Schwarz preconditioner, following [9].

In Table 4.2 we have compared the Schwarz preconditioner (P_S) without the coarse grid correction, the one augmented by the "classical" coarse operator, indicated by P_C, and the proposed preconditioners $P_{\mathrm{ACM},1}$ and $P_{\mathrm{ACM},2}$. Finally, we report the results obtained with the ASC preconditioner ASP-L-iluf.

Table 4.3. 2D Poisson problem. Comparison of different preconditioners. CPU-time in seconds to reduce the initial residual by a factor of 10^6.

	$M = 4$	$M = 9$	$M = 16$	$M = 25$
P_S	-	3.90	1.59	0.77
P_C	-	5.50	1.94	2.13
$P_{\text{ACM},1}$	-	5.10	2.08	1.68
$P_{\text{ACM},2}$	-	4.16	2.03	0.89
ASP-2-ilu0	12.64	3.94	3.34	2.12
ASP-4-ilu0	9.73	5.46	2.23	1.88
ASP-2-ilu1	8.04	3.87	1.80	2.11
ASP-2-ilu2	6.44	4.32	1.77	1.54

Table 4.4. 2D Poisson problem. $P_{\text{ACM},2}$. The table reports the number of iterations needed to reduce the initial residual by a factor of 10^6.

Local dimension of coarse space	$M = 9$	$M = 16$	$M = 25$
1	51	49	46
2	57	54	50
4	55	49	46
8	50	45	41
32	50	34	32

Table 4.5. 2D Poisson problem. $P_{\text{ACM},2}$. The table reports the CPU time (in seconds) needed to reduce the initial residual by a factor of 10^6.

Local dimension of coarse space	$M = 9$	$M = 16$	$M = 25$
1	4.16	2.03	0.89
2	4.28	1.60	0.98
4	4.22	1.55	0.96
8	4.16	1.54	0.88
32	4.17	1.52	0.94

The performance of the Schwarz method without any coarse correction degrades rapidly, demonstrating the poor scalability of the basic algorithm. Preconditioning with $P_{\text{ACM},1}$ has a very poor influence (probably also because of the relative small number of subdomains), while the 2-level version behaves much better. The ASC-type preconditioners show a very good scalability, even if the CPU times, reported in Table 4.3, are less interesting. Another important parameter is the dimension of the agglomeration coarse space, and it is analyzed in Tables 4.4 and 4.5. Note that increasing the dimension of the coarse space has a positive effect on the convergence. Even if the numerical experiments show that the "classic" preconditioner is in general more effective unless a "rich" local coarse space dimension is used in the agglomeration procedure, we point out again the greater flexibility and generality of the latter.

4.5.2 The Compressible Euler Equations

Any standard spatial discretisation applied to the Euler equations leads eventually to a system of ODE in time, which may be written as

$$\frac{\mathrm{d}U}{\mathrm{d}t} = R(\mathbf{U}) , \tag{4.19}$$

where $\mathbf{U} = (U_1, U_2, \ldots, U_n)^T$ is the vector of unknowns with $U_i = U_i(t)$ and $R(U)$ the result of the spatial discretisation of the Euler fluxes. An implicit two-step scheme applied to (4.19), for instance a backward Euler method, yields

$$\mathbf{U}^{n+1} - \mathbf{U}^{n} = \Delta t R\left(\mathbf{U}^{n+1}\right) , \tag{4.20}$$

where Δt is a *diagonal matrix* of local time steps, i.e. $\Delta t = \mathrm{diag}(\Delta t_i, i = 1, \ldots, n)$. The non-linear problem (4.20) may be solved by employing a Newton iterative procedure, which computes successive approximations $\mathbf{U}_{(k+1)}$ of $\mathbf{U}^{n+1}$ by solving

$$\left[I + \Delta t \frac{\partial R}{\partial \mathbf{U}}\left(\mathbf{U}_{(k)}\right)\right]\left(\mathbf{U}_{(k+1)} - \mathbf{U}^{n}\right) = \mathbf{U}_{(k)} - \mathbf{U}^{n} - \Delta t R\left(\mathbf{U}_{(k)}\right) , \quad k = 0, \ldots ,$$

with $\mathbf{U}_{(0)} = \mathbf{U}^n$. We have reduced the original non-linear problem to the solution of a series of linear systems which will be finally tackled by the proposed parallel techniques.

Since we are considering steady state solutions, we have taken just a single iteration of the Newton procedure. Furthermore, we have taken an approximate Jacobian $\frac{\partial R}{\partial \mathbf{U}}$. More precisely, the Jacobian is the exact Jacobian of a first-order upwind spatial discretisation. This ease up the computation greatly. The resulting method is sometimes called pseudo-transient continuation [6].

For the numerical experiments at hand we have used the parallel version of the code THOR, developed at the Von Karman Institute. This code uses for the spatial discretisation the multidimensional upwind finite element scheme [7]. The solutions obtained on two of the presented test cases are illustrated in Figures 4.5 and 4.6.

The test cases are summarized in Table 4.6. For all of them, the starting solution is a constant vector, and the local CFL numbers are varied from 10 to 10^5, by multiply them at each time level by a factor of 2 until we reach CFL=10^5. The computations are stopped when the Euclidean norm of the density residual is less than 10^{-6}. As previously described, at each time level we have to solve a linear system. This is done using GMRES(60) in the case of Schwarz-type preconditioner or GMRESR if the ASC preconditioner is chosen, up to a tolerance on the relative residual $||r||/||r_0||$ of 10^{-6}. The starting solution is the zero vector. The Schwarz preconditioner P_S uses a minimal overlap and the local problems are solved inexactly using an ILU(0) decomposition. A first test case concerns the flow around a Falcon Aircraft. We have considered a free-stream Mach number of 0.45 and zero angles of yaw and attach. The mesh is formed by 45387 elements, corresponding to 226935 degrees of freedom.

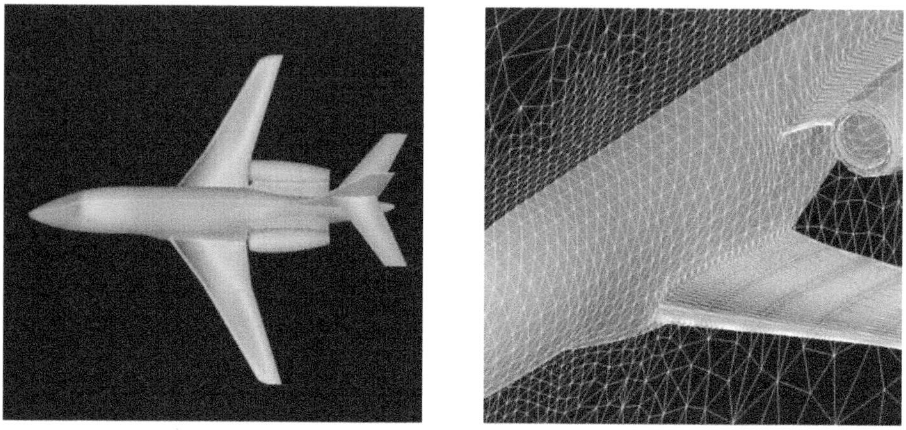

Fig. 4.5. Pressure coefficient contours for FALCON_45k. (For the color version, see Figure A.7 on page 470).

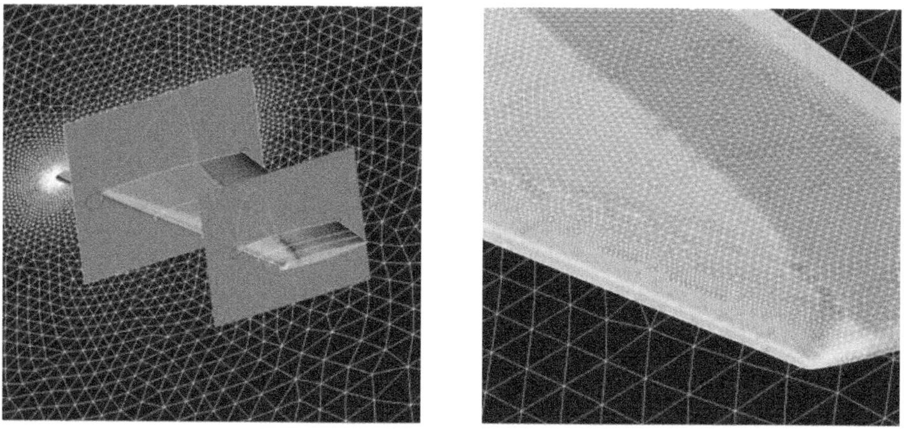

Fig. 4.6. Mach number contours for M6_316K. (For the color version, see Figure A.8 on page 470).

Figures 4.7 and 4.8 report the iterations to converge at each time level for different values of L, using 4, 8, 16 and 32 SGI-Origin 3000 processors for FALCON_45k. The time step in Equation (4.19) is increased exponentially. This quite common practice, here adopted in order to minimize the effect of "bad" starting solutions on the convergence of the Newton method, makes the first linear systems relatively well conditioned. The iterations to converge increase at each time level, to decrease again when the system is approaching the steady-state solution. We may notice that using 16 and 32 processors the ASC preconditioner cannot guarantee good performance, at least for some combinations of the number of levels L and CFL number. This may suggest to adapt the value of L to the CFL number. The elapsed CPU times are re-

Table 4.6. Main characteristics of the test cases.

name	M_∞	α	N_nodes	N_cells
FALCON_45k	0.45	1.0	45387	255944
M6_23k	0.84	3.06	23008	125690
M6_42k	0.84	3.06	42305	232706
M6_94k	0.84	3.06	94493	666569
M6_316k	0.84	3.06	316275	1940182

Table 4.7. FALCON_45k. CPU-time (in seconds) for ASC preconditioner, using different values of L.

N_procs	ASC-1-ilu0	ASC-2-ilu0	ASC-4-ilu0	ASC-8-ilu0
4	2542.4	2401.7	**2393.2**	3319.7
8	925.5	**897.6**	1406.6	1423.2
16	863.7	753.7	**561.6**	707.2
32	443.8	332.1	**248.6**	398.6

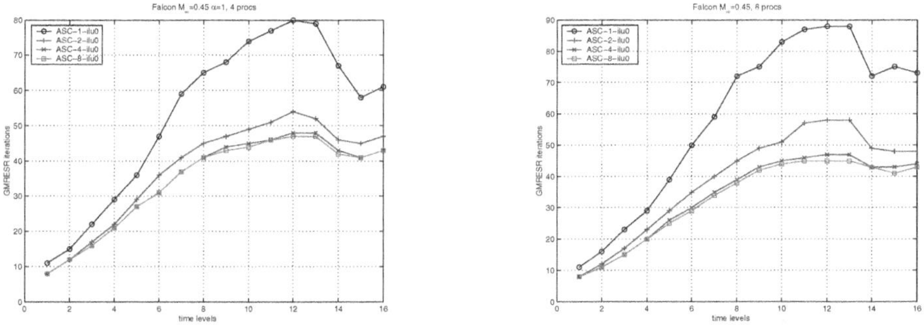

Fig. 4.7. FALCON_45k. Convergence history with 4 (left) and 8 processors (right) with the ASC preconditioner, for different values of L. (For the color version, see Figure A.9 on page 471).

ported in Table 4.7, where we have highlighted the best performance. We may notice that one iteration of the nested solver is not enough to guarantee good convergence results, especially as the number of processors grows. At the same time, high values of L will decrease the performance. A value of about 4 seems a good compromise.

In Figure 4.9 we have reported the results obtained with the proposed Schwarz methodology and in particular the influence of the local dimension of the coarse space N_p for $P_{\mathrm{ACM},1}$ and $P_{\mathrm{ACM},2}$, using 16 MIPS 14000 processors. We recall again that N_p is the dimension of the coarse space on each processor. For low CFL numbers the value of N_p does not affect remarkably the convergence of GMRES, while as the CFL number increases the positive effect of an increasing coarse space becomes more evident, as we may notice in Figure 4.10. The two-level preconditioner seems a better choice from the point of view of both iterations to converge and CPU

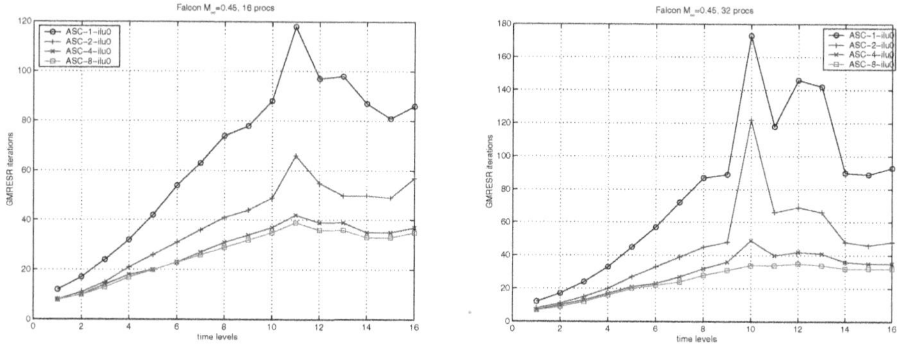

Fig. 4.8. FALCON_45k. Convergence history with 16 processors (left) and 32 processors (right) with the ASC preconditioner, for different values of L. (For the color version, see Figure A.10 on page 471).

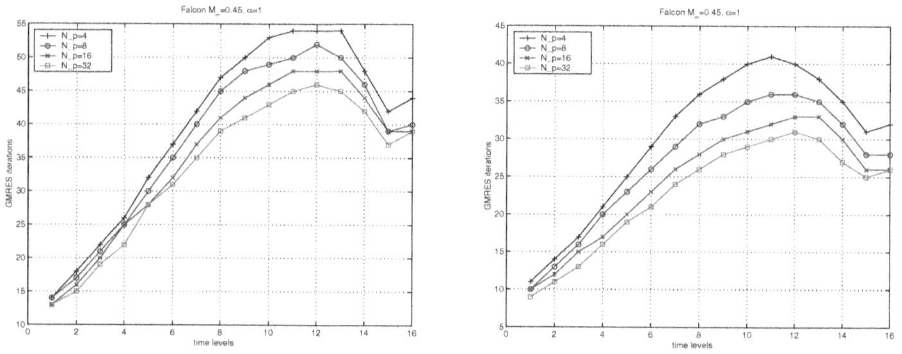

Fig. 4.9. FALCON_45k. Iterations to converge with different values of N_p for $P_{\mathrm{ACM},1}$ (left) and $P_{\mathrm{ACM},2}$ (right). (For the color version, see Figure A.11 on page 471).

time, especially for low values of N_p. Figure 4.11 shows a comparison among the Schwarz preconditioner without coarse correction P_S, the ASC preconditioner and the ACM preconditioner. Although better than P_S in terms of converge rate, ASC-2-ilu0 doesn't seem to be a suitable choice, as one may observe from the left picture of Figure 4.11, where we have plotted the time residual versus the CPU-time. On the contrary, both $P_{\mathrm{ACM},1}$ and $P_{\mathrm{ACM},2}$ are substantially better than P_S.

We have obtained similar results for the test case M6_94k, as reported in Figures 4.12 and 4.13. The CPU times are reported in Tables 4.8, 4.9 and 4.10 for M6_94K, and in table 4.11 for M6_316k.

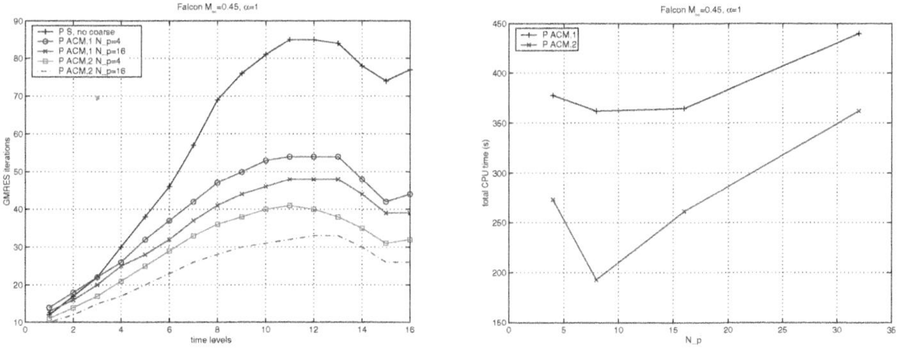

Fig. 4.10. FALCON_45k. Comparison among different preconditioners (left) and CPU time, in seconds (right). 16 SGI-Origin3000 processors. (For the color version, see Figure A.12 on page 472).

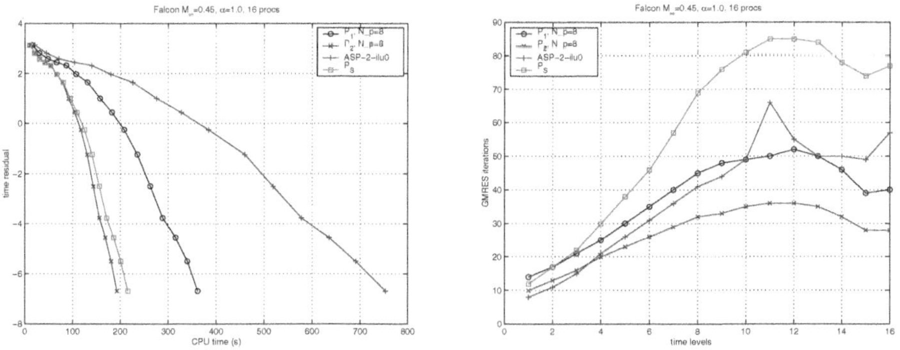

Fig. 4.11. FALCON_45k. Residual versus CPU-time (left) and iterations to converge at each time level (right), using 16 SGI-Origin3000 processors. (For the color version, see Figure A.13 on page 472).

Table 4.8. M6_94k. CPU-time (in seconds) for ASC preconditioner, using different values of the fixed iteration count L.

N_procs	ASC-2-ilu0	ASC-4-ilu0	ASC-8-ilu0
8	**1538.4**	1600.4	1859.9
16	544.8	**569.1**	1330.5
32	**248.5**	286.0	358.9

4.6 Conclusions

In this chapter we have presented a class of preconditioners based on the DD approach that are well suited for parallel implementation. A great variety of methods are in fact available in literature and we have chosen to focus our attention on

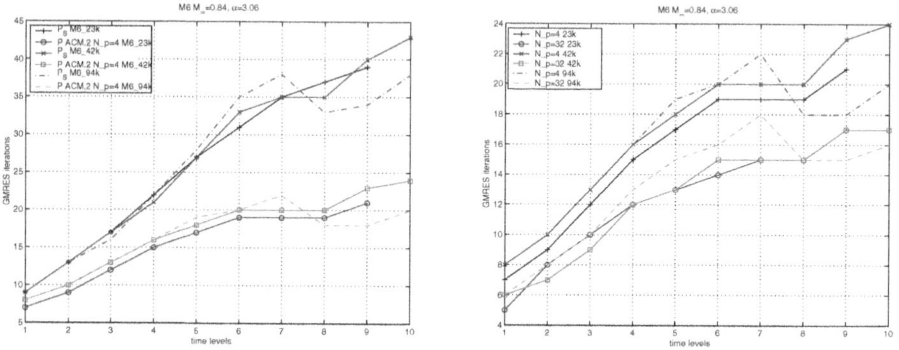

Fig. 4.12. M6_94k. Iterations to converge with P_S and $P_{\mathrm{ACM},2}$ (left), and iterations to converge with $P_{\mathrm{ACM},2}$ (right) using two different values of $N_p$and 16 processors. (For the color version, see Figure A.14 on page 473).

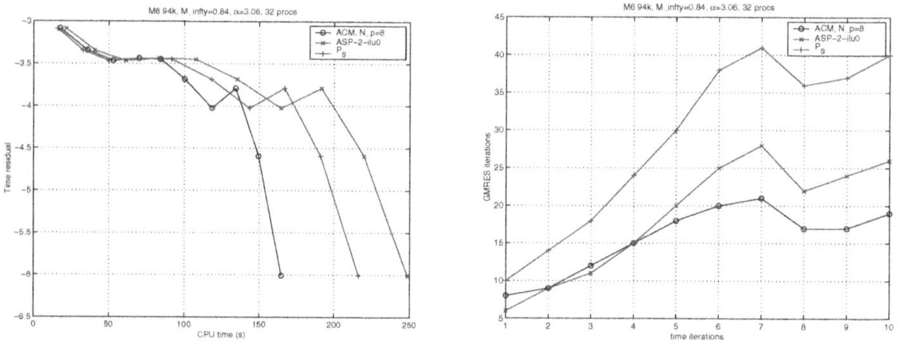

Fig. 4.13. M6_94k. Residual versus CPU-time (right) and iterations to converge at each time level (right), using 32 processors. (For the color version, see Figure A.15 on page 473).

Table 4.9. M6_94k. CPU-time to converge, using $P_{\mathrm{ACM},1}$ and varying the number of processors. Best results are highlighted.

N_procs	N_p=4	N_p=8	N_p=16	N_p=32
8	1008.2	**978.4**	1251.3	883.4
16	**502.5**	506.9	515.0	457.3
32	**208.0**	245.3	300.5	505.0

DD methods applied at the algebraic level, namely the Schur complement and the Schwarz algorithms. The reason being their generality and the fact that they are implemented in many available parallel linear algebra packages. We have illustrated mainly their use as preconditioners. Indeed, the first consideration we can make is that these methods are usually inefficient when used as solvers.

Table 4.10. M6_94k. CPU-time to converge, using $P_{\mathrm{ACM},2}$, and varying the number of processors. Best results are highlighted.

N_procs	N_p=4	N_p=8	N_p=16	N_p=32
8	934.8	945.6	**909.3**	925.6
16	458.6	**405.2**	413.9	442.6
32	164.4	**164.7**	181.4	515.6

Table 4.11. M6_316k. CPU-time (seconds) required to reach the steady state solution, using 32 processors. Comparison between Schwarz preconditioner without coarse grid, the ACM preconditioner, multiplicative version, and an approximated Schur complement preconditioner.

N_procs	P_S	$P_{\mathrm{ACM},2}$	ASC-4-ilu0
32	1524.2	1370.6	2691.3

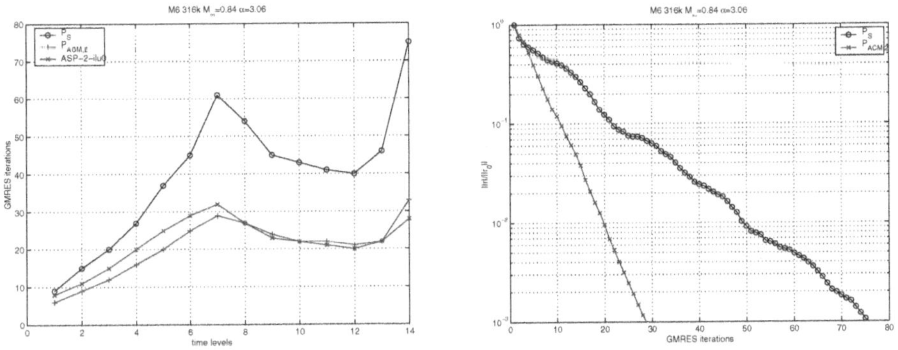

Fig. 4.14. M6_316k. Iterations at each time level (left) and converge history at the 14th time step (right). (For the color version, see Figure A.16 on page 473).

A clear cut comparison of the two is difficult as their performance is often problem dependent. As a general rule we may state that the approximate Schur complement system has generally a better preconditioning property at the price of an higher cost "per iteration". It performs better when the ratio unknowns/number of subdomains is "low". Otherwise, the computational cost linked to the solution of the internal problems (which in most cases scales with the square of the number of the local degrees of freedom) may degrade the effectiveness of the preconditioner. It may be attractive also if the ratio between computational and communication speed is high, for example when the processors are connected through a slow network. The smaller number of iterations to converge imply less communication, and in this case it may overcome the higher cost spent at local level.

The Schwarz preconditioner is often the matter of choice of many parallel linear algebra packages, because of its rather simple implementation. The minimal overlap variant is also rather attractive in term of memory usage. Yet it needs a coarse operator to obtain scalability. To this aim, here we have described an agglomeration procedure that has the advantage of generality and of a simple set up. The cost per iteration is smaller, since we need to solve the local problem only once. Yet its preconditioning properties are usually less marked, and this imply a slower convergence.

It is important to notice that all efficient DD preconditioners consist of a *local* and a *global* component. The local part, acts at the subdomain level and may possibly capture the coupling between neighboring subdomains through the interface nodes; the global part provides instead an overall communication among far away subdomains. In the Schur complement based methods, the global part is the solution of the Schur complement system itself, in the Schwarz technique this task is played by the coarse operator.

Acknowledgments

Part of this work is the result of a research carried out under the frame of the European project IDeMAS (contract number BRPR-CT97-0591). Other research agencies, such us the Italian CNR and MURST are also acknowledged for they financial support.

References

1. J. Bramble, J. Pasciak, and X. Zhang. Two-level preconditioners for 2nd order elliptic finite element problems. *East-West J. Numer. Math.*, 4:99–120, 1996.
2. T. Chan, B.Smith, and J. Zou. Overlapping Schwarz methods on unstructured meshes using non-matching coarse grids. *Numer. Math.*, 73:149–167, 1996.
3. T. Chan, S. Go, and J. Zou. Boundary treatments for multilevel methods on unstructured meshes. *SIAM J. Sci. Comput.*, 21(1):46–66, 1999.
4. T. Chan and T. Mathew. The interface probing technique in domain decomposition. *SIAM Journal on Matrix Analysis and Applications*, 13(1):212–238, 1992.
5. T. Chan and T. Mathew. Domain decomposition algorithms. *Acta Numerica*, pages 61–163, 1993.
6. T. Coffey, C. Kelley, and D. Keyes. Pseudotransient continuation and differential-algebraic equations. *SIAM Journal on Scientific Computing*, 25(2):553–569, 1996.
7. H. Deconinck, H. Paillère, R. Struijs, and P. Roe. Multidimensional upwind schemes based on fluctuaction splitting for systems of conservation laws. *Comput. Mech.*, 11:323–340, 1993.
8. M. Dryja, B. Smith, and O. Widlund. Schwarz analysis of iterative substructuring algorithms for elliptic problems in three dimensions. *SIAM J. Numer. Anal.*, 31(6):1662–1694, 1993.
9. M. Dryja and O. Widlund. Domain decomposition algorithms with small overlap. *SIAM J. Sci.Comput.*, 15(3):604–620, 1994.
10. C. Farhat and F. Roux. A method of finite element tearing and interconnecting and its parallel solution algorithm. *Internat. J. Numer. Meth. Engrg.*, 32:1205–1227, 1991.

11. A. Grama, A. Gupta, and V. Kumar. Isoefficiency: Measuring the scalability of parallel algorithms and architectures. *IEEE Parallel Distrib. Technol.*, 1:12–21, August 1993.
12. F. Hülsemann, M. Kowarschik, M. Mohr, and U. Rüde. Parallel geometric multigrid. In A. M. Bruaset and A. Tveito, editors, *Numerical Solution of Partial Differential Equations on Parallel Computers*, volume 51 of *Lecture Notes in Computational Science and Engineering*, pages 165–208. Springer-Verlag, 2005.
13. K. Hwang. *Advanced Computer Architecture: Parallelism Scalability, Programmability*. McGraw Hill, New York, 1993.
14. K. Hwang and Z. Xu. *Scalable Parallel Computing: Technology, Architecture, Programming*. McGraw-Hill, Inc., New York, NY, USA, 1998.
15. G. Karypis and V. Kumar. METIS: Unstructured graph partitining and sparse matrix ordering system. Technical Report 98-036, University of Minnesota, Department of Computer Science, 1998.
16. P. Lin, M. Sala, J. Shadid, and R. Tuminaro. Performance of fully-coupled algebraic multilevel domain decomposition preconditioners for incompressible flow and transport. *submitted to International Journal for Numerical Methods in Engineering*, 2004.
17. A. Quarteroni and A. Valli. *Domain Decomposition Methods for Partial Differential Equations*. Oxford University Press, Oxford, 1999.
18. Y. Saad. *Iterative Methods for Sparse Linear Systems*. Thompson, Boston, 1996.
19. M. Sala. Amesos 2.0 reference guide. Technical Report SAND-4820, Sandia National Laboratories, September 2004.
20. M. Sala. Analysis of two-level domain decomposition preconditioners based on aggregation. *Mathematical Modelling and Numerical Analysis*, 38(5):765–780, 2004.
21. B. Smith, P. Bjorstad, and W. Gropp. *Domain Decomposition, Parallel Multilevel Methods for Elliptic Partial Differential Equations*. Cambridge University Press, New York, 1996.
22. X.-H. Sun. Scalability versus execution time in scalable systems. *J. Parallel Distrib. Comput.*, 62(2):173–192, 2002.
23. X.-H. Sun and D. Rover. Scalability of parallel algorithm-machine combinations. *IEEE Parallel Distrib. Systems*, 5:599–613, June 1994.
24. P. L. Tallec. Domain decomposition methods in computational mechanics. *Computational Mechanics Advances*, 1:121–220, 1994.
25. A. Toselli and O. Widlund. *Domain Decomposition Methods - Algorithms and Theory*, volume 34 of *Springer Series in Computational Mathematics*. Springer-Verlag, New York, 2005.
26. H. van der Vorst. *Iterative Krylov Methods for Large Linear Systems*, volume 13 of *Cambridge Monographs on Applied and Computational Mathematics*. Cambridge University Press, Cambridge, 2003.
27. H. van der Vorst and C. Vuik. GMRESR: a family of nested GMRES methods. *Num. Lin. Alg. Appl.*, 1:369–386, 1994.
28. P. Vanek, J. Mandel, and M. Brezina. Algebraic multigrid based on smoothed aggregation for second and fourth order problems. *Computing*, 56:179–196, 1996.
29. U. M. Yang. Parallel algebraic multigrid methods - high performance preconditioners. In A. M. Bruaset and A. Tveito, editors, *Numerical Solution of Partial Differential Equations on Parallel Computers*, volume 51 of *Lecture Notes in Computational Science and Engineering*, pages 209–236. Springer-Verlag, 2005.

5

Parallel Geometric Multigrid

Frank Hülsemann[1], Markus Kowarschik[1], Marcus Mohr[2], and Ulrich Rüde[1]

[1] System Simulation Group, University of Erlangen, Germany
`[frank.huelsemann,markus.kowarschik,ulrich.ruede]@cs.fau.de`
[2] Department for Sensor Technology, University of Erlangen, Germany
`marcus.mohr@lse.eei.uni-erlangen.de`

Summary. Multigrid methods are among the fastest numerical algorithms for the solution of large sparse systems of linear equations. While these algorithms exhibit asymptotically optimal computational complexity, their efficient parallelisation is hampered by the poor computation-to-communication ratio on the coarse grids. Our contribution discusses parallelisation techniques for geometric multigrid methods. It covers both theoretical approaches as well as practical implementation issues that may guide code development.

5.1 Overview

Multigrid methods are among the fastest numerical algorithms for solving large sparse systems of linear equations that arise from appropriate discretisations of elliptic PDEs. Much research has focused and will continue to focus on the design of multigrid algorithms for a variety of application areas. Hence, there is a large and constantly growing body of literature. For detailed introductions to multigrid we refer to the earlier publications [7, 32] and to the comprehensive overview provided in [60]. A detailed presentation of the multigrid idea is further given in [10]. A long list of multigrid references, which is constantly being updated, can be found in the BibTEX file `mgnet.bib` [16].

Multigrid methods form a family of iterative algorithms for large systems of linear equations which is characterised by asymptotically optimal complexity. For a large class of elliptic PDEs, multigrid algorithms can be devised which require $\mathcal{O}(n)$ floating-point operations in order to solve the corresponding linear system with n degrees of freedom up to discretisation accuracy. In the case of parabolic PDEs, appropriate discretisations of the time derivative lead to series of elliptic problems. Hence, the application of multigrid methods to parabolic equations is straightforward.

Various other linear solvers such as Krylov subspace methods (e.g., the method of conjugate gradients and GMRES), for example, mainly consist of matrix-vector products as well as inner products of two vectors [23, 33]. Their parallel implementation is therefore quite straightforward. The parallelisation of multigrid algorithms tends to be more involved. This is primarily due to the necessity to handle problems of different mesh resolutions which thus comprise significantly varying numbers of unknowns.

In this chapter, we will focus on parallelisation approaches for geometric multigrid algorithms; see also [15, 45] and the tutorial on parallel multigrid methods by Jones that can be found at `http://www.mgnet.org`. We assume that each level of accuracy is represented by a computational grid which is distributed among the parallel resources (i.e., the processes) of the underlying computing environment. We further suppose that the processes communicate with each other via message passing.

Our chapter is structured as follows. In Section 5.2 we will present a brief and general introduction to geometric multigrid schemes. Section 5.3 describes elementary parallelisation techniques for multigrid algorithms. The case of multigrid methods for applications involving unstructured finite-element meshes is more complicated than the case of structured meshes and will be addressed subsequently in Section 5.4. Section 5.5 focuses on the optimisation of the single-node performance, which primarily covers the improvement of the utilisation of memory hierarchies. Advanced parallelisation approaches for multigrid will be discussed afterwards in Section 5.6. Conclusions will be drawn in Section 5.7.

5.2 Introduction to Multigrid

5.2.1 Overview

Generally speaking, all multigrid algorithms follow the same fundamental design principle. A given problem is solved by integrating different levels of resolution into the solution process. During this process, the contributions of the individual levels are combined appropriately in order to form the required solution.

In the classical sense, multigrid methods involve a hierarchy of computational grids of different mesh resolution and can therefore be considered geometrically motivated. This approach has led to the notion of *geometric multigrid (GMG)*. In contrast, later research has additionally addressed the development and the analysis of *algebraic multigrid (AMG)* methods, which target a multigrid-like iterative solution of linear systems without using geometric information from a grid hierarchy, but only the original linear system itself[3]. For an introduction to parallel AMG, see [65].

5.2.2 Preparations

We will first motivate the principles of a basic geometric multigrid scheme. For simplicity, we consider the example of a scalar elliptic boundary value problem

$$Lu = f \quad \text{in} \quad \Omega \ , \tag{5.1}$$

defined on the interval of unit length (i.e., $\Omega := (0,1)$), on the unit square (i.e., $\Omega := (0,1)^2$), or on the unit cube (i.e., $\Omega := (0,1)^3$). L denotes a second-order

[3] The term *algebraic multigrid* may thus appear misleading, since an ideal AMG approach would dispense with any computational grid.

linear elliptic differential operator, the solution of (5.1) is denoted as $u : \Omega \to \mathbb{R}$, and the function $f : \Omega \to \mathbb{R}$ represents the given right-hand side.

We assume Dirichlet boundary conditions only; i.e.,

$$u = g \quad \text{on} \quad \partial\Omega \ . \tag{5.2}$$

We concentrate on the case of an equidistant regular grid. As usual, we use h to denote the mesh width in each dimension. Hence, $n_{\text{dim}} := h^{-1}$ represents the number of sub-intervals per dimension. In the 1D case, the grid nodes are located at positions

$$\{x = ih;\ 0 \le i \le n_{\text{dim}}\} \subset [0,1] \ .$$

In the 2D case, they are located at positions

$$\{(x_1, x_2) = (i_1 h, i_2 h);\ 0 \le i_1, i_2 \le n_{\text{dim}}\} \subset [0,1]^2 \ ,$$

and in the 3D case, the node positions are given by

$$\{(x_1, x_2, x_3) = (i_1 h, i_2 h, i_3 h);\ 0 \le i_1, i_2, i_3 \le n_{\text{dim}}\} \subset [0,1]^3 \ .$$

Consequently, the grid contains $n_{\text{dim}} + 1$ *nodes* per dimension. Since the outermost grid points represent Dirichlet boundary nodes, the corresponding solution values are fixed. Hence, since u is a scalar function, our grid actually comprises $n_{\text{dim}} - 1$ *unknowns* per dimension.

Our presentation is general enough to cover the cases of finite differences as well as finite element discretisations involving equally sized line elements in 1D, square elements in 2D, and cubic elements in 3D, respectively. For the description of the core concepts, we focus on the case of *standard coarsening* only. This means that the mesh width H of any coarse grid is obtained as $H = 2h$, where h denotes the mesh width of the respective next finer grid. See [60] for an overview of alternative coarsening strategies such as red-black coarsening and semi-coarsening, for example. In Section 5.4, we turn to the question of multigrid methods on unstructured grids.

The development of multigrid algorithms is motivated by two fundamental and independent observations which we will describe in the following; the equivalence of the original equation and the residual equation as well as the convergence properties of basic iterative solvers.

5.2.3 The Residual Equation

A suitable discretisation of the continuous problem given by (5.1), (5.2) yields the linear system

$$A_h u_h = f_h \ , \tag{5.3}$$

where A_h denotes a sparse nonsingular matrix that represents the discrete operator. For the model case of a second-order finite difference discretisation of the negative Laplacian in 2D, A_h is characterised by the five-point stencil

$$\frac{1}{h^2}\begin{bmatrix} & -1 & \\ -1 & 4 & -1 \\ & -1 & \end{bmatrix} .$$

We refer to [60] for details including a description of this common stencil notation.

The exact solution of (5.3) is explicitly denoted as u_h^*, while u_h stands for an approximation to u_h^*. If necessary, we add superscripts to specify the iteration index; e.g., $u_h^{(k)}$ is used to denote the k-th iterate, $k \geq 0$. In the following, we further need to distinguish between approximations on grid levels with different mesh widths. Therefore, we use the indices h and H to indicate that the corresponding quantities belong to the grids of sizes h and H, respectively. The solutions of the linear systems under consideration represent function values located at discrete grid nodes. As a consequence, the terms *grid function* and *vector* are used interchangeably hereafter.

As usual, the *residual* r_h corresponding to the approximation u_h is defined as

$$r_h := f_h - A_h u_h \ . \tag{5.4}$$

The *(algebraic) error* e_h corresponding to the current approximation u_h is given by

$$e_h := u_h^* - u_h \ . \tag{5.5}$$

From these definitions, we obtain

$$A_h e_h = A_h (u_h^* - u_h) = A_h u_h^* - A_h u_h = f_h - A_h u_h = r_h \ ,$$

which relates the current error e_h to the current residual r_h. Hence, the *residual (defect) equation* reads as

$$A_h e_h = r_h \ . \tag{5.6}$$

Note that (5.6) is equivalent to (5.3), and the numerical solution of both linear systems is equally expensive since they involve the same system matrix A_h. The actual motivation for these algebraic transformations is not yet obvious and will be provided subsequently. In the following, we will briefly review the convergence properties of elementary iterative schemes and then illustrate the multigrid principle.

5.2.4 Convergence Behaviour of Elementary Iterative Methods

There is a downside to all elementary iterative solvers such as Jacobi's method, the method of Gauß-Seidel, and SOR. Generally speaking, when applied to large sparse linear systems arising in the context of numerical PDE solution, they cannot efficiently reduce the *slowly oscillating (low-frequency, smooth)* discrete Fourier components of the algebraic error. However, they often succeed in efficiently eliminating the *highly oscillating (high-frequency, rough)* error components [60].

This behaviour can be investigated analytically in detail as long as certain model problems (e.g., involving standard discretisations of the Laplacian) as well as the classical iterative schemes are used. This analysis is based on a decomposition of the initial error $e_h^{(0)}$. This vector is written as a linear combination of the eigenvectors of the corresponding iteration matrix. In the case of the model problems under consideration, these eigenvectors correspond to the discrete Fourier modes.

As long as standard model problems are considered, it can be shown that the spectral radius $\rho(M)$ of the corresponding iteration matrix M behaves like $1-\mathcal{O}(h^2)$ for the method of Gauß-Seidel, Jacobi's method, and weighted Jacobi. Similarly, $\rho(M)$ behaves like $1 - \mathcal{O}(h)$ for SOR with optimal relaxation parameter [57]. This observation indicates that, due to their slow convergence rates, these methods are hardly applicable to large problems involving small mesh widths h.

A closer look reveals that the smooth error modes, which cannot be eliminated efficiently, correspond to those eigenvectors of M which belong to the relatively large eigenvalues; i.e, to the eigenvalues close to 1[4]. This fact explains the slow reduction of low-frequency error modes. In contrast, the highly oscillating error components often correspond to those eigenvectors of M which belong to relatively small eigenvalues; i.e., to the eigenvalues close to 0. As we have mentioned previously, these high-frequency error components can thus often be reduced quickly and, after a few iterations only, the smooth components dominate the remaining error.

Note that whether an iterative scheme has this so-called *smoothing property* depends on the problem to be solved. For example, Jacobi's method cannot be used in order to eliminate high-frequency error modes quickly, if the discrete problem is based on a standard finite difference discretisation of the Laplacian, see [60]. In this case, high-frequency error modes can only be eliminated efficiently, if a suitable relaxation parameter is introduced; i.e., if the weighted Jacobi scheme is employed instead.

5.2.5 Aspects of Multigrid Methods

Coarse Grid Representation of the Residual Equation

As was mentioned in Section 5.2.4, many basic iterative schemes possess the smoothing property; within a few iterations only, the highly oscillating error components can often be eliminated and the smooth error modes remain. As a consequence, a coarser grid (i.e., a grid with fewer grid nodes) may be sufficiently fine to represent this smooth error accurately enough. Note that, in general, it is the algebraic error (and not the approximation to the solution of the original linear system itself) that becomes smooth after a few steps of an appropriate basic iterative scheme have been performed.

The observation that the error is smooth after a few iterations motivates the idea to apply a few iterations of a suitable elementary iterative method on the respective fine grid. This step is called *pre-smoothing*. Then, an approximation to the remaining smooth error can be computed efficiently on a coarser grid, using a coarsened representation of (5.6); i.e., the residual equation. Afterwards, the smooth error must

[4]Note that non-convergent iterative schemes involving iteration matrices M with $\rho(M) \geq 1$ may be used as smoothers as well.

be interpolated back to the fine grid and, according to (5.5), added to the current fine grid approximation in order to correct the latter.

In the simplest case, the coarse grid is obtained by standard coarsening; i.e., by omitting every other row of fine grid nodes in each dimension, cf. Section 5.2.2. This coarsening strategy results in an equidistant coarse grid with mesh width $H = 2h$. As usual, we use Ω^h and Ω^H to represent the fine grid and the coarse grid, respectively. Furthermore, we assume that n_h and n_H denote the total numbers of unknowns corresponding to the fine grid and the coarse grid, respectively. Note that standard coarsening reduces the number of unknowns by a factor of approximately 2^{-d}, where d is the dimension of the problem.

The coarse representation of (5.6), which is used to approximate the current algebraic fine grid error e_h, reads as

$$A_H e_H = r_H \ , \tag{5.7}$$

where $A_H \in \mathbb{R}^{n_H \times n_H}$ stands for the coarse grid operator and $e_H, r_H \in \mathbb{R}^{n_H}$ are suitable coarse grid representations of the algebraic fine grid error e_h and the corresponding fine grid residual r_h, respectively. Equation (5.7) must be solved for e_H.

Inter-Grid Transfer Operators

The combination of the fine grid solution process and the coarse grid solution process requires the definition of *inter-grid transfer operators*, which are necessary to map grid functions from the fine grid Ω^h to the coarse grid Ω^H, and vice versa. In particular, we need an *interpolation (prolongation) operator*

$$I_H^h : \mathbb{R}^{n_H} \rightarrow \mathbb{R}^{n_h} \ ,$$

which maps a coarse grid function to a fine grid function, as well as a *restriction operator*

$$I_h^H : \mathbb{R}^{n_h} \rightarrow \mathbb{R}^{n_H} \ ,$$

which maps a fine grid function to a coarse grid function. Note that, in the following, I_H^h and I_h^H are also used to denote the corresponding matrix representations of the interpolation and the restriction operators, respectively.

The restriction operator I_h^H is used to transfer the fine grid residual r_h to the coarse grid, yielding the right-hand side of the coarse grid representation (5.7) of the fine grid residual equation:

$$r_H := I_h^H r_h \ .$$

In the case of discrete operators A_h and A_H with slowly varying coefficients, typical choices for the restriction operator are *full weighting* or *half weighting*. These restriction operators compute weighted averages of the components of the fine grid function to be restricted. They do not vary from grid point to grid point. In 2D, for example, they are given as follows:

- Full weighting:

$$\frac{1}{16}\begin{bmatrix} 1 & 2 & 1 \\ 2 & 4 & 2 \\ 1 & 2 & 1 \end{bmatrix}_h^H$$

- Half weighting:

$$\frac{1}{8}\begin{bmatrix} 0 & 1 & 0 \\ 1 & 4 & 1 \\ 0 & 1 & 0 \end{bmatrix}_h^H$$

Here, we have used the common stencil notation for restriction operators. The entries of these stencils specify the weights for the values of the respective grid function, when transferred from the fine grid Ω^h to the corresponding coarser grid Ω^H. This means that the function value at any (interior) coarse grid node is computed as the weighted average of the function values at the respective neighbouring fine grid nodes, see [10] for example. Representations of the full weighting and the half weighting operators in 3D are provided in [60].

After the coarse representation e_H of the algebraic error has been determined, the interpolation operator is employed to transfer e_H back to the fine grid Ω^h:

$$\tilde{e}_h := I_H^h e_H \ .$$

We use $\tilde{e}_h$ to denote the resulting fine grid vector, which is an approximation to the actual fine grid error e_h. Ideally, the smoother would yield a fine grid error e_h which lies in the range of I_H^h such that it could be eliminated completely by the correction $\tilde{e}_h$.

A typical choice for the prolongation operator is *linear interpolation*. In 2D, for example, this constant operator is given as follows:

$$\left]\frac{1}{4}\begin{matrix} 1 & 2 & 1 \\ 2 & 4 & 2 \\ 1 & 2 & 1 \end{matrix}\right[_H^h$$

Here, we have employed the common stencil notation for interpolation operators. The entries of the interpolation stencils specify the weights for the values of the respective grid function, when prolongated from the coarse grid Ω^H to the corresponding finer grid Ω^h. This means that the function value at any (interior) coarse grid node is propagated to the respective neighbouring fine grid nodes using these weights [10]. A representation of the linear interpolation operator in 3D is again provided in [60].

If the stencil coefficients vary significantly from grid node to grid node, it may be necessary to employ operator-dependent inter-grid transfer operators, which do not just compute weighted averages when mapping fine grid functions to the coarse grid, and vice versa [1]. See also [8] for a discussion of how to select appropriate restriction and prolongation operators depending on the order of the differential operator L in (5.1).

Ideally, the high-frequency components dominate the fine grid error after the coarse grid representation e_H of the error has been interpolated to the fine grid Ω^h and added to the current fine grid approximation u_h; i.e., after the *correction*

$$u_h \leftarrow u_h + \tilde{e}_h$$

has been carried out. In addition, the interpolation of the coarse grid approximation e_H to the fine grid Ω^h usually even amplifies oscillatory error components. Therefore, a few further iterations of the smoother are typically applied to the fine grid solution u_h. This final step is called *post-smoothing*.

Note that the discrete operator and the stencils that represent the inter-grid transfer operators are often characterised by *compact stencils*. In 2D, this means that each of these stencils only covers the current node and its eight immediate neighbours in the grid. In a regular 3D grid, each interior node has 26 neighbours instead. This property of compactness simplifies the parallelisation of the multigrid components. We will return to this issue in Section 5.3 in the context of grid partitioning.

Coarse Grid Operators

The coarse grid operator A_H can be obtained by discretising the continuous differential operator L from (5.1) on the coarse grid Ω^H anew. Alternatively, A_H can be computed as the so-called *Galerkin product*

$$A_H := I_h^H A_h I_H^h \ . \tag{5.8}$$

An immediate observation of this choice is the following. If the fine grid operator A_h as well as the inter-grid transfer operators I_h^H and I_H^h are characterised by compact stencils (i.e., 3-point stencils in 1D, 9-point stencils in 2D, or 27-point stencils in 3D) the resulting coarse grid operator A_H will be given by corresponding compact stencils as well. In a multigrid implementation, this property enables the use of simple data structures and identical parallelisation strategies on all levels of the grid hierarchy, see Section 5.3.

Note that, if the restriction operator corresponds to the transpose of the interpolation operator (up to a constant factor), and if (5.8) holds, a symmetric fine grid operator A_h yields a symmetric coarse grid operator A_H. If A_h is even symmetric positive definite and the interpolation operator I_H^h has full rank, the corresponding coarse grid operator A_H will again be symmetric positive definite.

As a consequence, if the matrix corresponding to the finest grid of the hierarchy is symmetric positive definite (and therefore nonsingular) and, furthermore, both the inter-grid transfer operators and the generation of the coarse grid matrices are chosen appropriately, each of the coarse grid matrices will be symmetric positive definite as well. This property of the matrix hierarchy often simplifies the analysis of multigrid methods.

Algorithm 5.1 Recursive definition of the multigrid CGC V(ν_1,ν_2)-cycle.

1: Perform ν_1 iterations of the smoother on Ω^h (pre-smoothing):

$$u_h^{(k+\frac{1}{3})} \leftarrow S_h^{\nu_1}\left(u_h^{(k)}\right)$$

2: Compute the residual on Ω^h:

$$r_h \leftarrow f_h - A_h u_h^{(k+\frac{1}{3})}$$

3: Restrict the residual from Ω^h to Ω^H and initialise the coarse grid approximation:

$$f_H \leftarrow I_h^H r_h \;,\quad u_H \leftarrow 0$$

4: **if** Ω^H is the coarsest grid of the hierarchy **then**
5: Solve the coarse grid equation $A_H u_H = f_H$ on Ω^H exactly
6: **else**
7: Solve the coarse grid equation $A_H u_H = f_H$ on Ω^H approximately by (recursively) performing a multigrid V(ν_1,ν_2)-cycle starting on Ω^H
8: **end if**
9: Interpolate the coarse grid approximation (i.e., the error) from Ω^H to Ω^h:

$$\tilde{e}_h \leftarrow I_H^h u_H$$

10: Correct the fine grid approximation on Ω^h:

$$u_h^{(k+\frac{2}{3})} \leftarrow u_h^{(k+\frac{1}{3})} + \tilde{e}_h$$

11: Perform ν_2 iterations of the smoother on Ω^h (post-smoothing):

$$u_h^{(k+1)} \leftarrow S_h^{\nu_2}\left(u_h^{(k+\frac{2}{3})}\right)$$

Formulation of the Multigrid V-Cycle Correction Scheme

The previous considerations first lead to the two-grid *coarse grid correction (CGC)* V-cycle. This scheme assumes that, in each iteration, the coarse grid equation is solved exactly.

If, however, this linear system is still too large to be solved efficiently by using either a direct method or an elementary iterative method, the idea of applying a coarse grid correction (i.e., the idea of solving the corresponding residual equation on a coarser grid) can be applied recursively. This then leads to the class of multigrid schemes. See [10, 60], for example.

Algorithm 5.1 shows the structure of a single multigrid V(ν_1,ν_2)-cycle. The notation $S_h^{\nu}(\cdot)$ is introduced to indicate that ν iterations of an appropriate smoothing method are applied to the corresponding approximation on Ω^h. The parameters ν_1 and ν_2 denote the numbers of iterations of the smoother before and after the coarse grid correction, respectively. Typical values for ν_1 and ν_2 are 1, 2, or 3.

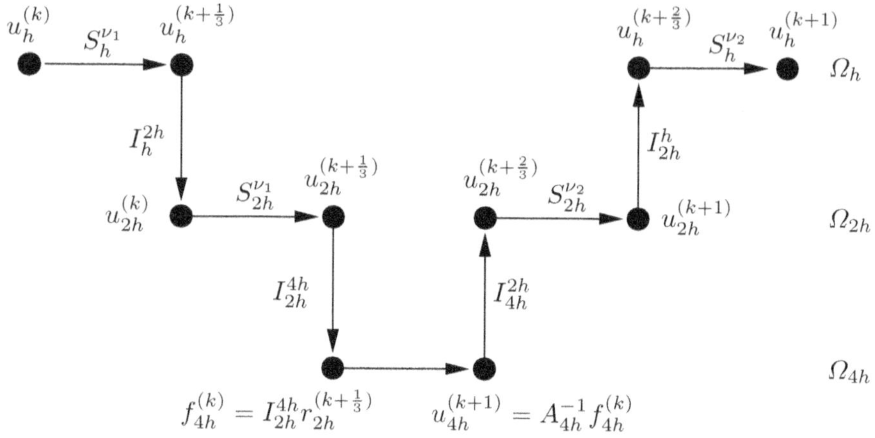

Fig. 5.1. Three-grid CGC V(ν_1,ν_2)-cycle.

Due to the recursive formulation of Algorithm 5.1, it is sufficient to distinguish between a fine grid Ω^h and a coarse grid Ω^H. When the recursive scheme calls itself in Step 7, the current coarse grid Ω^H becomes the fine grid of the next deeper invocation of the multigrid V-cycle procedure. Typically, $u_H := 0$ is used as initial guess on Ω^H. We assume that the initial guess and the right-hand side on the finest grid level as well as the matrices on all grid levels of the hierarchy have been initialised beforehand.

An additional parameter γ may be introduced in order to increase the number of multigrid cycles to be executed on the coarse grid Ω^H in Step 7 of Algorithm 5.1. This parameter γ is called the *cycle index*. The choice $\gamma := 1$ (as is implicitly the case in Algorithm 5.1) leads to multigrid V(ν_1,ν_2)-cycles, while different cycling strategies are possible. Another common choice is $\gamma := 2$, which leads to the multigrid W-cycle [60], see also Section 5.3.4. The names of these schemes are motivated by the order in which the various grid levels are visited during the multigrid iterations.

Figure 5.1 shows the algorithmic structure of a three-grid CGC V(ν_1,ν_2)-cycle. This figure illustrates the origin of the term V-cycle. We have used the level indices h, $2h$, and $4h$ in order to indicate that we generally assume the case of standard coarsening, see above.

Remarks on Multigrid Convergence Analysis

A common and powerful approach towards the quantitative convergence analysis (and the development) of multigrid methods is based on *local Fourier analysis (LFA)*. The principle of the LFA is to examine the impact of the discrete operators, which are involved in the two-grid or in the multigrid setting, by representing them in the basis of the corresponding Fourier spaces. The LFA ignores boundary conditions and, instead, assumes infinite grids [60, 62].

Alternatively, the convergence analysis of the two-grid scheme and the multigrid scheme can be based on the notions of *smoothing property* and *approximation property*, which have been introduced by Hackbusch [32]. As these names suggest, the smoothing property states that the smoother eliminates high-frequency error components without introducing smooth ones. In contrast, the approximation property states that the CGC performs efficiently; i.e., that the inverse of the coarse grid operator represents a reasonable approximation to the inverse of the fine grid operator. In comparison with the aforementioned LFA, the current approach only yields qualitative results.

The convergence analysis of multigrid methods reveals for example that, for W-cycle schemes applied to certain model problems, these algorithms behave asymptotically optimal. It can be shown that, for these cases, multigrid W-cycles only require $\mathcal{O}(N \log \epsilon)$ operations, where N denotes the number of unknowns corresponding to the finest grid level and ϵ stands for the required factor by which the norm of the algebraic error shall be improved, see [60].

Full Approximation Scheme

So far, we have considered the CGC scheme, or simply the *correction scheme (CS)*. This means that, in each multigrid iteration, any coarse grid is employed to compute an approximation to the error on the next finer grid.

Alternatively, the coarse grid can be used to compute an approximation to the fine grid solution instead. This approach leads to the *full approximation scheme/storage (FAS) method*. The FAS method is primarily used, if the discrete operator is nonlinear or if adaptive grid refinement is introduced. In the latter case, the finer grids may not cover the entire domain in order to reduce both memory consumption and computational work.

It can be shown that, for the nonlinear case, the computational efficiency of the FAS scheme is asymptotically optimal, as is the case for the aforementioned correction scheme. In addition, the parallelisation of the FAS method resembles the parallelisation of the correction scheme. See [7, 60] for details on the FAS method.

Nested Iteration and Full Multigrid

In most cases, the solution times of iterative methods (i.e., the numbers of iterations required to fulfil the given stopping criteria) can be reduced drastically by choosing suitable initial guesses. When applied recursively, the idea of determining an approximation on a coarse grid first and interpolating this approximation afterwards in order to generate an accurate initial guess on a fine grid leads to the principle of *nested iteration* [33].

The combination of nested iteration and the multigrid schemes we have described so far leads to the class of *full multigrid (FMG) methods*, which typically represent the most efficient multigrid algorithms. For typical problems, the computational work required to solve the discrete problem on the finest grid level up to discretisation accuracy is of order $\mathcal{O}(N)$ only. This results from the observation that, on each

Algorithm 5.2 Recursive formulation of the FMG scheme on Ω^h.

1: **if** Ω^h is the coarsest grid of the hierarchy **then**
2: Solve $A_h u_h = f_h$ on Ω^h exactly
3: **else**
4: Restrict the right-hand side from Ω^h to the next coarser grid Ω^H:

$$f_H \leftarrow I_h^H f_h$$

5: Solve $A_H u_H = f_H$ using FMG on Ω^H recursively
6: Interpolate the coarse grid approximation from Ω^H to Ω^h in order to obtain a good initial guess on Ω^h:

$$u_h^{(0)} \leftarrow \tilde{I}_H^h u_H$$

7: Improve the approximation on Ω^h by applying ν_0 multigrid iterations:

$$u_h \leftarrow \mathrm{MG}_{\nu_1,\nu_2}^{\nu_0}\left(u_h^{(0)}, A_h, f_h\right)$$

8: **end if**

level of the grid hierarchy, a constant number of V-cycles is sufficient to solve the corresponding linear system up to discretisation accuracy. See [8, 60] for details.

The FMG scheme generally starts on the coarsest level of the grid hierarchy. There, an approximation to the solution is computed and then interpolated to the next finer grid, yielding a suitable initial guess on this next finer grid. A certain number of multigrid cycles (either CGC-based or FAS-based) is applied to improve the approximation, before it is in turn interpolated to the next finer grid, and so on.

As Algorithm 5.2 shows, the FMG scheme can be formulated recursively. The linear system on the coarsest level of the grid hierarchy is assumed to be solved exactly. Since the approximations which are mapped from coarse to fine grids in Step 6 are not necessarily smooth and potentially large, it is commonly recommended to choose an interpolation operator $\tilde{I}_H^h$ of sufficiently high order [60].

Depending on the actual problem, the multigrid method applied in Step 7 of Algorithm 5.2 may either be based on the CGC scheme or on the FAS method. It may involve either V(ν_1,ν_2)-cycles or W(ν_1,ν_2)-cycles. The notation we have introduced in Step 7 is supposed to indicate that ν_0 multigrid cycles are performed on the linear system involving the matrix A_h and the right-hand side f_h, starting with the initial guess $u_h^{(0)}$ which has been determined previously by interpolation in Step 6.

Note that, in order to compute an approximation to the actual solution on each of the coarse grids, it is necessary to appropriately represent the original right-hand side; i.e., the right-hand side on the finest grid level. These coarse grid right-hand sides are needed whenever the corresponding coarse grid level is visited for the first time and, in the case of the FAS method, during the multigrid cycles as well. They can either be determined by successively restricting the right-hand side from the finest grid (as is the case in Algorithm 5.2, Step 4) or, alternatively, by discretising the continuous right-hand side (i.e., the function f in (5.1)) on each grid level anew [60].

5.3 Elementary Parallel Multigrid

In this section, we will introduce the basic concepts used in parallelising the geometric multigrid method.

5.3.1 Grid Partitioning

We begin with an outline of the parallelisation of a standard (geometric) multigrid method, as introduced in Section 5.2. In the simplest case, computations are performed on a hierarchy of L grids of $m_l \times n_l$ $(\times o_l)$ grid lines for 2D (3D) problems and for $1 \leq l \leq L$. A vertex-centred discretisation will associate unknowns with the grid vertices and the grid edges represent data dependencies (though not necessarily all) induced by the discrete equations on level l when applying a stencil operation.

Our parallel machine model is motivated by current cluster architectures. In particular, we assume a distributed memory architecture and parallelisation by message passing. The most common message passing standard, today, is the *message passing interface (MPI)*; see [29, 30], for example. Each compute node in this setting may itself be a (shared memory) multiprocessor.

In the following, we will adopt standard message passing terminology and use the notion of a *process* rather than a *processor*. In a typical cluster environment each processor of a compute node will execute one process. The message passing paradigm, however, also allows for situations where several processes are executed by a single processor. In a parallel environment with a number of compute nodes that is significantly smaller than the number of unknowns, the typical approach to parallelise any grid-based algorithm is to split the grid into several parts or *sub-grids* and assign each sub-grid and all of its grid nodes to one process. This approach is for obvious reasons denoted as *grid partitioning* or *domain partitioning*; see for example [41].

Note that grid (domain) partitioning is sometimes also referred to as *domain decomposition*. This terminology is actually misleading, since domain decomposition denotes a special solution approach. Classical domain decomposition algorithms include a subdomain solution phase in which each process solves a problem that is completely independent of the problems in the remaining sub-domains. The solution of the original problem posed on the global domain is then obtained by iteratively adapting the local problem on any sub-domain based on the solutions computed by other processes on the neighbouring sub-domains. The solution of the local problems does not induce a need for communication between the processes. This is restricted to the outer iteration in the adaptation phase. If the coupling is done in an appropriate way, the local solutions will converge to the global solution restricted to the local sub-domains. Thus, domain decomposition actually is a concept to design a new parallel algorithm using existing sequential ones as building blocks. For further details, we refer to [19].

Grid (domain) partitioning, in contrast, denotes a strategy to parallelise an existing sequential grid-based algorithm. Here we will consider the parallelisation of the (geometric) multigrid algorithm. In this case, there exist no independent local problems on the individual sub-domains and the execution of the algorithm itself induces the need to communicate data between the different processes.

Before we consider this point of interdependence further, we briefly introduce some notation. Let us denote by Ω^l the grid on level l, by Ω^l_k its k-th sub-grid, by $n^l_j, 1 \leq j \leq N^l$ a node of Ω^l, and by p^l_k the process responsible for sub-grid Ω^l_k.

As was illustrated in Section 5.2, all operations in a structured geometric multigrid method can in principle be expressed with the help of stencil operators that are applied to a grid function. Such stencils often have a compact support. Hence, when they are locally applied at a certain node $n^l_j \in \Omega^l_k$, this only involves values of the grid function at nodes that are neighbours of the node n^l_j. If the node n^l_j in question is located inside the sub-grid Ω^l_k, then the application of a stencil can be performed by process p^l_k independently. However, if n^l_j is lying in the vicinity of the sub-domain boundary, then the application of a stencil may involve nodes that belong to neighbouring sub-domains and are thus not stored by process p^l_k.

The straightforward solution to this problem is to let p^l_k query its neighbours for the node values involved whenever it requires them. This, however, must be ruled out for efficiency reasons. Instead, the following approach is typically employed.

Each sub-domain Ω^l_k is augmented by a layer of so-called *ghost nodes* that surround it. This layer is denoted as *overlap region* but other names such as *halo* are also often used. The ghost nodes in the overlap region correspond to nodes in neighbouring sub-domains that p^l_k must access in order to perform computations on Ω^l_k. The width of the overlap region is therefore determined by the extent of the stencil operators involved. Figures 5.2 and 5.3 give two examples; one for a structured and one for an unstructured grid. Note the difference in the size of the four subdomains in Figure 5.2. This will result in some load imbalance, but is inevitable for the $(2^k + 1) \times (2^m + 1)$ grids typically employed in geometric multigrid applications. In Section 5.6.2 we will present a partitioning approach that avoids this difficulty.

The ghost nodes around each subdomain can be considered as a kind of read-only nodes. This means that process p^l_k will never change the respective function values directly by performing computations on them, but will only access their values when performing computations on its own nodes.

Keeping the values in the overlap region up-to-date requires communication with the neighbouring processes. As an example, Figure 5.4 shows a strategy for updating the overlap regions in a partitioning with four sub-grids and also demonstrates how "diagonal" communication (in this example communication between the lower-left and top-right process as well as between the top-left and bottom-right process) can implicitly be avoided. The precise details of such an update and its frequency depend on the multigrid component involved and will be considered in the following sections.

When it can be assured that the function value at a ghost node g^l_j in the overlap region of Ω^l_k is always identical to the function value at its master node n^l_j when it is read by p^l_k, the computation will yield the same result as in the sequential case. One of the major challenges in designing parallel geometric multigrid methods is to

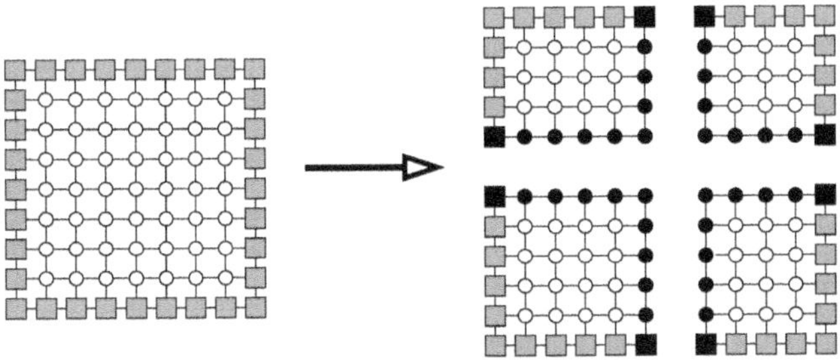

Fig. 5.2. Partitioning of a structured grid into four sub-domains with an overlap region of width one. We distinguish: inner nodes (*white circles*), boundary nodes (*grey squares*) and ghost nodes (*black circles / squares*).

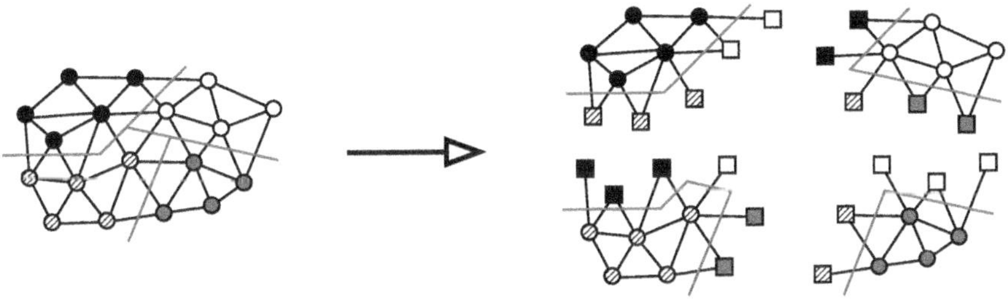

Fig. 5.3. Partitioning of a unstructured grid into four sub-domains with an overlap region of width one. Ghost nodes are coloured according to their corresponding master nodes.

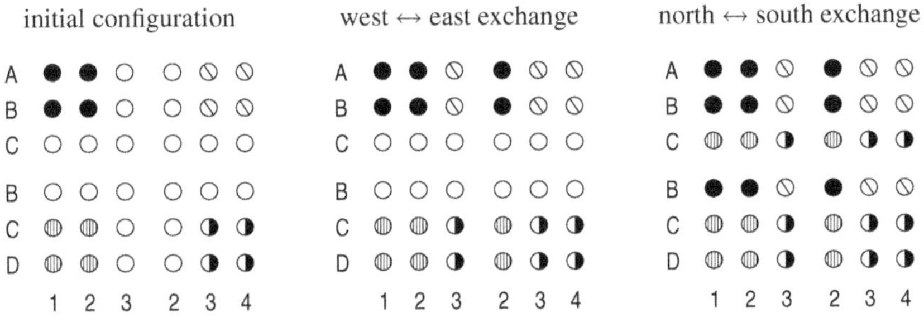

Fig. 5.4. Example update of ghost node values at the intersection of four sub-domains.

employ appropriate multigrid components and to develop suitable implementations, such that this requirement is met with as little communication costs as possible, while on the other side retaining the fast convergence speed and high efficiency achieved by sequential multigrid algorithms.

Another important question is how to choose the grid partitioning. Since the size of the surface of a sub-grid determines the number of ghost nodes in the overlap region, it is directly related to the amount of data that must be transferred and thus to the time spent with communication. This *communication time* is a consequence of the parallel processing. This is complemented by the *computation time*, which comprises all operations that a sequential program would have to carry out for the local data set. In our application, this time is dominated by floating-point operations, hence the name.

As a high computation-to-communication ratio is important for achieving reasonable parallel efficiency, one thus typically strives for a large volume-to-surface ratio when devising the grid partitioning. With structured grid applications, this initially led to the preference of 2D over 1D partitioning for 2D problems and 3D over 2D or 1D partitioning for 3D simulations.

In light of the increasing gap between CPU speed and main memory performance (cf. Section 5.5) the focus has started to change. Not only the amount of data to be communicated is now taken into account, but also the run-time costs for collecting and rearranging them. Here, lower dimensional splittings such as 1D partitionings in 3D, for example, can yield performance benefits. See [53], for example. They may further be advantageous in the case of the application of special smoothers such as line smoothers, cf. Section 5.6.1.

5.3.2 Parallel Smoothers

In this section, we will describe the parallelisation of the classical point-based smoothers that are typically used in geometric multigrid methods. More sophisticated smoothers are considered in Section 5.6.1. Since, in the context of multigrid methods, smoothers always operate on the individual grid levels, we drop the l superscript in this section.

Parallelisation with Read-Only Ghost Nodes

We start our exposition with the simplest smoother, the weighted Jacobi method given by

$$u^{\text{new}}(n_k) = u^{\text{old}}(n_k) - \frac{\omega}{\sigma(0,n_k)} \left[f(n_k) - \sum_{j \in \mathcal{K}(n_k)} \sigma(j,n_k)\, u^{\text{old}}(n_{k+j}) \right] , \tag{5.9}$$

where $\mathcal{K}(n_k)$ denotes the support of the stencil representing the discrete operator at node n_k, $\sigma(j, n_k)$ is the stencil weight for the value at node n_{k+j}, and u and f denote the approximate solution and the discrete right-hand side, respectively. Equation (5.9) implies that the new iterate u^{new} at each node is computed based solely on values of the previous iterate u^{old}. Thus, if the values of u^{old} are up-to-date at the ghost nodes of the overlap region, each process can perform one Jacobi sweep on its sub-domain independently of the remaining processes. Once this is done, one communication phase is required to update the overlap regions again in order to prepare for the next sweep or the next phase of the multigrid algorithm.

Let us now turn to Gauß-Seidel smoothing and its weighted variant; the successive over-relaxation (SOR). The formula for this smoother is given by

$$u^{\text{new}}(n_k) = u^{\text{old}}(n_k) - \frac{\omega}{\sigma(0,n_k)} \left[f(n_k) - \sum_{j \in \mathcal{K}(n_k)} \sigma(j,n_k)\, v(n_{k+j}) \right], \quad (5.10)$$

where the Gauß-Seidel method is obtained for $\omega = 1$. In (5.10), we have $v(n_{k+j}) = u^{\text{old}}(n_{k+j})$, if the node n_{k+j} has not yet been updated during the computation, and $v(n_{k+j}) = u^{\text{new}}(n_{k+j})$, otherwise. Thus, (5.10) shows that in the case of SOR smoothing the ordering in which the nodes are updated may influence the properties and the qualities of the smoother. This turns out to be true and details can be found in [60], for instance.

In a parallel setting this inherently sequential dependency causes some difficulty. For example, assume a splitting of a regular grid into four sub-domains as shown in Figure 5.2. Choosing a lexicographic ordering of the nodes for the update and starting the Gauß-Seidel sweep from the lower left would imply that only the process responsible for the lower left sub-domain could perform any computations until the update reaches the top left sub-domain, and so on. As a consequence, three processes would be idle all the time. A different ordering is therefore required in the parallel setting.

Fortunately, the ordering chosen for most sequential multigrid applications is not a lexicographic one anyway. Instead, a so-called red-black or checker-board ordering is frequently used. This ordering often has superior properties in the multigrid context. It is based on a splitting of the grid nodes into two groups depending on their positions; a red group and a black one. For the typical 5-point stencil representing the discrete Laplacian, see Section 5.2.3, the nodes in each of the groups are completely independent of each other. A red node only depends on black nodes for the update, and vice versa. The advantage of this splitting for parallelisation is that each subdomain can update its local red nodes using values at its black nodes (both local and ghost), without the need for communication during the update. Once this is done, one communication phase is required in order to update the values at the red ghost nodes. Afterwards, the update of the black nodes can proceed, again independently, followed by a second communication phase for updating the black ghost nodes.

Compared to the Jacobi smoother, the SOR smoother thus requires one additional communication step. The total amount of data that needs to be transmitted remains the same, however. For larger discretisation stencils, more colours are needed; a compact 9-point-stencil requires a splitting into four sets of nodes, for example.

Trading Computation for Communication

Another possibility for parallelising the red-black SOR smoother and smoothers with similar dependency patterns is to discard the read-only property of the ghost nodes and to trade computation for communication. Assume that we perform a V(2,0)-cycle, for example. In the first approach, four communication steps are required for the two pre-smoothing steps. Let us denote by the k-th generation of ghost nodes all nodes that are required to update in the SOR sweep the nodes in the (k-1)-th generation, with the nodes in a sub-domain belonging to generation 0. Broadly speaking, this definition implies that one layer of ghost nodes is added for each SOR iteration step that is meant to be performed without communication between the partitions.

If we extend the overlap region to include all ghost nodes up to fourth generation, we can perform the four partial SOR sweeps without communication. This is possible, since in the overlap region the same values are computed as in the neighbouring sub-domains. We must take into account, however, that each partial sweep invalidates one generation of ghost nodes; they cannot be updated anymore because their values would start to differ from those at the master nodes that they should mirror.

This second approach increases both the overlap region (and thus the amount of transferred data) as well as the computational work. Therefore, its advantage strongly depends on the parallel architecture employed for the simulation. In the first place, it is the ratio of the costs of floating-point computation, assembly of data into a message, transfer of a message depending on its size, and initialising a communication step between processes that are important here. Moreover, the size of the individual sub-domains plays an essential role as well. We will return to the latter aspect in Section 5.3.4.

Hybrid Smoothers

Let us conclude this subsection by mentioning another general concept for parallelising multigrid smoothers, which is known under the notion of *hybrid smoothing* [43]. See also [65] in this respect. The underlying idea is simple. One does not try to parallelise the chosen smoother, maintaining all data dependencies. Instead, this smoother is applied on each sub-domain independently of the other ones, and the values at the ghost nodes are only updated after each sweep. Hence, this scheme corresponds to an inexact block-Jacobi method. The impact of this approach on the smoothing property and the convergence speed of the multigrid method is, of course, strongly problem-dependent. It can be quite negative, though, and one must be cautious with this approach. However, recently approaches have been developed to improve this concept by the use of suitable weighting parameters; see [65] for more details.

5.3.3 Parallel Transfer Operators

As was introduced in Section 5.2, the use of grids of different resolutions in geometric multigrid necessitates the use of inter-grid transfer operators. These allow to restrict a grid function from the function space associated with a fine grid Ω^l to the space associated with a coarser grid Ω^{l-1} and to prolongate a grid function from Ω^{l-1} to Ω^l.

Both restriction and prolongation operators can typically be expressed by compact local stencils and, as with the Jacobi smoother of the previous section, the order in which the nodes are treated does not influence the final result. Using an overlap region of sufficient width, the multigrid transfer operators can be parallelised easily. Note, however, that using a vertex-centred discretisation, the subdomain Ω_k^{l-1} and Ω_k^l may cover different areas. An example is given in Figure 5.5, where a 1D prolongation by (piece-wise) linear interpolation is sketched. The overlap regions of the

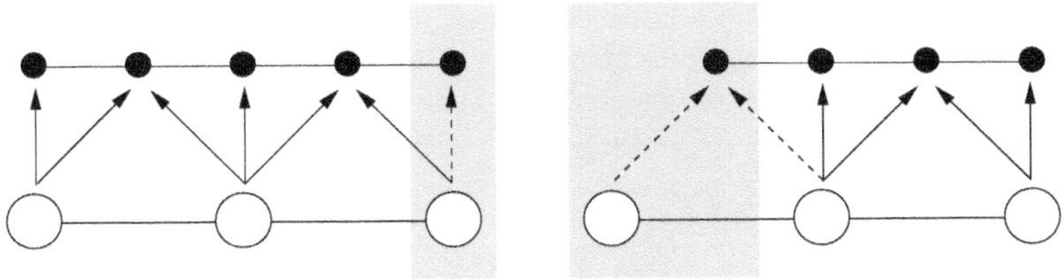

Fig. 5.5. Prolongation at the interface between two sub-domains in 1D.

two sub-domains are shaded in grey. Thus, some care must be taken when performing prolongation and restriction operations at the interfaces of the sub-domains.

5.3.4 Parallel Multigrid Cycles

Multigrid methods may employ different cycling strategies, the most prominent being the V- and the W-cycle. In principle, a parallel multigrid method can also be used with different cycling strategies. However, again some extra issues come into play in the parallel setting. One of the questions in this respect is that of parallel efficiency. As was already mentioned in Section 5.3.1, the latter is coupled to the ratio of computation time to communication time. In order to illustrate how this ratio changes from one grid level to another, we have to determine the dominating factors for the two parts.

In our numerically intensive case, it is safe to assume that the computation time is linearly proportional to the number of nodes in a sub-domain Ω_k^l. For the communication part, we ignore for the moment the impact of message latencies and assume that the communication time is linearly proportional to the number of ghost nodes in the overlap region of Ω_k^l. In practice, the latter can be well approximated by a multiple of the number of nodes on the sub-domain boundary. Thus, the ratio of communication time to computation time is directly proportional to the surface-to-volume ratio of the sub-domain.

As an example, Table 5.1 presents the values of the surface-to-volume ratio for typical square sub-domains in 2D and cubic sub-domains in 3D. We assume a standard coarsening by doubling the mesh width in each dimension, see Figure 5.6, and an overlap region of width 1. The example clearly shows that the surface-to-volume ratio on coarser grids is less favourable than on finer grids. Similar results hold for other partition geometries.

Depending on the representation of the discrete operator, partition sizes of a million or more unknowns for a scalar PDE are not uncommon. In 3D, the surface-to-volume ratio for a single grid algorithm operating on a partition consisting of one cube with two million unknowns is in the order of 5%. Put differently, there are twenty times more compute nodes than ghost nodes. However, on clusters with high performance processors but slow networks, even this factor of 20 may not be sufficient to hide the time for data exchange behind the time for the floating-point operations by using the technique of overlapping communication and computation

Table 5.1. Examples of surface-to-volume ratios in two and three spatial dimensions. l denotes the grid level in the hierarchy, $B(l)$ the number of ghost nodes for a sub-domain Ω_k^l, $V(l)$ the number of nodes in Ω_k^l and r is the quotient of $B(l)$ divided by $V(l)$.

	2D			3D		
l	$B(l)$	$V(l)$	r	$B(l)$	$V(l)$	r
1	8	1	8.00	26	1	26.0
2	16	9	1.77	98	27	3.63
3	32	49	0.65	386	343	1.13
4	64	225	0.28	1,538	3,375	0.46
5	128	961	0.13	6,146	29,791	0.21
6	256	3,969	0.06	24,578	250,047	0.10
7	512	16,129	0.03	98,306	2,048,383	0.05
8	1024	65,025	0.02	393,218	16,581,375	0.02
9	2048	261,121	0.01	1,572,866	133,432,831	0.01
10	4096	1,046,529	0.004	6,291,458	1,070,599,167	0.006

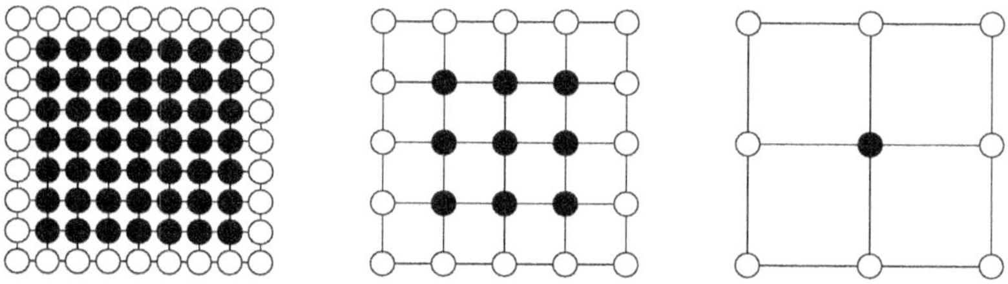

Fig. 5.6. Example of full coarsening in 2D. Hollow circles indicate ghost nodes, full circles denote the unknowns in the local partition.

discussed in Section 5.6.3. In 2D, the surface-to-volume ratio of a square shaped partition with two million unknowns is more favourable than in 3D at the same number of unknowns.

In the case of multigrid algorithms, the network performance becomes even more important. One aim of the sequential multigrid algorithm is to perform as little work as possible on the finest grid and to do as much work as possible on the coarser levels. The assumption behind this idea is that operations such as smoothing steps are much cheaper in terms of computing time on coarser grids than on finer ones. However, this is not necessarily valid in the parallel setting. Given that network data transfer is still significantly slower than accesses to main memory, it is safe to assume that, for instance, smoothing the 27 interior points on level 2 in a cube takes less time than communicating the 26 ghost values for the single unknown on grid level 1. Thus, in this case, it is less time-consuming to keep working on a finer level than to delegate the work to a coarser level, provided the work on the finer level is similarly efficient to solve the problem.

Depending on the processor and network specifications, the same reasoning may actually hold true for other levels as well, so that it may well be faster to stop the multigrid recursion at level three or level four, say, or to use fewer processes for the computations on the coarser grid levels. The problem then is to find an appropriate solver for the remaining coarsest level in terms of computation and communication.

Let us further examine the communication demands of the grid hierarchy in a parallel multigrid method from another perspective. Consider a 3D scalar elliptic PDE on a grid with $N = n^3$ unknowns. As can be seen from Section 5.2, the computational cost for a single V- ($\gamma = 1$) or W-cycle ($\gamma = 2$) in this case is of order $\mathcal{O}(N) = \mathcal{O}(n^3)$. The latter can be obtained from estimating the geometric series

$$n^3 + \gamma^1(n/2)^3 + \gamma^2(n/4)^3 + \gamma^3(n/8)^3 + \dots$$

in combination with a linear relationship of the number of nodes per level to the computational work on that level. Assuming that the domain is partitioned into a grid of $p^3 = P$ sub-domains of (nearly) equal size, each of these has $\mathcal{O}((n/p)^3)$ nodes, and therefore a number of ghost nodes of order $\mathcal{O}((n/p)^2)$. The *total* data volume of communication for a V- or W-cycle is then given by

$$\mathcal{V}^{3D}_{\text{comm}} = \mathcal{O}\left(p^3\,\frac{n^2}{p^2}\,(1+\gamma^1/4+\gamma^2/16+\gamma^3/64+\dots)\right) = \mathcal{O}(p\,n^2)\ .$$

This consideration demonstrates that, even asymptotically (i.e., for $n \to \infty$), the aggregate costs of communication of a 3D multigrid V- or W-cycle differ from that of a single data exchange between the partitions on the finest grid only by a constant factor. In 2D, however, one obtains

$$\mathcal{V}^{2D}_{\text{comm}} = \mathcal{O}\left(p^2\,\frac{n}{p}\,(1+\gamma^1/2+\gamma^2/4+\gamma^3/8+\dots)\right)\ .$$

Thus, asymptotically the argument in 2D only holds for the V-cycle, where we have $\mathcal{V}^{2D}_{\text{comm}} = \mathcal{O}(p\,n)$. In contrast, the W-cycle leads to $\mathcal{V}^{2D}_{\text{comm}} = \mathcal{O}(p\,n\log(n))$. In practice, of course, other effects will influence the run-time behaviour making the situation more complicated. The most noticeable additional effect on many current system architectures may be the startup cost *(latency)* of communication. When messages are small, as is the case when data is exchanged between sub-domains on coarse grids, then the time for initiating a message exchange may be so large that it cannot be neglected anymore. In contrast to the volume of communication, the number of messages is independent of the sub-domain size and thus will grow with $\log(n)$ for a standard multigrid method that employs all levels of the hierarchy. Also the commonly applied technique of overlapping communication and computation, see Section 5.6.3, that tries to reduce overall run-time requires that there is enough computational work behind which communication can be hidden. This of course becomes problematic, when sub-domains get too small on coarser grids. Additionally, the aggregate data volume for all processes says little about the time needed for communication. Depending on the network topology and allocation of the processes, network jams may occur, or bandwidth may depend on message sizes, etc.

Finally, it is the run-time of a parallel multigrid algorithm that is of major interest to the common user. The latter may be as much determined by the cost of communication as by the cost of computation, and finding ways to reduce communication cost may be more important than reducing the cost of computation.

As a consequence, the typical rule of thumb in parallel multigrid is to prefer cycling strategies that spend less time on coarser grids (e.g., V-cycles) to those that spend more time on coarser grids (e.g., W-cycles) as far as parallel efficiency is concerned. However, a concrete application problems may still require the use of W-cycles for its increased robustness.

The question of a suitable parallel multigrid cycling strategy is closely related to the question of how deep the grid hierarchy should be chosen. In multigrid, one often tries to coarsen the mesh as far as possible, in the extreme even to a level with only one unknown. It turns out that, in the sequential case, this is in fact a competitive strategy, not alone since the exact solution of the coarsest grid problem in this case can be performed with very small costs. In light of the above discussion, however, the question is whether it is sensible to have grid levels where the number of compute nodes exceeds that of grid nodes. One might even ask whether a grid level leading to sub-domains with a large ratio of ghost nodes to interior nodes is desirable. As often with parallel algorithms, there is no universal answer to these questions, since the best approach depends on the actual parallel environment and the application at hand.

However, two general strategies exist. One is known under the notion of *coarse grid agglomeration*. The idea here is to unite the sub-domains of different processes on one process once the ratio between interior and boundary nodes falls below a certain threshold. While this leaves processes idle and induces a significant communication requirement at those points in the multigrid cycle where one descends to or ascends from a grid level on which coarse grid agglomeration occurs, it can nevertheless be advantageous, since it reduces the communication work for performing operations (smoothing, residual computation, coarse grid correction, etc.) on this level and the lower ones. The extreme of this approach is to collect all sub-domains on one process on a certain level, and to perform a sequential multigrid algorithm on the grid levels below.

Unfortunately, while this technique is mentioned in nearly all publications on parallel multigrid, there appears, at least to our best knowledge, to exist no publication that thoroughly investigates the approach and gives the user some advice on how and when this should be done depending on the underlying parallel architecture and the specific multigrid algorithm.

Another strategy is to use a so-called *U-cycle*. The idea here is to confine oneself to a comparatively flat multigrid hierarchy, i.e. the number of grid levels is chosen such that the coarsest grid is still comparatively fine. While this induces the need to compute an exact solution for the coarsest grid problem already on a global grid with a larger number of unknowns, doing so can on one hand improve the convergence speed, which can alleviate the additional costs. On the other hand, one can employ another parallel solution method, such as a parallel sparse direct solver or a parallel preconditioned Krylov subspace method for this purpose. See e.g. [64] for a closer examination of this approach.

Generally, parallel multigrid designers should critically question what they are optimising for. Implementing a U-cycle which stops already at a very fine coarsest grid and simply solves the coarsest grid equations by a sufficient number of calls to the smoother may lead to excellent parallel efficiency and impressive aggregate Mflop/s rates[5]. However, in terms of run-time this will seldom be a high performance approach, since the coarse grid solver is of course inefficient and needs many iterations which incurs computation and communication time. This shows that all arguments about parallel efficiency in the multigrid context must be considered with some care.

At this point, it can also be pointed out that similar arguments should be considered when comparing multigrid algorithms with other iterative linear solvers. If, for example, a preconditioner achieves a condition number which depends logarithmically on the system size of a 3D problem (as is the case for some of the more advanced domain decomposition algorithms, for instance), then the CG solver will need $\mathcal{O}(\log n)$ iterations. Assuming that the preconditioner requires one next neighbour data exchange, the solver has an overall communication volume of order $\mathcal{O}(p\, n^2 \log n)$, which is asymptotically worse than for an equivalent multigrid method, while being asymptotically equivalent in the number of messages to be sent.

Summarising the argument in this section, we see that multigrid may make it difficult to obtain good parallel efficiency, but this should not misguide anyone to try to use too simplistic variants or be misinterpreted as necessarily leading to poor run-time performance. Asymptotically, multigrid is not only optimal in (sequential) computational complexity, but using grid partitioning, it also requires asymptotically only close to the minimal amount of communication.

5.3.5 Experiments

The computational results in this section have the very simple aim to illustrate the potential of parallel geometric multigrid methods and to show that, on suitable architectures, parallel multigrid can be designed to be extremely efficient both with respect to absolute timings and in terms of parallel efficiency. The test case is a Poisson problem with Dirichlet boundary conditions in three space dimensions. Though numerically no challenge, this is a hard test problem with respect to efficiency, since the ratio of computation to communication is very low and thus it is not trivial to achieve good speedup results.

For the scalability experiment, the problem domains consist of $9N$ unit cubes with N being the number of processes in the computation. Seven refinement levels are used in the grid hierarchy. In the case of the speed-up experiment, the number of computational cells in the problem domain remains constant, of course. The L-shaped problem domain consists of 128 unit cubes and six refinement levels are

---

[5] 1 Mflop/s = 10^6 floating-point operations per second, 1 Gflop/s = 10^9 floating-point operations per second.

CPU	Dof $\times 10^6$	Time in (s)
64	1179.48	44
128	2359.74	44
256	4719.47	44
512	9438.94	45
550	10139.49	48

(a)

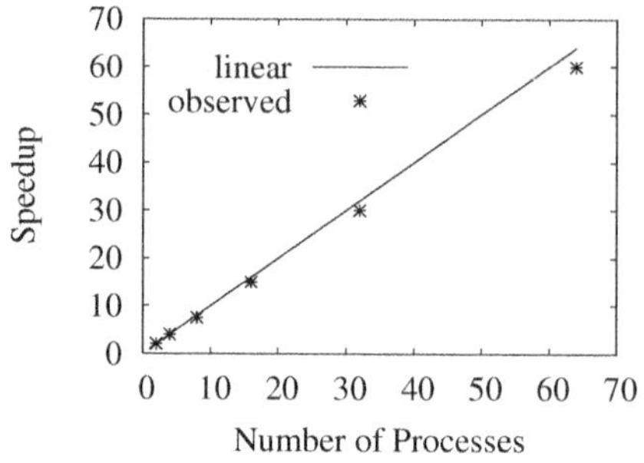

(b)

Fig. 5.7. Parallel performance results in 3D: (**a**) Scalability experiments using a Poisson problem with Dirichlet boundary conditions on an L-shaped domain. *Dof* denotes the degrees of freedom in the linear system, the timing results given refer to the wall clock time for the solution of the linear system using a full multigrid solver. (**b**) Speedup results for the same Poisson problem.

generated by repeated subdivision. The problem domain is distributed to 2, 4, 8, 16, 32, and 64 processes. The single processor computing time is estimated by dividing the two processor execution time by two.

The algorithmic components of the multigrid method are well established. We employ a variant of the red-black Gauß-Seidel iteration as smoother and full weighting and trilinear interpolation as the inter-grid transfer operators. The cycling strategy is a full multigrid method in which we perform two V(2,2)-cycles on each grid level before prolongating the current approximation to the next finer grid in the hierarchy. A discretisation by trilinear finite elements results in a 27-point stencil in the interior of the domain.

The experiments were carried out on the Hitachi SR8000 supercomputer at the Leibniz Computing Centre in Munich. The machine is made up of SMP nodes with eight processors each and a proprietary high speed network between the nodes. The program runs with an overall performance, including startup, reading the grid file and all communication, of 220 Mflop/s per process, which yields an agglomerated node performance of 1.76 Gflop/s out of the theoretical nodal peak performance of 12 Gflop/s. To put the solution time into perspective, we note that the individual processors have a theoretical peak performance of 1.5 Gflop/s, which is not much compared to the 6 Gflop/s and more of currently available architectures, such as e.g. the Intel Pentium IV 3.6 GHz, the Intel ItaniumII 1.5 GHz or the IBM PowerPC 970 2.2GHz. Nevertheless, a linear algebraic system with more than 10^{10} unknowns distributed over 550 processes and as many processors is solved in less than 50 seconds. For information concerning the design principles and the data structures of this multigrid code, we refer to [36].

The scalability results owe much to the ability of the full multigrid algorithm to arrive at the result with a fixed number of cycles, independent of the problem size, cf. Sect 5.2.5. One might think that the speedup experiment represents a harder test,

Table 5.2. Scale up results for parallel V(2,2)-cycle in 3D: *Dof* are the degrees of freedom, *Time* is the wall clock solution time for the linear system.

CPU	Dof $\times 10^6$	Time in (s)	V-cycles
8	133.4	78	9
16	267.1	81	10
32	534.7	95	11

as the amount of communication over the network increases, while the amount of computations per process decreases. However, as shown in Figure 5.7, the behaviour is close to optimal. In the experiment, an L-shaped domain consisting of 128 cubes is distributed to 2, 4, 8, 16, 32, and 64 processes. Each cube is regularly subdivided six times. The same Poisson problem is solved using the same multigrid algorithm as before.

Even the straightforward V-cycle can handle large problems in acceptable time frames, as the following experiment shows. We solve again a 3D Poisson problem with Dirichlet boundary conditions again on the Hitachi SR8000, but this time we employ the 7-point finite difference stencil on problem domains $\Omega_1 = [0, 2] \times [0, 2] \times [0, 2]$, partitioned into 8 hexahedra, $\Omega_2 = [0, 4] \times [0, 2] \times [0, 2]$, partitioned into 16 hexahedra, $\Omega_3 = [0, 4] \times [0, 4] \times [0, 2]$, partitioned into 32 hexahedra. On each grid level, we perform two pre- and two post-smoothing steps. The fact that the number of V-cycles in Tab. 5.2 increases with the problem size is a consequence of the stopping criteria that imposes an absolute limit on the l_2-norm of the residual vector. Still, the V-cycle multigrid manages to solve a system with more than $500 \cdot 10^6$ unknowns in 95 seconds, in this case on 32 processors.

5.4 Parallel Multigrid for Unstructured Grid Applications

5.4.1 Single Grid Aspects

In this section, we examine the added complications that arise when parallelising a multigrid solver on unstructured grids. We first consider the single grid case before turning to the multigrid setting.

From a sufficiently abstract point of view, the steps in the parallelisation of operations on an unstructured grid are the same as in the structured case considered so far. First, the grid has to be partitioned, then the necessary ghost points have to be determined. After that, it has to be worked out for each ghost value from which partition the information can be retrieved before finally the communication structures between the partitions are set up.

Fortunately, the commonly used programs to perform the partitioning step, such as *Metis* or *ParMetis* [39, 40], are applicable to structured and unstructured grids alike. As in the structured case, the discretisation scheme determines which ghost values are needed around the partition boundary. One major difference between the two grid types concerns the number of neighbouring partitions with which a given partition has to exchange information. In the structured case, the number of neighbours is bounded independently of the number of partitions, whereas in the unstructured case, this number is not bounded a priori. For operations such as a matrix vector product or an inner product between two distributed vectors, this of course affects the number of messages that have to be sent in order to keep the data synchronised across partitions, but the basic functionality is the same as in the structured case. However, for inherently sequential operations, such as the common Gauß-Seidel smoother in multigrid schemes, efficient parallelisation strategies that minimise the number of messages are more difficult to construct. The problem is linked to the one of the colouring of the data dependency graph between the partitions. Given that this graph depends on the distribution of the unstructured grid, it is only known at run-time. Hence, any parallelisation strategy has to be formulated and implemented in a general fashion, which is much more complicated than in the structured case, where the neighbourhood relationships are known at compile-time. For further details on parallel smoothing in the unstructured case we refer the reader to [65], where this problem is considered in detail, since parallel algebraic multigrid faces the same challenge.

5.4.2 Generation of Grid Hierarchies

Turning now to multigrid methods on unstructured grids, the first problem concerns the construction of the grid hierarchy. The common approach to generating the hierarchy of nested approximation spaces needed for standard geometric multigrid on unstructured grids consists in carrying out several steps of repeated regular refinement as depicted in Figure 5.8. The reason for the popularity of this approach lies in the complexity and difficulty of any alternative. Coarsening a given unstructured grid so that the resulting approximation spaces are nested is often impossible. Hence, one would have to work on non-nested spaces. Multigrid methods for such a setting are much more complicated than those for their nested counterparts. In summary, although it is possible to construct geometric multigrid methods on hierarchies of repeatedly coarsened unstructured grids, this option is rarely chosen in practice.

Concerning the parallelisation, the simplest approach is to perform the grid distribution only once, on the unstructured input grid, and then to let the finer levels inherit the communication pattern from the coarsest grid. However, this strategy will amplify any kind of load imbalance present in the distribution of the initial grid. The other extreme approach is to partition each grid level individually. This avoids the load balancing problems of the other method at the cost of a more complex initialisation phase and potentially more complicated communication patterns in the multigrid algorithm. Which approach yields the smaller run-time depends on the concrete problem and the CPU and network specifications. We refer to [58] for a discussion of load balancing issues.

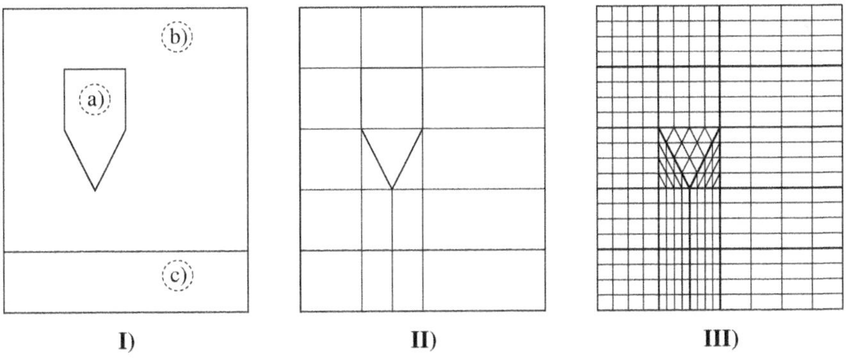

Fig. 5.8. Non-destructive testing example, from left to right: **I)** the problem domain with a) the coil that generates a magnetic field, b) air surrounding the coil, and c) the material to be tested. The material parameters (magnetic permeability) of the three different components differ from each other, but are assumed to be constant within each component (homogeneous materials). **II)** the coarse, unstructured grid that represents the problem geometry and **III)** the grid after two global regular subdivision steps

5.4.3 Hierarchical Hybrid Grids

The repeated refinement of an unstructured input grid described above opens up possibilities for performance improvement over standard unstructured implementations. The main observation is that, after a number of refinement steps, the resulting grids are still unstructured globally, but exhibit regular features locally.

We illustrate the generation of these locally regular structures using a simplified problem domain that arose in an electromagnetic field problem. Consider the setting in Figure 5.8.

For the representation of the discrete operator on the grid in Figure 5.8 II), it is appropriate to employ sparse matrix storage schemes. However, the situation on the grid in Figure 5.8 III) is different. Within each cell of the unstructured input grid, the repeated refinement has resulted in regular patches. In the interior of each coarse grid cell, the grid is regular and the material parameter is constant. This implies that one stencil suffices to represent the discrete operator inside such a regular region.

The main idea is now to turn operations on the refined, globally unstructured grid into a collection of operations on block-structured parts where possible and resort to unstructured operations only where necessary. Provided a sufficiently high level of refinement or, put differently, provided that the regular regions are sufficiently large in comparison to the remaining parts, this approach combines an improved single node floating-point performance with the geometric flexibility of unstructured grids, at least at the input level. For more details, see [5, 35].

Assuming a vertex-based discretisation, the discrete operator can be expressed by a stencil with constant shape in each regular region of the refined grids as the neighbourhood relationship does not change. In some cases, the entries of the stencil are also constant over the object. Inside such a region, the representation of the operator can be reduced to one single stencil. Both properties, constant stencil shape and constant stencil entries, help to improve the performance of operations involving the discrete operator. The scale of the improvement depends on the computer architecture.

On general unstructured grids, the discrete operator is usually stored in one of the many sparse matrix formats. If one uses such general formats in operations such as the matrix-vector product or a Gauß-Seidel iteration, one usually has to resort to indirect indexing to access the required entries in the vector. Being able to represent the discrete operator in the regular regions by stencils with fixed shapes, we can express the memory access pattern for the operation explicity through index arithmetic and thus enable the compiler to analyse and optimise the memory accesses better. In the sparse matrix case, the memory access pattern is known at run-time, in our setting it is known at compile-time, at least for a significant subset of all points.

On the Hitachi SR8000 supercomputer at the Leibniz Computing Centre in Munich, the change from indirect indexing to index arithmetic improves the Mflop/s performance of a Gauß-Seidel iteration on a single processor from around 50 Mflop/s for a CRS (compressed row storage) implementation to 300 Mflop/s for the stencil-based computation. These values were obtained with a 27-point finite element discretisation inside a refined hexahedron on a processor with a theoretical peak performance of 1500 Mflop/s. On this high memory bandwidth architecture, the explicit knowledge about the structure of the operations results in a six-fold performance improvement. Many cache-based architectures do not offer such a high bandwidth to main memory. Given that a new stencil has to be fetched for each unknown, the memory traffic to access the stencil values slows down the computations on these machines. As an example for such machines, we consider an Intel Pentium 4 processor with 2.4 GHz clock speed, 533 MHz front side bus and dual channel memory access. Our CRS-based Gauß-Seidel iteration runs at 190 Mflop/s on a machine with a theoretical peak performance of 4800 Mflop/s[6]. With the fixed stencil shape implementation, we observe a performance between 470 and 490 Mflop/s, depending on the problem size, which is 2.5 times more than for the standard unstructured one. In all cases, the problem size did not fit into the caches on the machines. In the case of constant stencil entries for an element, the advantage of the structured implementation is even clearer. Instead of fetching 27 stencil values from memory for each unknown, the same stencil is applied to all unknowns in the element. This obviously reduces the amount of memory traffic significantly. On the Hitachi SR8000, such a constant coefficient Gauß-Seidel iteration achieves 884 (out of 1500) Mflop/s on a hexahedron with 199^3 unknowns. Compared to the standard unstructured implementation, this amounts to a speed up factor of 17.5. Again, on the commodity architecture Pentium 4, the improvement is less impressive. For a problem of the same size, the constant coefficient iteration runs at 1045 Mflop/s, which is 5.5 times faster than its unstructured counterpart. In short, the hierarchical hybrid approach works well for the setting for which it was derived. This is the case when the input grid is com-

[6]The results on the PC architecture were obtained with the version 3.3.3 of the GNU compiler collection.

paratively coarse and has to be subdivided several times for accuracy reasons. Then the resulting grid hierarchy is well suited for the above mentioned approach. However, when, due to a complicated geometry, the input grid has to be very fine so that only few refinement steps are either necessary or possible, then the proposed data structures are not advantageous.

5.5 Single-Node Performance

5.5.1 Overview

In order to increase the run-time performance of any parallel numerical application, it is essential to address two related optimisation issues, each of which requires intimate knowledge in both the algorithm and the architecture of the parallel computing platform. Firstly, it is necessary to minimise the parallelisation overhead itself. This optimisation target represents the primary focus of this paper.

Secondly, it is essential to exploit the individual parallel resources as efficiently as possible; i.e., by achieving the highest possible performance on each node in the parallel environment. This is especially true for distributed memory systems found in clusters based on off-the-shelf workstations communicating via fast interconnection networks.

5.5.2 Memory Hierarchy Optimisations

According to Moore's law from 1975, the number of transistors on a silicon chip will double every 12 to 24 months. This prediction has already proved remarkably accurate for almost three decades. It has led to an average increase in CPU speed of approximately 55% every year. In contrast, DRAM speed has evolved rather slowly. Main memory latency and memory bandwidth have only been improving by about 5% and 10% per year, respectively [34]. The *International Technology Roadmap for Semiconductors*[7] predicts that this trend will continue further on and the gap between CPU speed and main memory performance will grow for more than another decade until technological limits will be reached. Therefore, today's computer architectures commonly employ *memory hierarchies* in order to hide both the relatively low main memory bandwidth as well as the rather high latency of main memory accesses.

A typical memory hierarchy covers the CPU registers, up to three levels of cache, and main memory. Cache memories are commonly based on fast semiconductor SRAM technology. They are intended to contain copies of main memory blocks to speed up accesses to frequently needed data. The reason why caches can substantially reduce program execution time is the principle of *locality of references*. This principle is empirically established and states that most programs do not access all code or data uniformly. Instead, recently used data as well as data that are stored close to the currently referenced data in address space are very likely to be accessed

[7] See the `http://public.itrs.net` web site.

in the near future. These properties are referred to as *temporal* and *spatial* locality, respectively [34].

Only if the hierarchical memory architecture is respected by the code, can efficient program execution (in terms of arithmetic operations per time unit) be expected. Unfortunately, current optimising compilers are not able to synthesise chains of complicated cache-based code transformations. Hence, they rarely deliver the performance expected by the users and much of the tedious and error-prone work concerning the tuning of the memory efficiency (particularly the utilisation of the cache levels) is thus left to the software developer. Typical cache optimisations techniques cover both *data layout transformations* as well as *data access transformations*.

Data layout optimisations aim at enhancing code performance by improving the arrangement of the data in address space. On one hand, such techniques can be applied to change the mapping of array data to the cache frames, thereby reducing the number of cache conflict misses. This is achieved by a layout transformation called *array padding* [54]. On the other hand, data layout optimisations can be applied to increase spatial locality. They can be used to reduce the size of the working set of a process; i.e., the number of virtual pages which are referenced alternatingly and should therefore be kept in main memory [34]. Furthermore, data layout transformations can be introduced in order to increase the reuse of cache blocks (or cache lines) once they have been loaded into cache. Since cache blocks are always transferred as a whole and contain several data items that are arranged next to each other in address space, it is reasonable to aggregate data items in address space which are likely to be referenced within a short period of time. This is primarily accomplished by the application of *array merging* [34].

In contrast, data access optimisations change the order in which iterations in a loop nest are executed. These transformations primarily strive to improve both spatial and temporal locality. Moreover, they can also expose parallelism and make loop iterations vectorisable. Typical examples of data access transformations are *loop interchange* and *loop blocking (loop tiling)*, see [2]. Loop interchange reverses the order of loops in a loop nest, thereby reducing the *strides* of array-based computations; i.e., the step sizes at which the corresponding arrays are accessed. This implies an improved reuse of cache blocks and thus causes an increase in spatial locality. In contrast, loop blocking is primarily used to improve temporal locality by enhancing the reuse of data in cache and reducing the number of cache capacity misses. Tiling a single loop replaces it by a pair of loops. The inner loop of the new loop nest traverses a block of the original iteration space with the same increment as the original loop. The outer loop traverses the original iteration space with an increment equal to the size of the block which is traversed by the inner loop. Thus, the outer loop feeds blocks of the whole iteration space to the inner loop which then executes them step by step.

Performance experiments emphasize the effectiveness of cache-based transformations. Depending on the properties of the underlying problem (2D/3D, constant/variable coefficients, etc.), the application of appropriate cache optimisation techniques has revealed significant speedups of up to a factor of 5 on common cache-based architectures. The impact of these techniques is typically examined by applying suitable *code profiling tools* that are usually based on platform-specific performance hardware such as dedicated counter registers [34].

For further details on cache optimisation techniques for numerical computations, we refer to [42, 61] and to the research in our *DiME*[8] project. In particular, we refer to [18] for an overview of cache optimisation techniques for multigrid methods.

5.5.3 Optimising for SMP Nodes

It is often the case that the individual nodes of a parallel computing environment consist of parallel architectures themselves. For example, many of today's workstation clusters are composed of so-called *symmetric multiprocessors (SMPs)*. SMP nodes are shared memory machines commonly consisting of two, four, or eight CPUs that access the local shared memory modules through fast interconnects such as a crossbar switches [34]. Lower levels of cache may be shared as well.

Within each SMP node, the parallel execution of the code is based on a shared memory programming model using *thread parallelism*. Thread parallelism can either be introduced automatically by parallelising compilers or by explicit programming; e.g., by using *OpenMP* directives [12, 13]. The individual nodes, however, typically communicate with each other using a distributed memory programming paradigm such as message passing.

The conclusions from the previous section carry over from single-CPU nodes to SMP nodes. Obviously, in the latter case, the efficient utilisation of the memory hierarchies is crucial for run-time performance as well. However, the manual introduction of cache-based transformations into thread-parallel code using OpenMP typically turns out to be even more tedious and error-prone.

5.6 Advanced Parallel Multigrid

In this section we will present a selection of extensions of the basic parallel multigrid method, as described in Section 5.3.

5.6.1 Parallelisation of Specific Smoothers

Multigrid methods come in many variants as required by specific applications. The treatment of anisotropies in particular is important in practice and has been studied extensively in the literature, see e.g. [60] and the references cited therein. In order to maintain full multigrid efficiency in the presence of mild anisotropies, one can use SOR smoothers with special weights [66], which in terms of a parallel implementation poses no additional difficulties. However, in the case of strong anisotropies, the ideal multigrid efficiency can only be maintained by either the use of non-standard coarsening strategies such as semi-coarsening, see [60], or the use of more advanced smoothers.

[8] See the `http://www10.informatik.uni-erlangen.de/dime` web site.

Fundamentally, the stronger coupling of the unknowns in a specific direction, as it is the case in anisotropic problems, must be observed in the algorithm. One way to accomplish this is to use so-called *line-smoothers*. These basically operate in the same fashion as the point smoothers described in Section 5.3.2 with the difference that the unknowns belonging to all nodes of a complete grid line are relaxed concurrently. In order to do this, a (small) linear system of equations, most often with tri-diagonal or banded structure, must be solved for each line. In view of a grid partitioning, this is uncritical as long as these lines of dependencies do not intersect sub-domain interfaces. If the tri-diagonal systems must be distributed across processes, then the solution of these systems must be distributed accordingly, where the usual starting point is the *cyclic reduction* algorithm; see [21], for example.

In the literature, many variants and extensions have been described, of which we mention only a few. Line smoothing is usually applied in a so-called *zebra order*, for example, akin to the red-black ordering for Gauß-Seidel and SOR, to simplify parallelisation. Having such a set of tri-diagonal systems which can be solved independently may also be used to parallelise (or vectorise) the simultaneous solution. This may lead to better parallel efficiency (or better vectorisation), since it avoids the need to devise parallel strategies for the inherent dependencies in the tri-diagonal systems.

Unfortunately, anisotropies aligned with the grid lines of a structured grid are just the simplest case. More complicated forms of anisotropies will require more complex smoothing strategies, such as alternating the direction of the line smoothers. This, however, again leads to problems with the parallelisation, since the lines of dependencies will then necessarily intersect sub-domain boundaries.

Finally, it should be mentioned that some 3D applications not only exhibit lines of strongly coupled unknowns, but planes of strong coupling. Smoothers adapted to this situation will treat all the unknowns in such a plane simultaneously. An efficient technique for solving the resulting linear systems is to revert to a 2D multigrid method [59]. Again, it may be necessary to do this in alternating plane directions, and each of the 2D multigrid algorithms may need line smoothing in order to be efficient. Any such strategy is naturally problem-dependent. Hence, it is difficult to develop any generally usable, robust multigrid method based on these techniques.

Besides anisotropies, the treatment of convection-diffusion equations (with dominating convection) is of high practical interest. In this case, the multigrid theory is still much less developed. One algorithmic multigrid approach to convection dominated PDEs is based on smoothers with *downstream relaxation*, where the smoother is designed such that it observes the (sequential) data dependencies along the characteristics (that is streamlines) of the flow. Again, a parallelisation of such a smoother is easy when the domains are split parallel to the lines of dependency, but unfortunately this is far from trivial to accomplish in most practical situations where the streamlines may change direction throughout the domain or — in the case of nonlinear equations — may depend on the flow itself. In these cases, no generally applicable rules for the parallelisation exist, but the best strategy depends on the application.

Another smoothing approach successfully applied to both of the above problems is to employ variants of the *incomplete LU decomposition (ILU)* method; See [56, 60,

63], for example. It has been shown that ILU smoothers lead to robust and efficient multigrid methods in the sequential setting, at least for 2D problems. The 3D case, however, remains problematic with respect to efficiency. Only for problems with dominating directions efficient 3D ILU smoothers exist so far, see the remarks in [60]. From the parallel point of view, the problem is that, while ILU algorithms are in principle point-based smoothers, they are intrinsically sequential and therefore difficult to parallelise, the more so, since their quality depends even stronger on the ordering of unknowns than this is the case for the SOR method. For more details on parallel ILU smoothers, see the references in [65], for example.

At the end of this subsection, we wish to give a word of warning. For any algorithm designer, it is tempting to simply neglect certain complexities that may arise from an efficient treatment of either anisotropies or dominating convection. This may be the case for sequential multigrid and is of course even more tempting when all the difficulties of parallelisation need to be addressed. In this case, algorithm designers often modify the algorithm slightly to simplify the parallel implementation. For example, the tridiagonal systems of line smoothers can simply be replaced by a collection of smaller tridiagonal systems as dictated by sub-domain boundaries, thus neglecting the dependencies across the sub-domains in the smoother. These smaller systems can be solved in parallel, benefiting parallel efficiency and simplifying the implementation. Such a simplified method will still converge because it is embedded in the overall multigrid iteration. In the case of only few sub-domains and moderate anisotropy, this may in fact lead to a fully satisfactory solver.

However, if the physics of the problem and the mathematical model really dictate a global dependency along the lines of anisotropy, then such a simplified treatment which does not fully address this feature will be penalised; the convergence rate will deteriorate with an increasing number of sub-domains to the point that the benefit of using a multigrid method is completely lost. Eventually there is no way to cheat the physics and the resulting mathematical properties of the problem. Multigrid methods have the disadvantage (or is this an advantage?) that they mercilessly punish any disregard for the underlying physics of the problem. Optimal multigrid performance with the typical convergence rates of 0.1 per iteration will only be achieved, if all the essential features of the problem are treated correctly — and this may be not easy at all, especially in parallel.

5.6.2 Alternative Partitioning Approaches

In this section, we will briefly introduce three different concepts that can be seen as alternatives to the standard grid partitioning approach described in Section 5.3.1.

Additive and Overlapping Storage

The grid partitioning introduced in Section 5.3.1 assumed that each node of the grid belongs to a unique sub-domain and ghost nodes were employed to handle sub-domain dependencies. In this subsection, we will briefly discuss an approach to grid partitioning that strikes a different path. For a detailed analysis, see [17, 31, 37]. It can most easily be explained from a finite element point of view.

In the first approach to grid partitioning, nodes uniquely belong to sub-domains *(node-oriented decomposition)*. Thus, the elements are intersected by sub-domain boundaries. In the second approach, one assigns each element to a unique sub-domain *(element-oriented decomposition)*. In this case, nodes on the sub-domain boundaries belong to more than one sub-domain. This introduces the question of how to store the values of a grid-function (i.e., an FEM vector) at these nodes. One combines two different schemes. The first one, denoted as *overlapping* storage, assumes that each process stores the function values at the respective nodes. In this case, the connection between the local vector v_k on process p_k and the global vector v can be expressed by a boolean matrix M_k as $v_k = M_k v$.

In the second scheme, the global function value $v(n_j)$ at a node n_j on the boundary is split between all sub-domains to which n_j belongs. If s_p is the number of sub-domains, then the global function v can be obtained from the local sub-domain functions v_k via

$$v = \sum_{k=1}^{s_p} M_k^T v_k \quad .$$

This is denoted as *adding* storage. Note that the conversion of an adding type vector to a vector of overlapping type requires communication, while the reverse operation can be performed without communication. In a similar fashion, one can define operators of adding and of overlapping type. This can be imagined in the following way. For an adding type operator a process stores all stencils for nodes in its sub-grid as vectors of adding type and analogously for an operator of overlapping type.

The application of an adding type operator to an overlapping type vector can be performed without communication and will result in a vector of adding type. Under some constraints regarding the dependency pattern of the respective operator, the application of an operator of overlapping type to a vector of either adding or overlapping type can also be performed without communication and results in a vector of the original type. Proofs as well as a detailed analysis of the limiting conditions can be found in [31].

The element-oriented decomposition in combination with the above storage concept can be used to parallelise a multigrid method in the following fashion. One stores the approximate solutions and the coarse grid corrections as vectors of overlapping type and the right hand sides and residuals as vectors of adding type. The transfer operators are stored in overlapping fashion, while the discrete differential operator is stored as operator of adding type.

Under these assumptions (and assuming that the dependency patterns of the operators fulfil the limiting conditions) one can show that none of the multigrid components prolongation, restriction, computation of the residual, and coarse grid correction step requires any communication. The only place where communication is required is the smoothing process. Here, typically the update to the old approximate can be computed as vector of adding type, which must be converted to overlapping storage before adding it to the old approximate. As is the case for node-oriented decompositions, each smoothing step will thus require one communication step. In the latter approach, however, restriction and prolongation will typically incur the need for a communication step in order to update the residual resp. the approximate solution at the ghost nodes, before the interpolation operation itself. Compared to a node-oriented decomposition there is, thus, a significant reduction in the number of messages and the amount of data that must be exchanged.

Another interesting aspect, though less general, is the following. Assume that our multigrid method employs a hierarchy of regular grids Ω^l composed of $(2^{m_l} + 1) \times (2^{n_l} + 1)$ nodes. As was mentioned in Section 5.3.3, the use of a node-oriented decomposition will lead to a hierarchy, where sub-grids on different levels cover different areas. With the element-oriented decomposition this problem does not arise. The sub-grid boundaries coincide on all levels. Furthermore, we will obtain a perfect load balance, as far as the number of grid nodes in each sub-grid is concerned. This is not the case for standard grid partitioning, see Figure 5.2 in this respect.

The reduced number of places in the algorithm, where communication occurs, as well as the symmetric view of the data, i.e. a node does not change its type from local to ghost, when going from one partition to the other, can also be seen as an advantage with respect to the implementation of the multigrid algorithm.

In summary, we think that the element oriented decomposition, if it is applicable, is a sincere challenger to the classical node-oriented decomposition approach.

Full Domain Partitioning

The partitioning schemes presented so far result in each process storing its patch (sub-grid) of the global grid plus some data from the immediate neighbouring patches. In this case, on the finer grid levels, each process knows its part of the global grid, but no process works on the whole problem domain. In the *Full Domain Partition* approach, in short *FuDoP*, proposed by Mitchell [46, 47, 48], each process starts from a coarse grid representing the whole problem domain and then adaptively refines the grid until the resolution is sufficiently accurate in its area of responsibility. As a result, each process computes a solution for the whole domain, albeit with a high accuracy only in its patch of the global domain on a grid that becomes increasingly coarse the further the distance to that patch.

The advantage of this approach is that existing serial, adaptive codes can be reused in a parallel context. In [46], a V-cycle multigrid method is presented that requires communication on the finest and on the coarsest level only, but not on the levels in between. However, on the downside, this approach requires all to all communication as now the grids on any two processes overlap. Whether this approach shows run-time advantages over standard partitioning schemes depends on the number of processes, the number of grid levels and, as usual, on the network characteristics.

Bank and Holst [3] promote a similar technique for parallel grid generation to limit load imbalances in parallel computations on adaptively refined grids. These authors also stress that existing adaptive components can be employed in a parallel setting with comparable ease.

Space-Filling Curves

The issues of partitioning and load balancing in the context of parallel adaptive multigrid have also been addressed through the use of so-called *space-filling curves*. A space-filling curve represents a mapping Φ of the unit interval $I := [0,1]$ to the space $\mathbb{R}^d$, $d = 2,3$, such that the image $\Phi(I)$ has a positive measure. See [28, 67], for example. The computational domain Ω is supposed to be a subset of $\Phi(I)$ such that all nodes of the adaptively refined discretisation grid are elements of $\Phi(I)$ and thus passed by the space-filling curve. At this point we should mention the following two aspects. The first one is that *space-filling curves* are typically constructed as the limit of a family of recursively defined functions. In practice no real space-filling curve, but only a finite approximation from such a family is employed in the method. Secondly, the assumption that the curve passes through all nodes of the grid is not a restrictive one. The space-filling curve by its nature can be chosen such that it completely covers a domain in 2D / 3D. Choosing a fine enough approximation will thus fulfil the assumption.

The fundamental advantage of the partitioning approach based on space-filling curves is that it allows for inexpensive load balancing. The idea, namely, is to use the inverse mapping of Φ and to divide the unit interval I into partitions that approximately contain the same numbers of pre-images of grid nodes. These partitions of I are then mapped to the available processes. Therefore, the computational load is balanced exactly. As a consequence of this decomposition approach, the original d-dimensional graph partitioning problem, which can be shown to be NP-hard [52], is approximated by a 1D problem that is easy to solve.

However, it has been demonstrated that the resulting partitioning of the actual d-dimensional computational grid can be suboptimal and may involve much more communication overhead than necessary. For a comprehensive presentation, we again point to [67] and the references therein.

5.6.3 Reduced and Overlapped Communication

Overlapped Communication

A technique to improve the parallel efficiency of any parallel algorithm is to overlap communication and computation, see also [41] in this context. In parallel multigrid, this can work as follows. Consider the Jacobi smoother discussed in Section 5.3.2 and assume that we are using a grid partitioning with ghost nodes. One sweep of the Jacobi method does not require any communication. However, after the sweep, the values at the ghost nodes must be updated. Remember that the order in which the nodes are treated during the Jacobi sweep does not play a role. Thus, we are free to alter it to our taste. By first updating all nodes at the boundary of a sub-grid we can overlap communication and computation, since the update of the ghost nodes can be performed while the new approximate solution is computed for nodes in the interior of the sub-grid. This idea directly carries over to other multigrid components that work in the same fashion; i.e., prolongation and restriction. More sophisticated smoothers, such as red-black SOR, for example, can also be treated in this way. Though, the order of treating nodes may become more intricate.

Reduced Communication

While overlapping communication with computation can significantly reduce the costs for communication, there further exist other more radical approaches. At this point, we briefly want to mention an algorithm introduced by Brandt and Diskin in [9, 14]. Their parallel multigrid method abandons communication on several of the finest levels of the grid hierarchy. In a standard parallel version of multigrid, this is the place where the largest chunk of communication occurs, as far as the amount of transferred data is concerned. In this respect, their approach differs from the ones discussed in Section 5.3.4 which addressed efficiency problems on coarser grids arising from their less favourable surface-to-volume ratio.

The core idea of the algorithm by Brandt and Diskin can be traced back to the *segmental-refinement-type* procedures originally proposed to overcome storage problems on sequential computers [8]. It can briefly be described as follows. The basic parallelisation is again done by grid partitioning, where each sub-grid is augmented by an extended overlap region in the same fashion as mentioned in Section 5.3.2. However, communication between processes is completely restricted to the subdomains on the coarsest grid level. Here, communication between the processes is required to form a common coarsest grid problem.

The overlap region fulfils two purposes. On one hand, if an appropriate relaxation scheme such as red-black Gauß-Seidel is chosen, this buffer slows the propagation of errors due to inexact values at the interfaces. On the other hand, in multigrid, the coarse grid correction typically introduces some high-frequency errors on the fine grid. Since values at the interfaces cannot be smoothed, the algorithm cannot eliminate these components. But in elliptic problems high-frequency components decay quickly. Hence, the overlap region keeps these errors from affecting the inner values too much.

It is obvious that the algorithm will in most cases not be able to produce an exact solution of the discrete problem. However, if one is solving a PDE problem, it is actually the continuous solution one is really interested in. Since the latter is represented by the discrete solution only up to a discretisation error, a discrete solution will be valuable, as long as its algebraic error remains in the same order as the discretisation error. For a detailed analysis of the communication pattern and the possible benefits of this approach, see [14, 50, 49].

5.6.4 Alternative Parallel Multigrid Algorithms

As was mentioned in Section 5.2.5, standard multigrid methods traverse the grid hierarchy sequentially, typically either in the form of W-cycles or V-cycles. We already discussed this inherently sequential aspect of multigrid in Section 5.3.4 and considered some standard approaches for dealing with it. Here, we want to briefly mention some further alternatives that have been devised.

The BPX Variant of Multigrid

In this context, the BPX algorithm (as proposed in [6]) and its analysis (see also [51] and the references therein) is of special importance. The BPX algorithm is usually used as a preconditioner for a Krylov subspace method, but it shares many features with the multigrid method. In fact, it can be regarded as an *additive* variant of the multigrid method.

In classical multigrid, as discussed so far, each grid level contributes its correction sequentially. If expressed by operators, this results in a representation of the multigrid cycle as a product of operators corresponding to the corrections on each level. Here, each level needs the input from the previous level in the grid hierarchy; therefore, it results in a multiplicative structure. For example, on the traversal to coarser grids, the smoothing has to be applied first, before the residuals for the coarser grids can be computed.

In BPX in contrast, all these corrections are computed *simultaneously* and are then added. Formally, this corresponds to a sum of the correction operators on each level and, consequently, it seems that the need for a sequential treatment of the levels has been avoided. Since this approach will only result in a convergent overall correction, the BPX algorithm is usually not used by itself, but instead employed as preconditioner. In this role, it can be shown to be an asymptotically optimal preconditioner. Thus, when combined with Krylov subspace acceleration, BPX results in an asymptotically optimal algorithm.

Unfortunately, the hope of the original authors that their algorithm would show better parallel properties per se (the original paper [6] was entitled "Parallel Multilevel Preconditioners") has only partially become true, since a closer look reveals that the BPX algorithm must internally compute a hierarchical sum of contributions from all the grid levels. To be more precise, the residuals have to be summed according to the grid transfer operators from the finest level to any of the auxiliary coarser levels. The restriction of the residual to the coarsest level will usually require intermediate quantities that are equivalent to computing the restriction of the residual to all intermediate levels. Therefore, computing these restrictions in parallel does not only create duplicated work, it also does not reduce the time needed to traverse the grid hierarchy in terms of a faster parallel execution. The coarsest level still requires a hierarchical traversal through all intermediate levels. Consequently, the BPX algorithm essentially requires the same sequential treatment of the grid hierarchy as the conventional multiplicative multigrid algorithm.

The above argument is of course theoretically motivated. In practice, other considerations may be essential and may change the picture. For example, the BPX algorithm may still be easier to implement in parallel, or it may have advantages in a particular adaptive refinement situation. For a comparison, see also [4, 27, 38].

Point-Block Algorithm and the Fully Adaptive Multigrid Method (FAMe)

The invention of the additive variant of multilevel algorithms, however, has shown that the strict sequential treatment of the levels is unnecessary and this has spurred a number of other ideas. For example the point-block algorithm in [24, 26] and the representation of all levels of the grid hierarchy in a single system, see [25, 55], relies on the analysis of additive multilevel systems.

For the parallelisation of a multilevel method, this construction and analysis can be exploited by realizing that the traversal of the grid hierarchy and the processing of the individual domain partitions can be decoupled.

In the point-block algorithm, it is exploited that each node of the finest grid may be associated with several coarser levels, and therefore it may belong to several discrete equations corresponding to these levels. It is now possible to set up this system of equations for each node. In a traversal through all fine grid nodes, one can now solve this system for each node. The resulting algorithm becomes a block relaxation method, with blocks of 1 to $\log(n)$ unknowns, corresponding to the number of levels to which a node belongs, and can be shown to have asymptotically optimal complexity. Imposing a grid partitioning, this algorithm can be parallelised. Note that, since the *multiple stencils* may extend deep into neighbouring partitions, the communication is more complicated.

Possibly the most far-reaching result towards an asynchronous execution of multilevel algorithms is the *fully adaptive multigrid* method of [55]. Here an active set strategy is proposed together with a meta-algorithm which permits many different concurrent implementations. In particular, it is possible to design a multilevel algorithm such that the traversal between the levels in the sub-domains can be performed completely asynchronously, as long as the essential data dependencies are being observed. The meta algorithm gives rigorous criteria which of these data dependencies must be observed. One possible realization is then to monitor the data dependencies dynamically at run-time within the algorithm. Any such dependency which extends across process boundaries need only be activated, if it is essential for the convergence of the algorithm, but it can be delayed until, for instance, enough such data have accumulated to amortise the startup cost.

The algorithmic framework of the fully adaptive multigrid method permits many different parallel implementations and includes as a special case the classical parallel multigrid algorithms as well as the point-block method.

Multiple Coarse Grid Algorithms and Concurrent Methods

For the sake of completeness, we also want to mention the algorithms by Fredrickson and McBryan [20], Chan and Tuminaro [11], as well as Gannon and van Rosendale [22]. While these methods also address the topic of the sequentiality of cycling through the multigrid hierarchy and the loss of parallel efficiency on coarser grids, they were designed primarily with massively parallel systems in mind. Such systems are denoted as *fine-grain*; i.e., the number p of processes is on the same order as the number N of unknowns. In contrast, we have considered *coarse-grain* parallel systems so far; i.e., systems with $p \ll N$. Nevertheless, the methods deserve to be mentioned here, since they complement the aforementioned ideas of this section.

The Chan-Tuminaro and the Gannon-van Rosendale algorithms both belong to the class of *concurrent methods*. Their idea, not quite unlike the one of the BPX or the point-block algorithm, is to break up the sequential cycling structure and to allow for a concurrent treatment of the different grid levels; hence the name concurrent methods. The basic approach is to generate independent sub-problems for the different grid levels by projection onto orthogonal sub-spaces. The algorithms differ in the way this decomposition is performed and the way solutions are combined again. For details, we refer to the original publications cited above.

The algorithm of Fredrickson and McBryan follows a completely different approach. In a fine-grain setting the sequential cycling structure of a standard multigrid method will lead to a large number of idle processes while coarser grids are treated. The Fredrickson–McBryan algorithm retains the sequential cycling structure, and instead tries to take advantage of these idle processes. Opposed to standard multigrid, the method does not employ a single grid to compute a coarse grid correction, but composes on each level several coarse grid problems. Ideally, the additional information obtained from these *multiple coarse grids* can be used to improve the convergence rate of the multigrid method, thus improving not only parallel efficiency, but also actual run-time. Indeed, an impressive improvement in convergence speed can be demonstrated for selected applications, which lead the authors to denote this approach as *parallel superconvergent multigrid* (PSMG).

While all three algorithms can in principle also be employed in a coarse-grain setting, the (theoretical) considerations in [44] strongly indicate that from a pure run-time perspective this is very unlikely to pay-off. Even in a fine-grain environment, the additional work induced by the methods seems to clearly over-compensate for the gain in convergence speed and/or improved parallelism.

5.7 Conclusions

The parallelisation of multigrid algorithms is a multifaceted field of ongoing research. Due to the difficulties resulting from the decreasing problem sizes on the coarse grids, parallel multigrid implementations are often characterized by relatively poor parallel efficiency when compared to competing methods. Nevertheless, it is the time required to solve a problem that finally matters. From this perspective, suitably designed multigrid methods are often by far superior to alternative elliptic PDE solvers.

Since the multigrid principle leads to a large variety of multigrid algorithms for different applications, it is impossible to derive general parallelisation approaches that address all multigrid variants. Instead, as we have pointed out repeatedly, the choice of appropriate multigrid components and their efficient parallel implementation is highly problem-dependent. As a consequence, our contribution can definitely not be complete and should be understood as an introductory overview to the development of parallel multigrid algorithms.

References

1. R. E. Alcouffe, A. Brandt, J. E. Dendy, and J. W. Painter. The multi–grid methods for the diffusion equation with strongly discontinuous coefficients. *SIAM J. Sci. Stat. Comput.*, 2:430–454, 1981.
2. R. Allen and K. Kennedy. *Optimizing Compilers for Modern Architectures.* Morgan Kaufmann Publishers, San Francisco, California, USA, 2001.
3. R. E. Bank and M. Holst. A new paradigm for parallel adaptive meshing algorithms. *SIAM J. Sci. Comput.*, 22(4):1411–1443, 2000.
4. P. Bastian, W. Hackbusch, and G. Wittum. Additive and multiplicative multi-grid — a comparison. *Computing*, 60(4):345–364, 1998.
5. B. Bergen and F. Hülsemann. Hierarchical hybrid grids: data structures and core algorithms for multigrid. *Numerical Linear Algebra with Applications*, 11:279–291, 2004.
6. J. H. Bramble, J. E. Pasciak, and J. Xu. Parallel multilevel preconditioners. *Math. Comp.*, 55:1–22, 1990.
7. A. Brandt. Multi–level adaptive solutions to boundary–value problems. *Math. Comp.*, 31:333–390, 1977.
8. A. Brandt. *Multigrid techniques: 1984 guide with applications to fluid dynamics.* GMD–Studien Nr. 85. Gesellschaft für Mathematik und Datenverarbeitung, St. Augustin, 1984.
9. A. Brandt and B. Diskin. Multigrid solvers on decomposed domains. In *Domain Decomposition Methods in Science and Engineering: The Sixth International Conference on Domain Decomposition*, volume 157 of *Contemporary Mathematics*, pp. 135–155, Providence, Rhode Island, 1994. American Mathematical Society.
10. W. Briggs, V. Henson, and S. McCormick. *A Multigrid Tutorial.* SIAM, 2. edition, 2000.
11. T. F. Chan and R. S. Tuminaro. Analysis of a parallel multigrid algorithm. In J. Mandel, S. F. McCormick, J. E. Dendy, C. Farhat, G. Lonsdale, S. V. Parter, J. W. Ruge, and K. Stüben, editors, *Proceedings of the Fourth Copper Mountain Conference on Multigrid Methods*, pp. 66–86, Philadelphia, 1989. SIAM.
12. R. Chandra, L. Dagum, D. Kohr, D. Maydan, J. McDonald, and R. Menon. *Parallel Programming in OpenMP.* Morgan Kaufmann, 2001.
13. L. Dagum and R. Menon. OpenMP: An industry-standard API for shared-memory programming. *IEEE Comp. Science and Engineering*, 5(1):46–55, 1998.
14. B. Diskin. Multigrid Solvers on Decomposed Domains. Master's thesis, Department of Applied Mathematics and Computer Science, The Weizmann Institute of Science, 1993.
15. C. C. Douglas. A review of numerous parallel multigrid methods. In G. Astfalk, editor, *Applications on Advanced Architecture Computers*, pp. 187–202. SIAM, Philadelphia, 1996.
16. C. C. Douglas and M. B. Douglas. MGNet Bibliography. Department of Computer Science and the Center for Computational Sciences, University of Kentucky, Lexington, KY, USA and Department of Computer Science, Yale University, New Haven, CT, USA, 1991–2002 (last modified on September 28, 2002); see `http://www.mgnet.org/mgnet-bib.html`.
17. C. C. Douglas, G. Haase, and U. Langer. *A Tutorial on Elliptic PDE Solvers and their Parallelization.* SIAM, 2003.
18. C. C. Douglas, J. Hu, M. Kowarschik, U. Rüde, and C. Weiß. Cache optimization for structured and unstructured grid multigrid. *Elect. Trans. Numer. Anal.*, 10:21–40, 2000.
19. L. Formaggia, M. Sala, and F. Saleri. Domain decomposition techniques. In A. M. Bruaset and A. Tveito, editors, *Numerical Solution of Partial Differential Equations on Parallel Computers*, volume 51 of *Lecture Notes in Computational Science and Engineering*, pp. 135–163. Springer-Verlag, 2005.

20. P. O. Frederickson and O. A. McBryan. Parallel superconvergent multigrid. In S. F. McCormick, editor, *Multigrid Methods: Theory, Applications, and Supercomputing*, volume 110 of *Lecture Notes in Pure and Applied Mathematics*, pp. 195–210. Marcel Dekker, New York, 1988.
21. T. L. Freeman and C. Phillips. *Parallel numerical algorithms.* Prentice Hall, New York, 1992.
22. D. B. Gannon and J. R. Rosendale. On the structure of parallelism in a highly concurrent pde solver. *J. Parallel Distrib. Comput.*, 3:106–135, 1986.
23. A. Greenbaum. *Iterative Methods for Solving Linear Systems.* SIAM, 1997.
24. M. Griebel. Grid– and point–oriented multilevel algorithms. In W. Hackbusch and G. Wittum, editors, *Incomplete Decompositions (ILU) – Algorithms, Theory, and Applications*, Notes on Numerical Fluid Mechanics, pp. 32–46. Vieweg, Braunschweig, 1993.
25. M. Griebel. Multilevel algorithms considered as iterative methods on semidefinite systems. *SIAM J. Sci. Stat. Comput.*, 15:547–565, 1994.
26. M. Griebel. Parallel point–oriented multilevel methods. In *Multigrid Methods IV, Proceedings of the Fourth European Multigrid Conference, Amsterdam, July 6-9, 1993*, volume 116 of *ISNM*, pp. 215–232, Basel, 1994. Birkhäuser.
27. M. Griebel and P. Oswald. On the abstract theory of additive and multiplicative Schwarz algorithms. *Numer. Math.*, 70:163–180, 1995.
28. M. Griebel and G. W. Zumbusch. Hash-storage techniques for adaptive multilevel solvers and their domain decomposition parallelization. In J. Mandel, C. Farhat, and X.-C. Cai, editors, *Proceedings of Domain Decomposition Methods 10, DD10*, number 218 in Contemporary Mathematics, pp. 279–286, Providence, 1998. AMS.
29. W. Gropp, E. Lusk, N. Doss, and A. Skjellum. A high-performance, portable implementation of the MPI message passing interface standard. *Parallel Computing*, 22(6):789–828, Sept. 1996.
30. W. Gropp, E. Lusk, and A. Skjellum. *Using MPI, Portable Parallel Programming with the Mesage-Passing Interface.* MIT Press, second edition, 1999.
31. G. Haase. *Parallelisierung numerischer Algorithmen für partielle Differentialgleichungen.* B. G. Teubner Stuttgart – Leipzig, 1999.
32. W. Hackbusch. *Multigrid Methods and Applications*, volume 4 of *Computational Mathematics.* Springer–Verlag, Berlin, 1985.
33. W. Hackbusch. *Iterative Solution of Large Sparse Systems of Equations*, volume 95 of *Applied Mathematical Sciences.* Springer, 1993.
34. J. Hennessy and D. Patterson. *Computer Architecture: A Quantitative Approach.* Morgan Kaufmann Publisher, Inc., San Francisco, California, USA, 3. edition, 2003.
35. F. Hülsemann, B. Bergen, and U. Rüde. Hierarchical hybrid grids as basis for parallel numerical solution of PDE. In H. Kosch, L. Böszörményi, and H. Hellwagner, editors, *Euro-Par 2003 Parallel Processing*, volume 2790 of *Lecture Notes in Computer Science*, pp. 840–843, Berlin, 2003. Springer.
36. F. Hülsemann, S. Meinlschmidt, B. Bergen, G. Greiner, and U. Rüde. *gridlib* – a parallel, object-oriented framework for hierarchical-hybrid grid structures in technical simulation and scientific visualization. In *High Performance Computing in Science and Engineering, Munich 2004. Transactions of the Second Joint HLRB and KONWIHR Result and Reviewing Workshop*, pp. 37–50, Berlin, 2004. Springer.
37. M. Jung. On the parallelization of multi–grid methods using a non–overlapping domain decomposition data structure. *Appl. Numer. Math.*, 23(1):119–137, 1997.

38. M. Jung. Parallel multiplicative and additive multilevel methods for elliptic problems in three–dimensional domains. In B. H. V. Topping, editor, *Advances in Computational Mechanics with Parallel and Distributed Processing*, pp. 171–177, Edinburgh, 1997. Civil–Comp Press. Proceedings of the EURO–CM–PAR97, Lochinver, April 28 - May 1, 1997.
39. G. Karypis and V. Kumar. A fast and highly quality multilevel scheme for partitioning irregular graphs. *SIAM J. Sci. Comput.*, 20(1):359–392, 1999.
40. G. Karypis and V. Kumar. Parallel multilevel k-way partition scheme for irregular graphs. *SIAM Review*, 41(2):278–300, 1999.
41. C. Körner, T. Pohl, U. Rüde, N. Thürey, and T. Zeiser. Parallel lattice boltzmann methods for cfd applications. In A. M. Bruaset and A. Tveito, editors, *Numerical Solution of Partial Differential Equations on Parallel Computers*, volume 51 of *Lecture Notes in Computational Science and Engineering*, pp. 439–466. Springer-Verlag, 2005.
42. M. Kowarschik. *Data Locality Optimizations for Iterative Numerical Algorithms and Cellular Automata on Hierarchical Memory Architectures.* PhD thesis, Lehrstuhl für Informatik 10 (Systemsimulation), Institut für Informatik, Universität Erlangen-Nürnberg, Erlangen, Germany, July 2004. SCS Publishing House.
43. H. Lötzbeyer and U. Rüde. Patch-adaptive multilevel iteration. *BIT*, 37:739–758, 1997.
44. L. R. Matheson and R. E. Tarjan. Parallelism in multigrid methods: how much is too much? *Int. J. Paral. Prog.*, 24:397–432, 1996.
45. O. A. McBryan, P. O. Frederickson, J. Linden, A. Schuller, K. Solchenbach, K. Stuben, C.-A. Thole, and U. Trottenberg. Multigrid methods on parallel computers — a survey of recent developments. *Impact Comput. Sci. Eng.*, 3:1–75, 1991.
46. W. F. Mitchell. A parallel multigrid method using the full domain partition. *Elect. Trans. Numer. Anal.*, 6:224–233, 1997.
47. W. F. Mitchell. The full domain partition approach to distributing adaptive grids. *Appl. Numer. Math.*, 26:265–275, 1998.
48. W. F. Mitchell. Parallel adaptive multilevel methods with full domain partitions. *App. Num. Anal. and Comp. Math.*, 1:36–48, 2004.
49. M. Mohr. Low Communication Parallel Multigrid: A Fine Level Approach. In A. Bode, T. Ludwig, W. Karl, and R. Wismüller, editors, *Proceedings of Euro-Par 2000: Parallel Processing*, volume 1900 of *Lecture Notes in Computer Science*, pp. 806–814. Springer, 2000.
50. M. Mohr and U. Rüde. Communication Reduced Parallel Multigrid: Analysis and Experiments. Technical Report 394, Institut für Mathematik, Universität Augsburg, 1998.
51. P. Oswald. *Multilevel Finite Element Approximation, Theory and Applications.* Teubner Skripten zur Numerik. Teubner Verlag, Stuttgart, 1994.
52. A. Pothen. Graph partitioning algorithms with applications to scientific computing. In D. E. Keyes, A. H. Sameh, and V. Venkatakrishnan, editors, *Parallel Numerical Algorithms*, volume 4 of *ICASE/LaRC Interdisciplinary Series in Science and Engineering*. Kluwer Academic Press, 1997.
53. M. Prieto, I. Llorente, and F. Tirado. A Review of Regular Domain Partitioning. *SIAM News*, 33(1), 2000.
54. G. Rivera and C.-W. Tseng. Data Transformations for Eliminating Conflict Misses. In *Proc. of the ACM SIGPLAN Conf. on Programming Language Design and Implementation*, Montreal, Canada, 1998.
55. U. Rüde. *Mathematical and Computational Techniques for Multilevel Adaptive Methods*, volume 13 of *Frontiers in Applied Mathematics*. SIAM, Philadelphia, 1993.
56. Y. Saad. *Iterative Methods for Sparse Linear Systems*. SIAM, 2nd edition, 2003.

57. J. Stoer and R. Bulirsch. *Numerische Mathematik 2*. Springer, 4. edition, 2000.
58. J. D. Teresco, K. D. Devine, and J. E. Flaherty. Partitioning and dynamic load balancing for the numerical solution of partial differential equations. In A. M. Bruaset and A. Tveito, editors, *Numerical Solution of Partial Differential Equations on Parallel Computers*, volume 51 of *Lecture Notes in Computational Science and Engineering*, pp. 55–88. Springer-Verlag, 2005.
59. C.-A. Thole and U. Trottenberg. Basic smoothing procedures for the multigrid treatment of elliptic 3D–operators. *Appl. Math. Comput.*, 19:333–345, 1986.
60. U. Trottenberg, C. W. Oosterlee, and A. Schüller. *Multigrid*. Academic Press, London, 2000.
61. C. Weiß. *Data Locality Optimizations for Multigrid Methods on Structured Grids*. PhD thesis, Lehrstuhl für Rechnertechnik und Rechnerorganisation, Institut für Informatik, Technische Universität München, Munich, Germany, Dec. 2001.
62. R. Wienands and C. W. Oosterlee. On three-grid fourier analysis for multigrid. *SIAM J. Sci. Comput.*, 23(2):651–671, 2001.
63. G. Wittum. On the robustness of ILU–smoothing. *SIAM J. Sci. Stat. Comput.*, 10:699–717, 1989.
64. D. Xie and L. Scott. The Parallel U–Cycle Multigrid Method. In *Virtual Proceedings of the 8th Copper Mountain Conference on Multigrid Methods*, 1997. Available at `http://www.mgnet.org`.
65. U. M. Yang. Parallel algebraic multigrid methods - high performance preconditioners. In A. M. Bruaset and A. Tveito, editors, *Numerical Solution of Partial Differential Equations on Parallel Computers*, volume 51 of *Lecture Notes in Computational Science and Engineering*, pp. 209–236. Springer-Verlag, 2005.
66. I. Yavneh. On red-black SOR smoothing in multigrid. *SIAM J. Sci. Comput.*, 17:180–192, 1996.
67. G. Zumbusch. *Parallel Multilevel Methods — Adaptive Mesh Refinement and Loadbalancing*. Advances in Numerical Mathematics. Teubner, 2003.

6

Parallel Algebraic Multigrid Methods – High Performance Preconditioners

Ulrike Meier Yang

Center for Applied Scientific Computing, Lawrence Livermore National Laboratory, Box 808, L-560, Livermore, CA 94551, USA

umyang@llnl.gov

Summary. The development of high performance, massively parallel computers and the increasing demands of computationally challenging applications have necessitated the development of scalable solvers and preconditioners. One of the most effective ways to achieve scalability is the use of multigrid or multilevel techniques. Algebraic multigrid (AMG) is a very efficient algorithm for solving large problems on unstructured grids. While much of it can be parallelized in a straightforward way, some components of the classical algorithm, particularly the coarsening process and some of the most efficient smoothers, are highly sequential, and require new parallel approaches. This chapter presents the basic principles of AMG and gives an overview of various parallel implementations of AMG, including descriptions of parallel coarsening schemes and smoothers, some numerical results as well as references to existing software packages.

6.1 Introduction

The development of multilevel methods was a very important step towards being able to solve partial differential equations fast and efficiently. One of the first multilevel methods was the multigrid method. Multigrid builds on a relaxation method, such as the Gauß-Seidel or the Jacobi method, and was prompted by the discovery, that while these relaxation schemes efficiently damp high frequency errors, they make little or no progress towards reducing low frequency errors. If however one moves the problem to a coarser grid previously low frequency errors turn now into high frequency errors and can be damped efficiently. If this procedure is recursively applied, one obtains a method with a computational cost that depends only linearly on the problem size.

There are two basic multigrid approaches: geometric and algebraic. In geometric multigrid [28], the geometry of the problem is used to define the various multigrid components. Algebraic multigrid (AMG) methods use only the information available in the linear system of equations and are therefore suitable to solve problems on more complicated domains and unstructured grids. AMG was first introduced in the 80s [7, 8, 5, 48]. Since then, a lot of research has been done and many new variants have been developed, e.g. smoothed aggregation [58, 57], AMGe [11], spectral AMGe [14], to just name a few. Whole books have been written on this topic [39, 54]. A good tutorial on multigrid, including AMG, is [12]. A very detailed introduction on AMG is [51].

Focus of this chapter is not an overview of all existing AMG methods (there would not be enough space), but a presentation of the basic idea of AMG, the challenges that come with a parallel implementation of AMG and how to overcome them, as well as the impact this challenge has had on AMG.

With the advent of parallel computers one has sought parallel algorithms. When vector computers where developed in the 70s, it was important to develop algorithms that operated on long vectors to make good use of pipelines or vector registers. For parallel computers with tens or hundreds of processors and slow intercommunication it was necessary to be able to partition an algorithm into big independent pieces in order to avoid large communication cost. Multigrid methods, due to their decreasing levels, did not appear to be good candidates for these machines. However with the development of high performance computer with tens or hundreds of thousands of processors it has become very important to develop scalable algorithms, and therefore multigrid methods as well as the application of multilevel techniques in other algorithms have become very popular in parallel computing.

In this chapter, the concept of AMG as well as various parallel variations will be described. While most of AMG can be parallelized in a straightforward way, the coarsening algorithm and the smoother are more difficult to parallelize. Therefore a large portion of this chapter is devoted to parallel approaches for these components. Additionally, interpolation as well as a few numerical results are presented. Finally, an overview of various parallel software packages is given that contain algebraic multigrid or multilevel codes.

6.2 Algebraic Multigrid - Concept and Description

We begin by outlining the basic principles and techniques that comprise AMG. Detailed explanations may be found in [48]. Consider a problem of the form

$$Au = f, \tag{6.1}$$

where A is an $n \times n$ matrix with entries a_{ij}. For convenience, the indices are identified with grid points, so that u_i denotes the value of u at point i, and the grid is denoted by $\Omega = \{1, 2, \ldots, n\}$. In any multigrid method, the central idea is that "smooth error," e, that is not eliminated by relaxation must be removed by coarse-grid correction. This is done by solving the residual equation $Ae = r$ on a coarser grid, then interpolating the error back to the fine grid and using it to correct the fine-grid approximation by $u \leftarrow u + e$.

Using superscripts to indicate level number, where 1 denotes the finest level so that $A^1 = A$ and $\Omega^1 = \Omega$, the components that AMG needs are as follows:

1. "Grids" $\Omega^1 \supset \Omega^2 \supset \ldots \supset \Omega^M$ with subsets:

Set of coarse points or C-points $C^k, k = 1, 2, \dots M-1$,
Set of fine points or F-points $F^k, k = 1, 2, \dots M-1$.

2. Grid operators $A^1, A^2, \dots, A^M$.
3. Grid transfer operators:
   Interpolation $P^k, k = 1, 2, \dots M-1$,
   Restriction $R^k, k = 1, 2, \dots M-1$.
4. Smoothers $S^k, k = 1, 2, \dots M-1$.

These components of AMG are constructed in the first step, known as the *setup phase*.

AMG Setup Phase:

1. Set $k = 1$.
2. Partition Ω^k into disjoint sets C^k and F^k.
   a) Set $\Omega^{k+1} = C^k$.
   b) Define interpolation P^k.
3. Define R^k (often $R^k = (P^k)^T$).
4. Set $A^{k+1} = R^k A^k P^k$.
5. Set up S^k, if necessary.
6. If Ω^{k+1} is small enough, set $M = k+1$ and stop. Otherwise, set $k = k+1$ and go to step 2.

Once the setup phase is completed, the *solve phase*, a recursively defined cycle, can be performed as follows:

Algorithm: $MGV(A^k, R^k, P^k, S^k, u^k, f^k)$.
If $k = M$, solve $A^M u^M = f^M$ with a direct solver.
Otherwise:
Apply smoother S^k μ_1 times to $A^k u^k = f^k$.
Perform coarse grid correction:
Set $r^k = f^k - A^k u^k$.
Set $r^{k+1} = R^k r^k$.
Apply $MGV(A^{k+1}, R^{k+1}, P^{k+1}, S^{k+1}, e^{k+1}, r^{k+1})$.
Interpolate $e^k = P^k e^{k+1}$.
Correct the solution by $u^k \leftarrow u^k + e^k$.
Apply smoother S^k μ_2 times to $A^k u^k = f^k$.

The algorithm above describes a V(μ_1, μ_2)-cycle, other more complex cycles such as W-cycles can be found in [12].

Coarse grid selection and interpolation must go hand in hand and affect each other in many ways. There are basically two measures which give an indication about the quality of the AMG method, both need to be considered and are important, although depending on the user's priorities one might be more important than the other. The first one, the *convergence factor*, gives an indication on how fast the method converges, i.e. how many iterations are needed to achieve the desired accuracy, the second one, *complexity*, affects the number of operations per iteration and the memory usage.

There are two types of complexities that need to be considered: the *operator complexity* and the *average stencil size*. The operator complexity C_{op} is defined as the quotient of the sum of the numbers of nonzeroes of the matrices on all levels, $A^k, k = 1, ..., M$, divided by the number of nonzeroes of the original matrix $A^1 = A$. This measure indicates how much memory is needed. If memory usage is a concern, it is important to keep this number small. It also affects the number of operations per cycle in the solve phase. Small operator complexities lead to small cycle times. The average stencil size $s(A^k)$ is the average number of coefficients per row of A^k. While stencil sizes of the original matrix are often small, it is possible to get very large stencil sizes on coarser levels. Large stencil sizes can lead to large setup times, even if the operator complexity is small, since various components, particularly coarsening and to some degree interpolation, require that neighbors of neighbors are visited and so one might observe superlinear or even quadratic growth in the number of operations when evaluating the coarse grid or the interpolation matrix. Large stencil sizes can also increase parallel communication cost, since they might require the exchange of larger sets of data.

Both convergence factors and complexities need to be considered when defining the coarsening and interpolation procedures, as they often affect each other; increasing complexities can improve convergence, and small complexities lead to a degradation in convergence. The user needs therefore to decide his/her priority. Note that often a degradation in convergence due to low complexity can be overcome or diminished by using the AMG method as a preconditioner for a Krylov method like conjugate gradient, GMRES, BiCGSTAB, etc.

Many parts of AMG can be parallelized in a straightforward way, since they are matrix or vector operations. This is generally true for the evaluation of the interpolation or prolongation matrix as well as the generation of the triple matrix product $R^k A^k P^k$. Certainly all those operations require communication among processors and data exchange in some way, however all of this can be done in a straightforward way. There are however two components which present potentially a serious challenge, the coarsening routine as well as the relaxation routine. The original coarsening routine as described in [48] as well as the basic aggregation procedure are inheritantly sequential. Also, the relaxation routine used in general is the Gauß-Seidel algorithm, which is also sequential in nature. In the following sections, the various components, coarse grid selection, interpolation and smoothing are described.

6.3 Coarse Grid Selection

Before describing any parallel coarsening schemes, we will describe various sequential coarsening schemes, since most parallel schemes build on these.

6.3.1 Sequential Coarsening Strategies

There are basically two different ways of choosing a coarse grid. The first approach (which can be found e.g. in [48, 51]) strives to separate all points i into either coarse points or C-points, which will be taken to the next level, and fine points or F-points, which will be interpolated by the C-points. The second approach, coarsening by aggregation or agglomeration ([58]), accumulates aggregates which will be the coarse "points" for the next level.

"Classical" Coarsening

Since most likely not all matrix coefficients are equally important for the determination of the coarse grids, one should only consider those matrix entries that are sufficiently large. We introduce the concept of *strong influence* and *strong dependence*. A point i depends strongly on j or j strongly influences i if

$$-a_{ij} \geq \theta \max_{k \neq i}(-a_{ik}). \tag{6.2}$$

Note that this definition was originally motivated by the assumption that A is a symmetric M-matrix, i.e. a matrix that is positive definite and off-diagonally nonpositive, it can however formally be applied to more general matrices. The size of the *strength threshold* θ can have a significant influence on complexities, particularly stencil size, as well as convergence, as is demonstrated in Section 6.6. In the classical coarsening process (which we will denote Ruge-Stüben or RS coarsening) the attempt is made to fulfill the following two conditions:

(C1): For each point j that strongly influences an F-point i, j is either a C-point or it strongly depends on a C-point k that also strongly influences i.

(C2): The C-points should be a maximal independent subset of all points, i.e. no two C-points are connected to each other, and if another C-point is added the independence is lost.

(C1) is designed to insure the quality of interpolation, while (C2) is designed to restrict the size of the coarse grids. In general, it is not possible to fulfill both conditions, therefore (C1) is enforced, while (C2) is used as a guideline. We will show in Section 6.4 why (C1) is important.

RS coarsening consists of two passes and is illustrated in Figure 6.1 for a 4×4-grid. In the first pass, each point i is assigned a measure λ_i, which equals the number of points that are strongly influenced by i. Then a point with a maximal λ_i (there usually will be several) is selected as the first coarse point. Now all points that strongly depend on i become F-points. For all points that strongly influence these new F-points, λ_j is incremented by the number of new F-points that j strongly influences in order to increase j's chances of becoming a C-point. This process is repeated until all points are either C- or F-points.

Since this first pass does not guarantee that condition (C1) is satisfied, it is followed by a second pass, which examines all strong $F-F$ connections for common coarse neighbors. If (C1) is not satisfied new C-points are added. This is illustrated in the second part of Figure 6.1, where solid lines denote strong $F-F$ connections that do not satisfy condition (C1).

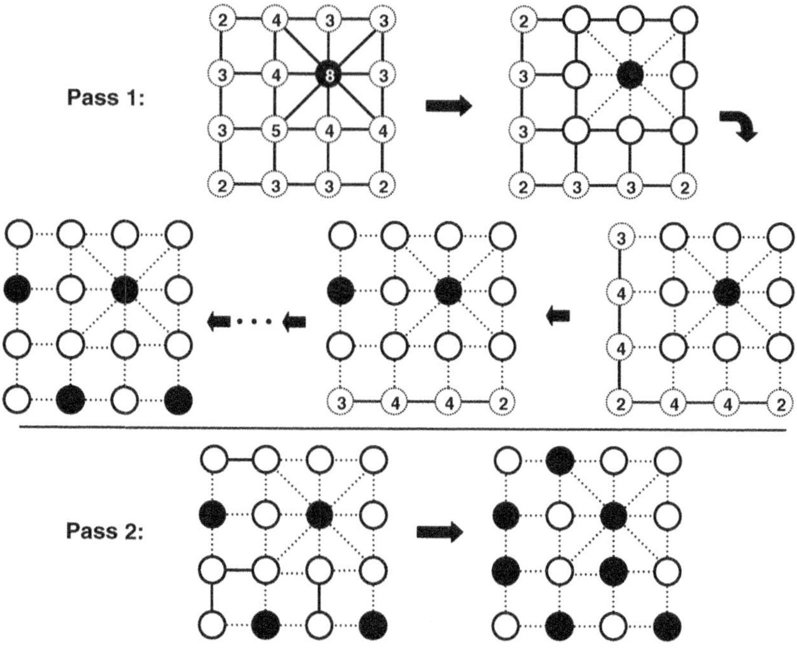

Fig. 6.1. RS coarsening. Black points denote C-points, white points with solid border denote F-points, white points with dotted border denote undetermined points. Pass 2: solid lines denote strong $F-F$ connections that do not satidfy condition (C1).

Aggressive Coarsenings

Experience has shown [51] that often the second pass generates too many C-points, causing large complexities and inefficiency. Therefore condition (C1) has been modified to the following.

(C1′): Each F-point i needs to strongly depend on at least one C-point j.

Now just the first pass of the RS coarsening fulfills this requirement. This method leads to better complexities, but worse convergence.

Even though this approach often decreases complexities significantly, complexities can still be quite high and require more memory than desired. Allowing C-points to be even further apart leads to *aggressive coarsening*. This is achieved by the following new definition of strength: A variable i is *strongly n-connected along a path of length* l to a variable j, if there exists a sequence of variables $i_0, i_1, \ldots i_l$, with $i = i_0$ and $j = i_l$ and i_k strongly connected (as previously defined) to i_{k+1} for $k = 0, \ldots, l-1$. A variable i is *strongly n-connected w.r.t.* (p, l) to a variable j, if at least p paths of lengths $\leq l$ exist such that i is strongly n-connected to j along each of these paths. This can be most efficiently implemented by applying the first pass of RS coarsening twice, the first time as described in the previous section, the

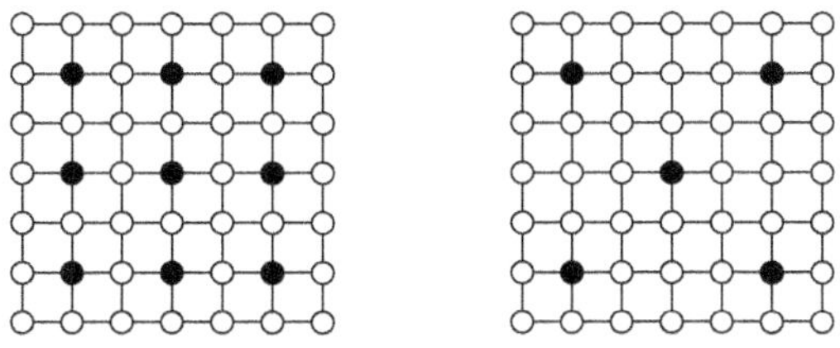

Fig. 6.2. Various sequential coarsenings for a 5-point Laplacian.

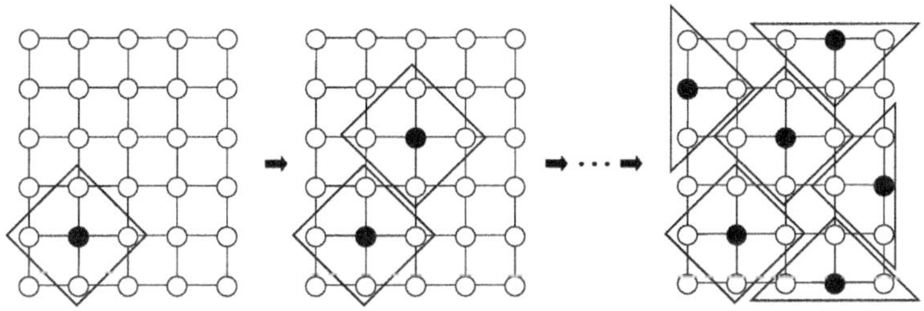

Fig. 6.3. Coarsening by aggregation. Black points denote root points, boxes and triangles denote aggregates.

second time by defining strong n-connectivity w.r.t. (p, l) only between the resulting C-points (via neighboring F-points). For further details see [51]. The result of applying aggressive A2 coarsening, i.e. choosing $p = 2$ and $l = 2$, and aggressive A1 coarsening, i.e. $p = 1$ and $l = 2$, to the 5-point Laplacian on a 7×7-grid is illustrated in Figure 6.2.

Coarsening by Aggregation

For the aggregation scheme, a different concept of strength is used. Here only matrix coefficients a_{ij} are considered, if they fulfill the following condition:

$$|a_{ij}| > \theta \sqrt{|a_{ii} a_{jj}|}. \tag{6.3}$$

An aggregate is defined by a root point i and its neighborhood, i.e. all points j, for which a_{ij} fulfills (6.3). Now the basic aggregation procedure consists of the following two phases. In the first pass, a root point is picked that is not adjacent to any existing aggregate. This procedure is repeated until all unaggregated points are adjacent to an aggregate. It is illustrated in Figure 6.3 for a 5-point Laplacian on a 6×5-grid. In the second pass, all remaining unaggregated points are either integrated into already existing aggregates or used to form new aggregates. Since root points are connected by paths of length of at least 3, this approach leads to fast coarsening and small complexities. While aggregation is fundamentally different from classical coarsening, many of the same concerns arise. In particular, considerable care must be taken within the second pass when deciding to create new aggregates and what points should be placed into already existing aggregates. If too many aggregates are created in this phase, complexities grow. If aggregates are enlarged too much or have highly irregular shapes, convergence rates suffer.

6.3.2 Parallel Coarsening Strategies

There are various approaches of parallelizing the coarse grid selection schemes described in the previous section, which are described in the following subsections.

Decoupled Coarsening Schemes

The most obvious approach to parallelize any of the coarsening schemes described in the previous section is to partition all variables into subdomains, assign each processor a subdomain, coarsen the variables on each subdomain using any of the methods described above, and find a way of dealing with the variables that are located on the processor boundaries.

The easiest option is to just coarsen independently on each subdomain while ignoring the processor boundaries. Such an approach is the most efficient one, since it requires no communication, but will most likely not produce a very good coarse grid. The decoupled RS coarsening, which will be denoted by RS0 coarsening, generally violates condition (C1) by generating strong $F - F$ connections without common coarse neighbors (see Figure 6.5a, which shows the coarse grid generated by RS0 coarsening on 4 processors; black points denote C-points, while white and gray points denote F-points.) and often leads to poor convergence, see Section 6.6 and [26]. While in practice this approach might lead to fairly good results for coarsening by aggregation [55], it can produce many aggregates near processor boundaries that are either smaller or larger than an ideal aggregate and so lead to larger complexities or have a negative effect on convergence. Another disadvantage of this approach is that it cannot have fewer coarse points or aggregates than processors. In the case of thousands of processors this leads to a large grid and a large system on the coarsest level and might be inefficient to solve using a direct solver. Ways to overcome this problem are described at the end of this section.

Coupled Coarsening Strategies

If we want to improve RS0 coarsening, we need to find ways to deal with the variables that are located on the processor boundaries. This starts with a decision about whether one wants to compute the measures locally or globally, i.e. whether one wants to include off processor influences. The use of global measures can improve the coarsening and convergence, but is in general not enough to fulfill condition (C1). One possible way of treating this problem is — after one has performed a first and a second pass on each processor independently — to perform a third pass only on the processor boundary points which will add further C-points and thus ensure that condition (C1) is fulfilled. This approach is called RS3 coarsening and can be found in [26]. It is illustrated in Figure 6.5a for a 5-point Laplacian on a 10×10-grid on 4 processors. The black points denote the C-points that have been generated in the first (and second) pass, the gray points denote the C-points generated in the third pass. One of the disadvantages of this approach is that this can generate C-point clusters on the boundaries, thus increasing stencil sizes at the boundaries where one would like to avoid those, in order to keep communication cost low.

In the coupled aggregation method, aggregates are first built on the boundary. This step is not completely parallel. When there are no more unaggregated points adjacent to an aggregate on the processor boundaries, one can proceed to choose aggregates in the processor interiors, which can be done in parallel. In the third phase unaggregated points on the boundaries and in the interior are swept into local aggregates. Finally, if there are any remaining points, new local aggregates are formed. This process yields significantly better aggregates and does not limit the coarseness of grids to the number of processors, see [55].

Parallel Independent Set Coarsenings

A completely parallel approach is suggested in [17, 26]. It is based on parallel independent set algorithms as described by Luby and Jones and Plassman in [38, 32]. This algorithm, the CLJP (Cleary-Luby-Jones-Plassman) coarsening, begins by generating global measures as in RS coarsening, and then adding a random number between 0 and 1 to each measure, thus making them distinctive. It is now possible to find unique local maxima. The algorithm, which is illustrated for a small example on a 4×4-grid in Figure 6.4, proceeds as follows: If i is a local maximum, make i a C-point, eliminate the connections to all points j that influence i and decrement j's measure. (Thus instead of immediately making C-point neighbors F-points, we increase their likelihood of becoming F-points. This models the two passes of the RS coarsening into one pass.) Further for all points j that depend on i, remove its connection to i and examine all points k that depend on j on whether they also depend on i. If i is a common neighbor for both k and j decrement the measure of j and remove the edge connecting k and j from the graph. If a measure gets smaller than 1, the point associated with it becomes an F-point. The advantage of this procedure is, assuming one uses the same global set of random numbers, that it is completely independent of the number of processors, a feature that is desired by some users. It also facilitates debugging. The additional advantage of this procedure is that it does not require the existence of a coarse point in each processor as the coarsening schemes above and thus coarsening does not slow down on the coarser levels. While this approach works fairly well on truly unstructured grids, it often leads to C-point clusters and fairly high complexities, see Figure 6.5b. These appear to be caused by the enforcement of condition (C1).

To reduce operator complexities, while keeping the property of being independent of the number of processors, a new algorithm, the PMIS coarsening [19], has been developed that is more comparable to using one pass of the RS coarsening.

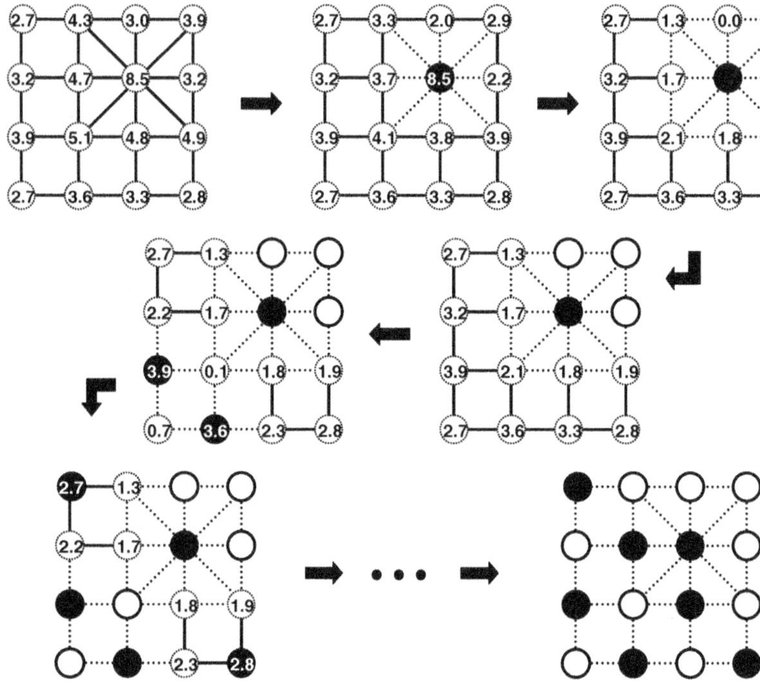

Fig. 6.4. CLJP coarsening. Black points are C-points, white points with solid border F-points, white points with dotted border undetermined points.

While it does not fulfill condition (C1), it fulfills condition (C1′). PMIS coarsening begins just as the CLJP algorithm with distinctive global measures, and sets local maxima to be C-points. Then points that are influenced by C-points are made F-points, and are eliminated from the graph. This procedure will continue until all points are either C- or F-points. For an illustration of the PMIS coarsening applied to a 5-point Laplacian see Figure 6.5c.

Another aggregation scheme suggested in [55] is also based on a parallel maximally independent set algorithm. Since the goal is to find an initial set of aggregates with as many points as possible with the restriction that no root point can be adjacent to an existing aggregate. Therefore maximizing the number of aggregates is equivalent to finding the largest number of root points such that the distance between any two root points is at least three. This can be accomplished by applying a parallel maximally independent set (MIS) algorithm, e.g. the asynchronous distributed memory algorithm ADMMA [1], to the square of the matrix in the first phase of the coupled aggregation scheme.

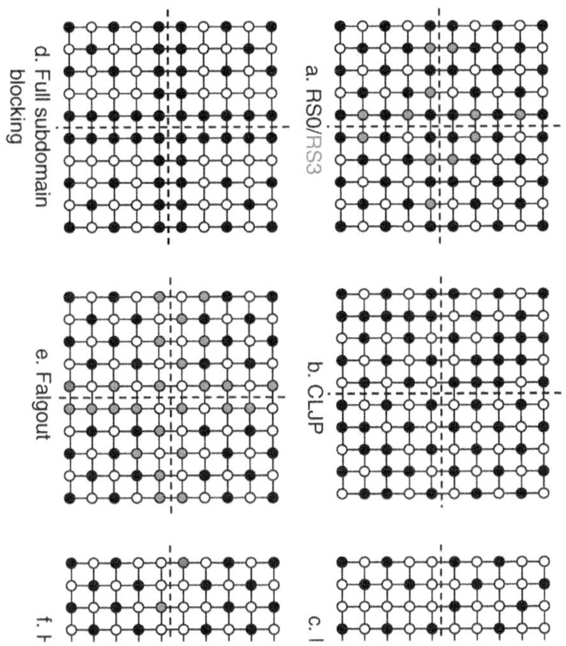

Fig. 6.5. Various parallel coarsenings of a 5-point Laplacian on a 10×10-grid using 4 processors. White points are F-points, black points are C-points, gray points are C-points generated during special boundary treatments.

Subdomain Blocking

Another parallel approach is *subdomain blocking* [35]. Here, coarsening starts with the processor boundaries, and one then proceeds to coarsen the inside of the domains. *Full subdomain blocking* is performed by making all boundary points coarse and then coarsening into the interior of the subdomain using any coarsening scheme one wishes to use, such as one pass of RS coarsening or any of the aggressive coarsening schemes. The disadvantage of this scheme is that it generates far too many C-points on the boundary, which can cause problems on the coarser grids. For an illustration see Figure 6.5d. A method, which avoids this problem, is *minimum subdomain blocking*. This approach uses standard coarsening on the boundaries and then coarsens the interior of the subdomains.

Combination Approaches and Miscellaneous

Another option which has shown to work quite well for structured problems is the following combination of the RS and the CLJP coarsening which is based on an idea by Falgout [26]. This coarsening starts out as RS0 coarsening. It then uses the C-points that have been generated in the first step and are located in the interior of each processor as the first independent set (i.e. they will all remain C-points) and feeds them into the CLJP-algorithm. The resulting coarsening, which satisfies condition (C1), fills the boundaries with further C-points and possibly adds a few in the interior of the subdomains, see Figure 6.5e. A more aggressive scheme, which satisfies condition (C1′), and uses the same idea, is the HMIS coarsening [19]. It performs only the first pass of RS0 coarsening to generate the first independent set, which then is used by the PMIS algorithm. Figure 6.5f shows its application to a 5-point Laplacian on 4 processors. The black C-points are the first independent set, which have been generated by the first pass of the RS0 coarsening, whereas the gray C-points have been determined by the application of the PMIS coarsening.

Another approach is to color the processors so that subdomains of the same color are not connected to each other. Then all these subdomains can be coarsened independently. This approach can be very inefficient since it might lead to many idle processors. An efficient implementation that builds on this approach can be found in [33]. Here the number of colors is restricted to n_c, i.e. processors with color numbers higher than n_c are assigned the color n_c. Good results were achieved using only two colors on the finest level, but allowing more colors on the coarser levels.

Dealing with the Coarser Levels

One of the difficulties that arises when parallelizing multilevel schemes is the treatment of the coarser levels. Since many AMG schemes use the sequential approach on each processor and this often requires at least one coarse point or aggregate on each processor, coarsening will slow down on the coarser grids leading to a coarsest grid of the size of at least the number of processors. This could be thousands of unknowns and lead to a very inefficient coarse grid solve, and potentially prevent scalability. There are various possibilities to deal with this situation. Via aggregation one can combine the contents of various processors, when a certain size is achieved. This approach also coarsens previous processor boundaries and thus deal with cluster of coarse points in these areas. One disadvantage of this approach is that it requires a lot of communication and data transfer across processors.

Another possibility to deal with a slowdown in coarsening is to switch to a coarsening scheme that does not require C-points on each processor, such as CLJP or PMIS, when coarsening slows.

6.4 Interpolation

In this section, we will consider the construction of the interpolation operator. The interpolation of the error at the F-point i takes the form

$$e_i = \sum_{j \in C_i} w_{ij} e_j \tag{6.4}$$

where w_{ij} is an interpolation weight determining the contribution of the value e_j in the final value e_i, and C_i is the subset of C-points whose values will be used to interpolate a value at i.

In classical AMG the underlying assumption is that algebraically smooth error corresponds to having very small residuals; that is the error is smooth when the residual $r = f - Au \approx 0$. Since the error, e, and the residual are related by $Ae = r$, smooth error has the property $Ae \approx 0$. Let i be an F-point to which we wish to interpolate, N_i the neighborhood of i, i.e. the set of all points, which influence i. Then the ith equation becomes

$$a_{ii}e_i + \sum_{j \in N_i} a_{ij}e_j = 0. \tag{6.5}$$

Now, "classical" interpolation as described in [48] proceeds by dividing N_i into the set of coarse neighbors, C_i, the set of strongly influencing neighbors, F_i^s, and the set of weakly influencing neighbors, F_i^w. Using those distinctions as well as condition (C1), which guarantees that a neighbor in F_i^s is also strongly influenced by at least one point in C_i yields the following interpolation formula

$$w_{ij} = -\frac{1}{a_{ii} + \sum_{k \in F_i^w} a_{ik}} \left(a_{ij} + \sum_{k \in F_i^s} \frac{a_{ik}a_{kj}}{\sum_{m \in C_i} a_{km}} \right). \tag{6.6}$$

Obviously, this interpolation formula fails whenever (C1) is violated, since there would be no m and one would divide by zero. One can somewhat remedy this by including elements of F_i^s that violate (C1) in F_i^w, but this will affect the quality of the interpolation and lead to worse convergence. Nevertheless often good results can be achieved with this interpolation, if the resulting AMG method is used as a preconditioner for a Krylov method, see [19]. One advantage of this interpolation formula is that it only involves immediate neighbors and thus is easier to implement in parallel, since it requires only one layer of ghost points located on a neighbor processor.

Another interpolation formula which also requires only immediate neighbors, and can be used when (C1) is violated, but leads in general to worse convergence rates is *direct interpolation*:

$$w_{ij} = -\left(\frac{\sum_{k \in N_i} a_{ik}}{\sum_{l \in C_i} a_{il}} \right) \frac{a_{ij}}{a_{ii}}. \tag{6.7}$$

It is easy to implement sequentially as well as in parallel.

However, if one needs an interpolation that yields lower convergence rates, a better approach is *standard interpolation*, which uses an extended neighborhood. For all points $j \in F_i^s$, one substitutes e_j in (6.5) by $-\sum_{k \in N_j} a_{jk}e_k/a_{jj}$. This leads to a new formula

$$\hat{a}_{ii}e_i + \sum_{j \in \hat{N}_i} \hat{a}_{ij}e_j = 0, \quad \hat{N}_i = \{j \neq i : \hat{a}_{ij} \neq 0\}. \tag{6.8}$$

The interpolation weights are then defined as in the direct interpolation by replacing a by $\hat{a}$ and N by $\hat{N}$.

If one uses any of the aggressive coarsening schemes, it is necessary to use long range interpolation, such as *multipass interpolation*, in order to achieve reasonable convergence. Multipass interpolation starts by deriving interpolation weights using direct interpolation for all fine points that are influenced by coarse points (which are connected by a path of length 1). In the following pass it evaluates weights using the same approach as in standard interpolation for points that are influenced by those points for which interpolation weights have already been determined and which are connected by a path of length 1. This process is repeated until weights have been obtained for all remaining points. One disadvantage of multipass interpolation is that since it is based on direct interpolation it often converges fairly slowly. Therefore it has been suggested to improve this interpolation formula by using an a posteriori Jacobi relaxation and applying it to the interpolation operator as follows

$$P_{FC}^{(n)} = (I - D_{FF}^{-1} A_{FF}) P_{FC}^{(n-1)} - D_{FF}^{-1} A_{FC}, \tag{6.9}$$

where P_{FC} and $(A_{FF} \quad A_{FC})$ consist of those rows of the matrices P and A that refer to the F-points only, and D_{FF} is the diagonal matrix with the diagonal of A_{FF}. Of course using one or more sweeps of Jacobi interpolation will increase the number of nonzeroes of P and also the complexity of the AMG method. Therefore often *interpolation truncation* is used, which truncates those elements in the interpolation operator that are absolutely smaller than a chosen *truncation factor* τ, leading to smaller stencil sizes.

Parallel implementation of long range interpolation can be tedious, since due to the long ranges it involves several layers of off processor points. Some of these points might even be located in processors that are not neighbor processors according to the chosen data structure. Therefore this approach might require expensive communication. One way to avoid this problem has been suggested in [51]. There, interpolation proceeds only into the interior of the subdomain and not across boundaries. While this approach can be implemented very efficiently on a parallel computer, it can cause convergence problems.

Aggregation methods use a different type of interpolation. At first a tentative interpolation operator $(P^k)^{(0)}$ is created using one or more seed vectors. These seed vectors should correspond to components that are difficult to smooth. The tentative interpolation interpolates the seed vectors perfectly by ensuring that all of them are in the range of the interpolation operator. For Poisson problems the entries are defined as follows:

$$(P^k)_{ij}^{(0)} = \begin{cases} 1 & \text{if point } i \text{ is contained in aggregate } j \\ 0 & \text{otherwise,} \end{cases} \tag{6.10}$$

For specific applications such as elasticity problems, more complicated tentative prolongators can be derived based on rigid body motions. Once the tentative prolongator is created it is smoothed via a damped Jacobi iteration

$$(P^k)^{(n)} = (I - \omega (D^k)^{-1} A^k)(P^k)^{(n-1)}, n = 0, \ldots, \tag{6.11}$$

where D^k is the diagonal matrix with the diagonal of A^k. For a more detailed description of this method see [58, 57].

6.5 Smoothing

One important component of algebraic multigrid is the smoother. A good smoother will reduce the oscillatory error components, whereas the 'smooth' error is transferred to the coarser grids. Although the classical approach of AMG focused mainly on the Gauß-Seidel method, the use of other iterative solvers has been considered. Gauß-Seidel has proven to be an effective smoother for many problems, however its main disadvantage is its sequential nature. For elasticity problems, Schwarz smoothers have shown to be extremely efficient, here again the most efficient ones are multiplicative Schwarz smoothers, which are highly sequential.

The general definition of a smoother S applied to a system $Au = f$ is

$$e_{n+1} = Se_n \textbf{ or } u_{n+1} = Su_n + (I - S)A^{-1}f, \tag{6.12}$$

where $e_n = u_n - u$ denotes the error. Often S is the iteration matrix of an iterative solver $S = I - Q^{-1}A$, where Q is a matrix that is part of a splitting $Q + (Q - A)$ of A, e.g. Q is the lower triangular part of A for the Gauß-Seidel method. Other approaches such as polynomial smoothers or approximate inverse set Q^{-1} to be an approximation of A^{-1} that can easily be evaluated. Often iterative schemes that are used as smoothers are also presented as

$$u_{n+1} = u_n + Q^{-1}(f - Au_n). \tag{6.13}$$

6.5.1 Parallel Relaxation Schemes

There are various conventional relaxation schemes that are already parallel, such as the Jacobi or the block Jacobi algorithm. Here Q is the diagonal matrix (or block diagonal matrix) with the diagonal (or block diagonal) elements of A. These relaxation schemes require in general a smoothing or relaxation parameter for good convergence, as discussed in the next subsection. Another equally parallel related algorithm that in general leads to better convergence than Jacobi relaxation, is C-F Jacobi relaxation, where first the variables associated with C-points are relaxed, then the F-variables. In algebraic multigrid C-F Jacobi is used on the downward cycle, and F-C Jacobi, i.e. relax the F-variables before the C-variables, on the upward cycle.

6.5.2 Hybrid Smoothers and the Use of Relaxation Parameters

The easiest way to implement any smoother in parallel is to just use it independently on each processor, exchanging boundary information after each iteration. We will call such a smoother a *hybrid smoother*. Using the terminology of (6.13), for a computer with p processors Q would be a block diagonal matrix with p diagonal blocks $Q_k, k = 1, ..., p$. For example, if one applies this approach to Gauß-Seidel, Q_k are lower triangular matrices (we call this particular smoother hybrid Gauß-Seidel; it has also been referred to as Processor Block Gauß-Seidel [3]). While this approach is easy to implement, it has the disadvantage of being more similar to a block Jacobi method, albeit worse, since the block systems are not solved exactly. Block Jacobi methods can converge poorly or even diverge unless used with a suitable damping parameter. Additionally, this approach is not scalable, since the number of blocks increases with the number of processors and with it the number of iterations increases. In spite of this, good results can be achieved by setting $Q = (1/\omega)\tilde{Q}$ and choosing a suitable relaxation parameter ω. Finding good parameters is not easy and made even harder by the fact that in a multilevel scheme one deals with a new system on each level, which requires new parameters. It is therefore important to find an automatic procedure to evaluate these parameters. Such a procedure has been developed for symmetric positive problems and smoothers in [59] using convergence theory for regular splittings. A good smoothing parameter for a positive symmetric matrix A is $\omega = 1/\lambda_{max}(\tilde{Q}^{-1}A)$, where $\lambda_{max}(M)$ denotes the maximal eigenvalue of M. A good estimate for this value can be obtained by using a few relaxation steps of Lanczos or conjugate gradient preconditioned with $\tilde{Q}$. This procedure can be applied to any symmetric positive definite hybrid smoother, such as hybrid symmetric Gauß-Seidel, Jacobi, Schwarz smoothers or symmetric positive definite variants of sparse approximate inverse or incomplete Cholesky smoothers.

6.5.3 Multicoloring Approaches

Another approach to parallelize Gauß-Seidel or similar smoothers such as block Gauß-Seidel or multiplicative Schwarz smoothers is to color subsets of points in such a way that subsets of the same color are independent of each other and can be processed in parallel. There are various parallel coloring algorithms available [32], however use of those as smoothers has shown to be inefficient, since often they generate too many colors, particularly on the coarser levels. Another approach is to color the processors, which will give some parallelism, but is overall not very efficient, since it leads to a large number of processors being idle most of the time.

One truly efficient implementation of a multicolor Gauß-Seidel method is described in [2, 3]. Here the nodes are ordered in such a way that interior nodes are processed while waiting for communication necessary to process boundary nodes. The algorithm first colors the processors and uses an ordering of the colors to receive an ordering of the processors. Each processor partitions its nodes into interior and boundary nodes, which are further divided into boundary nodes that require only communication with higher processors, those that communicate only with lower processors and the remaining boundary nodes. Interior nodes are also divided into smaller sets which are determined by taking into account computational cost to ensure that updates of boundary points occur at roughly the same time and thus idle time is minimized. For further details see [2].

6.5.4 Polynomial Smoothers

Another very parallel approach which promises to be also scalable is the use of polynomial smoothers. Here Q^{-1} in (6.13) is chosen to be a polynomial $p(A)$ with

$$p(A) = \sum_{0 \le j \le m} \alpha_j A^j. \tag{6.14}$$

To use this iteration as a multigrid smoother, the error reduction properties of $q(A) = I - p(A)A$ must be complementary to those of the coarse grid correction. One way to achieve this is by splitting the eigenvalues of A into low energy and high energy groups. The ideal smoother is then given by a Chebyshev polynomial that minimizes over the range that contains the high energy eigenvalues subject to the constraint $q(0) = 1$. If one knows the two eigenvalues that define the range, it is easy to compute the coefficients of the polynomials via a simple recursion formula.

Another polynomial smoother is the MLS (multilevel smoother) polynomial smoother. It is based on a combined effect of two different smoothing procedures, which are constructed to complement each other on the range of coarse grid correction. The two procedures are constructed so that their error propagation operators have certain optimum properties. The precise details concerning this polynomial can be found in [10] where this smoother was first developed in conjunction with the smoothed aggregation method.

The advantage of these methods is that they are completely parallel and their parallelism is independent of the number of processors. The disadvantage is that they require the evaluation of eigenvalues. Both require the maximal eigenvalue of a matrix and the Chebyshev polynomial smoother also the lower range of the high frequency eigenvalues. However, if these smoothers are used in the context of smoothed aggregation, which is usually the case, the maximal eigenvalues are already available since they are needed for the smoothing of the interpolation. Further details on this topic can be found in [3].

6.5.5 Approximate Inverse, Parallel ILU and More

Obviously, one can use any other parallel solver or preconditioner as a smoother. Q^{-1} can be chosen as any approximation to A^{-1}, e.g. a sparse approximate inverse. The use of approximate inverses in the context of multilevel methods is described in [53, 13, 40]. ParaSails, a very efficient parallel implementation of an approximate inverse preconditioner, approximates Q by minimizing the Frobenius norm of $I - QA$ and uses graph theory to predict good sparsity patterns for Q [15, 16]. Other options for smoothers are incomplete LU factorizations with $Q = \tilde{L}\tilde{U}$, where $\tilde{L}$ and $\tilde{U}$ are sparse approximations of the actual lower and upper triangular factors of A. There are various good parallel implementations available such as Euclid [30, 31], which obtains scalable parallelism via local and global reorderings, or PILUT [34], a parallel ILUT factorization. It is also possible to use conjugate gradient as a smoother.

6.6 Numerical Results

This section gives a few numerical results to illustrate some of the effects described in the previous sections. We apply various preconditioned AMG methods to 2-

dimensional (2D) and 3-dimensional discretizations (3D) of the Laplace equation

$$-\Delta u = f, \tag{6.15}$$

with homogeneous Dirichlet boundary conditions on a unit square or unit cube. We use the codes BoomerAMG and MLI from the *hypre* library. BoomerAMG is mostly built on the "classical" AMG method and provides the coarsening algorithms: RS0, RS3, CLJP, Falgout, PMIS and HMIS. It deals with a slowdown in coarsening by switching to CLJP, and uses Gaussian elimination on the coarsest level, which is at most of size 9. It uses a slightly modified form of the "classical" interpolation described in Section 6.4 [26, 19]. MLI is an aggregation code. It uses the fast parallel direct solver SuperLU [52] on the coarsest level, which in our experiments is chosen to be at least 1024, the number of processors in our largest test run, and smaller than 4100. MLI's coarsening strategy is decoupled aggregation. We consider two variants: aggregation (AG), which uses the unsmoothed interpolation operator $(P^k)^{(0)}$ in (6.10), and smoothed aggregation (SA), which applies one step of smoothed Jacobi to $(P^k)^{(0)}$. For all methods, hybrid symmetric Gauß-Seidel smoothing was used. While RS3, Falgout and CLJP work almost as well as stand-alone solvers for the first two test problems, all other methods are significantly improved when accelerated by a Krylov method. Therefore in all of our experiments, they were used as preconditioners for GMRES(10).

All test problems were run on the Linux cluster MCR at Lawrence Livermore National Laboratory. We use the following notations in the tables:

- p: number of processors,
- θ: strength threshold as defined in (6.2) for RS0, RS3, CLJP, Falgout, PMIS and HMIS, as defined in (6.3) for AG and SA, (Section 6.3),
- τ: interpolation truncation factor (Section 6.4),
- C_{op}: operator complexity (Section 6.2),
- s_{avg}: maximal average stencil size, i.e. $\max_{1 \leq k \leq M} s(A^k)$, (Section 6.2),
- $\#its$: number of iterations,
- t_{setup}: setup time in seconds,
- t_{solve}: time of solve phase in seconds,
- t_{total}: total time in seconds.

The first test problem is a 2D Laplace problem with a 9-point discretization on a unit square. We kept the number of grid points fixed at 122,500 (350×350) on each processor, while increasing the number of processors and the overall problem size. The setup and total times for runs using 1, 4, 16, 64, 256 and 1024 processors are presented in Figure 6.6. For the aggregation based solvers AG and SA, $\theta = 0$ was chosen. For almost all other solvers, $\theta = 0.25$ was used, since this choice led to the best performance. RS0, however, performed significantly better with $\theta = 0$. Table 6.1 contains complexities and numbers of iterations for the 4 processor and 1024 processor runs. It shows that operator complexities are constant across an increasing number of processors, while stencil sizes increase for RS0, RS3, Falgout and SA. The aggregation methods obtain the best operator complexities, while CLJP's operator

Table 6.1. Complexities and numbers of iterations for GMRES(10) preconditioned with various AMG methods.

	$p = 4$			$p = 1024$		
Method	C_{op}	s_{avg}	$\#its$	C_{op}	s_{avg}	$\#its$
RS0	1.33	11	14	1.33	18	172
RS3	1.34	20	6	1.36	69	7
Falgout	1.34	16	6	1.35	32	8
CLJP	1.92	32	8	1.96	36	9
PMIS	1.25	11	24	1.23	11	42
HMIS	1.33	9	10	1.33	12	19
AG	1.13	9	37	1.13	9	93
SA	1.13	9	10	1.13	27	21

complexities are the highest. The overall best stencil sizes are obtained by AG, HMIS and PMIS. The most significant growth in stencil size occurs for RS3, apparently caused by the addition of the third pass on the boundaries. The effect of this stencil growth can clearly be seen in Figure 6.6a. AG and RS0 have the largest number of iterations, showing the influence of the decoupled coarsening. It also turns out that increasing the strength threshold for RS0 leads quickly to disastrous convergence results: for $\theta = 0.25$, $p = 1024$, RS0 needs more than 700 iterations to converge. While SA also uses decoupled coarsening, its interpolation operator is significantly improved by the use of one weighted Jacobi iteration, leading to faster convergence.

The results in Figure 6.6 show that the overall fastest method for this problem is Falgout, closely followed by HMIS and CLJP. While RS3 has slightly better convergence, the fact that its average stencil sizes increase faster with increasing number of processors than those of the other coarsenings, leads to faster increasing setup times and worse total times than Falgout, HMIS and CLJP. The smallest setup time is obtained by RS0, which requires no communication during the coarsening phase, followed by PMIS and HMIS, which benefit from their small stencil sizes. SA's setup times are larger than those of AG due to the smoothing of the interpolation operator. Note that while the results for the aggregation methods demonstrate good complexities, and thus low memory usage, they are not necessarily representative of smoothed aggregation in general or other implementations of this method. The use of more sophisticated parallel aggregation schemes, a different treatment of the coarsest level and the use of other smoothers might yield better results.

The second test problem is a 7-point finite difference discretization of the 3-dimensional Laplace equation on a unit cube using $40 \times 40 \times 40$ points per processor. For three-dimensional test problems, it becomes more important to consider complexity issues. While the choice of $\theta = 0.25$ often still leads to best convergence, complexities can become so large for this choice that setup times and memory requirements might be unreasonable. The effect of using different strength thresholds on complexities and convergence is illustrated in Table 6.2 for Falgout-GMRES(10) for 288 processors. Based on the results of this experiment we are choosing $\theta = 0.75$

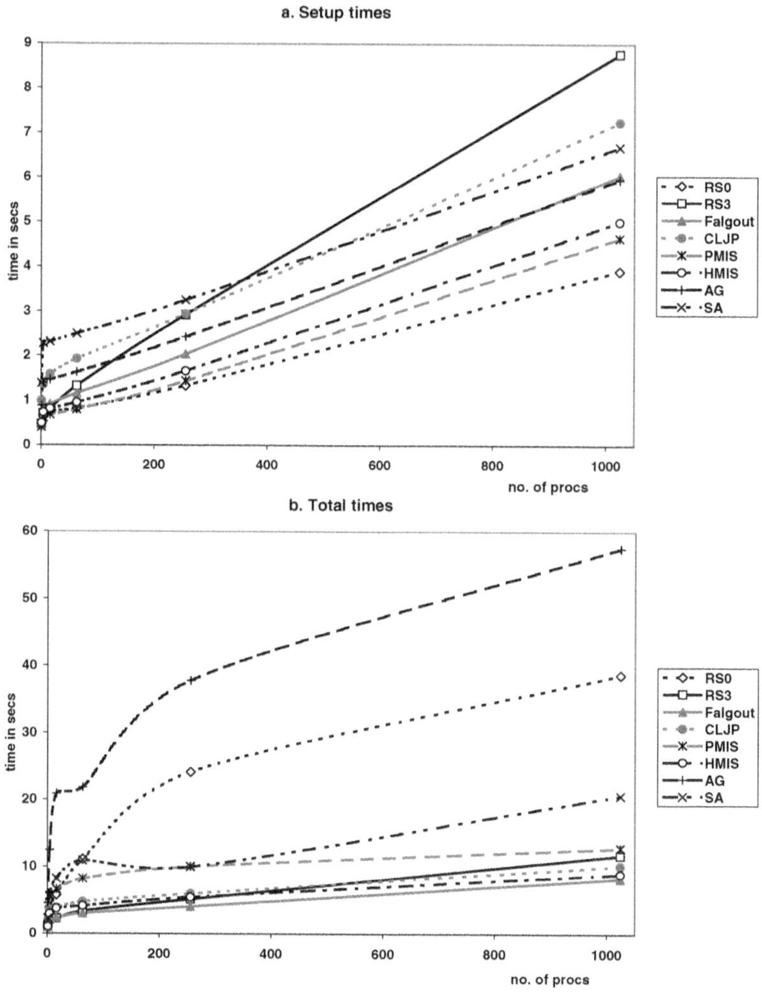

Fig. 6.6. Setup and total times for a 2-dimensional Laplacian problem with a 9-point stencil for an increasing number of processors and increasing problem size with a fixed number of grid points (350×350) per processor.

for our further experiments. Another useful tool to decrease stencil size is interpolation operator truncation. Increasing the truncation factor τ also decreases s_{avg}, while only slightly decreasing C_{op}. Best timings when choosing $\theta = 0$ and increasing τ were achieved for $\tau = 0.3$. In general, this choice leads to larger stencil sizes than choosing $\theta = 0.75$ and $\tau = 0$, but better convergence, as can be seen in Table 6.3. When increasing the number of processors and with it the overall problem size, operator complexities and stencil sizes initially increase noticeably, particularly for RS3, Falgout and CLJP. They, however, cease growing for large problem sizes. The complexities presented in Table 6.3 are therefore close to those that can be obtained for

Table 6.2. Effect of strength threshold on Falgout-GMRES(10) for a 7-point finite difference discretization of the 3-dimensional Laplacian on 216 processors with 40^3 degrees of freedom per processor.

θ	C_{op}	s_{avg}	$\#its$	t_{setup}	t_{solve}	t_{total}
0.00	3.26	458	6	21.82	2.67	24.49
0.25	6.32	3199	5	96.77	7.93	104.70
0.50	5.32	538	7	18.60	5.97	24.57
0.75	6.08	232	10	11.05	6.73	17.78

Table 6.3. AMG-GMRES(10) with different coarsening strategies applied to a 7-point finite difference discretization of the 3-dimensional Laplacian on 512 processors with 40^3 degrees of freedom per processor.

Method	θ	τ	C_{op}	s_{avg}	$\#its$	t_{setup}	t_{solve}	t_{total}
RS0	0.50	0.0	3.70	170	14	6.69	4.67	11.36
RS3	0.75	0.0	6.18	457	9	19.35	5.83	25.18
	0.00	0.3	3.52	756	6	42.83	5.69	48.52
CLJP	0.75	0.0	13.38	119	16	14.94	13.23	28.17
	0.00	0.3	4.46	232	9	17.94	2.80	20.74
Falgout	0.75	0.0	6.12	237	10	13.79	5.88	19.67
	0.00	0.3	3.22	275	7	19.14	2.56	21.70
PMIS	0.00	0.0	2.09	49	20	4.69	4.29	8.98
HMIS	0.00	0.0	2.75	61	13	5.15	3.44	8.59
AG	0.08	0.0	1.25	16	36	3.22	11.74	14.96
SA	0.08	0.0	1.75	172	13	4.96	5.90	10.86

larger problem sizes. Operator complexities are overall larger than in the previous test problem, particularly for the CLJP coarsening, when no interpolation truncation is applied. Best operator complexities are obtained for the aggregation based schemes, followed by PMIS and HMIS. The best overall timings are achieved by HMIS. RS0 performs very well for this problem for this particular parameter choice, however if applied to the 7-point 3D problem using 39^3 instead of 40^3 unknowns per processor, $p = 1024$, C_{op} is twice, $\#its$ is four times and t_{total} is three times as large as the values reported in Table 6.3, while the other coarsenings are affected to a much lesser degree by the change in system size.

Finally, in Table 6.4 we present results for the 3D Laplace problem on the unit cube using an unstructured finite element discretization. Note that operator complexities are overall lower here. RS0 is performing the worst due to a large number of iterations. The best timings are achieved by PMIS, followed by HMIS and SA. Interestingly enough, while CLJP's operator complexities were much larger than Falgout's and convergence was slightly worse for the structured test problems, it performs slightly better than Falgout for this truly unstructured problem. A similar

Table 6.4. AMG-GMRES(10) with different coarsening strategies applied to an unstructured finite element discretization of the 3-dimensional Laplacian on 288 processors with approx. 20,000 degrees of freedom per processor.

Method	θ	C_{op}	s_{avg}	$\#its$	t_{setup}	t_{solve}	t_{total}
RS0	0.75	2.51	47	86	3.55	15.53	19.08
RS3	0.75	2.65	56	41	4.87	11.69	16.56
CLJP	0.75	2.71	76	18	5.79	6.33	12.12
Falgout	0.75	2.84	77	21	6.42	7.20	13.62
PMIS	0.25	1.46	53	22	2.41	3.60	6.01
HMIS	0.25	1.60	61	21	2.91	3.64	6.55
AG	0.00	1.06	18	54	2.13	12.22	14.35
SA	0.00	1.24	102	18	2.73	5.68	8.61

effect can be observed for HMIS and PMIS. Results for a variety of test problems applied to various coarsening schemes in BoomerAMG can be found in [26, 19]. Numerical test results for various elasticity problems comparing BoomerAMG and MLI can be found in [9].

6.7 Software Packages

There are various software packages for parallel computers which contain algebraic multilevel methods. This section contains very brief descriptions of these codes. Unless specifically mentioned otherwise, these packages are open source codes, and information on how to obtain them is provided below.

6.7.1 hypre

The software library *hypre* [24] is being developed at Lawrence Livermore National Laboratory and can be downloaded from [29]. One of its interesting features are its conceptual interfaces [24, 22], which are described in further detail in [23] It contains various multilevel preconditioners, including the geometric multigrid codes SMG and PFMG [21],the AMG code BoomerAMG and the smoothed aggregation code MLI, as well as sparse approximate inverse and parallel ILU preconditioners.

6.7.2 LAMG

LAMG is a parallel algebraic multigrid code, which has been developed at Los Alamos National Laboratory. It makes extensive use of aggressive coarsening, such as described in [51]. It has shown to give extremely scalable results on upto 3500 processors. Further information on the techniques used in LAMG can be found in [33]. LAMG is not an open source code.

6.7.3 ML

ML [27] is a massively parallel algebraic multigrid solver library for sparse linear systems. It contains various parallel multigrid methods, including smoothed aggregation, a version of classic AMG and a special algebraic multigrid solver for Maxwell's equations. The smoothed aggregation code offers all the parallel coarsening options for aggregation based AMG described above and among other features polynomial and multi-colored Gauß-Seidel smoothers. It is being developed at Sandia National Laboratories and can be downloaded from [43].

6.7.4 pARMS

The library pARMS contains parallel algebraic recursive multilevel solvers. These solvers are not algebraic multigrid methods in the sense described above, but are algebraic multilevel solvers which rely on a recursive multi-level ILU factorization. Further details on these methods can be found in [49, 36] The pARMS library has been developed at the University of Minnesota and is available at [44].

6.7.5 PEBBLES

PEBBLES (Parallel and Element Based grey Box Linear Equation solver) is an algebraic multigrid package for solving large sparse, symmetric positive definite linear equations which arise from finite element discretizations of elliptic PDEs of second order [25]. It is a research code with element preconditioning, different interpolation and coarsening schemes and more that is being developed at the University in Linz. More information on the package and how to obtain the code is available at [45].

6.7.6 PHAML

PHAML (Parallel Hierarchical Adaptive Multilevel solvers) is a parallel code for the solution of general second order linear self-adjoint elliptic PDEs. It uses a finite element method with linear, quadratic or cubic elements over triangles. The adaptive refinement and multigrid iteration are based on a hierarchical basis formulation [42]. For further information and to download the software package which is being developed at the National Institute for Science and Technology (NIST) see [46].

6.7.7 Prometheus

Prometheus is a parallel multigrid library that has been developed originally at the University of California at Berkeley. It contains a multigrid solver for PDE's on finite element generated unstructured grids, but also a smoothed aggregation component named ATLAS. Prometheus is available at [47].

6.7.8 SAMGp

SAMGp (Algebraic Multigrid Methods for Systems) is a parallel software library which has been developed at Fraunhofer SCAI. It uses subdomain blocking as a parallel coarsening scheme [35] and allows for various different coarsening schemes, such as aggressive coarsening, making it very memory efficient. SAMGp is a commercial code. Information on how to purchase it as well as a user's manual can be found at [50].

6.7.9 SLOOP

A parallel object oriented library for solving sparse linear systems named SLOOP [18, 41] is being developed at CEA (Commisariat a l'Energie Atomique) in France. SLOOP's primary goal is to provide a friendly user interface to the parallel solvers and ease the integration for new preconditioners and matrix structures. It contains various parallel preconditioners: algebraic multigrid, approximate inverse and incomplete Cholesky. This library is currently not an open source code, but there are plans to change this in the near future.

6.7.10 UG

UG (Unstructured Grids) is a parallel software package that is being developed at the University of Heidelberg particularly for the solution of PDEs on unstructured grids [4]. It has various features, including a parallel AMG code, adaptive local grid refinement and more. For further information and a copy of the code see [56].

6.8 Conclusions and Future Work

Overall, there are many efficient parallel implementations of algebraic multigrid and multilevel methods. Various parallel coarsening schemes, interpolation procedures, parallel smoothers as well as several parallel software packages have been briefly described. There has truly been an explosion of research and development in the area of algebraic multilevel techniques for parallel computers with distributed memories. Even though we have tried to cover as much information as possible, there are still various interesting approaches that have not been mentioned. One of those approaches that shows a lot of promise is the concept of compatible relaxation. This was originally suggested by Achi Brandt [6]. Much research has been done in this area. Although many theoretical results have been obtained [20, 37], we are not aware of an efficient implementation of this algorithm to this date. However, once this has been formulated, compatible relaxation holds much promise for parallel computation. Since the smoother is used to build the coarse grid, use of a completely parallel smoother (e.g. C-F Jacobi relaxation) will lead to a parallel coarsening algorithm.

References

1. M. Adams. A parallel maximal independent set algorithm. In *Proceedings of the 5th Copper Mountain Conference on Iterative Methods*, 1998.
2. M. Adams. A distributed memory unstructured Gauss-Seidel algorithm for multigrid smoothers. In *ACM/IEEE Proceedings of SC2001: High Performance Networking and Computing*, 2001.
3. M. Adams, M. Brezina, J. Hu, and R. Tuminaro. Parallel multigrid smoothing: polynomial versus Gauss-Seidel. *Journal of Computational Physics*, 188:593–610, 2003.
4. P. Bastian, K. Birken, K. Johannsen, S. Lang, n. Neuß, H. Rentz-Reichert, and C. Wieners. UG: a flexible software toolbox for solving partial differential euations. *Computing and Visualization in Science*, 1:27–40, 1997.
5. A. Brandt. Algebraic multigrid theory: The symmetric case. *Appl. Math. Comp.*, 19:23–56, 1986.
6. A. Brandt. General highly accurate algebraic coarsening schemes. *Electronic Transactions on Numerical Analysis*, 10:1–20, 2000.
7. A. Brandt, S. McCormick, and J. Ruge. Algebraic multigrid (AMG) for automatic multigrid solutions with application to geodatic computations. Technical report, Institute for Computational Studies, Fort Coolins, CO, 1982.
8. A. Brandt, S. McCormick, and J. Ruge. Algenbraic multigrid (AMG) for sparse matrix equations. In D. Evans, editor, *Sparsity and Its Applications*. Cambridge University Press, 1984.
9. M. Brezina, C. Tong, and R. Becker. Parallel algebraic multigrids for structural mechanics. *SIAM Journal of Scientific Computing*, submitted, 2004. Also available as LLNL technical report UCRL-JRNL-204167.
10. M. Brezina. Robust iterative solvers on unstructured meshes. Technical report, University of Colorado at Denver, 1997. Ph.D.thesis.
11. M. Brezina, A. J. Cleary, R. D. Falgout, V. E. Henson, J. E. Jones, T. A. Manteuffel, S. F. McCormick, and J. W. Ruge. Algebraic multigrid based on element interpolation (AMGe). *SIAM J. Sci. Comput.*, 22(5):1570–1592, 2000. Also available as LLNL technical report UCRL-JC-131752.
12. W. Briggs, V. Henson, and S. McCormick. *A multigrid tutorial*. SIAM, Philadelphia, PA, 2000.
13. O. Bröker and M. Grote. Sparse approximate inverse smoothers for geometric and algebraic multigrid. *Applied Numerical Mathematics*, 41:61–80, 2002.
14. T. Chartier, R. D. Falgout, V. E. Henson, J. Jones, T. Manteuffel, S. McCormick, J. Ruge, and P. Vassilevski. Spectral AMGe (ρAMGe). *SIAM Journal on Scientific Computing*, 25:1–26, 2003.
15. E. Chow. A priori sparsity patterns for parallel sparse approximate inverse preconditioners. *SIAM J. Sci. Comput.*, 21(5):1804–1822, 2000. Also available as LLNL Technical Report UCRL-JC-130719 Rev.1.
16. E. Chow. Parallel implementation and practical use of sparse approximate inverses with a priori sparsity patterns. *Int'l J. High Perf. Comput. Appl.*, 15:56–74, 2001. Also available as LLNL Technical Report UCRL-JC-138883 Rev.1.
17. A. J. Cleary, R. D. Falgout, V. E. Henson, and J. E. Jones. Coarse-grid selection for parallel algebraic multigrid. In *Proc. of the Fifth International Symposium on: Solving Irregularly Structured Problems in Parallel*, volume 1457 of *Lecture Notes in Computer Science*, pp. 104–115, New York, 1998. Springer–Verlag. Held at Lawrence Berkeley National Laboratory, Berkeley, CA, August 9–11, 1998. Also available as LLNL Technical Report UCRL-JC-130893.

18. L. Colombet, G. Meurant, et al. Manuel utilisateur de la bibliotheque (SLOOP) 3.2 SLOOP 3.2 users manual. Technical report, CEA/DIF/DSSI/SNEC, 2004.
19. H. De Sterck, U. M. Yang, and J. Heys. Reducing complexity in parallel algebraic multigrid preconditioners. *SIAM Journal on Matrix Analysis and Applications*, submitted, 2004. Also available as LLNL technical report UCRL-JRNL-206780.
20. R. Falgout and P. Vassilevski. On generalizing the AMG framework. *SIAM Journal on Numerical Analysis*, to appear, 2003. Also available as LLNL technical report UCRL-JC-150807.
21. R. D. Falgout and J. E. Jones. Multigrid on massively parallel architectures. In E. Dick, K. Riemslagh, and J. Vierendeels, editors, *Multigrid Methods VI*, volume 14 of *Lecture Notes in Computational Science and Engineering*, pp. 101–107, Berlin, 2000. Springer. Proc. of the Sixth European Multigrid Conference held in Gent, Belgium, September 27-30, 1999. Also available as LLNL technical report UCRL-JC-133948.
22. R. D. Falgout, J. E. Jones, and U. M. Yang. Conceptual interfaces in hypre. *Future Generation Computer Systems*, to appear, 2003. Also available as LLNL technical report UCRL-JC-148957.
23. R. D. Falgout, J. E. Jones, and U. M. Yang. The design and implementation of *hypre*, a library of parallel high performance preconditioners. In A. M. Bruaset and A. Tveito, editors, *Numerical Solution of Partial Differential Equations on Parallel Computers*, volume 51 of *Lecture Notes in Computational Science and Engineering*, pp. 267–294. Springer-Verlag, 2005.
24. R. D. Falgout and U. M. Yang. *hypre*: a library of high performance preconditioners. In P. Sloot, C. Tan., J. Dongarra, and A. Hoekstra, editors, *Computational Science - ICCS 2002 Part III*, volume 2331 of *Lecture Notes in Computer Science*, pp. 632–641. Springer–Verlag, 2002. Also available as LLNL Technical Report UCRL-JC-146175.
25. G. Haase, M. Kuhn, and S. Reitzinger. Parallel algebraic multigrid methods on distributed memory computers. *SIAM Journal on Scientific Computing*, 24:410–427, 2002.
26. V. E. Henson and U. M. Yang. BoomerAMG: a parallel algebraic multigrid solver and preconditioner. *Applied Numerical Mathematics*, 41:155–177, 2002. Also available as LLNL technical report UCRL-JC-141495.
27. J. Hu, C. Tong, and R. Tuminaro. ML 2.0 smoothed aggregation user's guide. Technical Report SAND2001-8028, Sandia National Laboratories, 2002.
28. F. Hülsemann, M. Kowarschik, M. Mohr, and U. Rüde. Parallel geometric multigrid. In A. M. Bruaset and A. Tveito, editors, *Numerical Solution of Partial Differential Equations on Parallel Computers*, volume 51 of *Lecture Notes in Computational Science and Engineering*, pp. 165–208. Springer-Verlag, 2005.
29. hypre: High performance preconditioners. `http://www.llnl.gov/CASC/hypre/`.
30. D. Hysom and A. Pothen. Efficient parallel computation of ILU(k) preconditioners. In *Proceedings of SuperComputing 99*. ACM, November 1999. published on CDROM, ISBN #1-58113-091-0, ACM Order #415990, IEEE Computer Society Press Order # RS00197.
31. D. Hysom and A. Pothen. A scalable parallel algorithm for incomplete factor preconditioning. *SIAM J. Sci. Comput.*, 22(6):2194–2215, 2001.
32. M. Jones and P. Plassman. A parallel graph coloring heuristic. *SIAM J. Sci. Comput.*, 14:654–669, 1993.
33. W. Joubert and J. Cullum. Scalable algebraic multigrid on 3500 processors. Technical Report Technical Report No. LAUR03-568, Los Alamos National Laboratory, 2003.

34. G. Karpis and V. Kumar. Parallel threshold-based ILU factorization. Technical Report 061, University of Minnesota, Department of Computer Science/Army HPC Research Center, Minneapolis, MN 5455, 1998.
35. A. Krechel and K. Stüben. Parallel algebraic multigrid based on subdomain blocking. *Parallel Computing*, 27:1009–1031, 2001.
36. Z. Li, Y. Saad, and M. Sosonkina. pARMS: a parallel version of the algebraic recursive multilevel solver. *Numerical Linear Algebra with Applications*, 10:485–509, 2003.
37. O. Livne. Coarsening by compatible relaxation. *Numerical Linear Algebra with Applications*, 11:205–228, 2004.
38. M. Luby. A simple parallel algorithm for the maximal independent set problem. *SIAM J. on Computing*, 15:1036–1053, 1986.
39. S. F. McCormick. *Multigrid Methods*, volume 3 of *Frontiers in Applied Mathematics*. SIAM Books, Philadelphia, 1987.
40. G. Meurant. A multilevel AINV preconditioner. *Numerical Algorithms*, 29:107–129, 2002.
41. G. Meurant. Numerical experiments with parallel multilevel preconditioners on a large number of processors. *SIAM Journal on Matrix Analysis and Applications*, submitted, 2004.
42. W. Mitchell. Unified multilevel adaptive finite element methods for elliptic problems. Technical Report UIUCDCS-R-88-1436, Department of Computer Science, University of Illinois, Urbana, IL, 1988. Ph.D. thesis.
43. ML: A massively parallel algebraic multigrid solver library for solving sparse linear systems. `http://www.cs.sandia.gov/~tuminaro/ML_Description.html`.
44. pARMS: Parallel algebraic recursive multilevel solvers. `http://www-users.cs.umn.edu/~saad/software/pARMS/`.
45. PEBBLES: Parallel and elment based grey box linear equation solver. `http://www.numa.uni-linz.ac.at/Research/Projects/pebbles.html`.
46. PHAML: The parallel hierarchical adaptive multilevel project. `http://math.nist.gov/phaml/`.
47. Prometheus. `http://www.cs.berkeley.edu/~madams/prometheus/`.
48. J. W. Ruge and K. Stüben. Algebraic multigrid (AMG). In S. F. McCormick, editor, *Multigrid Methods*, volume 3 of *Frontiers in Applied Mathematics*, pp. 73–130. SIAM, Philadelphia, PA, 1987.
49. Y. Saad and B. Suchomel. ARMS: an algebraic recursive multilevel solver for general sparse linear systems. *Numerical Linear Algebra with Applications*, 9:359–378, 2002.
50. SAMGp: Algebraic multigrid methods for systems. `http://www.scai.fraunhofer.de/samg.htm`.
51. K. Stüben. Algebraic multigrid (AMG): an introduction with applications. In U. Trottenberg, C. Oosterlee, and A. Schüller, editors, *Multigrid*. Academic Press, 2001.
52. SuperLU. `http://acts.nersc.gov/superlu/`.
53. W.-P. Tang and W. L. Wan. Sparse approximate inverse smoother for multigrid. *SIAM J. Matrix Anal. Appl.*, 21:1236–1252, 2000.
54. U. Trottenberg, C. Oosterlee, and A. Schüller. *Multigrid*. Academic Press, 2001.
55. R. Tuminaro and C. Tong. Parallel smoothed aggregation multigrid: aggregation strategies on massively parallel machines. In J. Donnelley, editor, *Supercomputing 2000 Proceedings*, 2000.
56. UG: A flexible software toolbox for solving partial differential equations. `http://cox.iwr.uni-heidelberg.de/~ug/index.html`.
57. P. Vanek, M. Brezina, and J. Mandel. Convergence of algebraic multigrid based on smoothed aggregation. *Numerische Mathematik*, 88:559–579, 2001.

58. P. Vaněk, J. Mandel, and M. Brezina. Algebraic multigrid based on smoothed aggregation for second and fourth order problems. *Computing*, 56:179–196, 1996.
59. U. M. Yang. On the use of relaxation parameters in hybrid smoothers. *Numerical Linear Algebra with Applications*, 11:155–172, 2004.

7

Parallel Mesh Generation

Nikos Chrisochoides

Computer Science Department, College of William and Mary, Williamsburg, VA 23185, USA

Division of Applied Mathematics, Brown University, 182 George Street, Providence, RI 02912, USA

nikos@cs.wm.edu

Summary. Parallel mesh generation is a relatively new research area between the boundaries of two scientific computing disciplines: computational geometry and parallel computing. In this chapter we present a survey of parallel unstructured mesh generation methods. Parallel mesh generation methods decompose the original mesh generation problem into smaller subproblems which are meshed in parallel. We organize the parallel mesh generation methods in terms of two basic attributes: (1) the sequential technique used for meshing the individual subproblems and (2) the degree of coupling between the subproblems. This survey shows that without compromising in the stability of parallel mesh generation methods it is possible to develop parallel meshing software using off-the-shelf sequential meshing codes. However, more research is required for the efficient use of the state-of-the-art codes which can scale from emerging chip multiprocessors (CMPs) to clusters built from CMPs.

7.1 Introduction

This chapter presents a survey of parallel unstructured mesh generation methods based on three widely used techniques: Delaunay [39], Advancing Front [66], and Edge Subdivision [59]. Parallel methods for quadrilateral [6] and hexahedral [54] mesh generation as well as block structured [93, 20, 90] and structured adaptive mesh refinement [1] methods are not reviewed in this chapter.

Parallel mesh generation procedures in general decompose the original 2-dimensional (2D) or 3-dimensional (3D) mesh generation problem into N_s smaller subproblems which are solved (i.e., meshed) concurrently using P processors. The subproblems can be formulated to be either tightly coupled [60, 56, 78], partially coupled [55, 31, 19] or even decoupled [38, 79, 52]. The coupling of the subproblems determines the intensity of the communication and the amount/type of synchronization required between the subproblems.

The challenges in parallel mesh generation methods are: to maintain *stability* of the parallel mesher (i.e., retain the quality of finite elements generated by state-of-the-art sequential codes) and at the same time achieve 100% *code re-use* (i.e., leverage the continuously evolving and fully functional off-the-shelf sequential meshers) without substantial deterioration of the *scalability* of the parallel mesher. In this chapter we review parallel mesh generation methods having in mind these three requirements.

We build on top of previous work [30, 39] where parallel mesh generation methods are classified in terms of the *way* and the *order* the artificial boundary surfaces (interfaces) of the subproblems are meshed. Specifically, in [31, 39] existing parallel methods are classified in three categories: (i) methods that first mesh (either in parallel [55] or sequentially [79]) the interfaces of the subproblems and then mesh in parallel the individual subproblems, (ii) methods that first solve the meshing problem in each of the subproblems in parallel and then mesh the interfaces so that the global mesh is conforming [36], and (iii) methods that simultaneously mesh and improve the interfaces as they mesh the individual subproblems [25, 19, 26].

In this chapter we organize the parallel mesh generation methods in terms of two basic attributes. First, the sequential technique used for meshing the individual subproblems: *(1) Delaunay, (2) Advancing Front, and (3) Edge Subdivision*. Second, the degree of coupling between the subproblems: *(a) tightly-coupled, (b) partially-coupled, and (c) decoupled methods*.

7.2 Domain Decomposition Approaches

Parallel mesh generation methods use a sequential pre-processing step for the data partitioning problem with the exception of [47, 48]. The data are partitioned using either the continuous domain which is decomposed into *subdomains* (see Figure 7.1, left) or a discrete approximation (i.e., an initial coarser mesh) of the domain which is decomposed into *submeshes* (see Figure 7.1, right). The internal boundaries between the subdomains or submeshes (S_i) are called *interfaces* or *separators* (∂S_i). In both cases the number of generated subdomains or submeshes (N_s) can be significantly greater than the number of processors P (*over-decomposition*). Over-decomposition was introduced in parallel computing in mid 80s. It is used to hide communication latency in message passing [50] and to mask information dissemination, decision making and data migration costs in dynamic load balancing [21].

The domain decomposition (DD) problem in parallel mesh generation is defined as follows:

$$lfs(\Omega) \le lfs(S_i) \;\; i = 1, N_s \tag{7.1}$$

$$\min_{i=1,N_s} \frac{|\partial S_i|}{|S_i|} \tag{7.2}$$

$$\partial S_i \text{ form ``good'' angles between each other and the boundary } \partial\Omega. \tag{7.3}$$

where $lfs(\Omega)$ and $lfs(S_i)$ are the local feature size [85] of the original domain Ω and the subdomains (or submeshes) S_i, respectively. The $|\partial S_i|$ denotes the length (in 2D) and surface (in 3D) of the interfaces while the $|S_i|$ denotes the area (in 2D) and volume (in 3D) of the subdomains (or submeshes) S_i.

Fig. 7.1. Domain decomposition of the continuous geometry [52] and the discrete geometry [17] of a cross section of a rocket pipe. (For the color version, see Figure A.17 on page 474).

Continuous Domain Decomposition

The continuous domain decomposition methods partition the region Ω into subdomains $\Omega_i, i = 1, N_s$. There are two types of continuous DD methods. The first and most popular approach is based on quadtree/octree methods [31, 56, 58]. The octree methods utilize an octree structure for the decomposition of Ω into blocks (octants). The octants along with a description of the external boundary $\partial\Omega$ define the subdomains. Another class of continuous DD methods [52] is based on auxiliary structures like the Medial Axis [9, 68, 91] so that the subdomains Ω_i have no new features like small angles between the separators and the separators and external boundary [52].

Continuous DD approaches are attractive because they refine the individual subdomains by re-using existing well tested and fine-tuned sequential codes on each subdomain independently. However, independence in mesh refinement and high code re-use in some cases come at a price. The polyhedral surfaces which arise due to the decomposition of the initial mesh impose additional constraints on the execution of sequential meshing algorithms in each of the subdomains. Poorly generated interface surfaces can affect the termination of meshing algorithms and the quality of the elements. Moreover, the artificially imposed interfaces can affect the mesh gradation.

Discrete Domain Decomposition

The Discrete DD methods partition an initial coarse mesh (usually a boundary conforming mesh), D into a number of simply-connected submeshes $D_i, i = 1, N_s$ while they try to minimize the surface-to-volume ratio for each of the submeshes. Usually a coarse mesh is generated on a high-performance workstation using sequential mesh generators. The partitioning of a coarse mesh is performed either sequentially or in parallel using generic graph partitioning libraries like Metis/Parallel Metis [80] and Chaco [42]. Also, there are mesh partitioning libraries like Domain Decomposer [22, 23], Zoltan [33], Drama [3], Plum [64], and Jove [87] (to mention

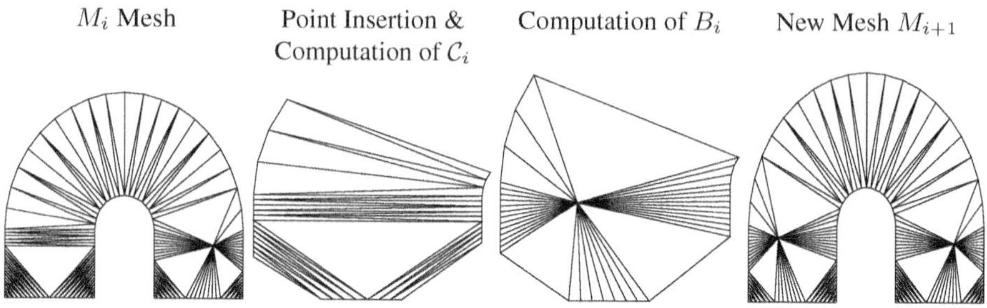

Fig. 7.2. Bowyer-Watson kernel starts with a mesh M_i (left), computes the cavity (center left) of a newly inserted point, triangulates the cavity (center right) and updates the mesh into M_{i+1} (right).

a few) which extend and customize the generic data partitioning techniques for FEM calculations.

7.3 Parallel Mesh Generation Methods

In this section we review parallel mesh generation methods which are based on Delaunay triangulation in Section 7.3.1, Advancing Front Technique in Section 7.3.2, and Edge Subdivision methods in Section 7.3.3.

7.3.1 Delaunay Based Methods

There are many approaches to generate Delaunay meshes [39], we focus on methods based on Bowyer-Watson [10, 96] kernel which can lead to: (1) more efficient parallel implementations due to easier optimizations for improving data locality and (2) simpler and more efficient data structures. The Bowyer-Watson (BW) kernel is described in Figure 7.2 and the loop bellow:

Algorithm 1 (BW($M_o, p_1, ..., p_n$)).

```
Input: Mesh M_o an initial mesh and a set of n points
  for i = 1, n
     Compute the cavity of C_i of the point p_i
     Compute the ball B_i of point p_i
     M_{i+1} = M_i - C_i + B_i
  endfor
Output: A new mesh M = M_{n+1}
```

where the cavity C of a point p is defined as the set of all triangles whose circumcircle includes p; the ball B of a point p is defined as set of new triangles defined by the point p and the vertices of the boundary of its cavity [39].

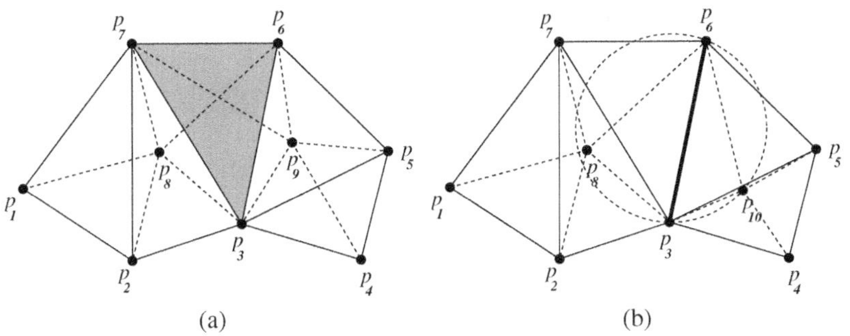

Fig. 7.3. (a) Intersection of two cavities and (b) two cavities share an edge; solid lines represent the edges of the initial triangulation, and dashed lines edges created by the insertion of p_8, p_9, and p_{10}.

The challenge, for parallel mesh generation methods based on the BW kernel, is to maintain the following loop invariant: M_i *is conformal and Delaunay, for* $i = 1, n$. Figure 7.3 depicts two cases where the concurrent point insertion violates the loop invariant. First, the cavities intersect i.e., there is a triangle $\triangle p_3p_6p_7 \in \mathcal{C}(p_8) \cap \mathcal{C}(p_9)$, then concurrent insertion of p_8 and p_9 results in a non-conformal mesh. Second, the cavities share an edge in 2D (or a face in 3D), an edge p_3p_6 is shared by $\mathcal{C}(p_8) = \{\triangle p_1p_2p_7, \triangle p_2p_3p_7, \triangle p_3p_6p_7\}$ and $\mathcal{C}(p_{10}) = \{\triangle p_3p_5p_6, \triangle p_3p_4p_5\}$, then the new triangle $\triangle p_3p_{10}p_6$ can have point p_8 inside its circircle, thus, violating the Delaunay property.

The focus of this section is on parallel mesh generation methods that address this challenge. There is a number of parallel Delaunay and triangulation methods like the MIMD method in [92] and the HPF implementation in [14] which target parallel programming paradigms no longer in use for practical purposes. Other methods [27, 62, 63, 41] also contributed in shaping up this author's directions and work in parallel mesh generation and implicitly contribute in this chapter.

In [7] the authors describe a divide-and-conquer projection-based algorithm for constructing in parallel 2D Delaunay triangulations of a set of given points. The method extends to 3D, but its implementation is quite complex. The goal in parallel mesh generation, though, is to refine an existing mesh by inserting new points i.e., the set of points in the final mesh is not known in advance.

In [46, 48] the authors extended [7] for parallel 2D mesh generation which further eliminates the sequential step for an initial mesh, but does not address the issue of code re-use. The method in [46, 48] is partially coupled.

In [35] the authors define the points x and y as independent if the closures of their *prestars* (or *cavities*) are disjoint. The approach in [35] does not provide a way to schedule the concurrent insertion of points whose cavity closures are disjoint.

In [88] the authors presented the first theoretical analysis of the complexity of parallel Delaunay refinement algorithms. However, the assumption in [88] is that the global mesh is completely retriangulated each time a set of independent points is inserted. In [89] the authors developed a more practical algorithm.

In the rest of this section we describe five different practical (i.e., they have been implemented) parallel Delaunay mesh generation methods. These methods formulate the subproblems to be: (1) tightly coupled, (2) decoupled, and (3) partially coupled.

Tightly Coupled Methods

A straight forward approach to parallel computing is based on identifying some partial order among the computations of well understood and successful sequential kernels and then in a brute-force fashion use message passing or threads to implement the computations on distributed and shared memory parallel machines, respectively. This approach leads to the tightly coupled method presented in [60] for parallel guaranteed quality Delaunay mesh generation.

Parallel Optimistic Delaunay Meshing (PODM) Method

In [60] the authors presented the first provable 3D parallel guaranteed quality Delaunay mesh generation method for polyhedral domains. PODM is based on discrete domain decomposition, but it is not constrained by the interfaces of the submeshes. The algorithm guarantees the stability by simultaneously re-partitioning and refining the interface surfaces and volume of the submeshes [26] —refinement due to a point insertion might extend across subproblem (or submesh) boundaries. The extension of a cavity beyond the interfaces is a source of intensive communication. However, PODM can tolerate most of the communication by concurrently refining other regions of the submeshes while it waits for remote data to arrive. Unfortunately, the concurrent refinement can create a number of inconsistencies in the mesh (see Figure 7.3). These inconsistencies are resolved at the cost of setbacks (or rollbacks [44]) and thus we call this method Parallel Optimistic Delaunay Meshing method. Setbacks is a source of major algorithm and code re-structuring (due to overlapping cavities) and they lead to zero code re-use. Unfortunately, the overlapping of the cavities becomes even more complex when they are near the external boundary, where a certain order of inserted points needs to be maintained due to encroachment rules that are used to maintain and prove the quality of the elements and thus satisfy the stability requirement.

Figure 7.4a depicts a cavity which extends beyond the submesh interfaces (because two of the cavity BHGFAC triangles $t \in M_1$ and $t^* \in M_2$ are non-local to submesh M_0) in order to guarantee the quality of the mesh. The extension of the cavity beyond the interfaces is a source of intensive communication. However, as Figure 7.4b shows PODM can tolerate the communication by concurrently refining other regions (e.g. compute a new cavity ABCDE) of the submeshes while it waits for remote data (e.g. the partially completed cavity BCAF) to arrive (eg. rest of the cavity BFGH). Unfortunately, the concurrent refinement can lead the violation of the loop invariant by creating non-conforming meshes and/or the violation of the Delaunay property as is the case in Figure 7.4a where the point P_j is within the circumcenter of $\triangle P_l CA$ which is a newly created triangle from the triangulation of the cavity (BHGFAC) that corresponds to the point P_1. These violations are resolved at

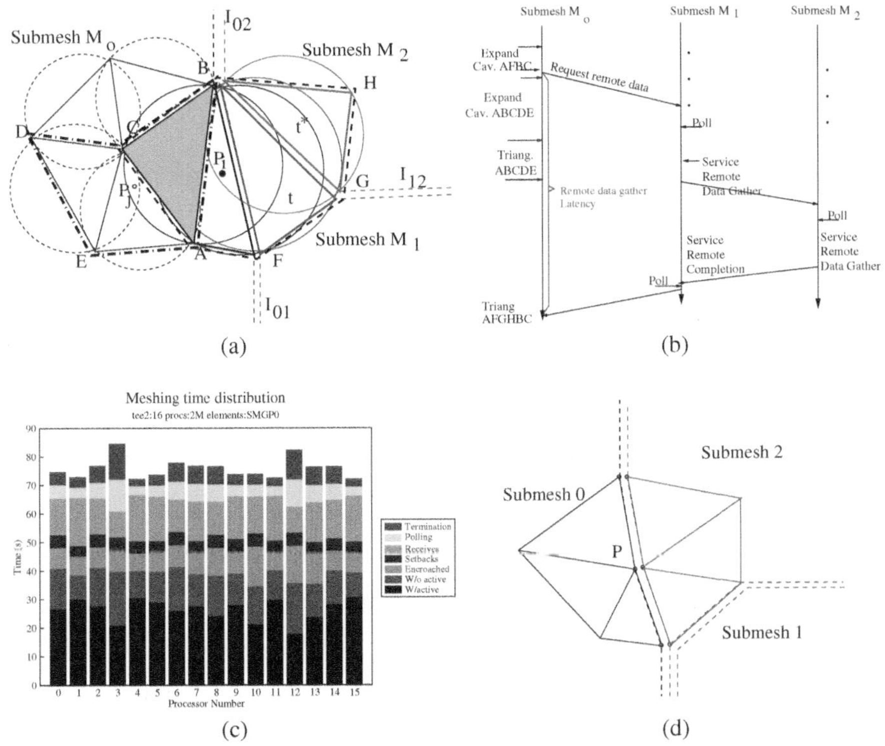

Fig. 7.4. a) cavity extension beyond submesh interfaces, b) time diagram with concurrent point insertion, c) a breakdown of execution time for PODM, and finally d) the refinement of a cavity with simultaneous distribution of the newly created elements. (For the color version, see Figure A.18 on page 475).

the cost of setbacks and frequent message polling shown in the performance graph of Figure 7.4c. With some additional communication cost the PODM becomes domain decomposition independent and moreover re-distributes new elements as they are generated (Figure 7.4d).

In summary, PODM does not depend on good domain decompositions before, during and after parallel meshing at the cost of being labor intensive approach. PODM is a stable and tightly coupled method, with zero code re-use.

Decoupled Methods

In [38, 52] the authors present two approaches which achieve 100% code re-use and eliminate communication and synchronization. Both approaches rely on continuous domain decomposition and decouple the individual subdomains (subproblems) so

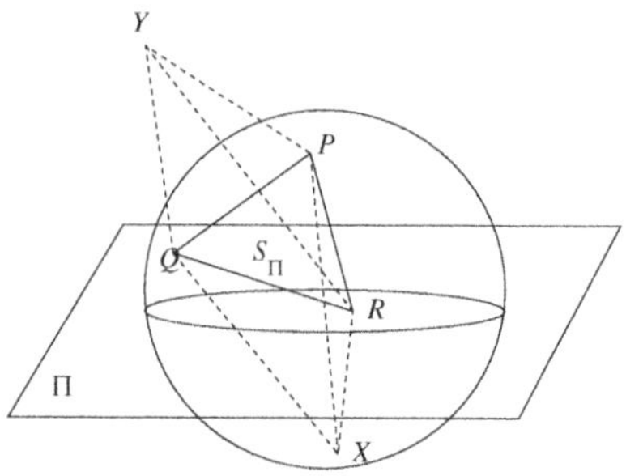

Fig. 7.5. The circumcenter of the face △PQR lies on the plane Π which helps define a separator S_Π. Note that △PQR $\in S_\Pi$.

that they can be meshed independently. Earlier, in [8] the authors presented similar approach for the parallel triangulation of a set of fixed points.

Parallel Projective Delaunay Meshing

The Parallel Projective Delaunay Meshing (P^2DM) method [38] starts by sequentially meshing the external surfaces of the geometry and by pre-computing domain separators whose facets are Delaunay-admissible (i.e., the precomputed interface faces of the separators will appear in the final Delaunay mesh). The separators decompose the continuous domain into subdomains which are meshed in parallel using a sequential Delaunay mesh generation method on each of the processors.

The basic idea for computing Delaunay-admissible separators can be explained easier in the context of the parallel triangulation of a convex hull for a set of points $S \in \mathcal{R}^3$ [36, 30]. The convex hull of a set of points S is decomposed in two subdomains by computing a Delaunay admissible separator as follows: First, the position of a surface (in practice a plane Π) is computed using an Inertia Axis Decomposition method [33]. The plane Π decomposes the convex hull of S into two almost equal pieces (in terms of points). Then the algorithm finds all faces (P,Q,R) $\in R^3$ (see Figure 7.5) for which there is an empty sphere whose center lies on the plane Π and passes through the points P,Q,R. These faces constitute a polyhedral separator S_Π which decomposes the domain into two subdomains assuming that the corresponding tetrahedra PQRX and PQRY contain the centers of their respective circumscribed spheres i.e., the quality of the initial mesh around the separators is very good which requires substantial refinement around the separators. In [38] it is shown that the faces of the polyhedral separator S_Π will appear in the final Delaunay triangulation of the convex hull. The generalization of the idea to complex geometries is possible, however it is much more difficult and it is explained in [38].

It is possible that the pre-constructed separators can not be Delaunay-admissible [38] and the whole process has to start from the beginning. This is a very difficult problem which for 2D has been solved in [52] using a different approach.

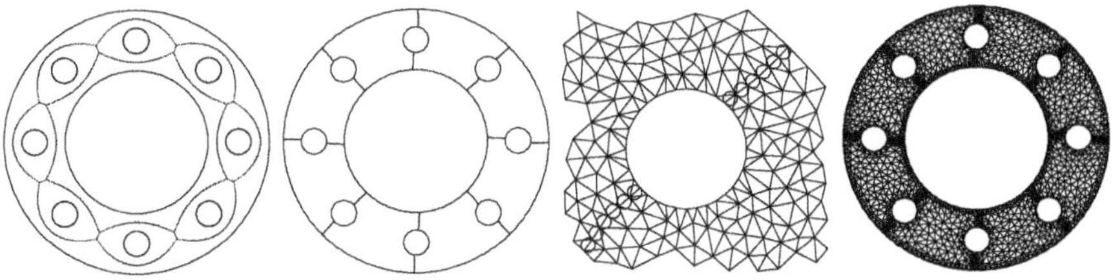

(a) (b) (c) (d)

Fig. 7.6. The Medial Axis Transformation (a) which in turn is used to achieve high quality domain decomposition (b). For PD^3 the interfaces of the subdomains are refined (c) in a preprocessing step in order to decouple the subdomains which are refined independently (d).

Parallel Delaunay Domain Decoupling PD^3 Method

The PD^3 method [52] like P^2DM is based on continuous domain decomposition. PD^3, for the domain decomposition of 2D geometries, uses medial axis of the domain and relies on the following simple geometric property [52]:

Lemma 1. *Let $MA(\Omega)$ be the medial axis of Ω and b a contact point of $c \in MA(\Omega)$. The angles formed by the segment cb and the tangent of the boundary $\partial\Omega$ at b are at least $\pi/2$.*

The medial axis of a domain Ω is approximated by Voronoi points of a discretization of the domain. Figure 7.6a depicts the medial axis approximation and a 8-way partition (b) for the same geometry. The level of the discretization of the boundary determines the quality of the approximation of the medial axis. However, the goal in [52] is not to approximate accurately the medial axis, but to obtain good angles from the separator. Therefore, the criteria for the discretization of the domain are determined from the quality of the angles formed between the separators and the external boundary of the domain [53].

After the decomposition of the domain (see Figure 7.6b), PD^3 constructs a “zone” around the interfaces of the submeshes. The “zone” consist of the union of all diametral circles of the interface edges (see Figure 7.6c). The interfaces of the subdomains are refined using the lfs of the original domain. This leads into an overrefinement of the final distributed mesh. Experimental data from PD^3 (see Table 7.1) suggest that the overrefinement is not as high as one could expect. However, the authors of [52] are working on a new approach which will use adaptive domain decomposition [53] and different lfs for different interfaces of the subdomains. This method is expected to reduce overrefinement of the interfaces and produce well graded meshes [51].

In [52] the authors prove that sequential Delaunay meshers will not insert any new points within a zone around the subdomain interfaces i.e., the sequential Delaunay meshing on the individual submeshes can terminate without inserting any new points on the interfaces and thus eliminate communication and modifications of the sequential codes. This way, the problem of parallel meshing is reduced into a

Table 7.1. Overrefinement data as we increase the number of subdomains for the decomposition of a cross section of a rocket pipe model.

Subs	1	16	32	64	128
Elms :	21,016,403	21,016,857	21,018,522	21,030,711	21,044,689
ORef.Elms/Sub	0	28	66	379	299

"proper" domain decomposition and a discretization of interfaces. However, the construction of decompositions that can decouple the mesh is a challenging problem, since its solution is based on medial axis which is very expensive and difficult to construct (even to approximate) for complex 3-dimensional geometries [40, 83, 29].

Partially Coupled Methods

The parallel tightly-coupled and decoupled methods we have seen so far address some of the parallel mesh generation requirements we described in Section 7.1. For example, PODM is a 3D stable and domain decomposition independent, but it is zero code re-use with high communication method; while P^2DM and PD^3 address the code re-use and communication issues, but their applicability in 3D is limited by the Delaunay-admissible and domain decomposition problem, respectively. In the rest of this section we present two partially coupled methods that make an attempt to balance trade-offs between all three requirements and the domain decomposition problem at the cost of some communication.

Parallel Constrained Delaunay Meshing (PCDM) Method

In order to address the communication and synchronization problem in [19], the authors developed the PCDM which is asynchronous and can reduce the variable and unpredictable communication patterns to irregular but bulk communication.

The PCDM [19] is based on the Constrained Delaunay Triangulation [18] and a discrete DD method. Each submesh is treated as an independent mesh defined by external boundary (if any) and/or constrained edges which are the edges of the interfaces between any pair of adjacent submeshes.

Intuitively, the constrained Delaunay triangulation is as close as one can get to the Delaunay triangulation given that one needs to preserve certain (constrained) edges and internal boundaries. It has been shown in [18] that the constrained internal edges do not affect the quality of the resulting mesh more than the edges and faces that define the external boundary. However, one might be able to identify such boundaries (interfaces for the PDCM) in the resulting mesh by noting the way in which triangle edges are aligned. Using the idea of a constrained Delaunay mesh generation one can introduce in the mesh artificial constrained edges (interfaces) which decompose the mesh into submeshes and can be meshed almost independently.

By the definition of the constrained Delaunay mesh, points inserted on one side of an interface have no effect on triangles on the other side; thus, no synchronization is required during the element creation process. In addition, communication between

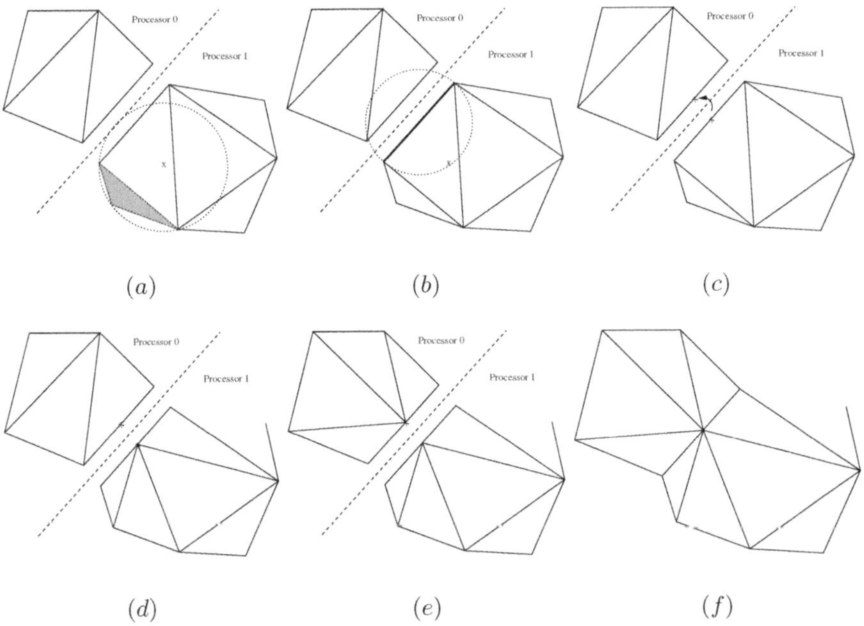

Fig. 7.7. Processor P1 inserts a new point (a) which is encroaching upon an interface edge (b). Then P1 discards the new point and inserts the midpoint of the encroached edge (c) while at the same time it sends a request to split the same interface edge on processor P0. Processor P0 computes the cavity of the midpoint (d). The triangulation of the cavities (e) and (f) of the midpoint of the interface edge results in a new conforming and distributed Delaunay (in the CDT sense) triangulation which guarantees the quality of the elements.

submeshes is tremendously simplified: the only message between adjacent processes is of the form [19]: "Split this interface (i.e., constrained) edge" if a newly inserted point encroaches (see Fig. 7.7) upon an interface edge. Since interface edges are always split exactly in half, no additional information needs to be communicated.

The PCDM is an asynchronous with bulk communication and thus partially coupled method. Moreover, the number and size of messages can be reduced by message aggregation [17]. Although this optimization improves the performance of the PCDM it has its own problems when many "Split this interface edge" messages are delayed. This causes performance degradation due to: (1) the large number of accumulated messages which can consume memory, (2) redundant computation (by delaying messages), from neighboring processors which are unaware of each other's interface splits. In [17] these problems are addressed by a mechanism which adaptively changes the number of messages allowed to be aggregated before a low-level message is send.

However, code re-use remains a problem due to "Split this interface edge" message and optimizations required for reducing the fine-grain communication to a bulk and asynchronous message passing.

Parallel Delaunay Refinement (PDR) Method

The above tightly coupled (PODM) and partially coupled (PCDM) methods [61, 24, 19] require algorithm re-structuring and thus lead to completely new implementations for parallel Delaunay mesh generation. The implementation of sequential mesh generation codes is labor intensive and requires multi-disciplinary effort; it takes about ten to fifteen years to develop the algorithmic and software infrastructure for sequential industrial strength mesh generation libraries. Moreover, improvements in terms of quality, speed, and functionality are open ended and permanent which makes the task of delivering state-of-the-art parallel mesh generation codes much more difficult.

This problem is addressed by P^2DM and PD^3 in [38, 52], where two decoupling methods are presented in order to use (without modifications) optimized and fully functional sequential codes on each of the subproblems and eliminate communication and synchronization. However, P^2DM can suffer setbacks due to difficulty of constructing Delaunay-admissible separators and PD^3, for 3D geometries, is expected to be suffer high pre-processing overhead due the construction (or approximation) of the medial axis.

With PDR in [16, 15] the authors try to balance trade-offs between the data decomposition, communication and code re-use i.e., maintain stability and achieve high code re-use using a simple domain decomposition method at the cost of some communication. The key idea of the PDR method is based on the concurrent point insertion of more than two points without calculating their corresponding cavities ahead of time in order to decide whether they violate the conformity and Delaunay properties of the mesh. PDR accomplishes this objective by introducing for the first time a practical Delaunay-independence criterion for concurrent point insertion [16]:

Theorem 1 *Let $\bar{r}$ be the upper bound on triangle circumradius in the mesh and $p_i, p_j \in \Omega \subset \mathbb{R}^2$. Then if $\|p_i - p_j\| \geq 4\bar{r}$, then independent insertion of p_i and p_j will result in a mesh which is both conformal and Delaunay.*

Theorem 1 is applicable throughout the run of the algorithm, since the execution of the Bowyer-Watson kernel, either sequentially [10, 96] or in parallel [61], does not violate the condition that $\bar{r}$ is the upper bound on triangle circumradius in the entire mesh [16]. However, checking the inequality of the theorem, for every pair of candidate points, would be quite expensive task. In [16] the authors present a simple block domain decomposition scheme[1] which guarantees that any pair of points in non-adjacent cells are far apart no less than $4\bar{r}$. To enforce the $\bar{r}$ circumradius bound in the mesh they derive the following relation which allows the use of

[1]This scheme is based on a simple block decomposition for uniform mesh refinement [16] and octree decomposition for graded mesh refinement [15].

a standard sequential Delaunay refinement algorithm/software like Triangle [84] for preprocessing [16]:

Theorem 2 *If* $\bar{\rho}$ *and* $\bar{\Delta}$ *are upper bounds on triangle circumradius-to-shortest edge ratio and area, respectively, then* $\bar{r} = 2(\bar{\rho})^{3/2}\sqrt{\bar{\Delta}}$ *is an upper bound on triangle circumradius.*

7.3.2 Advancing Front Based Methods

All five parallel Delaunay methods we present in Section 7.3.1 maintain the stability of the parallel mesher. However, parallel finite element codes require only "good" quality of elements and the definition of quality depends on the field solver and varies from code to code. For example, in [79] although the stability is not guaranteed, it appears that the generated meshes are practical and of "good" quality. This raises the following two questions: Is the stability of parallel mesher important? Does the parallel mesh generation without the stability requirement become easier?

The answer to the first question depends on the upstream solver. Regarding the second question, our experience[2] suggests that even if we relax the stability criterion the problem of parallel mesh generation does not become easier. In fact, the termination problem (which is even more fundamental than the stability) becomes, for some cases, a very important issue. In some cases, subdomains or submeshes obtained from state-of-the-art partitioning libraries can not be meshed even by industrial strength advancing front sequential meshers. Parallel mesh smoothing techniques [57] are helpful, but do not work always.

There is a trade-off between the domain decomposition and the capability of the sequential mesher required to mesh the individual subdomains. A balance between the two is important not only for stability but even for termination. Two successful Parallel Advancing Front Techniques [56, 28] address this issue by what we refer to as *guided re-partitioning or shifting of the separators*. In [56] the authors present a tightly coupled method for shared memory machines and in [28] the authors present a partially coupled method for distributed memory machines. We review both methods in the rest of this section.

Tightly Coupled Methods

Löhner et al. in [56] revisit a partially coupled Parallel Advancing Front Technique (PAFT) they developed in [55] (see bellow) in order to address the termination, stability, and code re-use requirements. In [56] they address these issues by developing a PAFT for shared memory computers ($PAFT_{SM}$). However, instead of generating and partitioning a very fine-grain octree as in [58] on a single processor, for the whole geometry, they use an octree to identify the zones where elements can be introduced concurrently. They set the edge length of the smallest octree box to be an order of magnitude larger than the specified size of elements and they use the "shift

[2] From the implementation of a method similar to one appeared in [79].

and regrid" technique, but in a completely different way from the method in [32]. The PAFT$_{SM}$ is broken into two phases: (1) the AFT phases and (2) "shift" or as we call it here guided re-partitioning phase. At each AFT phase the active front expands and a new one is created. The process continues until the whole domain is meshed. The PAFT$_{SM}$ method synchronizes at the beginning of each AFT phase in order to sequentially refine and re-partition the global octree, for the new active front, whose leaves will be refined in parallel. The method is suitable for shared memory machines but can not be used in large-scale distributed memory parallel platforms, because of the global synchronization required between the mesh generation and re-partitioning phases.

The PAFT$_{SM}$ is stable and code re-use is achieved at the cost of global synchronization which is not expensive on shared memory machines.

Partially Coupled Methods

In [55] Lóhner et al. introduced the first 2D PAFT. The initial mesh is subdivided into submeshes using a discrete domain decomposition approach. Each submesh is further separated into an interior region and interface regions, where interface regions of a submesh are defined to be the set of elements that are adjacent to elements that belong to different submehses. The interior regions of each submesh are refined independently. The interface regions and then the corners are refined once all the interior and interface regions are meshed, respectively (a posteriori approach). The order of meshing interface and interior regions can change i.e., interfaces can be refined first (a priori approach) and the interior regions refined last [55]. The submeshes synchronize locally, because no new elements can be inserted in the interfaces and corner regions before the meshing of adjacent interior and interface regions, respectively. The pre-computed interface regions work well for AFT because they create buffer zones which fully decouple the interior regions of the submeshes.

Parallel Octree AFT (POAFT) Method

The 3D POAFT in [28], contrary to the PAFT in [55], is based on continuous domain decomposition method. The POAFT method generates a distributed coarse-grain octree using a divide-and-conquer algorithm. The terminal octants and the geometric model of a domain define the subdomains. The terminal octants of the octree are classified into: interior, interface, boundary, and complete. Interface octants have at least one adjacent octant which is not local. Boundary octants include mesh entities from the input surface mesh. Complete octants have no front faces in their volumes. The subdomains represented as subtrees (on each processor) which are refined further until their leaves reach to a predefined size to use tetrahedral meshing templates. The new octrees are repartitioned (using stop-and-repartition methods) in order to guarantee load balancing during the execution of meshing templates. After the redistribution of interior octants, mesh templates are applied so that the triangulations conform on both sides of interface-octant faces. The interior octants of one processor are independent of the interior octants in other subdomains and thus can be meshed in parallel. At this step code re-use is high, since the meshing templates of sequential octree meshing code [82] can be used on each processor. Existing scalar meshing templates for the interface octants can be used, but some communication will be required during the meshing process. Instead in [28] meshing templates were re-designed in order to guarantee conformity without compromising stability and eliminating communication. The potential for ambiguous splits of faces is addressed and resolved in [69].

Before the boundary octants are refined and meshed a re-partitioning might take place if it is necessary. Any parallel partitioning algorithm can be used; in [28] the parallel recursive inertia bisection method is applied. The meshing of boundary octants is a challenging task. Every processor applies a tree-based face removal procedure [28] in order to connect the input surface mesh with the mesh of the interior octants. The face removal (from the active front) is a basic operation in AFT and it consists of connecting a front mesh face to a target mesh vertex which is drawn from a "neighborhood" of the face [28]. In the parallel face removal, portion of the "neighborhood" might be on a remote processor and a target vertex can not be found locally; in this case the face removal is postponed. This will create unmeshed regions between the terminal interface boundary octants and input surface mesh. In [28] active terminal and boundary interface octants are repartitioned so that the remaining unmeshed "neighborhoods" become local and thus the face removal becomes a local operation. This permits code re-use. This process is repeated until there are no unmeshed regions. The "guided" repartitioning is a very challenging problem.

7.3.3 Edge Subdivision Based Methods

Parallel Edge Subdivision (PES) methods have been used successfully for both 2D domains [97, 45] and 3D geometries [13, 28, 65, 78]. PES methods use discrete DD for data decomposition and their termination and stability does not depend on the geometric properties of the submeshes. Once a coarse mesh is partitioned into submeshes, the individual submeshes are refined in parallel by splitting tetrahedra using sequential subdivision techniques. The longest-edge bisection method [72, 76] is the most commonly used for parallel refinement/derefinement [97, 45, 13, 78]. In 2D an element is refined into two triangles by adding an edge defined by the longest-edge midpoint and its opposite vertex, while in 3D an element is refined into two tetrahedra by adding a triangle defined by the longest-edge midpoint and its two opposite vertices. The longest-edge bisection technique is attractive because it simplifies the management of intermediate non-conforming points throughout the process. With the introduction of terminal-edges in [78] this management is localized in a similar way the cavity localizes the computation of Delaunay based methods.

Like all parallel mesh generation methods PES refinement methods should satisfy all three requirements we listed in the introduction of this chapter. Existing PES methods address some of these requirements successfully and have the potential to meet all the requirements in the future. In [12] the authors present a termination proof, for parallel longest-edge bisection algorithms, using Dijkstra's general termination algorithm [34, 5]. Moreover, they prove the stability and even show that the mesh refined in parallel is identical to a sequentially generated mesh.

The scalability of PES methods depends on the way they address the *refinement collision*: *more than one processor split concurrently two different copies of the same interface edge*. Other factors that affect the scalability is the choice of dynamic load balancing methods and the degree of code re-use. For example, frequent use of stop-and-repartition methods due to global barrier operations can deteriorate the scalability of computationally inexpensive parallel mesh generation methods [2]. In general parallel mesh generation methods that do not take advantage of highly optimized sequential codes have difficulty to demonstrate good scalability against the best sequential codes.

In [65] it has been shown that 100% code re-use is possible at the cost of 10% overhead by putting a wrapper around the sequential data structure in order to handle data distribution and remote memory accesses. Communication is another aspect of parallel codes that affects scalability. In [28] the authors present a number of subdivision templates that can be used to decouple the refinement on different processors and thus eliminate communication completely.

The main challenge in PES methods is the *collision refinement* problem. In order to achieve mesh conformity and correctness the interface faces between the submeshes should be subdivided the same way from all submeshes that share them. Thus interface edges that are subdivided in one submesh are marked to be subdivided from all other submeshes that share them. This causes communication which is handled by sending, at the end of the refinement of the interface faces, a message to submeshes that share refined faces and edges. Based on the communication and synchronization requirements for handling the refinement collision problem, the PES methods are classified into three categories: tightly coupled methods [78], partially coupled methods [45, 28, 13, 65] and decoupled methods [74].

Tightly Coupled Methods

The 3D Parallel Terminal-Edge (PTE) method described in [78] is an inherently decoupled method. However, the PTE method in [78] is implemented as a coupled method. In [74] a new design and implementation is presented so that the stability and code re-use requirements are satisfied while the global synchronization for maintaining the global name of all bisected edges is eliminated.

Parallel Coupled Terminal-Edge (PCTE) Method

In this paragraph we will refer to the tightly coupled implementation of the PTE method as PCTE. The PCTE method is based on longest-edge bisection approach introduced by Rivara [71, 72, 76]. Triangles/ tetrahedra are refined by bisecting their longest-edge. The longest-edge bisection algorithm requires the management of sequences of intermediate non-conforming mesh points throughout the refinement process. This complicates its parallel implementation because it requires some synchronization in order to handle the collision refinement and global name of newly inserted vertices, both are required to maintain the conformity of the distributed mesh.

The PCTE method [78], although it requires zero communication between processors, relies on a central processor for global name-assignment of new mesh points. The use of the central processor limits the scalability of the method for more than 60 processors and reduces the speed (tetrahedra per second) of the method by an order of magnitude. However, in [74] the authors present a decoupled method and implementation which takes full advantage of the terminal-edge algorithm introduced in [78]. The terminal-edge of a longest edge propagation path of t, Lepp(t), is the longest-edge between all the edges involved in Lepp(t) including the boundary of the Lepp polygon [73, 75, 77]. We review this method at the end of this Section.

Partially Coupled Methods

Partially coupled methods resolve inconsistencies during the collision refinement by processing interface edges in 2D (or faces in 3D) using independent sets of elements [45] and by breaking the mesh refinement process into two phases [28, 13, 65]: computation (actual refinement of elements) and communication (exchange of information about the newly created points and elements due to refinement of interfaces).

Parallel Independent Set Method

In [45] the refinement of a 2D mesh takes place in phases (refinement of one independent set at a time). This guarantees the conformity of the mesh and the elimination of the collision refinement problem, since non-local adjacent elements never refine interface edges concurrently and the processors are always aware of bisections of their interface edges. Specifically, the authors in [45] use a vertex-based partition of a 2D mesh to generate P submeshes, where P is the number of processors. Then all non-local adjacent elements (i.e., elements that share an edge) and adjacent vertices to the elements and vertices of submeshes are computed to create a layer of "ghost" mesh entities which are used to minimize communication in the independent set (IS) phase. The distributed memory implementation of the IS phase in [45] computes a distributed independent set $I = \cup_{i=1}^{P} I_{M_i}$, where $I_{M_i} = I \cap M_i$ and M_i is a submesh, as follows: a triangle $t_a \in I_{M_i}$ if: $\forall t_b \in adj(t_a)$ and one of the following three holds (1) $t_a, t_b \in M_i$, (2) $\rho(t_a) > \rho(t_b)$, and (3) t_b is not a marked triangle for refinement, where adj(t) is the set of adjacent triangles of t, and $\rho(t)$ is a unique random number assigned to each element in the mesh $M_i,\ i = 1, P$. Note that due to the ghost elements, the communication for checking the above conditions is eliminated. The algorithm requires communication only for: (a) the update of the bisections of ghost elements, and (b) a global reduction operation for termination. Both take place at the end of the refinement of an independent set. These two types of communication make the algorithm partially coupled, since experimental data in [45] indicate that the number of refinement phases (or loop iterations) is small (10 to 20) as the number of processors and the size of the mesh increase to 200 processors and a million elements, respectively.

Parallel Alternate Bisection Method

DeCougny et. al [28] addresses the collision refinement problem by using, first, alternate bisection on the interface faces then by applying region subdivision templates on the rest of the tetrahedra. After the mesh faces are subdivided, it is possible to create non-conforming interface edges on the interfaces. The non-conforming interface edges are sent to the corresponding adjacent submeshes that are refined by different processors. This will start a new mesh face subdivision followed by a communication phase, until no mesh faces need to be subdivided. Upon termination of face subdivisions, the mesh is conforming across the interfaces and then a region subdivision using sequential templates is applied in parallel to the rest of the interior tetrahedra.

Parallel Nested Elements Method

Castaños and Savage [13] have parallelized the non-conforming longest edge bisection algorithm both in 2D and 3D. In this case the refinement propagation implies the creation of sequences of non-conforming edges that can cross several submeshes involving several processors. This also means the creation of non-conforming interface edges which is particularly complex to deal with in 3D. To perform this task each processor P_i iterates between a no-communication phase (where refinement propagation between processors is delayed) and an interprocessor communication phase. Different processors can be in different phases during the refinement process, their termination is coordinated by a central processor P_0. The subdivision of an interface edge might leads to either a non-conforming edge or to a conforming edge, but the creation of different copies (one per subdomain) of its midpoint. However, after the communication phase a remote cross reference for each newly created interface edge midpoint along with *nested elements* information guarantee a unique logical name for these newly created vertices [11].

Decoupled Methods

The PTE method [74] in addition to the terminal-edge of a Lepp(t) takes full advantage of the terminal-star, which is the set of tetrahedra that share a terminal-edge. The terminal-star can play the same role in PTE the cavity plays in PCDM. Contrary to the method in [13] the terminal-star refinement algorithm completely avoids the management of non-conforming edges both in the interior of the submeshes and in the inter-subdomain interface. This eliminates the communication among subdomains and thus processors. Similarly to Castaños et al. the terminal-star method can terminate using a single processor as coordinator for adaptive mesh refinement i.e., when a global stopping criterion like the minimum-edge length of terminal-edges is not used.

The decoupled PTE algorithm and its implementation lead to an order of magnitude performance improvements compared to a previous tightly coupled implementation [74] of the same algorithm. Although the algorithm is theoretically scalable, our performance data indicate the contrary; the reason is the work-load imbalances and heterogeneity of the clusters we use. We will address these two issues in Section 7.5.

	Mesh Generation Technique		
Coupling	Deluanay	Advancing Front	Edge Bisection
Tight	PODM	$PAFT_{SM}$	PCTE
Partial	PCDT, PDR	POAFT	PIS, PNE
None	P^2DM, PD^3	PAFT	PTE

Fig. 7.8. Taxonomy of Parallel Mesh Generation Methods.

7.4 Taxonomy

The taxonomy in Figure 7.8 helps to clarify basic similarities and differences between parallel tetrahedral meshing methods. The taxonomy is based on the two attributes we used to classify the methods reviewed in this chapter: (i) the basic sequential meshing technique used for each subproblem and (ii) the degree of coupling between the subproblems. The coupling (i.e., the amount of communication and synchronization between the subproblems) is determined by the degree of dependency between the subproblems.

7.5 Implementation

The complexity of implementing efficient parallel mesh generation codes arises from the dynamic, data-dependent, and irregular computation and communication patterns of the algorithms. This inherent complexity, when combined with challenges from using primitive tools for communication like message-passing libraries [86, 4], makes the development of parallel mesh generation codes even more time-consuming and error-prone.

In the rest of the section we focus on dynamic load balancing issue. The scientific computing community has developed application-specific runtime libraries and software systems [3, 33, 49, 64, 67, 98] for dynamic load balancing. These systems are designed to support the development of parallel multi-phase applications which are computationally-intensive and consist of phases that are separated by computations such as the global error estimation. In these cases the load-balancing is accomplished by dynamically repartitioning the data after a global synchronization [95]. Throughout this chapter we call this approach to load balancing the *stop-and-repartition* method.

The stop-and-repartition approaches are good for loosely-synchronous applications like iterative PDE solvers, however they are not well-suited for applications such as adaptive mesh generation and refinement. Because for asynchronous and not

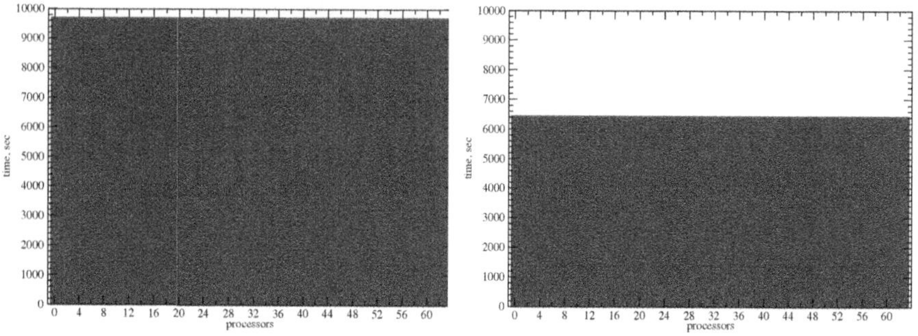

Fig. 7.9. Execution time of PAFT on Whirlwind subcluster with 64 homogeneous processors for the simplified human brain model without load balancing (left) and with load balancing using PREMA (right). The final mesh in both cases is 1.2 billion tetrahedrons.

computation-intensive applications the global synchronization overhead can overwhelm the benefits from load balancing. This problem is exacerbated as the number of processors in the parallel system grows. In order to address this issue, the authors in [2] developed a *Parallel Runtime Environment for Multi-computer Applications* (PREMA).

PREMA is a software library which provides a set of tools to application developers via a concise and intuitive interface. It supports single-sided communication primitives which conform to the *active messages* paradigm [94], a global namespace, message forwarding mechanisms to cope with object/data migration and a preemptive dynamic load balancing [2].

Performance Evaluation

In the rest of this section we present some performance data that show the effects of two sources of imbalance: (1) work-load due to geometric complexity of the subdomains/submeshes, and (2) processor heterogeneity. The experimental study was performed on Sciclone [81] cluster at the College of William and Mary which consists of many different heterogeneous subclusters. We have used three subclusters: (1) Whirlwind subcluster which consists of 64 single-cpu Sun Fire V120 nodes (650 MHz, 1 GB RAM), (2) Tornado which consists of 32 dual-cpu Sun Ultra 60 nodes (360 MHz, 512 MB RAM) and (3) Typhoon which consists of 64 single-cpu Sun Ultra 5 nodes (333 MHz, 256 MB RAM). The models we used are: (i) a cross-section of the rocket pipe (see Figure 7.1) and (ii) a simplified model of a human brain (see Figure 7.11, left).

Figure 7.9 shows the impact of dynamic load balancing on the performance of PAFT on the human brain model. The work-load imbalances are due to differences in the geometric complexity of the submeshes. The PAFT with dynamic load balancing (using PREMA) took 1.7 hours to generate the 1.2 billion elements while without dynamic load balancing it took 2.7 hours. The dynamic load balancing improved the

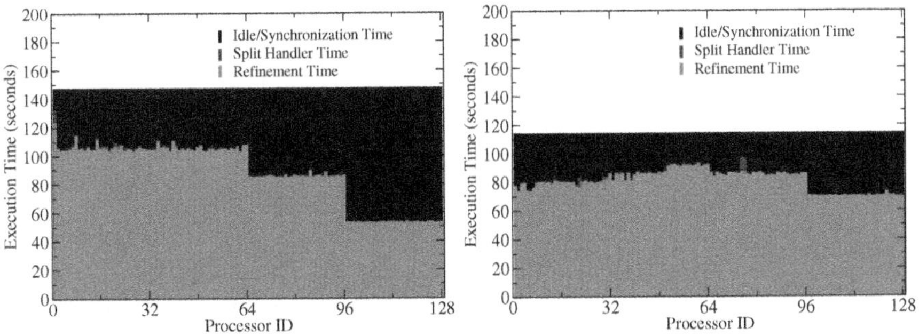

Fig. 7.10. The execution time of PCDM for the cross section of the rocket pipe whose data are equidistributed on 128 heterogenous processors; without load balancing (left) and with load balancing using PREMA (right).

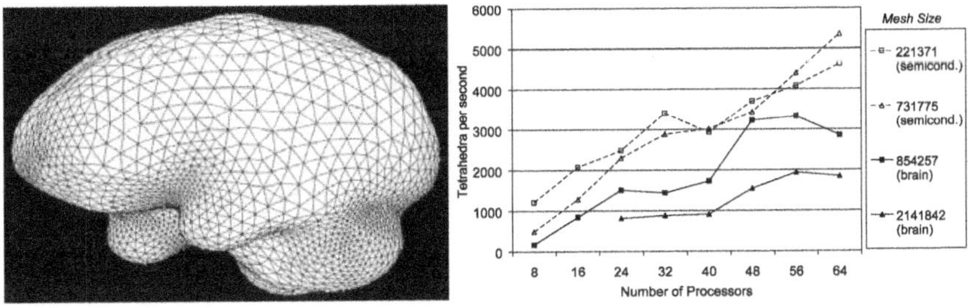

Fig. 7.11. Surface of the tetrahedral mesh for a simplified model of a human brain generated from an advancing front method [43].

performance of PAFT by more than 30%. Sequentially using Solidmesh [37], it takes three days one hour and 27 minutes, by executing the subdomains one at a time.

Figure 7.10 shows the impact of dynamic load balancing (using PREMA) on the performance of the PCDM for the cross section of rocket pipe. Although we used state-of-the-art ab-initio data partition methods for equidistributing the data and thus the computation among all 128 processors, the imbalances are due to heterogeneity of the three different clusters; the first 64 processors are from Typhoon (slowest cluster), the next 32 processors are from Tornado and the last 32 processors are from Whirlwind (the fastest cluster). Again, the dynamic load balancing (using PREMA) improved the performance of parallel mesh generation by 23%.

Finally, the data from Figure 7.11 and Table 7.2 indicate the impact of work load imbalances due to: (1) the differences in the work-load of submeshes and (2) heterogeneity of processors using the PTE method. Figure 7.11(right), shows that the speed of the PTE method is substantially lower, for the brain model (see Figure 7.11, left), due to work-load imbalances; while for a more regular geometry (the semiconductor test case [74]), the PTE speed is almost twice higher, because of better load balancing

Table 7.2. PTE speed (in tetrahedra per second) for the simplified human brain model using min-edge = 2.0. The final mesh is about 2.4 million tetrahedra.

Processors	8	16	32	48	64	96
Whirlwind	5427	9920	16195	21890	29035	23571
Tornado	3852	7025	11408	15526	20312	23571

due to more uniform point distribution. Also, Table 7.2 indicates a 19% slowdown in the PTE's speed once we increase the number of processors from 64 to 96 using additional 32 slower processors, despite the fact the PTE is a scalable method. Finally, a comparison between the speed data from the Figure 7.11 (right) and Table 7.2, for the brain model, indicate that the coupling (i.e., global synchronization) in the PCTE method slows down the speed of the code by an order of magnitude.

These data (and some more from [24, 2, 17, 74]) suggest that the tightly coupling methods should be used as a last resort. In addition, these data suggest that work-load imbalances are no longer a problem and it should not limit our creativity in the second round of our search for practical and effective parallel mesh generation methods. Runtime software systems like PREMA [2] can handle work-load imbalances quite successfully.

7.6 Future Directions

It takes about ten to fifteen years to develop the algorithmic and software infrastructure for *sequential industrial strength mesh generation libraries*. Moreover, improvements in terms of quality, speed, and functionality are open ended and permanent which makes the task of delivering state-of-the-art parallel mesh generation codes even more difficult.

This survey demonstrates that without compromising in the stability of parallel mesh generation methods it is possible for all three mesh generation classes of techniques to develop parallel meshing software using off-the-shelf sequential meshing codes.

An area with immediate high benefits to parallel mesh generation is domain decomposition. The DD problem as it is posed in Section 7.2 is still open for 3D geometries and its solution will help to deliver stable and scalable methods that rely on off-the-shelf mesh generation codes for Delaunay and Advancing Front Techniques. The edge subdivision methods are independent off the domain decomposition.

A longer term goal should be the development of both theoretical and software frameworks like PDR to implement new mesh generation methods which can: (1) take advantage of multicore architectures with more than two hardware contexts for the next generation of high-end workstations and (2) scale without any substantial implementation costs for clusters of high-end workstations.

Finally, a long term investment to parallel mesh generation is to attract the attention of mathematicians with open problems in mesh generation and broader impact in mathematics. For example, develop theoretical frameworks able to prove the correctness of single threaded guaranteed quality Delaunay theory in the context of partial order [70].

Acknowledgment

This work is supported by the IBM professorship at Brown University, the Alumni Memorial Endowment at the College of William and Mary, and National Science Foundation (NSF) grants: CNS-0312980, ANI-0203974, ACI-0085969, Career Award CCR-0049086, EIA-9972853, EIA-9726388. The experimental work was performed using computational facilities at the College of William and Mary which were enabled by grants from Sun Microsystems, the NSF, and Virginia's Commonwealth Technology Research Fund. Andrey Chernikov made many insightful recommendations and Leonidas Linardakis made useful comments for the Delaunay methods. Most of the figures and tables are the result of my collaboration with my students from the College of William and Mary Kevin Barker, Andrey Chernikov, Andriy Fedorov, Brian Holinka, Leonidas Linardakis, from Notre Dame Demian Nave, and from Universidad de Chile Carlo Calderon and Daniel Pizarro. The anonymous referee for his many useful comments.

References

1. S. Baden, N. Chrisochoides, D. Gannon, and M. Norman, editors. *Structured Adaptive Mesh Refinement Grid Methods*. Springer-Verlag, 1999.
2. K. Barker, A. Chernikov, N. Chrisochoides, and K. Pingali. A load balancing framework for adaptive and asynchronous applications. *IEEE Transactions on Parallel and Distributed Systems*, 15(2):183–192, Feb. 2004.
3. A. Basermann, J. Clinckemaillie, T. Coupez, J. Fingberg, H. Digonnet, R. Ducloux, J. Gratien, U. Hartmann, G. Lonsdale, B. Maerten, D. Roose, and C. Walshaw. Dynamic load-balancing of finite element applications with the drama library. *Applied Mathematical Modeling*, 25:83–98, 2000.
4. A. Belguelin, J. Dongarra, A. Geist, R. Manchek, S. Otto, and J. Walpore. PVM: Experiences, current status, and future direction. In *Supercomputing '93 Proceedings*, pp. 765–766, 1993.
5. D. Bertsekas and J. Tsitsiklis. *Parallel and Distributed Computation: Numerical Methods*. Prentice Hall, 1989.
6. T. B.H.V. and B. Cheng. Parallel adaptive quadrilateral mesh generation. *Computers and Structures,*, 73:519–536, 1999.
7. G. E. Blelloch, J. Hardwick, G. L. Miller, and D. Talmor. Design and implementation of a practical parallel Delaunay algorithm. *Algorithmica*, 24:243–269, 1999.
8. G. E. Blelloch, G. L. Miller, and D. Talmor. Developing a practical projection-based parallel Delaunay algorithm. In *12th Annual Symposium on Computational Geometry*, pp. 186–195, 1996.
9. H. Blum. A transformation for extracting new descriptors of shape. In *Models for the Perception of speech and Visual Form*, pp. 362–380. MIT Press, 1967.

10. A. Bowyer. Computing Dirichlet tesselations. *Computer Journal*, 24:162–166, 1981.
11. J. G. Castaños and J. E. Savage. The dynamic adaptation of parallel mesh-based computation. In *SIAM 7th Symposium on Parallel and Scientific Computation*, 1997.
12. J. G. Castaños and J. E. Savage. Parallel refinement of unstructured meshes. In *Proceedings of the IASTED, International Conference of Parallel and Distributed Computing and Systems*, 1999.
13. J. G. Castaños and J. E. Savage. PARED: a framework for the adaptive solution of PDEs. In *8th IEEE Symposium on High Performance Distributed Computing*, 1999.
14. M.-B. Chen, T. R. Chuang, and J.-J. Wu. Efficient parallel implementations of near Delaunay triangulation with high performance Fortran. *Concurrency: Practice and Experience*, 16(12), 2004.
15. A. N. Chernikov and N. P. Chrisochoides. Parallel guaranteed quality planar Delaunay mesh generation by concurrent point insertion. In *14th Annual Fall Workshop on Computational Geometry*, pp. 55–56. MIT, Nov. 2004.
16. A. N. Chernikov and N. P. Chrisochoides. Practical and efficient point insertion scheduling method for parallel guaranteed quality Delaunay refinement. In *Proceedings of the 18th annual international conference on Supercomputing*, pp. 48–57. ACM Press, 2004.
17. A. N. Chernikov, N. P. Chrisochoides, and L. P. Chew. Design of a parallel constrained Delaunay meshing algorithm, 2005.
18. L. P. Chew. Constrained Delaunay triangulations. *Algorithmica*, 4(1):97–108, 1989.
19. L. P. Chew, N. Chrisochoides, and F. Sukup. Parallel constrained Delaunay meshing. In *ASME/ASCE/SES Summer Meeting, Special Symposium on Trends in Unstructured Mesh Generation*, pp. 89–96, Northwestern University, Evanston, IL, 1997.
20. N. Chrisochoides. An alternative to data-mapping for parallel iterative PDE solvers: Parallel grid generation. In *Scalable Parallel Libraries Conference*, pp. 36–44. IEEE, 1993.
21. N. Chrisochoides. Multithreaded model for load balancing parallel adaptive computations. *Applied Numerical Mathematics*, 6:1–17, 1996.
22. N. Chrisochoides, C. Houstis, E.N.Houstis, P. Papachiou, S. Kortesis, and J. Rice. Domain decomposer: A software tool for partitioning and allocation of PDE computations based on geometry decomposition strategies. In *4th International Symposium on Domain Decomposition Methods*, pp. 341–357. SIAM, 1991.
23. N. Chrisochoides, E. Houstis, and J. Rice. Mapping algorithms and software environment for data parallel PDE iterative solvers. *Special issue of the Journal of Parallel and Distributed Computing on Data-Parallel Algorithms and Programming*, 21(1):75–95, 1994.
24. N. Chrisochoides and D. Nave. Parallel Delaunay mesh generation kernel. *Int. J. Numer. Meth. Engng.*, 58:161–176, 2003.
25. N. Chrisochoides and F. Sukup. Task parallel implementation of the Bowyer-Watson algorithm. In *Proceedings of Fifth International Conference on Numerical Grid Generation in Computational Fluid Dynamics and Related Fields*, 1996.
26. N. P. Chrisochoides. A new approach to parallel mesh generation and partitioning problems. *Computational Science, Mathematics and Software*, pp. 335–359, 2002.
27. P. Cignoni, D. Laforenza, C. Montani, R. Perego, and R. Scopigno. Evaluation of parallelization strategies for an incremental Delaunay triangulator in E^3. *Concurrency: Practice and Experience*, 7(1):61–80, 1995.
28. H. D. Cougny and M. Shephard. Parallel refinement and coarsening of tetrahedral meshes. *Int. J. Meth. Eng.*, 46(1101-1125), 1999.
29. T. Culver. *Computing the Medial Axis of a Polyhedron Reliably and Efficiently*. PhD thesis, The University of North Carolina at Chapel Hill, 2000.
30. H. de Cougny and M. Shephard. *CRC Handbook of Grid Generation*, chapter Parallel unstructured grid generation, pp. 24.1–24.18. CRC Press, Inc., 1999.

31. H. de Cougny and M. Shephard. Parallel volume meshing using face removals and hierarchical repartitioning. *Comp. Meth. Appl. Mech. Engng.*, 174(3-4):275–298, 1999.
32. H. L. de Cougny, M. S. Shephard, and C. Ozturan. Parallel three-dimensional mesh generation on distributed memory mimd computers. *Engineering with Computers*, 12:94–106, 1995.
33. K. Devine, B. Hendrickson, E. Boman, M. S. John, and C. Vaughan. Design of dynamic load-balancing tools for parallel applications. In *Proc. of the Int. Conf. on Supercomputing*, Santa Fe, May 2000.
34. E. W. Dijkstra and C. Sholten. Termination detection for diffusing computations. *Inf. Proc. Lettres*, 11, 1980.
35. H. Edelsbrunner and D. Guoy. Sink-insertion for mesh improvement. In *Proceedings of the Seventeenth Annual Symposium on Computational Geometry*, pp. 115–123. ACM Press, 2001.
36. P. J. Frey and P. L. George. *Mesh Generation: Applications to Finite Element.* Hermis; Paris, 2000.
37. J. Gaither, D. Marcum, and B. Mitchell. Solidmesh: A solid modeling approach to unstructured grid generation. In *7th International Conference on Numerical Grid Generation in Computational Field Simulations*, 2000.
38. J. Galtier and P. L. George. Prepartitioning as a way to mesh subdomains in parallel. In *Special Symposium on Trends in Unstructured Mesh Generation*, pp. 107–122. ASME/ASCE/SES, 1997.
39. P. L. George and H. Borouchaki. *Delaunay Triangulation and Meshing: Applications to Finite Element.* Hermis; Paris, 1998.
40. H. N. Gürsoy. *Shape interrogation by medial axis transform for automated analysis.* PhD thesis, Massachusetts Institute of Technology, 1989.
41. J. C. Hardwick. Implementation and evaluation of an efficient 2D parallel Delaunay triangulation algorithm. In *Proceedings of the 9th Annual ACM Symposium on Parallel Algorithms and Architectures,*, 1997.
42. B. Hendrickson and R. Leland. The Chaco user's guide, version 2.0. Technical Report SAND94-2692, Sandia National Laboratories., 1994.
43. A. M. S. Ito, Yasushi and B. K. Soni. Reliable isotropic tetrahedral mesh generation based on an advancing front method. In *Proceedings 13th International Meshing Roundtable, Williamsburg, VA, Sandia National Laboratories*, pp. 95–106, 2004.
44. D. A. Jefferson. Virtual time. In *ACM Transactions on Programming Languages and Systems*, volume 7, pp. 404–425, July 1985.
45. M. T. Jones and P. E. Plassmann. Parallel algorithms for the adaptive refinement and partitioning of unstructured meshes. In *Proceedings of the Scalable High-Performance Computing Conference*, 1994.
46. C. Kadow. Adaptive dynamic projection-based partitioning for parallel Delaunay mesh generation algorithms. In *SIAM Workshop on Combinatorial Scientific Computing*, Feb. 2004.
47. C. Kadow and N. Walkington. Design of a projection-based parallel Delaunay mesh generation and refinement algorithm. In *Fourth Symposium on Trends in Unstructured Mesh Generation*, July 2003. `http://www.andrew.cmu.edu/user/sowen/usnccm03/agenda.html`.
48. C. M. Kadow. *Parallel Delaunay Refinement Mesh Generation.* PhD thesis, Carnegie Mellon University, May 2004.
49. S. Kohn and S. Baden. Parallel software abstractions for structured adaptive mesh methods. *Journal of Par. and Dist. Comp.*, 61(6):713–736, 2001.

50. R. Konuru, J. Casas, R. Prouty, S. Oto, and J. Walpore. A user level process package for pvm. In *Proceedings of Scalable High-Performance Computing Conferene*, pp. 48–55. IEEE, 1997.
51. L. Linardakis and N. Chrisochoides. Parallel Delaunay domain decoupling method for non-uniform mesh generation. *SIAM Journal on Scientific Computing*, 2005.
52. L. Linardakis and N. Chrisochoides. Parallel domain decoupling Delaunay method. *SIAM Journal on Scientific Computing*, in print, accepted Nov. 2004.
53. L. Linardakis and N. Chrisochoides. Medial axis domain decomposition method. *ACM Trans. Math. Software*, To be submited, 2005.
54. R. Lober, T. Tautges, and R. Cairncross. The parallelization of an advancing-front, all-quadrilateral meshing algorithm for adaptive analysis. In *4th International Meshing Roundtable*, pp. 59–70, October 1995.
55. R. Löhner, J. Camberos, and M. Marshal. *Unstructured Scientific Computation on Scalable Multiprocessors (Eds. Piyush Mehrotra and Joel Saltz)*, chapter Parallel Unstructured Grid Generation, pp. 31–64. MIT Press, 1990.
56. R. Löhner and J. R. Cebral. Parallel advancing front grid generation. In *Proceedings of the Eighth International Meshing Roundtable*, pp. 67–74, 1999.
57. F. Lori, M. Jones, and P. Plassmann. An efficient parallel algorithm for mesh smoothing. In *Proceedings 4th International Meshing Roundtable*, pp. 47–58, 1995.
58. R. P. M. Saxena. Parallel FEM algorithm based on recursive spatial decomposition. *Computers and Structures*, 45(9-6):817–831, 1992.
59. S. N. Muthukrishnan, P. S. Shiakolos, R. V. Nambiar, and K. L. Lawrence. Simple algorithm for adaptative refinement of three-dimensionalfinite element tetrahedral meshes. *AIAA Journal*, 33:928–932, 1995.
60. D. Nave, N. Chrisochoides, and L. P. Chew. Guaranteed: quality parallel Delaunay refinement for restricted polyhedral domains. In *SCG '02: Proceedings of the eighteenth annual symposium on Computational geometry*, pp. 135–144. ACM Press, 2002.
61. D. Nave, N. Chrisochoides, and L. P. Chew. Guaranteed–quality parallel Delaunay refinement for restricted polyhedral domains. *Computational Geometry: Theory and Applications*, 28:191–215, 2004.
62. T. Okusanya and J. Peraire. Parallel unstructured mesh generation. In *5th International Conference on Numerical Grid Generation on Computational Field Simmulations*, pp. 719–729, April 1996.
63. T. Okusanya and J. Peraire. 3D parallel unstructured mesh generation. In S. A. Canann and S. Saigal, editors, *Trends in Unstructured Mesh Generation*, pp. 109–116, 1997.
64. L. Oliker and R. Biswas. Plum: Parallel load balancing for adaptive unstructured meshes. *Journal of Par. and Dist. Comp.*, 52(2):150–177, 1998.
65. L. Oliker, R. Biswas, and H. Gabow. Parallel tetrahedral mesh adaptation with dynamic load balancing. *Parallel Computing Journal*, pp. 1583–1608, 2000.
66. S. Owen. A survey of unstructured mesh generation. Technical report, ANSYS Inc., 2000.
67. M. Parashar and J. Browne. Dagh: A data-management infrastructure for parallel adaptive mesh refinement techniques. Technical report, Dept. of Comp. Sci., Univ. of Texas at Austin, 1995.
68. N. Patrikalakis and H. Gürsoy. Shape interrogation by medial axis transform. In *Design Automation Conference (ASME)*, pp. 77–88, 1990.
69. P. Pebay and D. Thompson. Parallel mesh refinement without communication. In *Proceedings of International Meshing Roundtable*, pp. 437–443, 2004.
70. S. Prassidis and N. Chrisochoides. A categorical approach for parallel Delaunay mesh generation, July 2004.

71. M. C. Rivara. Algorithms for refining triangular grids suitable for adaptive and multigrid techniques. *International Journal for Numerical Methods in Engineering*, 20:745–756, 1984.
72. M. C. Rivara. Selective refinement/derefinement algorithms for sequences of nested triangulations. *International Journal for Numerical Methods in Engineering*, 28:2889–2906, 1989.
73. M. C. Rivara. New longest-edge algorithms for the refinement and/or improvement of unstructured triangulations. *International Journal for Numerical Methods in Engineering*, 40:3313–3324, 1997.
74. M.-C. Rivara, C. Calderon, D. Pizarro, A. Fedorov, and N. Chrisochoides. Parallel decoupled terminal-edge bisection algorithm for 3D meshes. *(Invited) Engineering with Computers*, 2005.
75. M. C. Rivara, N. Hitschfeld, and R. B. Simpson. Terminal edges Delaunay (small angle based) algorithm for the quality triangulation problem. *Computer-Aided Design*, 33:263–277, 2001.
76. M. C. Rivara and C. Levin. A 3D refinement algorithm for adaptive and multigrid techniques. *Communications in Applied Numerical Methods*, 8:281–290, 1992.
77. M. C. Rivara and M. Palma. New lepp algorithms for quality polygon and volume triangulation: Implementation issues and practical behavior. In *In Trends unstructured mesh generationi Eds: S. A. Cannan . Saigal, AMD*, volume 220, pp. 1–8, 1997.
78. M.-C. Rivara, D. Pizarro, and N. Chrisochoides. Parallel refinement of tetrahedral meshes using terminal-edge bisection algorithm. In *13th International Meshing Roundtable*, Sept. 2004.
79. R. Said, N. Weatherill, K. Morgan, and N. Verhoeven. Distributed parallel Delaunay mesh generation. *Computer Methods in Applied Mechanics and Engineering*, (177):109–125, 1999.
80. K. Schloegel, G. Karypis, and V. Kumar. Parallel multilevel diffusion schemes for repartitioning of adaptive meshes. Technical Report 97-014, Univ. of Minnesota, 1997.
81. Sciclone cluster project. Last accessed, March 2005. `http://www.compsci.wm.edu/SciClone/`.
82. M. Shephard and M. Georges. Automatic three-dimensional mesh generation by the finite octree technique. *International Journal for Numerical Methods in Engineering*, 32:709–749, 1991.
83. E. C. Sherbrooke. *3-D shape interrogation by medial axial transform*. PhD thesis, Massachusetts Institute of Technology, 1995.
84. J. Shewchuk. Triangle: Engineering a 2D quality mesh generator and Delaunay triangulator. In *Proceedings of the First workshop on Applied Computational Geometry*, pp. 123–133, Philadelphia, PA, 1996.
85. J. R. Shewchuk. *Delaunay Refinement Mesh Generation*. PhD thesis, Carnegie Mellon University, 1997.
86. M. Snir, S. Otto, S. Huss-Lederman, and D. Walker. MPI the complete reference. MIT Press, 1996.
87. A. Sohn and H. Simon. Jove: A dynamic load balancing framework for adaptive computations on an SP-2 distributed memory multiprocessor, 1994. Technical Report 94-60, Dept. of Comp. and Inf. Sci., New Jersey Institute of Technology, 1994.
88. D. A. Spielman, S.-H. Teng, and A. Üngör. Parallel Delaunay refinement: Algorithms and analyses. In *Proceedings of the Eleventh International Meshing Roundtable*, pp. 205–217, 2001.

89. D. A. Spielman, S.-H. Teng, and A. Üngör. Time complexity of practical parallel Steiner point insertion algorithms. In *Proceedings of the sixteenth annual ACM symposium on Parallelism in algorithms and architectures*, pp. 267–268. ACM Press, 2004.
90. M. Stuti and A. Moitra. Considerations of computational optimality in parallel algorithms for grid generation. In *5th International Conference on Numerical Grid Generation in Computational Field Simulations*, pp. 753–762, 1996.
91. T. Tam, M. Price, C. Armstrong, and R. McKeag. Computing the critical points on the medial axis of a planar object using a Delaunay point triangulation algorithm.
92. Y. A. Teng, F. Sullivan, I. Beichl, and E. Puppo. A data-parallel algorithm for three-dimensional Delaunay triangulation and its implementation. In *SuperComputing*, pp. 112–121. ACM, 1993.
93. J. F. Thompson, B. K. Soni, and N. P. Weatherill. *Handbook of Grid Generation*. CRC Press, 1999.
94. T. von Eicken, D. Culler, S. Goldstein, and K. Schauser. Active messages: A mechanism for integrated communication and computation. In *Proceedings of the 19th Int. Symp. on Comp. Arch.*, pp. 256–266. ACM Press, May 1992.
95. C. Walshaw, M. Cross, and M. Everett. Parallel dynamic graph partitioning for adaptive unstructured meshes. *Journal of Par. and Dist. Comp.*, 47:102–108, 1997.
96. D. F. Watson. Computing the n-dimensional Delaunay tesselation with application to Voronoi polytopes. *Computer Journal*, 24:167–172, 1981.
97. R. Williams. *Adaptive parallel meshes with complex geometry*. Numerical Grid Generation in Computational Fluid Dynamics and Related Fields, 1991.
98. X. Yuan, C. Salisbury, D. Balsara, and R. Melhem. Load balancing package on distributed memory systems and its application particle-particle and particle-mesh (P3M) methods. *Parallel Computing*, 23(10):1525–1544, 1997.

Part III

Parallel Software Tools

8

The Design and Implementation of *hypre*, a Library of Parallel High Performance Preconditioners

Robert D. Falgout, Jim E. Jones, and Ulrike Meier Yang

Center for Applied Scientific Computing, Lawrence Livermore National Laboratory, P.O. Box 808, L-561, Livermore, CA 94551, USA

`[rfalgout,jjones,umyang]@llnl.gov`

Summary. The *hypre* software library provides high performance preconditioners and solvers for the solution of large, sparse linear systems on massively parallel computers. One of its attractive features is the provision of *conceptual interfaces*. These interfaces give application users a more natural means for describing their linear systems, and provide access to methods such as geometric multigrid which require additional information beyond just the matrix. This chapter discusses the design of the conceptual interfaces in *hypre* and illustrates their use with various examples. We discuss the data structures and parallel implementation of these interfaces. A brief overview of the solvers and preconditioners available through the interfaces is also given.

8.1 Introduction

The increasing demands of computationally challenging applications and the advance of larger more powerful computers with more complicated architectures have necessitated the development of new solvers and preconditioners. Since the implementation of these methods is quite complex, the use of high performance libraries with the newest efficient solvers and preconditioners becomes more important for promulgating their use into applications with relative ease.

The *hypre* library [16, 22] has been designed with the primary goal of providing users with advanced scalable parallel preconditioners. Multigrid preconditioners are a major focus of the library. Issues of robustness, ease of use, flexibility and interoperability have also been important. It can be used both as a solver package and as a framework for algorithm development. Its object model is more general and flexible than most current generation solver libraries [10]. The *hypre* library also provides several of the most commonly used solvers, such as conjugate gradient for symmetric systems or GMRES for nonsymmetric systems to be used in conjunction with the preconditioners. The code is open source and available for download from the web [22].

A unique and important design innovation in *hypre* is the notion of *conceptual (linear system) interfaces*. These interfaces make it possible to provide an array of powerful solvers that have largely not been available before in linear solver library form. By allowing users to access *hypre* in the way they naturally think about their problems, these interfaces ease the coding burden and provide extra application information required by certain solvers. For example, application developers that use structured grids typically think of their linear systems in terms of stencils and grids, so an interface that involves stencils and grids is more natural. Such an interface also makes it possible to supply solvers like geometric multigrid that take advantage of structure. In addition, the use of a particular conceptual interface does not preclude users from building more general sparse matrix structures (e.g., compressed sparse row) and using more traditional solvers (e.g., incomplete LU factorization) that do not use the additional application information. In fact, the construction of different underlying data structures is done internally by *hypre* and requires almost no changes to user code. The conceptual interfaces currently implemented include stencil-based structured and semi-structured interfaces, a finite-element based unstructured interface, and a traditional linear-algebra based interface.

The primary focus of this paper is on the design and implementation of the conceptual interfaces in *hypre*. The paper is organized as follows. The first two sections are of general interest. We begin in Section 8.2 with an introductory discussion of conceptual interfaces and point out the advantages of matching the linear solver interface with the natural concepts (grids, stencils, elements, etc.) used in the application code discretization. In Section 8.3, we discuss *hypre*'s object model, which is built largely around the notion of an operator. Sections 8.4 through 8.7 discuss specific conceptual interfaces available in *hypre* and include various examples illustrating their use. These sections are intended to give application programmers an overview of each specific conceptual interface. We then discuss some implementation issues in Section 8.8. This section may be of interest to application programmers, or more likely, others interested in linear solver code development on large scale parallel computers. The next two sections are aimed at application programmers potentially interested in using *hypre*. Section 8.9 gives a brief overview of the solvers and preconditioners currently available in *hypre*, and Section 8.10 contains additional information on how to obtain and build the library. The paper concludes with some comments on future plans to enhance the library.

8.2 Conceptual Interfaces

Each application to be implemented lends itself to natural ways of thinking of the problem. If the application uses structured grids, a natural way of formulating it would be in terms of grids and stencils, whereas for an application that uses unstructured grids and finite elements it is more natural to access the preconditioners and solvers via elements and element stiffness matrices. Consequently, the provision of different conceptual views of the problem being solved (*hypre*'s so-called conceptual interfaces) facilitates the use of the library.

Conceptual interfaces also decrease the coding burden for users. The most common interface used in linear solver libraries today is a linear-algebraic one. This

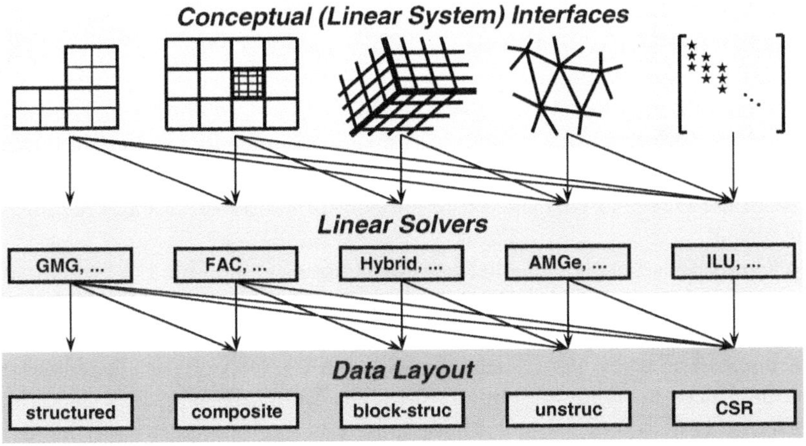

Fig. 8.1. Graphic illustrating the notion of conceptual interfaces.

interface requires that the user compute the mapping of their discretization to row-column entries in a matrix. This code can be quite complex; for example, consider the problem of ordering the equations and unknowns on the composite grids used in structured AMR codes. The use of a conceptual interface merely requires the user to input the information that defines the problem to be solved, leaving the forming of the actual linear system as a library implementation detail hidden from the user.

Another reason for conceptual interfaces—maybe the most compelling one—is that they provide access to a large array of powerful scalable linear solvers that need the extra application information beyond just the matrix. For example, geometric multigrid (GMG) [21] cannot be used through a linear-algebraic interface, since it is formulated in terms of grids. Similarly, in many cases, these interfaces allow the use of other data storage schemes with less memory overhead and provide for more efficient computational kernels.

Figure 8.1 illustrates the idea behind conceptual interfaces (note that the figure is not intended to represent the current state within *hypre*, although it is highly representative). The level of generality increases from left to right. On the left are specific interfaces with algorithms and data structures that take advantage of more specific information. On the right are more general interfaces, algorithms and data structures. Note that the more specific interfaces also give users access to general solvers like algebraic multigrid (AMG) [35] or incomplete LU factorization (ILU). The top row shows various concepts: structured grids, composite grids, unstructured grids, or just matrices. In the second row, various solvers and preconditioners are listed. Each of these requires different information from the user, which is provided through the conceptual interfaces. For example, GMG needs a structured grid and can only be used with the leftmost interface. AMGe [5], an algebraic multigrid method, needs finite element information, whereas general solvers can be used with any interface.

The bottom row contains a list of data layouts or matrix-vector storage schemes that can be used for the implementation of the various algorithms. The relationship between linear solver and storage scheme is similar to that of interface and linear solver. One minor difference, however, is that solvers can appear more than once. Consider ILU, for example. It is easy to imagine implementing ILU in terms of a structured-grid data layout. In the figure, such a solver would appear in the leftmost box (with the GMG solver) since it requires information about the structure in the linear system.

In *hypre*, four conceptual interfaces are currently supported: a structured-grid interface, a semi-structured-grid interface, a finite-element interface, and a linear-algebraic interface. For the purposes of this paper, we will refer to these interfaces by the names `Struct`, `semiStruct`, `FEI`, and `IJ`, respectively; the actual names of the interfaces in *hypre* are slightly different. Similarly, when interface routines are discussed below, we will not strictly adhere to the prototypes in *hypre*, as these prototypes may change slightly in the future in response to user needs. Instead, we will focus on the basic design components and basic use of the interfaces, and refer the reader to the *hypre* documentation [22] for current details.

Users of *hypre* only need to use one conceptual interface in their application code. The best choice of interface depends on the application, but it is usually better to use a more specific one that utilizes as much application information as possible (e.g., something further to the left in Figure 8.1). Access to more general solvers and data structures are still possible and require almost no changes to the user code. Note, finally, that *hypre* does not determine the parallel partitioning of the problem, but builds the internal parallel data structures (often quite complicated) according to the partitioning provided by the user.

8.3 Object Model

In this section, we discuss the basic object model for *hypre*, illustrated in Figure 8.2. This model has evolved since the version presented in [10], but the core design ideas have persisted. We will focus here on the primary components of the *hypre* model, but encourage the reader to see [10] for a discussion of other useful design ideas also utilized in the library.

Note that, although *hypre* uses object-oriented (OO) principles in its design, the library is actually written in C, which is not an object-oriented language (the PETSc library [3, 4, 26] employs a similar approach). The library also provides an interface for Fortran, still another non-OO language. In addition, the next generation of interfaces in *hypre* are being developed through a tool called Babel [2], which makes it possible to generate interfaces automatically for a wide variety of languages (e.g., C, C++, Fortran77, Fortran90, Python, and Java) and provides support for important OO concepts such as polymorphism. An object model such as the one in Figure 8.2 is critical to the use of such a tool.

Central to the design of *hypre* is the use of multiple inheritance of interfaces. An *interface* is simply a collection of routines that can be invoked on an object, e.g.,

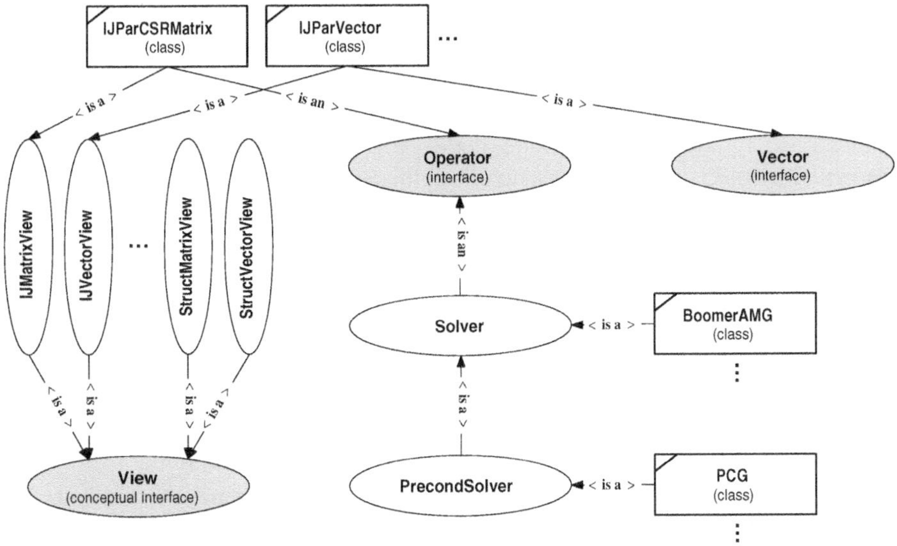

Fig. 8.2. Illustration of the *hypre* object model. Ovals represent abstract interfaces and rectangles are concrete classes. The `View` interface encapsulates the conceptual interface idea described in Section 8.2.

a matrix-vector multiply routine (for readers familiar with OO languages, the term interface is the same as in Java, and the same as an abstract base class in C++). The base interfaces in the *hypre* model are: `View`, `Operator`, and `Vector` (note that we use slightly different names here than those used in the *hypre* library). The `View` interface represents the notion of a conceptual interface as described in Section 8.2. This interface serves as our means of looking at (i.e., "viewing") the data in matrix and vector objects. Specific views such as `IJMatrixView` are inherited from, or *extend*, the base interface. That is, `IJMatrixView` contains the routines in `View`, plus routines specific to its linear-algebraic way of looking at the data.

The `Vector` interface is fairly standard across most linear solver libraries. It consists of basic routines such as `Dot()` (vector inner product) and `Axpy()` (vector addition).

The `Operator` interface represents the general notion of operating on vectors to produce an output vector, and primarily consists of the routine `Apply()`. As such, it unifies many of the more common mathematical objects used in linear (and even nonlinear) solver libraries, such as matrices, preconditioners, iterative methods, and direct methods. The `Solver` and `PrecondSolver` interfaces are extensions of the `Operator` interface. Here, `Apply()` denotes the action of solving the linear system, i.e., the action of "applying" the solution operator to the right-hand-side vector. The two solver interfaces contain a number of additional routines that are common to iterative methods, such as `SetTolerance()` and `GetNumIterations()`. But, the main extension is the addition of the `SetOperator()` routine, which defines the operator in the linear system to be solved. The `PrecondSolver` extends the

Operator interface further with the addition of the SetPreconditioner() routine.

The extensive use of the Operator interface is a novel feature of the *hypre* object model. In particular, note that both Solver and PrecondSolver are also Operator interfaces. Furthermore, the SetOperator() interface routine takes as input an object of type Operator, and the SetPreconditioner() routine takes as input an object of type Solver (an Operator). The latter fact is of interest. In *hypre*, there is currently no specific preconditioner type. Instead, preconditioning is considered to be a *role* played by solvers. The ability to use solvers for preconditioning is available in many other packages, but most employ additional design constructs to achieve this. For example, the linear algebra part of the Diffpack package [27, 28] utilizes a special class that bridges solvers and preconditioners.

Another fundamental design feature in the *hypre* library is the separation of the View and Operator interfaces in matrix classes such as IJParCSRMatrix. In this particular class, both interfaces are present, and the underlying data storage is a parallel compressed sparse row format. This format is required in solvers like BoomerAMG. However, the inherited IJMatrixView interface is generic, which allows users to write generic code for constructing matrices. As a result, the underlying storage format can be changed (giving access to potentially different solvers) by modifying only one or two lines of code.

The separation of the View and Operator interfaces also makes it possible for objects to support one functionality without supporting the other. The classic example is the so-called matrix-free linear operator (or matrix-free preconditioner), which is an object that performs matrix-vector multiplication, but does not offer access to the coefficients of the matrix. This has proven to be a useful technique in many practical situations.

For the OO-aware reader and users of *hypre*, we comment that Figure 8.2 is an accurate model for the current Babel interface in *hypre*, but it is not fully reflective of the native C interface. In particular, much of the polymorphism in the model is currently not implemented. Also, the native View interface follows the "Builder" design pattern discussed in [19] instead of the model in the figure, but the difference is trivial in practice. Finally, we remark that since *hypre* is moving toward using Babel for all of its interfaces in the future, the so-called native C interface mentioned here (and its shortcomings) will soon be moot.

8.4 The Structured-Grid Interface (Struct)

The Struct interface is appropriate for scalar applications on structured grids with a fixed stencil pattern of nonzeros at each grid point. It provides access to *hypre*'s most efficient scalable solvers for scalar structured-grid applications, the geometric multigrid methods SMG and PFMG. The user defines the *grid* and the *stencil*; the matrix and right-hand-side vector are then defined in terms of the grid and the stencil.

The Struct grid is described via a global d-dimensional *index space*, i.e. via integer singles in 1D, tuples in 2D, or triples in 3D (the integers may have any value,

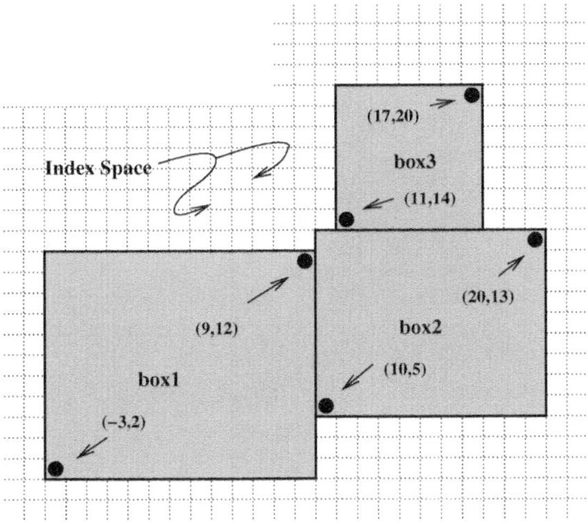

Fig. 8.3. A box is a collection of abstract cell-centered indices, described by its minimum and maximum indices. Here, three boxes are illustrated.

positive or negative). The global indices are used to discern how data is related spatially, and how it is distributed across the parallel machine. The basic component of the grid is a *box*: a collection of abstract cell-centered indices in index space, described by its "lower" and "upper" corner indices (see Figure 8.3). The scalar grid data is always associated with cell centers, unlike the more general `semiStruct` interface which allows data to be associated with box indices in several different ways. Each process describes the portion of the grid that it "owns", one box at a time. Note that it is assumed that the data has already been distributed, and that it is handed to the library in this distributed form.

The stencil is described by an array of integer indices, each representing a relative offset (in index space) from some gridpoint on the grid. For example, the geometry of the standard 5-pt stencil can be represented in the following way:

$$\begin{bmatrix} & (0,1) & \\ (-1,0) & (0,0) & (1,0) \\ & (0,-1) & \end{bmatrix}. \tag{8.1}$$

After the grid and stencil are defined, the matrix coefficients are set using the `MatrixSetBoxValues()` routine with the following arguments: a box specifying where on the grid the stencil coefficients are to be set; a list of stencil entries indicating which coefficients are to be set (e.g., the "center", "south", and "north" entries of the 5-point stencil above); and the actual coefficient values.

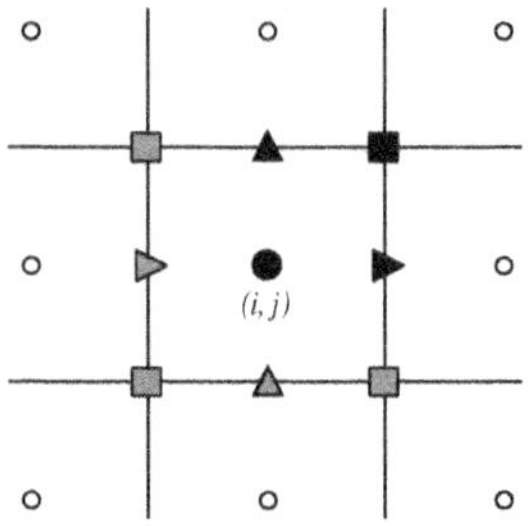

Fig. 8.4. Grid variables in *hypre* are referenced by the abstract cell-centered index to the left and down in 2D (and analogously in 3D). So, in the figure, index (i, j) is used to reference the variables in black. The variables in grey, although contained in the pictured cell, are not referenced by the (i, j) index.

8.5 The Semi-Structured-Grid Interface (`semiStruct`)

The `semiStruct` interface is appropriate for applications with grids that are mostly—but not entirely—structured, e.g. block-structured grids, composite grids in structured AMR applications, and overset grids. In addition, it supports more general PDEs than the `Struct` interface by allowing multiple variables (system PDEs) and multiple variable types (e.g. cell-centered, face-centered, etc.). The interface provides access to data structures and linear solvers in *hypre* that are designed for semi-structured grid problems, but also to the most general data structures and solvers.

The `semiStruct` grid is composed out of a number of structured grid *parts*, where the physical inter-relationship between the parts is arbitrary. Each part is constructed out of two basic components: boxes (see Section 8.4) and *variables*. Variables represent the actual unknown quantities in the grid, and are associated with the box indices in a variety of ways, depending on their types. In *hypre*, variables may be cell-centered, node-centered, face-centered, or edge-centered. Face-centered variables are split into x-face, y-face, and z-face, and edge-centered variables are split into x-edge, y-edge, and z-edge. See Figure 8.4 for an illustration in 2D.

The `semiStruct` interface uses a *graph* to allow nearly arbitrary relationships between part data. The graph is constructed from stencils plus some additional data-coupling information set by the `GraphAddEntries()` routine. Another method for relating part data is the `GridSetNeighborbox()` routine, which is particularly suited for block-structured grid problems. Several examples are given in the following sections to illustrate these concepts.

8.5.1 Block-Structured Grids

In this section, we describe how to use the `semiStruct` interface to define block-structured grid problems. We will do this primarily by example, paying particular attention to the construction of stencils and the use of the `GridSetNeighborbox()` interface routine.

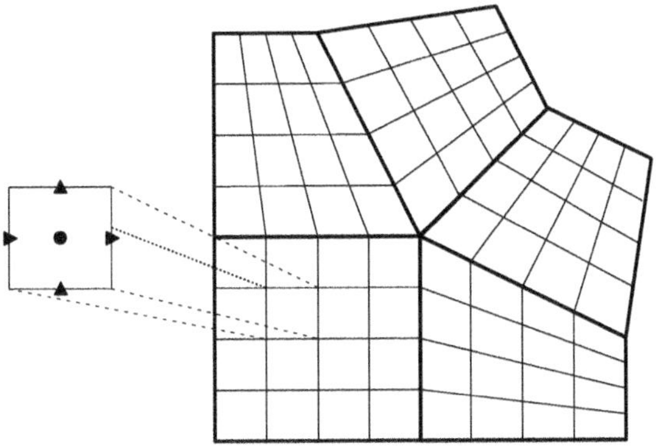

Fig. 8.5. Block-structured grid example with five logically-rectangular blocks and three variables types: cell-centered, x-face, and y-face.

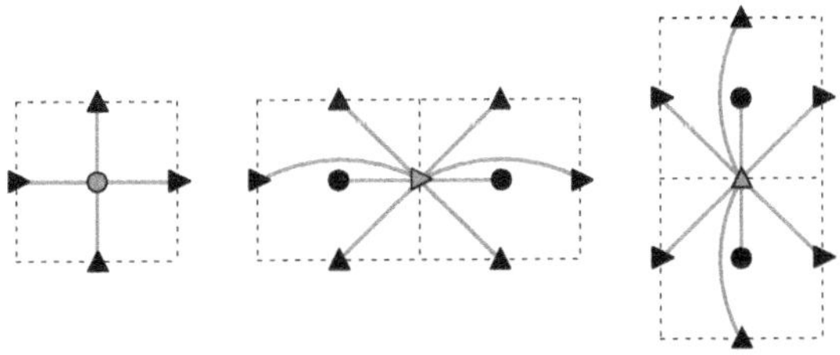

Fig. 8.6. Discretization stencils for the cell-centered (left), x-face (middle), and y-face (right) variables for the block-structured grid example in Figure 8.5.

Consider the solution of the diffusion equation

$$-\nabla \cdot (D\nabla u) + \sigma u = f \tag{8.2}$$

on the block-structured grid in Figure 8.5, where D is a scalar diffusion coefficient, and $\sigma \geq 0$. The discretization [30] introduces three different types of variables: cell-centered, x-face, and y-face. The three discretization stencils that couple these variables are given in Figure 8.6. The information in these two figures is essentially all that is needed to describe the nonzero structure of the linear system we wish to solve. Traditional linear solver interfaces require that this information first be translated to row-column entries of a matrix by defining a global ordering of the unknowns. This translation process can be quite complicated, especially in parallel. For example, face-centered variables on block boundaries are often replicated on two different processes, but they should only be counted once in the global ordering. In contrast, the `semiStruct` interface enables the description of block-structured grid problems in a way that is much more natural.

Two primary steps are involved for describing the above problem: defining the grid (and its associated variables) in Figure 8.5, and defining the stencils in Figure 8.6. The grid is defined in terms of five separate logically-rectangular parts as shown in Figure 8.7, and each part is given a unique label between 0 and 4. Each part

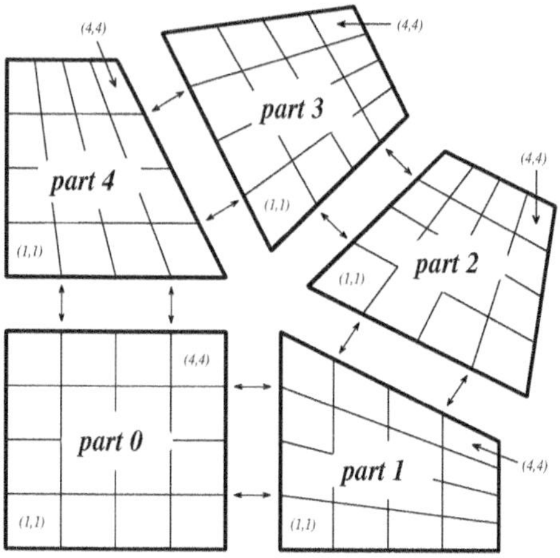

Fig. 8.7. Assignment of parts and indices to the block-structured grid example in Figure 8.5. In this example, the data for part p lives on process p.

consists of a single box with lower index $(1, 1)$ and upper index $(4, 4)$ (see Section 8.4), and the grid data is distributed on five processes such that data associated with part p lives on process p. Note that in general, parts may be composed out of arbitrary unions of boxes, and indices may consist of non-positive integers (see Figure 8.3). Also note that the `semiStruct` interface expects a domain-based data distribution by boxes, but the actual distribution is determined by the user and simply described (in parallel) through the interface.

For simplicity, we restrict our attention to the interface calls made by process 3. Each process describes through the interface only the grid data that it owns, so process 3 needs to describe the data pictured in Figure 8.8. That is, it describes part 3 plus some additional neighbor information that ties part 3 together with the rest of the grid. To do this, the `GridSetExtents()` routine is first called, passing it the lower and upper indices on part 3, $(1, 1)$ and $(4, 4)$. Next, the `GridSetVariables()` routine is called, passing it the three variable types on part 3: cell-centered, x-face, and y-face.

At this stage, the description of the data on part 3 is complete. However, the spatial relationship between this data and the data on neighboring parts is not yet defined. To do this, we need to relate the index space for part 3 with the index spaces of parts 2 and 4. More specifically, we need to tell the interface that the two grey boxes neighboring part 3 in Figure 8.8 also correspond to boxes on parts 2 and 4. To do this, two calls are made to the `GridSetNeighborbox()` routine. With this additional neighbor information, it is possible to determine where off-part stencil entries couple. Take, for example, any shared part boundary such as the boundary between parts 2 and 3. Along these boundaries, some stencil entries reach outside of the part. If no neighbor information is given, these entries are effectively zeroed out, i.e., they don't participate in the discretization. However, with the additional

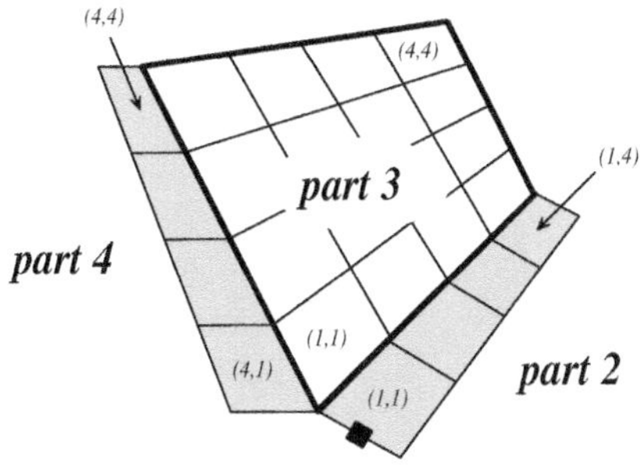

Fig. 8.8. Grid information given to the `semiStruct` interface by process 3 for the example in Figure 8.5. The shaded boxes are used to relate part 3 to parts 2 and 4.

neighbor information, when a stencil entry reaches into a neighbor box it is then coupled to the part described by that neighbor box information.

An important consequence of the use of the `GridSetNeighborbox()` routine is that it can declare variables on different parts as being the same. For example, consider the highlighted face variable at the bottom of Figure 8.8. This is a single variable that lives on both part 2 and part 1. Note that process 3 cannot make this determination based solely on the information in the figure; it must use additional information on other processes. Also note that a variable may be of different types on different parts. Take for example the face variables on the boundary of parts 2 and 3. On part 2 they are x-face variables, but on part 3 they are y-face variables.

The grid is now complete and all that remains to be done is to describe the stencils in Figure 8.6. For brevity, we consider only the description of the y-face stencil, i.e. the third stencil in the figure. To do this, the stencil entries are assigned unique labels between 0 and 8 and their "geometries" are described relative to the "center" of the stencil. This process is illustrated in Figure 8.9. Nine calls are made to the routine `StencilSetEntry()`. As an example, the call that describes stencil entry 5 in the figure is given the entry number 5, the offset $(-1, 0)$, and the identifier for the x-face variable (the variable to which this entry couples). Recall from Figure 8.4 the convention used for referencing variables of different types. The geometry description uses the same convention, but with indices numbered relative to the referencing index $(0, 0)$ for the stencil's center.

With the above, we now have a complete description of the nonzero structure for the matrix. Using the `MatrixSetValues()` routine in a manner similar to what is described in Sections 8.4 and 8.5.2, the matrix coefficients are then easily set. See the *hypre* documentation [22] for details.

An alternative approach for describing the above problem through the interface is to use the `GraphAddEntries()` routine. In this approach, the five parts are explicitly "sewn" together by adding non-stencil couplings to the matrix graph (see Section 8.5.2 for more information on the use of this routine). The main downside to this approach for block-structured grid problems is that variables along block boundaries are no longer considered to be the same variables on the corresponding parts

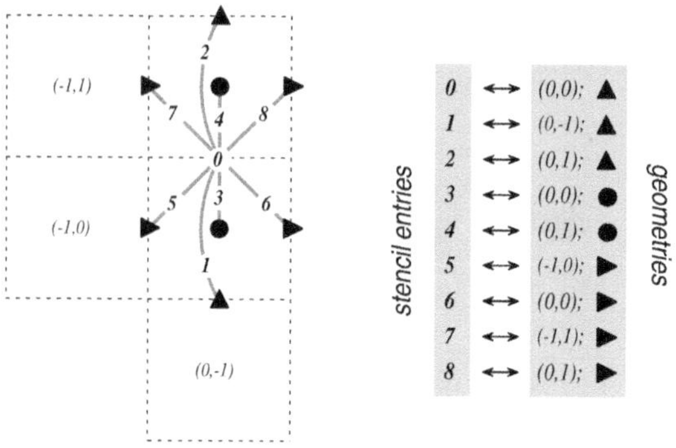

Fig. 8.9. Assignment of labels and geometries to the y-face stencil in Figure 8.6. Stencil geometries are described relative to the $(0, 0)$ index for the "center" of the stencil. The $(0, 0)$ index is located at the center of the cell just above the cell marked $(0, -1)$.

that share these boundaries. For example, the face variable at the bottom of Figure 8.8 would now represent two different variables that live on different parts. To "sew" the parts together correctly, we need to explicitly select one of these variables as the representative that participates in the discretization, and make the other variable a dummy variable that is decoupled from the discretization by zeroing out appropriate entries in the matrix.

8.5.2 Structured Adaptive Mesh Refinement

We now briefly discuss how to use the `semiStruct` interface in a structured AMR application. Consider Poisson's equation on the simple cell-centered example grid illustrated in Figure 8.10. For structured AMR applications, each refinement level should be defined as a unique part. There are two parts in this example: part 0 is the global coarse grid and part 1 is the single refinement patch. Note that the coarse unknowns underneath the refinement patch (gray dots in Figure 8.10) are not real physical unknowns; the solution in this region is given by the values on the refinement patch. In setting up the composite grid matrix [29] for *hypre* the equations for these "dummy" unknowns should be uncoupled from the other unknowns (this can easily be done by setting all off-diagonal couplings to zero in this region).

In the example, parts are distributed across the same two processes with process 0 having the "left" half of both parts. For simplicity, we discuss the calls made by process 0 to set up the composite grid matrix. First to set up the *grid*, process 0 will make `GridSetVariables()` calls to set the one variable type, cell-centered, and will make `GridSetExtents()` calls to identify the portion of the grid it owns for each part: *part 0*, $(1, 1) \times (2, 4)$; *part 1*, $(6, 6) \times (7, 9)$. Recall that there is no rule relating the indexing of different parts. Also note that there is no way to define hierarchies in the interface. In *hypre*, this additional information is passed directly into the solvers that need it, e.g., in the Fast Adaptive Composite Grid method (FAC).

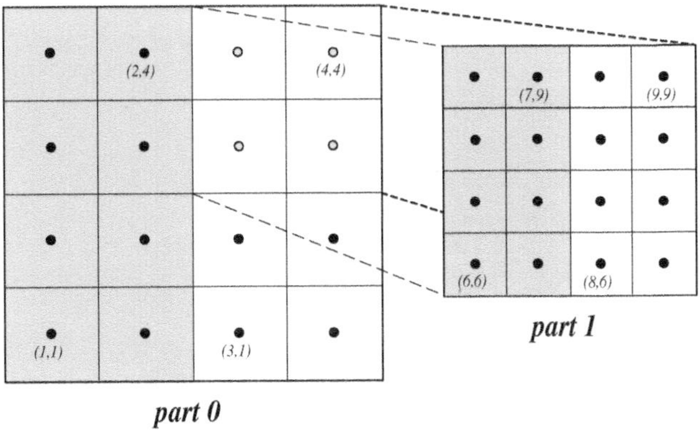

Fig. 8.10. Structured AMR grid example with one refinement patch (part 1) in the upper-right quadrant of the coarse grid (part 0). Shaded regions correspond to process 0, unshaded to process 1. The grey dots are dummy variables.

Next, the *stencil* is set up. In this example we are using a finite volume approach resulting in the standard 5-point stencil (8.1) in both parts.

The grid and stencil are used to define all intra-part coupling in the graph, the non-zero pattern of the composite grid matrix. The inter-part coupling at the coarse-fine interface is described by `GraphAddEntries()` calls. This coupling in the composite grid matrix is typically the composition of an interpolation rule and a discretization formula. In this example, we use a simple piecewise constant interpolation, i.e. the solution value at any point in a coarse cell is equal to the solution value at the cell center. Then the flux across a portion of the coarse-fine interface is approximated by a difference of the solution values on each side. As an example, consider the illustration in Figure 8.11. Following the discretization procedure above results in an equation for the variable at cell $(6, 6)$ involving not only the stencil couplings to $(6, 7)$ and $(7, 6)$ on part 1 but also non-stencil couplings to $(2, 3)$ and $(3, 2)$ on part 0. These non-stencil couplings are described by `GraphAddEntries()` calls. The syntax for this call is simply the part and index for both the variable whose equation is being defined and the variable to which it couples. After these calls, the non-zero pattern of the matrix (and the graph) is complete. Note that the "west" and "south" stencil couplings simply "drop off" the part, and are effectively zeroed out.

The remaining step is to define the actual numerical values for the composite grid matrix. This can be done by either `MatrixSetValues()` calls to set entries in a single equation, or by `MatrixSetBoxValues()` calls to set entries for a box of equations in a single call. The syntax for the `MatrixSetValues()` call is a part and index for the variable whose equation is being set and an array of entry numbers identifying which entries in that equation are being set. The entry numbers may correspond to stencil entries or non-stencil entries.

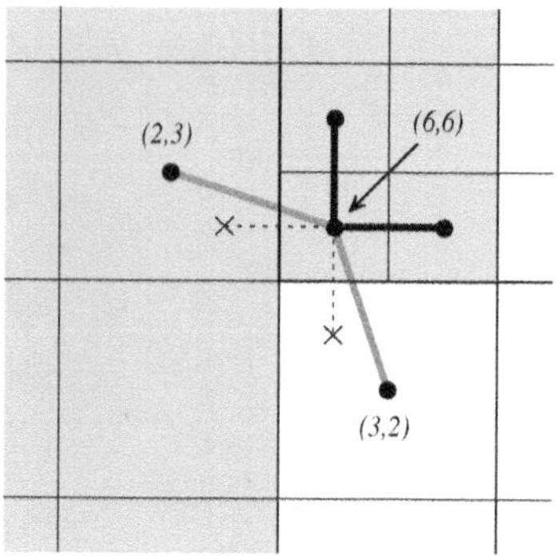

Fig. 8.11. Coupling for equation at corner of refinement patch. Black lines (solid and broken) are stencil couplings. Gray line are non-stencil couplings.

8.6 The Finite Element Interface (`FEI`)

The finite element interface is appropriate for users who form their systems from a finite element discretization. The interface mirrors typical finite element data structures. Though this interface is provided in *hypre*, its definition was determined elsewhere [11]. A brief summary of the actions required by the user is given below.

The use of this interface to build the underlying linear system requires two phases: initialization and loading. During the initialization phase, the structure of the finite-element data is defined. This requires the passing of control data that defines the underlying element types and solution fields; data indicating how many aggregate finite-element types will be utilized; element data, including element connectivity information to translate finite-element nodal equations to systems of sparse algebraic equations; control data for nodes that need special handling, such as nodes shared among processes; and data to aid in the definition of any constraint relations local to a given process. These definitions are needed to determine the underlying matrix structure and allocate memory for the load step.

During the loading phase, the structure is populated with finite-element data according to standard finite-element assembly procedures. Data passed during this step includes: boundary conditions (essential, natural, and mixed boundary conditions); element stiffness matrices and load vectors, passed as aggregate element set abstractions; and constraint relations, defined in terms of nodal algebraic weights and tables of associated nodes.

For a more detailed description with specific function call definitions, the user is referred to [11], the web site [17] which contains information on newer versions of the `FEI`, as well as the *hypre* user manual. The current `FEI` version used in *hypre* is version 2.x, and comprises additional features not defined in [11].

8.7 The Linear-Algebraic Interface (`IJ`)

The `IJ` interface is the traditional linear-algebraic interface. Here, the user defines the right hand side and the matrix in the general linear-algebraic sense, i.e. in terms of row and column indices. This interface provides access only to the most general data structures and solvers and as such should only be used when none of the grid-based interfaces is applicable.

As with the other interfaces in *hypre*, the `IJ` interface expects to get the data in distributed form. Matrices are assumed to be distributed across p processes by contiguous blocks of rows. That is, the matrix must be blocked as follows:

$$\begin{pmatrix} A_0 \\ A_1 \\ \vdots \\ A_{p-1} \end{pmatrix}, \tag{8.3}$$

where each submatrix A_k is "owned" by a single process k. A_k contains the rows $n_k, n_k + 1, ..., n_{k+1} - 1$, where n_0 is arbitrary (n_0 is typically set to either 0 or 1).

First, the user creates an empty `IJ` matrix object on each process by specifying the row extents, n_k and $n_{k+1} - 1$. Next, the object type needs to be set. The object type determines the underlying data structure. Currently only one data structure, the ParCSR matrix data structure, is available. However, work is underway to add other data structures. Additional data structures are desirable for various reasons, e.g. to be able to link to other packages, such as PETSc [3, 4, 26]. Also, if additional matrix information is known, more efficient data structures are possible. For example, if the matrix is symmetric, it would be advantageous to design a data structure that takes advantage of symmetry. Such an approach could lead to a significant decrease in memory usage. Another data structure could be based on blocks and thus make better use of the cache. Small blocks could naturally occur in matrices derived from systems of PDEs, and be processed more efficiently in an implementation of the nodal approach for systems AMG.

After setting the object type, the user can give estimates of the expected row sizes to increase efficiency. Providing additional detail on the nonzero structure and distribution of the matrix can lead to even more efficiency, with significant savings in time and memory usage. In particular, if information is given about how many column indices are "local" (i.e., between n_k and $n_{k+1} - 1$), and how many are not, both the ParCSR and PETSc matrix data structures can take advantage of this information during construction.

The matrix coefficients are set via the `MatrixSetValues()` routine, which allows a great deal of flexibility (more than its typical counterpart routines in other linear solver libraries). For example, one call to this routine can set a single coefficient, a row of coefficients, submatrices, or even arbitrary collections of coefficients. This is accomplished with the following parameters, describing which matrix coefficients are being set:

- `nrows` (scalar integer): the number of rows,

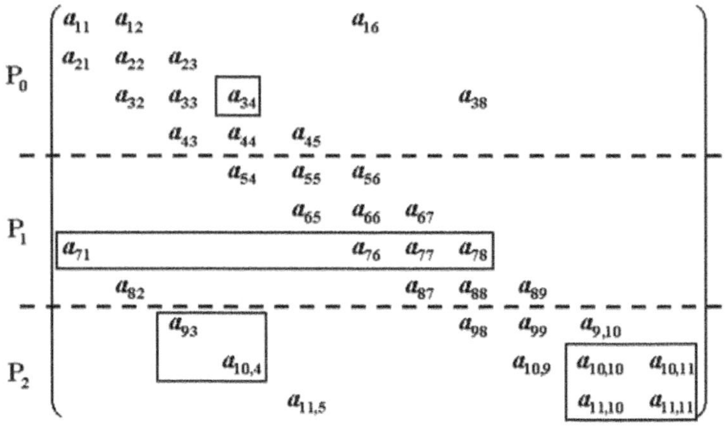

Fig. 8.12. An example of a ParCSR matrix, distributed across 3 processes.

- `ncols` (array of integers): the number of coefficients in each row,
- `rows` (array of integers): the global indices of the rows,
- `cols` (array of integers): the column indices of each coefficient,
- `values` (array of doubles): the actual values of the coefficients.

It is also possible to add values to the coefficients with the `MatrixAddValues()` call.

Figure 8.12 illustrates this for the example of an 11×11-matrix, distributed across 3 processes. Here, rows 1–4 reside on process 0, rows 5–8 on process 1 and rows 9–11 on process 2. We now describe how to define the above parameters to set the coefficients in the boxes in Figure 8.12.

On process 0, only one element a_{34} is to be set, which requires the following parameters: `nrows` = 1, `ncols` = [1], `rows` = [3], `cols` = [4], `values` = $[a_{34}]$. On process 1, the third (local) row is defined with the parameters: `nrows` = 1, `ncols` = [4], `rows` = [7], `cols` = [1,6,7,8], `values` = $[a_{71}, a_{76}, a_{77}, a_{78}]$. On process 2, the values contained in two submatrices are to be set. This can be done by two subsequent calls setting one submatrix at a time, or more generally, all of the values can be set in one call with the parameters: `nrows` = 3, `ncols` = [1, 3, 2], `rows` = [9, 10, 11], `cols` = [3, 4, 10, 11, 10, 11], `values` = $[a_{93}, a_{10,4}, a_{10,10}, a_{10,11}, a_{11,10}, a_{11,11}]$.

8.8 Implementation

This section discusses implementation issues of the various interfaces. It describes the data structures and discusses how to obtain neighborhood information. The focus is on those issues which impact performance on a large scale parallel computer. A more detailed analysis can be found in [14].

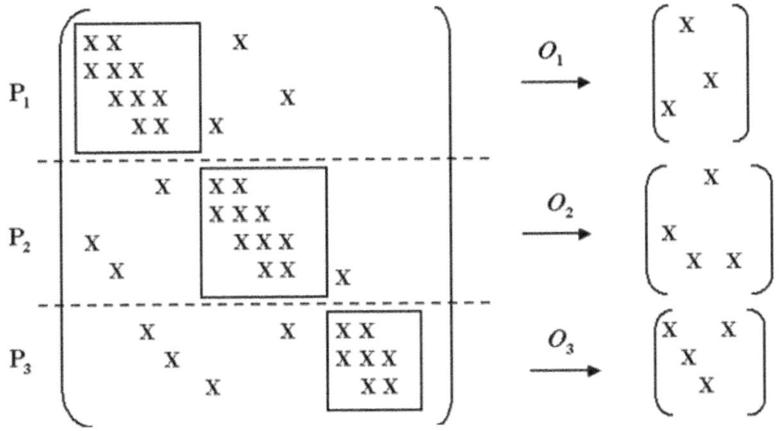

Fig. 8.13. An example of a ParCSR matrix, distributed across 3 processors. Matrices with local coefficients, D_1, D_2 and D_3, are shown as boxes within each processor. The remaining coefficients are compressed into the matrices O_1, O_2 and O_3.

8.8.1 `IJ` Data Structure and Communication Package

Currently only one data structure, the ParCSR matrix data structure, is available. It is similar to the parallel AIJ matrix format in PETSc [3, 4, 26], but is not the same. It is based on the sequential compressed sparse row (CSR) data structure.

A *ParCSR matrix* consists of p parts A_k, $k = 1, \ldots, p$ (see (8.3)), where A_k is stored locally on processor k. Each A_k is split into two matrices D_k and O_k. D_k is a square matrix of order $n_k \times n_k$, where $n_k = r_{k+1} - r_k$ is the number of rows residing on processor k. D_k contains all coefficients a_{ij}^k, with $r_k \leq i, j \leq r_{k+1} - 1$, i.e. column indices pointing to rows stored locally. The second matrix O_k contains those coefficients of A_k, whose column indices j point to rows that are stored on other processors with $j < r_k$ or $j \geq r_{k+1}$. Both matrices are stored in CSR format. Whereas D_k is a CSR matrix in the usual sense, in O_k, which in general is extremely sparse with many zero columns and rows, all non-zero columns are renumbered for greater efficiency. Thus, one needs to generate an array of length n_{O_k} that defines the mapping of local to global column indices, where n_{O_k} is the number of non-zero columns of O_k. We denote this array as COL_MAP_O_k.

An example of an 11×11 matrix that illustrates this data structure is given in Figure 8.13. The matrix is distributed across 3 processors, with 4 rows on processor 1 and processor 2, and 3 rows on processor 3. The 4×4 matrices D_1 and D_2 and the 3×3 matrix D_3 are illustrated as boxes. The remaining coefficients are compressed into the 4×3 matrix O_1 (with COL_MAP_O_1 = (5,6,8)), the 4×4 matrix O_2 (with COL_MAP_O_2 = (1,2,4,9)) and the 3×4 matrix O_3 (with COL_MAP_O_3 = (3,4,5,8)). Since often O_p is extremely sparse, efficiency can be further increased by introducing a row mapping that compresses zero rows by renumbering the non-zero rows.

For parallel computations it is necessary to generate the neighborhood information, which is assembled in a communication package. The communication package is based on the concept of what is needed for a matrix-vector multiplication. Let us consider the parallel multiplication of a matrix A with a vector $\mathbf{x}$. Processor k owns rows r_k through $r_{k+1}-1$ as well as the corresponding chunk of $\mathbf{x}$, $\mathbf{x}_k = (x_{r_k}, \ldots, x_{r_{k+1}-1})^T$. In order to multiply A with $\mathbf{x}$, processor k needs to perform the operation $A_k\mathbf{x} = D_k\mathbf{x}_k + O_k\tilde{\mathbf{x}}_k$, where $\tilde{\mathbf{x}}_k = (x_{col\_map\_O_k(1)}, \ldots, x_{col\_map\_O_k(n_{O_k})})^T$. While the multiplication of D_k and $\mathbf{x}_k$ can be performed without any communication, the elements of $\tilde{\mathbf{x}}_k$ are owned by the *receive processors* of k. Another necessary piece of information is the amount of data to be received by each processor. In general processor k owns elements of $\mathbf{x}$ that are needed by other processors. Consequently processor k needs to know the indices of the elements that need to be sent to each of its *send processors*.

In summary, the communication package on processor k consists of the following information:

- the IDs of the receive processors
- the size of data to be received by each processor
- the IDs of the send processors
- the indices of the elements that need to be sent to each send processor

Recall that each processor by design has initially only local information available, i.e. its own range and the rows of the matrix that it owns. In order to determine the communication package, it needs to get more information from other processors. There are various ways of dealing with this situation. We will overview two approaches here (for a more detailed analysis, see [14]).

The simpler approach is to allow each processor to own the complete partitioning information, which is easily communicated via MPI_ALLGATHER. From this it can determine its receive processors. This is a reasonable approach when dealing with computers with a few thousand processors. However, when using a supercomputer with 100,000 or more processors, this approach is expensive, requiring $O(p)$ computations and memory usage.

A better approach for this case would be to use a so-called distributed directory algorithm [31], which is a rendezvous algorithm that utilizes a concept we refer to as an assumed partitioning [14]. The idea is to assume a partitioning that is known by all processors, requires only $O(1)$ storage, and can be queried by an $O(1)$ function. In general, this function does not describe the actual partitioning, but should not be too far from it. This approach consists of two main steps. In the first step, the so-called distributed directory is set up. Here, each processor sends information about the data it owns in the actual partitioning to the processors responsible for that data in the assumed partitioning. Once this is done, each processor knows how the data in its piece of the assumed partitioning is distributed in the actual partitioning. Hence, it is now possible to answer queries through the distributed directory (in $O(\log p)$ time) about where data lives in the actual partitioning. Step two involves just such a set of queries. That is, each processor determines where its needed receive data lives in the assumed partitioning, then contacts the responsible processors for the needed receive processor information. In this step, it is necessary to use a distributed termination detection (DTD) algorithm to figure out when to stop servicing queries. This latter fact is the sole reason for the $O(\log p)$ cost (instead of $O(1)$) [14].

If matrix A has a symmetric structure, the receive processors are also send processors, and no further communication is necessary. However, this is different in the non-symmetric case. In the current implementation, the IDs of the receive processors and the amount of data to be obtained from each receive processor are communicated to all processors via a MPI_ALLGATHERV. For a moderate number of processors, even up to 1000, this is a reasonable approach, however it can become a potentially high cost if we consider 100,000 processors. This can be avoided by a using a DTD algorithm in a similar manner to step two above.

When each processor knows its receive and send processors, the remaining necessary information can be exchanged directly. For most PDE-based linear systems and data partitionings, the number of neighbors and the amount of data is independent of p. Hence, the computational complexity and the storage requirement is $O(1)$.

8.8.2 `Struct` Data Structure and Neighborhood Information

The underlying matrix data structure, *Struct matrix*, contains the following.

- *Struct grid*: describes the boxes owned by the processor (local boxes) as well as information about other nearby boxes (the actual data associated with these nearby boxes lives on other processors). Note that a box is stored by its "lower" and "upper" indices, called the box's *extents*.
- *Struct stencil*: an array of indices defining the coupling pattern in the matrix.
- *data*: an array of doubles defining the coupling coefficients in the matrix.

The corresponding vector data structure is similar except is has no stencil and the data array defines the vector values. In both the vector and matrix the data array is stored so that all values associated with a given box are stored contiguously. To facilitate parallel implementation of a matrix-vector product, the vector data array includes space for values associated with a box somewhat larger than the actual box; typically including one boundary layer of cells or ghost cells (see Figure 8.14). Assuming that the boxes are large, the additional storage of these ghost cells is fairly small as the boundary points also take only a small percentage of the total number of points. Some of these ghost cells may be part of other boxes, owned by either the same or a different processor. Determining communication patterns for updating ghost cells is the major task in implementing the `Struct` interface in a scalable manner.

Recall that in the interface, a given processor k is passed only information about the grid boxes that it owns. Determining how to update ghost cell values requires information about nearby boxes on other processors. This information is generated and stored when the Struct grid is assembled. Determining which processors own ghost cells is similar to the problem in the `IJ` interface of determining the receive processors. In the `IJ` case, this requires information about the global partitioning. In the `Struct` case, it requires information about the global grid.

The algorithm proceeds as follows. Here we let p denote the number of processors and b denote the total number of boxes in the grid (note $b \geq p$). First we

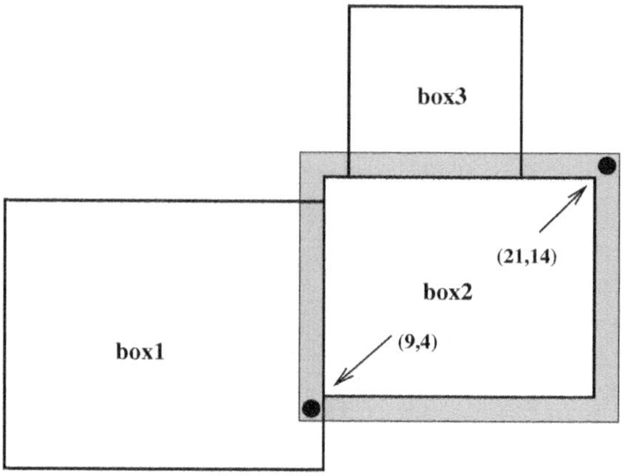

Fig. 8.14. For parallel computing, additional storage is allocated for cells nearby a box (ghost cells). Here, the ghost cells for BOX2 are illustrated.

accumulate information about the global grid by each processor sending the extents of its boxes to all other processors. As in the `IJ` case, this can be done using MPI_ALLGATHER with $O(\log p)$ operations. Memory usage is of order $O(b)$, since the global grid contains b boxes.

Once the global grid is known, each local box on processor k is compared to every box in the global grid. In this box-by-box comparison a *distance index* is computed which describes the spatial relationship in the index space of one box to another. This comparison of each local box to every global box involves $O(b)$ computations. Once the comparison is done, all global boxes within a specified distance (typically 2) from a local box are stored as part of a *neighborhood* data structure on processor k. Boxes not in this neighborhood can be deleted from processor k's description of the global grid. The storage requirement for the neighborhood is independent of p.

To perform the matrix-vector product, processor k must have up-to-date values in all ghost cells that will be "touched" when applying the matrix stencil at the cells owned by processor k. Determining these needed ghost cells is done by taking each box owned by the processor, shifting it by each stencil entry and finding the intersection of the shifted box with boxes in the neighborhood data structure. As an example, consider the same layout of boxes as before with each box on a different processor (see Figure 8.15). If the matrix has the 5-pt stencil (8.1), then shifting BOX2 by the "north" stencil entry and intersecting this with BOX3 produces one of the dark shaded regions labeled as a receive box. Using this procedure, a list of receive boxes and corresponding owner processors is generated. The procedure for determining the cells owned by processor k that are needed to update ghost cells on other processors is similar.

The current `Struct` interface implementation shares some of the same drawbacks as the current `IJ` interface implementation. The storage requirement in generating the neighborhood structure is $O(b)$ as the global grid is initially gathered

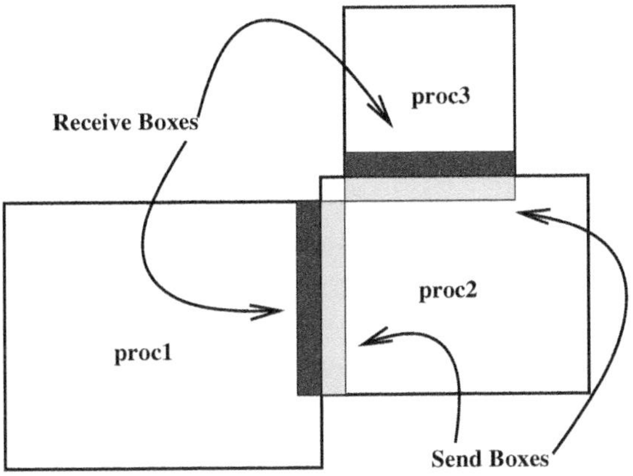

Fig. 8.15. The communication package for processor 2 contains send boxes (values owned by processor 2 needed by other processors) and receive boxes (values owned by other processors need by processor 2.)

to all processors and the box-by-box comparison to determine neighbors involves $O(b)$ operations, again note $b \geq p$. One possible approach to eliminate these drawbacks would be similar to the assumed partitioning approach described in Section 8.8.1. The idea is to have a function describing an assumed partitioning of the index space to processors and have this function available to all processors. Unlike the one-dimensional `IJ` partitioning, this partition would be d-dimensional. A processor would be able to determine its neighbors in the assumed partition in $O(1)$ computations and storage. A multi-phase communication procedure like that previously described for the `IJ` case could be used to determine the actual neighbors with $O(\log p)$ complexity.

8.8.3 `semiStruct` Data Structure

The `semiStruct` interface allows the user to choose from two underlying data structures for the matrix. One option is to use the ParCSR matrix data type discussed in Section 8.8.1. The second option is the *semiStruct matrix* data type which is based on a splitting of matrix non-zeros into structured and unstructured couplings $A = S + U$. The S matrix is stored as a collection of Struct matrices and the U matrix is stored as a ParCSR matrix. In our current implementation, the stencil couplings within variables of the same type are stored in S, all other couplings are stored in U. If the user selects the ParCSR data type, then all couplings are stored in U (i.e. $S = 0$.)

Since the `semiStruct` interface can use both Struct and ParCSR matrices, the issues discussed in the previous two sections impact its scalability as well. The major new issue impacting scalability is the need to relate the semi-structured description of unknowns and the global ordering of unknowns in the ParCSR matrix, i.e. the mapping $M(\text{PART,VAR,INDEX}) = \text{GLOBAL\_RANK}$. The implementation needs this

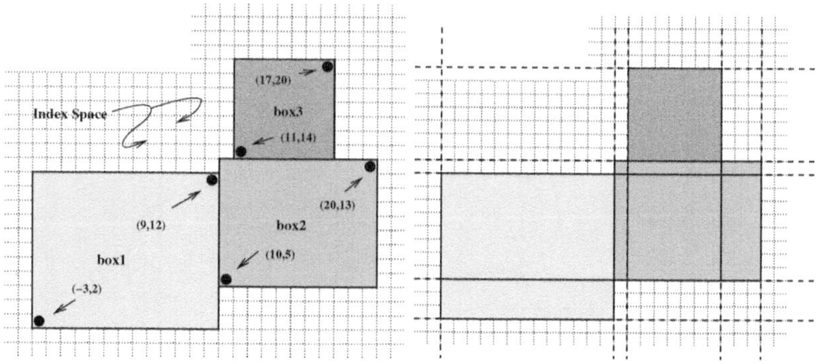

Fig. 8.16. The BOXMAP structure divides the index space into regions defined by cuts in each coordinate direction.

mapping to set matrix entries in U. The global ordering of unknowns is an issue internal to the `semiStruct` implementation; the user is not aware of this ordering, and does not need to be.

In our implementation of the semi-structured grid we include the concept of BOXMAP to implement this mapping. There is a BOXMAP for each variable on each part; the purpose is to quickly compute the global rank corresponding to a particular index. To describe the BOXMAP structure we refer to Figure 8.16. By cutting the index space in each direction by lines coinciding with boxes in the grid, the index space is divided into regions where each region is a either empty (not part of the grid) or is a subset of a box defining the grid. The data structure for the BOXMAP corresponds to a d-dimensional table of BOXMAPENTRIES. In three dimensions, BOXMAPENTRY$[i][j][k]$ contains information about the region bounded by cuts i and $i+1$ in the first coordinate direction, cuts j and $j+1$ in the second coordinate direction, cuts k and $k+1$ in the third coordinate direction. Among the information contained in BOXMAPENTRY is the first global rank (called *offset*) and the extents for the grid box which this region is a subset of. The global rank of any index in this region can be easily computed from this information.

The mapping $M(\text{PART}, \text{VAR}, \text{INDEX}) = \text{GLOBAL\_RANK}$ is computed by accessing the BOXMAP corresponding to PART and VAR, searching in each coordinate direction to determine which cuts INDEX falls between, retrieving the offset and box extents from the appropriate BOXMAPENTRY, and computing GLOBAL_RANK from this retrieved information. This computation has $O(1)$ (independent of number of boxes and processors) complexity except for the searching step. The searching is done by a simple linear search so worst case complexity is $O(b)$ since the number of cuts is proportional to the number of boxes. However, we retain the current position in the BOXMAP table, and in subsequent calls to the mapping function, we begin searching from this position. In most applications, subsequent calls will have map indices nearby the previous index and the search has $O(1)$ complexity. Further optimization is accomplished by retrieving BOXMAPENTRIES not for a single index but for an entire box of indices in the index space.

Table 8.1. Current solver availability via *hypre* conceptual interfaces.

	Conceptual Interfaces			
Solvers	`Struct`	`semiStruct`	`FEI`	`IJ`
Jacobi	x			
SMG	x			
PFMG	x			
Split		x		
MLI		x	x	x
BoomerAMG		x	x	x
ParaSails		x	x	x
PILUT		x	x	x
Euclid		x	x	x
PCG	x	x	x	x
GMRES	x	x	x	x
BiCGSTAB	x	x	x	x
Hybrid	x	x	x	x

The BoxMap structure does allow quick mapping from the semi-structured description to the global ordering of the ParCSR matrix, but it does have drawbacks: storage and computational complexity of initial construction. Since we store the structure on all processors, the storage costs are $O(b)$ where b is the global number of boxes (again b is at least as large as p, the number of processors). Constructing the structure requires knowledge of all boxes (accomplished by the MPI_Allgather with $O(\log p)$ operations and $O(b)$ storage as in the `Struct` case), and then scanning the boxes to define the cuts in index space (requiring $O(b)$ operations and storage.) As in the `IJ` and `Struct` cases, it may be possible to use the notion of an assumed partitioning of the index space to remove these potential scalability issues.

8.9 Preconditioners and Solvers

The conceptual interfaces provide access to several preconditioners and solvers in *hypre* (we will refer to them both as solvers in the sequel, except when noted). Table 8.1 lists the current solver availability. We expect to update this table continually in the future with the addition of new solvers in *hypre*, and potentially with the addition of solvers in other linear solver packages (e.g., PETSc). We also expect to update the `Struct` interface column, which should be completely filled in.

Great efforts have been made to generate highly efficient codes. Of particular concern has been the scalability of the solvers. Roughly speaking, a method is *scalable* if the time required to produce the solution remains essentially constant as both the problem size and the computing resources increase. All methods implemented here are generally scalable per iteration step, the multigrid methods are also scalable with regard to iteration count.

The solvers use MPI for parallel processing. Most of them have also been threaded using OpenMP, making it possible to run *hypre* in a mixed message-passing / threaded mode, of potential benefit on clusters of SMPs.

All of the solvers can be used as stand-alone solvers, except for ParaSails, Euclid, PILUT and MLI which can only be used as preconditioners. For most problems, it is recommended that one of the Krylov methods be used in conjunction with a preconditioner. The Hybrid solver can be a good option for time-dependent problems, where a new matrix is generated at each time step, and where the matrix properties change over time (say, from being highly diagonally dominant to being weakly diagonally dominant). This solver starts with diagonal-scaled conjugate gradient (DSCG) and automatically switches to multigrid-preconditioned conjugate gradient (where the multigrid preconditioner is set by the user) if DSCG is converging too slowly. SMG [33, 7] and PFMG [1, 15] are parallel semicoarsening methods, with the more robust SMG using plane smoothing and the more efficient PFMG using pointwise smoothing. The Split solver is a simple iterative method based on a regular splitting of the matrix into its "structured" and "unstructured" components, where the structured component is inverted using either SMG or PFMG. This is currently the only solver that takes advantage of the structure information passed in through the `semiStruct` interface, but solvers such as the Fast Adaptive Composite-Grid method (FAC) [29] will also be made available in the future. MLI [6] is a parallel implementation of smoothed aggregation [34]. BoomerAMG [20] is a parallel implementation of algebraic multigrid with various coarsening strategies [32, 12, 18, 13] and smoothers (including the conventional pointwise smoothers such as Jacobi, as well as more complex smoothers such as ILU, sparse approximate inverse and Schwarz). ParaSails [8, 9] is a sparse approximate inverse preconditioner. PILUT [25] and Euclid [23, 24] are ILU algorithms, where PILUT is based on Saad's dual-threshold ILU algorithm, and Euclid supports variants of ILU(k) as well as ILUT preconditioning.

After the matrix and right hand side are set up as described in the previous sections, the preconditioner (if desired) and the solver are set up, and the linear system can finally be solved. For many of the preconditioners and solvers, it might be desirable to choose parameters other than the default parameters, e.g. the strength threshold or smoother for BoomerAMG, a drop tolerance for PILUT, the dimension of the Krylov space for GMRES, or convergence criteria, etc. These parameters are defined using `Set()` routines. Once these parameters have been set to the satisfaction of the user, the preconditioner is passed to the solver with a `SetPreconditioner()` call. After this has been accomplished, the problem is solved by calling first the `Setup()` routine (this call may become optional in the future) and then the `Solve()` routine. When this has finished, the basic solution information can be extracted using a variety of `Get()` calls.

8.10 Additional Information

The *hypre* library can be downloaded by visiting the *hypre* home page [22]. It can be built by typing `configure` followed by `make`. There are several options that can be used with configure. For information on how to use those, one needs to type `configure --help`. Although *hypre* is written in C, it can also be called from Fortran. More specific information on *hypre* and how to use it can be found in the users manual and the reference manual, which are also available at the same URL.

8.11 Conclusions and Future Work

The introduction of conceptual interfaces in *hypre* gives application users a more natural means for describing their linear systems, and provides access to an array of powerful linear solvers that require additional information beyond just the matrix. Work continues on the library on many fronts. We highlight two areas: providing better interfaces and solvers for structured AMR applications and scaling up to 100,000's of processors.

As in the example in Section 8.5.2, the current `semiStruct` interface can be used in structured AMR applications. However, the user must explicitly calculate the coarse-fine coupling coefficients which are typically defined by the composition of two equations: a structured grid stencil coupling and an interpolation formula. A planned extension to the `semiStruct` interface would allow the user to provide these two equations separately, and the composition would be done inside the *hypre* library code. This extension would make the `semiStruct` interface more closely match the concepts used in AMR application codes and would further ease the coding burden for potential users. We are also finishing the implementation of a new FAC [29] solver in *hypre*. This is an efficient multigrid solver specifically tailored to structured AMR applications.

Another area of work is ensuring good performance on very large numbers of processors. As mentioned previously, the current implementations in *hypre* are appropriate for thousands of processors but do have places where, say, the storage needed is $O(p)$. These potential bottlenecks may be of real importance on machines with 100,000 processors. The crux of the problem is that the interfaces only provide local information and determining neighboring processors requires global information. We have mentioned the assumed partitioning approach as one way we are trying to overcome this hurdle.

Acknowledgments

This paper would not have been possible without the many contributions of the *hypre* library developers: Allison Baker, Edmond Chow, Andy Cleary, Van Henson, Ellen Hill, David Hysom, Mike Lambert, Barry Lee, Jeff Painter, Charles Tong and Tom Treadway. This work was performed under the auspices of the U.S. Department of

Energy by University of California Lawrence Livermore National Laboratory under contract No. W-7405-Eng-48.

References

1. S. F. Ashby and R. D. Falgout. A parallel multigrid preconditioned conjugate gradient algorithm for groundwater flow simulations. *Nuclear Science and Engineering*, 124(1):145–159, September 1996. Also available as LLNL Technical Report UCRL-JC-122359.
2. Babel: A language interoperability tool. `http://www.llnl.gov/CASC/components/`.
3. S. Balay, K. Buschelman, V. Eijkhout, W. Gropp, M. Knepley, L. McInnes, B. Smith, and H. Zhang. PETSc users manual. ANL-95/11-Revision 2.2.1. Technical report, Aronne National Laboratory, 2004.
4. S. Balay, W. D. Gropp, L. C. McInnes, and B. F. Smith. Efficient management of parallelism in object oriented numerical software libraries. In E. Arge, A. M. Bruaset, and H. P. Langtangen, editors, *Modern Software Tools in Scientific Computing*, pp. 163–202. Birkhauser Press, 1997.
5. M. Brezina, A. J. Cleary, R. D. Falgout, V. E. Henson, J. E. Jones, T. A. Manteuffel, S. F. McCormick, and J. W. Ruge. Algebraic multigrid based on element interpolation (AMGe). *SIAM J. Sci. Comput.*, 22(5):1570–1592, 2000. Also available as LLNL technical report UCRL-JC-131752.
6. M. Brezina, C. Tong, and R. Becker. Parallel algebraic multigrid for structural mechanics. *SIAM J. Sci. Comput.*, submitted, 2004. Also available as Lawrence Livermore National Laboratory technical report UCRL-JRNL-204167.
7. P. N. Brown, R. D. Falgout, and J. E. Jones. Semicoarsening multigrid on distributed memory machines. *SIAM J. Sci. Comput.*, 21(5):1823–1834, 2000. Special issue on the Fifth Copper Mountain Conference on Iterative Methods. Also available as LLNL technical report UCRL-JC-130720.
8. E. Chow. A priori sparsity patterns for parallel sparse approximate inverse preconditioners. *SIAM J. Sci. Comput.*, 21(5):1804–1822, 2000. Also available as LLNL Technical Report UCRL-JC-130719 Rev.1.
9. E. Chow. Parallel implementation and practical use of sparse approximate inverses with a priori sparsity patterns. *Int'l J. High Perf. Comput. Appl.*, 15:56–74, 2001. Also available as LLNL Technical Report UCRL-JC-138883 Rev.1.
10. E. Chow, A. J. Cleary, and R. D. Falgout. Design of the *hypre* preconditioner library. In M. Henderson, C. Anderson, and S. Lyons, editors, *Proc. of the SIAM Workshop on Object Oriented Methods for Inter-operable Scientific and Engineering Computing*, Philadelphia, PA, 1998. SIAM. Held at the IBM T.J. Watson Research Center, Yorktown Heights, New York, October 21-23, 1998. Also available as LLNL technical report UCRL-JC-132025.
11. R. L. Clay, K. D. Mish, I. J. Otero, L. M. Taylor, and A. B. Williams. An annotated reference guide to the finite-element interface (FEI) specification: version 1.0. Sandia National Laboratories report SAND99-8229, January 1999.
12. A. J. Cleary, R. D. Falgout, V. E. Henson, and J. E. Jones. Coarse-grid selection for parallel algebraic multigrid. In *Proc. of the Fifth International Symposium on: Solving Irregularly Structured Problems in Parallel*, volume 1457 of *Lecture Notes in Computer*

Science, pp. 104–115, New York, 1998. Springer–Verlag. Held at Lawrence Berkeley National Laboratory, Berkeley, CA, August 9–11, 1998. Also available as LLNL Technical Report UCRL-JC-130893.
13. H. De Sterck, U. M. Yang, and J. Heys. Reducing complexity in parallel algebraic multigrid preconditioners. *SIAM J. Matrix Anal. Appl.*, submitted, 2004. Also available as LLNL Technical Report UCRL-JRNL-206780.
14. R. Falgout, J. Jones, and U. M. Yang. Pursuing scalability for hypre's conceptual interfaces. *ACM Transaction on Mathematical Software*, submitted, 2003. Also available as Lawrence Livermore National Laboratory technical report UCRL-JP-200044.
15. R. D. Falgout and J. E. Jones. Multigrid on massively parallel architectures. In E. Dick, K. Riemslagh, and J. Vierendeels, editors, *Multigrid Methods VI*, volume 14 of *Lecture Notes in Computational Science and Engineering*, pp. 101–107, Berlin, 2000. Springer. Proc. of the Sixth European Multigrid Conference held in Gent, Belgium, September 27-30, 1999. Also available as LLNL technical report UCRL-JC-133948.
16. R. D. Falgout and U. M. Yang. *hypre*: a library of high performance preconditioners. In P. Sloot, C. Tan., J. Dongarra, and A. Hoekstra, editors, *Computational Science - ICCS 2002 Part III*, volume 2331 of *Lecture Notes in Computer Science*, pp. 632–641. Springer–Verlag, 2002. Also available as LLNL Technical Report UCRL-JC-146175.
17. The finite-element interface (FEI). `http://z.cz.sandia.gov/fei/`.
18. K. Gallivan and U. M. Yang. Efficiency issues in parallel coarsening schemes. Technical Report UCRL-ID-513078, Lawrence Livermore National Laboratory, 2003.
19. E. Gamma, R. Helm, R. Johnson, and J. Vlissides. *Design patterns: elements of reusable object-oriented software*. Addison-Wesley, 1995.
20. V. E. Henson and U. M. Yang. BoomerAMG: a parallel algebraic multigrid solver and preconditioner. *Applied Numerical Mathematics*, 41:155–177, 2002. Also available as LLNL technical report UCRL-JC-141495.
21. F. Hülsemann, M. Kowarschik, M. Mohr, and U. Rüde. Parallel geometric multigrid. In A. M. Bruaset and A. Tveito, editors, *Numerical Solution of Partial Differential Equations on Parallel Computers*, volume 51 of *Lecture Notes in Computational Science and Engineering*, pp. 165–208. Springer-Verlag, 2005.
22. *hypre*: High performance preconditioners. `http://www.llnl.gov/CASC/hypre/`.
23. D. Hysom and A. Pothen. Efficient parallel computation of ILU(k) preconditioners. In *Proceedings of Supercomputing 99*, New York, 1999. ACM. published on CDROM, ISBN #1-58113-091-0, ACM Order #415990, IEEE Computer Society Press Order # RS00197.
24. D. Hysom and A. Pothen. A scalable parallel algorithm for incomplete factor preconditioning. *SIAM J. Sci. Comput.*, 22(6):2194–2215, 2001.
25. G. Karypis and V. Kumar. Parallel threshold-based ILU factorization. Technical Report 061, University of Minnesota, Department of Computer Science/Army HPC Research Center, Minneapolis, MN 5455, 1998.
26. M. G. Knepley, R. F. Katz, and B. Smith. Developing a geodynamics simulator with petsc. In A. M. Bruaset and A. Tveito, editors, *Numerical Solution of Partial Differential Equations on Parallel Computers*, volume 51 of *Lecture Notes in Computational Science and Engineering*, pp. 413–438. Springer-Verlag, 2005.
27. H. P. Langtangen. *Computational Partial Differential Equations. Numerical Methods and Diffpack Programming*, volume 1 of *Texts in Computational Science and Engineering*. Springer, 2003. 2nd ed.
28. H. P. Langtangen and A. Tveito, editors. *Advanced Topics in Computational Partial Differential Equations. Numerical Methods and Diffpack Programming*, volume 33 of *Lecture Notes in Computational Science and Engineering*. Springer, 2003.

29. S. F. McCormick. *Multilevel Adaptive Methods for Partial Differential Equations*, volume 6 of *Frontiers in Applied Mathematics*. SIAM Books, Philadelphia, 1989.
30. J. Morel, R. M. Roberts, and M. J. Shashkov. A local support-operators diffusion discretization scheme for quadrilateral *r-z* meshes. *Journal of Computational Physics*, 144:17–51, 1998.
31. A. Pinar and B. Hendrickson. Communication support for adaptive communication. In *Proceedings of 10th SIAM Conference on Parallel Processing for Scientific computing*, 2001.
32. J. W. Ruge and K. Stüben. Algebraic multigrid (AMG). In S. F. McCormick, editor, *Multigrid Methods*, volume 3 of *Frontiers in Applied Mathematics*, pp. 73–130. SIAM, Philadelphia, PA, 1987.
33. S. Schaffer. A semi-coarsening multigrid method for elliptic partial differential equations with highly discontinuous and anisotropic coefficients. *SIAM J. Sci. Comput.*, 20(1):228–242, 1998.
34. P. Vaněk, J. Mandel, and M. Brezina. Algebraic multigrid based on smoothed aggregation for second and fourth order problems. *Computing*, 56:179–196, 1996.
35. U. M. Yang. Parallel algebraic multigrid methods - high performance preconditioners. In A. M. Bruaset and A. Tveito, editors, *Numerical Solution of Partial Differential Equations on Parallel Computers*, volume 51 of *Lecture Notes in Computational Science and Engineering*, pp. 209–236. Springer-Verlag, 2005.

9

Parallelizing PDE Solvers Using the Python Programming Language

Xing Cai and Hans Petter Langtangen

Simula Research Laboratory, P.O. Box 134, NO-1325 Lysaker, Norway

Department of Informatics, University of Oslo, P.O. Box 1080, Blindern, NO-0316 Oslo, Norway

[xingca,hpl]@simula.no

Summary. This chapter aims to answer the following question: Can the high-level programming language Python be used to develop sufficiently efficient parallel solvers for partial differential equations (PDEs)? We divide our investigation into two aspects, namely (1) the achievable performance of a parallel program that extensively uses Python programming and its associated data structures, and (2) the Python implementation of generic software modules for parallelizing existing serial PDE solvers. First of all, numerical computations need to be based on the special array data structure of the Numerical Python package, either in pure Python or in mixed-language Python-C/C++ or Python/Fortran setting. To enable high-performance message passing in parallel Python software, we use the small add-on package `pypar`, which provides efficient Python wrappers to a subset of MPI routines. Using concrete numerical examples of solving wave-type equations, we will show that a mixed Python-C/Fortran implementation is able to provide fully comparable computational speed in comparison with a pure C or Fortran implementation. In particular, a serial legacy Fortran 77 code has been parallelized in a relatively straightforward manner and the resulting parallel Python program has a clean and simple structure.

9.1 Introduction

Solving partial differential equations (PDEs) on a computer requires writing software to implement numerical schemes, and the resulting program should be able to effectively utilize the modern computer architecture. The computing power of parallel platforms is particularly difficult to exploit, so we would like to avoid implementing every thing from scratch when developing a parallel PDE solver. The well-tested computational modules within an existing serial PDE solver should in some way participate in the parallel computations. Most of the parallelization effort should be invested in dealing with the parallel-specific tasks, such as domain partitioning, load balancing, high performance communication etc. Such considerations call for a flexible, structured, and layered programming style for developing parallel PDE software. The object-oriented programming language C++, in particular, has obtained a considerable amount of success in this field, see e.g. [2, 3, 15, 4, 13].

In comparison with C++, the modern object-oriented programming language Python is known for its even richer expressiveness and flexibility. Moreover, the Python language simplifies interfacing legacy software written in the traditional compiled programming languages Fortran, C, and C++. This is a desired feature in parallelizing PDE solvers, because the serial computational modules of a PDE solver and existing software libraries may exist in different programming styles and languages. However, due to the computational inefficiency of its core language, Python has seldom been used for large-scale scientific computations, let alone parallelizations.

Based on the findings in [5], we have reasons to believe that Python is capable of delivering high performance in both serial and parallel computations. This assumes that the involved data structure uses arrays of the Numerical Python package [17]. Moreover, either the technique of vectorization must be adopted, or computation-intensive code segments must be migrated to extension modules implemented in a compiled language such as Fortran, C or C++. It is the purpose of the present chapter to show that the clean syntax and the powerful tools of Python help to simplify the implementation tasks of parallel PDE solvers. We will also show that existing serial legacy codes and software libraries can be re-used in parallel computations driven by Python programs.

The remainder of the chapter is organized as follows. Section 9.2 presents the most important ingredients in achieving high-performance computing in Python. Section 9.3 explains two parallelization approaches of relevance for PDE solvers, whereas Sections 9.4 gives examples of how Python can be used to code generic functions useful for parallelizing serial PDE solvers. Afterwards, Section 9.5 demonstrates two concrete cases of parallelization, and Section 9.6 summarizes the main issues of this chapter.

9.2 High-Performance Serial Computing in Python

To achieve good performance of any parallel PDE solver, the involved serial computational modules must have high performance. This is no exception in the context of the Python programming language. The present section therefore discusses the issue of high-performance serial computing in Python. Some familiarity with basic Python is assumed. We refer to [14] for an introduction to Python with applications to computational science.

9.2.1 Array-Based Computations

PDE solvers frequently employ one-dimensional or multi-dimensional arrays. Computations are normally done in the form of nested loops over values stored in the arrays. Such nested loops run very slowly if they are coded in the core language of Python. However, the add-on package *Numerical Python* [17], often referred to as NumPy, provides efficient operations on multi-dimensional arrays.

The NumPy package has a basic module defining a data structure of multi-dimensional contiguous-memory arrays and many associated efficient C functions.

Two versions of this module exist at present: `Numeric` is the classical module from the mid 1990s, while `numarray` is a new implementation. The latter is meant as a replacement of the former, and no further development of `Numeric` is going to take place. However, there is so much numerical Python code utilizing `Numeric` that we expect both modules to co-exist for a long time. We will emphasize the use of `Numeric` in this chapter, because many tools for scientific Python computing are at present best compatible with `Numeric`.

The allocation and use of NumPy arrays is very user friendly, which can be demonstrated by the following code segment:

```
from Numeric import arange, zeros, Float
n = 1000001
dx = 1.0/(n-1)
x = arange(0, 1, dx)          # x = 0, dx, 2*dx, ...
y = zeros(len(x), Float)      # array of zeros, as long as x
                              # as Float (=double in C) entries
from math import sin, cos     # import scalar math functions
for i in xrange(len(x)):      # i=0, 1, 2, ..., length of x-1
    xi = x[i]
    y[i] = sin(xi)*cos(xi) + xi**2
```

Here the `arange` method allocates a NumPy array and fills it with values from a start value to (but not including) a stop value using a specified increment. Note that a Python method is equivalent to a Fortran subroutine or a C/C++ function. Allocating a double precision real-value array of length `n` is done by the `zeros(n,Float)` call. Traversing an array can be done by a `for`-loop as shown, where `xrange` is a method that returns indices 0, 1, 2, and up to (but not including) the length of `x` (i.e., `len(x)`) in this example. Indices in NumPy arrays always start at 0.

Detailed measurements of the above `for`-loop show that the Python code is considerably slower than a corresponding code implemented in plain Fortran or C. This is due to the slow execution of `for`-loops in Python, even though the NumPy data arrays have an efficient underlying C implementation. Unlike the traditional compiled languages, Python programs are complied to byte code (like Java). Loops in Python are not highly optimized in machine code as Fortran, C, or C++ compilers would do. Such optimizations are in fact *impossible* due to the dynamic nature of the Python language. For example, there is no guarantee that `sin(xi)` does not return a new data type at some stage in the loop and thereby turning `y` into a new data type. (The only requirement of the new type is that it must be valid to subscript it as `y[i]`.) In the administering parts of a PDE code, this type of flexibility can be the reason to adopt Python, but in number crunching with large arrays the efficiency frequently becomes unacceptable.

The problem with slow Python loops can be greatly relieved through vectorization [16], i.e., expressing the loop semantics via a set of basic NumPy array operations, where each operation involves a loop over array entries efficiently implemented in C. As an improvement of the above Python `for`-loop, we now introduce a vectorized implementation as follows:

```
from Numeric import arange, sin, cos
n = 1000001
```

```
dx = 1.0/(n-1)
x = arange(0, 1, dx)          # x = 0, dx, 2*dx, ...
y = sin(x)*cos(x) + x**2
```

The expression `sin(x)` applies the sine function to each entry in the array `x`, the same for `cos(x)` and `x**2`. The loop over the array entries now runs in highly optimized C code, and the reference to a new array holding the computed result is returned. Consequently, there is no need to allocate `y` beforehand, because an array is created by the vector expression `sin(x)*cos(x) + x**2`. This vectorized version runs about 8 times faster than the pure Python version. For more information on using NumPy arrays and vectorization, we refer to [14, 5].

9.2.2 Interfacing with Fortran and C

Vectorized Python code may still run a factor of 3-10 slower than optimized implementations in pure Fortran or C/C++. This is not surprising because an expression like `sin(x)*cos(x)+x**2` involves three unary operations (`sin(x)`, `cos(x)`, `x**2`) and two binary operations (multiplication and addition), each begin processed in a separate C routine. For each intermediate result, a temporary array must be allocated. This is the same type of overhead involved in compound vector operations in many other libraries (Matlab/Octave, R/S-Plus, C++ libraries [20, 19]).

In these cases, or in cases where vectorization of an algorithm is cumbersome, computation-intensive Python loops can be easily migrated directly to Fortran or C/C++. The code migration can be easily achieved in Python, using specially designed tools such as F2PY [9]. For example, we may implement the expression in a simple loop in Fortran 77:

```
      subroutine someloop(x, n)
      integer n, i
      real*8 x(n), xi
Cf2py intent(in,out) x
      do i = 1, n
         xi = x(i)
         x(i) = sin(xi)*cos(xi) + xi**2
      end do
      return
      end
```

The only non-standard feature of the above Fortran subroutine is the command line starting with `Cf2py`. This helps the Python-Fortran translation tool F2PY [9] with information about input and output arguments. Note that `x` is specified as both an input and an output argument. An existing pure Fortran subroutine can be very easily transformed by F2PY to be invokable inside a Python program. In particular, the above `someloop` method can be called inside a Python program as

```
y = someloop(x)
# or
someloop(x)        # rely on in-place modifications of x
```

The "Python" way of writing methods is to have input variables as arguments and return all output variables. In the case of `someloop(x)`, the Fortran code overwrites the input array with values of the output array. The length of the `x` array is needed in the Fortran code, but not in the Python call because the length is available from the NumPy array `x`.

The variable `x` is a NumPy array object in Python, while the Fortran code expects a pointer to a contiguous data segment and the array size. The translation of Python data to and from Fortran data must be done in *wrapper code*. The tool F2PY can read Fortran source code files and automatically generate the wrapper code. It is the wrapper code that extracts the array size and the pointer to the array data segment from the NumPy array object and sends these to the Fortran subroutine. The wrapper code also converts the Fortran output data to Python objects. In the present example we may store the `someloop` subroutine in a file named `floop.f` and run F2PY like

```
f2py -m floop -c floop.f
```

This command creates an *extension module* `floop` with the wrapper code and the `someloop` routine. The extension module can be imported as any pure Python program.

Migrating intensive computations to C or C++ via wrapper code can be done similarly as explained for Fortran. However, the syntax has a more rigid style and additional manual programming is needed. There are tools for automatic generation of the wrapper code for C and C++, SWIG [8] for instance, but none of the tools integrate C/C++ seamlessly with NumPy arrays as for Fortran. Information on computing with NumPy arrays in C/C++ can be found in [1, 14].

9.3 Parallelizing Serial PDE Solvers

Parallel computing is an important technique for solving large-scale PDE problems, where multiple processes form a collaboration to solve a large computational problem. The total computational work is divided into multiple smaller tasks that can be carried out concurrently by the different processes. In most cases, the collaboration between the processes is realized as exchanging computing results and enforcing synchronizations. This section will first describe a subdomain-based framework which is particularly well-suited for parallelizing PDE solvers. Then, two approaches to parallelizing different types of numerical schemes will be explained, in the context of Python data structures.

9.3.1 A Subdomain-Based Parallelization Framework

The starting point of parallelizing a PDE solver is a division of the global computational work among the processes. The division can be done in several different ways. Since PDE solvers normally involve a lot of loop-based computations, one strategy is to (dynamically) distribute the work of every long loop among the processes. Such a strategy often requires that all the processes have equal access to all the data, and that a master-type process is responsible for (dynamically) carrying out the work load distribution. Therefore, shared memory is the most suitable platform type for this strategy. In this chapter, however, we will consider dividing the work based on subdomains. That is, a global solution domain Ω is explicitly decomposed into a set of subdomains $\cup\Omega_s = \Omega$, where there may be an overlap zone between each pair of neighboring subdomains. The part of the subdomain boundary that borders a neighboring subdomain, i.e., $\partial\Omega_s \backslash \partial\Omega$, is called the *internal boundary* of Ω_s. This decomposition of the global domain gives rise to a decomposition of the global data arrays, where subdomain s always concentrates on the portion of the global data arrays lying inside Ω_s. Such a domain and data decomposition results in a natural division of the global computational work among the processes, where the computational work on one subdomain is assigned to one process. This strategy suits both shared-memory and distributed-memory platforms.

In addition to ensuring good data locality on each process, the subdomain-based parallelization strategy also strongly promotes code re-use. One scenario is that parts of an existing serial PDE solver are used to carry out array-level operations (see Section 9.3.2), whereas another scenario is that an entire serial PDE solver works as a "black box subdomain solver" in Schwarz-type iterations (see Section 9.3.3). Of course, communication between subdomains must be enforced at appropriate locations of a parallel PDE solver. It will be shown in the remainder of this section that the needed communication operations are of a generic type, independent of the specific PDE problem. Moreover, Section 9.4 will show that Python is a well-suited programming language for implementing the generic communication operations as re-usable methods.

9.3.2 Array-Level Parallelization

Many numerical operations can be parallelized in a straightforward manner, where examples are (1) explicit time-stepping schemes for time-dependent PDEs, (2) Jacobi iterations for solving a system of linear equations, (3) inner-products between two vectors, and (4) matrix-vector products. The parallel version of these operations computes identical numerical results as the serial version. Therefore, PDE solvers involving only such numerical operations are readily parallelizable. Inside the framework of subdomains, the basic idea of this parallelization approach is as follows:

1. The mesh points of each subdomain are categorized into two groups: (1) points lying on the internal boundary and (2) the interior points (plus points lying on the physical boundary). Each subdomain is only responsible for computing values on its interior points (plus points lying on the physical boundary). The values on the internal boundary points are computed by the neighboring subdomains and updated via communication. (Such internal boundary points are commonly called *ghost points*.)
2. The global solution domain is partitioned such that there is an overlap zone of minimum size between two neighboring subdomains. For a standard second-order finite difference method this means one layer of overlapping cells (higher

order finite differences require more overlapping layers), while for finite element methods this means one layer of overlapping elements. Such an overlapping domain partitioning ensures that every internal boundary point of one subdomain is also located in the interior of at least one neighboring subdomain.

3. The parallelization of a numerical operation happens at the *array level*. That is, the numerical operation is concurrently carried out on each subdomain using its local arrays. When new values on the interior points are computed, neighboring subdomains communicate with each other to update the values on their internal boundary points.

Now let us consider a concrete example of parallelizing the following five-point stencil operation (representing a discrete Laplace operator) on a two-dimensional array `um`, where the computed results are stored in another array `u`:

```
for i in xrange(1,nx):
    for j in xrange(1,ny):
        u[i,j] = um[i,j-1] + um[i-1,j] \
                 -4*um[i,j] + um[i+1,j] + um[i,j+1]
```

The above Python code segment is assumed to be associated with a uniform two-dimensional mesh, having n_x cells (i.e., $n_x + 1$ entries) in the x-direction and n_y cells (i.e., $n_y + 1$ entries) in the y-direction. Note that the boundary entries of the `u` array are assumed to be computed separately using some given boundary condition (not shown in the code segment). We also remark that `xrange(1,nx)` covers indices 1, 2, and up to `nx-1` (i.e., not including index `nx`).

Suppose P processors participate in the parallelization. Let us assume that the P processors form an $N_x \times N_y = P$ lattice. A division of the global computational work naturally arises if we partition the interior array entries of `u` and `um` into P small rectangular portions, using imaginary horizontal and vertical cutting lines. That is, the $(n_x - 1) \times (n_y - 1)$ interior entries of `u` and `um` are divided into $N_x \times N_y$ rectangular portions. For a processor that is identified by an index tuple (l, m), where $0 \leq l < N_x$ and $0 \leq m < N_y$, $(n_x^l - 1) \times (n_y^m - 1)$ interior entries are assigned to it, plus one layer of boundary entries around the interior entries (i.e., one layer of overlapping cells). More precisely, we have

$$\sum_{l=0}^{N_x-1} (n_x^l - 1) = n_x - 1, \quad \sum_{m=0}^{N_y-1} (n_y^m - 1) = n_y - 1,$$

where we require that the values of n_x^0, n_x^1, ..., and $n_x^{N_x-1}$ are approximately the same as

$$(n_x - 1)/N_x + 1$$

for the purpose of achieving a good work load balance. The same requirement should also apply to the values of n_y^0, n_y^1, ..., and $n_y^{N_y-1}$.

To achieve a parallel execution of the global five-point stencil operation, each processor now loops over its $(n_x^l - 1) \times (n_y^m - 1)$ interior entries of `u_loc` as follows:

```
for i in xrange(1,loc_nx):
    for j in xrange(1,loc_ny):
        u_loc[i,j] = um_loc[i,j-1] + um_loc[i-1,j] \
                     -4*um_loc[i,j] \
                     + um_loc[i+1,j] + um_loc[i,j+1]
```

We note that the physical boundary entries of `u_loc` are assumed to be computed separately using some given boundary condition (not shown in the above code segment). For the internal boundary entries of `u_loc`, communication needs to be carried out. Suppose a subdomain has a neighbor on each of its four sides. Then the following values should be sent out to the neighbors:

- `u[1,:]` to the negative-x side neighbor,
- `u[nx_loc-1,:]` to the positive-x side neighbor,
- `u[:,1]` to the negative-y side neighbor, and
- `u[:,ny_loc-1]` to the positive-y side neighbor.

Correspondingly, the following values should be received from the neighbors:

- `u[0,:]` from the negative-x side neighbor,
- `u[nx_loc,:]` from the positive-x side neighbor,
- `u[:,0]` from the negative-y side neighbor, and
- `u[:,ny_loc]` from the positive-y side neighbor.

The *slicing* functionality of NumPy arrays is heavily used in the above notation. For example, `u[0,:]` refers to the first row of `u`, whereas `u[0,1:4]` means a one-dimensional slice containing entries `u[0,1]`, `u[0,2]`, and `u[0,3]` of `u` (not including `u[0,4]`), see [1]. The above communication procedure for updating the internal boundary entries is of a generic type. It can therefore be implemented as a Python method applicable in any array-level parallelization on uniform meshes. Similarly, the domain partitioning task can also be implemented as a generic Python method. We refer to the methods `update_internal_boundaries` and `prepare_communication` in Section 9.4.2 for a concrete Python implementation.

9.3.3 Schwarz-Type Parallelization

Apart from the numerical operations that are straightforward to parallelize, as discussed in the preceding text, there exist numerical operations which are non-trivial to parallelize. Examples of the latter type include (1) direct linear system solvers based on some kind of factorization, (2) Gauss-Seidel/SOR/SSOR iterations, and (3) incomplete LU-factorization based preconditioners for speeding up the Krylov subspace linear system solvers. Since these numerical operations are inherently serial in their algorithmic nature, a 100% mathematically equivalent parallelization is hard to implement.

However, if we "relax" the mathematical definition of these numerical operations, an "approximate" parallelization is achievable. For this purpose, we will adopt the basic idea of *additive Schwarz iterations* [7, 24, 10]. Roughly speaking, solving a PDE problem by a Schwarz-type approach is realized as an iterative procedure, where during each iteration the original problem is decomposed into a set of individual subdomain problems. The coupling between the subdomains is through enforcing an artificial Dirichlet boundary condition on the internal boundary of each subdomain, using the latest computed solutions in the neighboring subdomains. This mathematical strategy has shown to be able to converge toward correct global solutions for many types of PDEs, see [7, 24]. The convergence depends on a sufficient amount of overlap between neighboring subdomains.

Additive Schwarz Iterations

The mathematics of additive Schwarz iterations can be explained through solving the following PDE of a generic form on a global domain Ω:

$$\mathcal{L}(u) = f, \quad x \in \Omega, \tag{9.1}$$

$$u = g, \quad x \in \partial\Omega. \tag{9.2}$$

For a set of overlapping subdomains $\{\Omega_s\}$, the restriction of (9.1) onto Ω_s becomes

$$\mathcal{L}(u) = f, \quad x \in \Omega_s. \tag{9.3}$$

Note that (9.3) is of the same type as (9.1), giving rise to the possibility of re-using both numerics and software. In order to solve the subdomain problem (9.3), we have to assign the following boundary condition for subdomain s:

$$\begin{aligned} u &= g_s^{\text{artificial}} & x &\in \partial\Omega_s \backslash \partial\Omega, \\ u &= g, & x &\in \partial\Omega, \end{aligned} \tag{9.4}$$

where the artificial Dirichlet boundary condition $g_s^{\text{artificial}}$ is provided by the neighboring subdomains in an iterative fashion. More specifically, we generate on each subdomain a series of approximate solutions $u_{s,0}$, $u_{s,1}$, $u_{s,2}$, ..., where during the kth Schwarz iteration we have

$$u_{s,k} = \tilde{\mathcal{L}}^{-1} f, \quad x \in \Omega_s, \tag{9.5}$$

and the artificial condition for the kth iteration is determined as

$$g_{s,k}^{\text{artificial}} = u_{\text{glob},k-1}|_{\partial\Omega_s \backslash \partial\Omega}, \quad u_{\text{glob},k-1} = \text{composition of all } u_{s,k-1}. \tag{9.6}$$

The symbol $\tilde{\mathcal{L}}^{-1}$ in (9.5) indicates that an approximate local solve, instead of an exact inverse of $\mathcal{L}$, may be sufficient. Solving the subdomain problem (9.5) in the kth Schwarz iteration implies that $u_{s,k}$ attains the values of $g_{s,k}^{\text{artificial}}$ on the internal boundary, where the values are provided by the neighboring subdomains. We note that the subdomain local solves can be carried out independently, thus giving rise to parallelism. At the end of the kth additive Schwarz iteration, the conceptually existing global approximate solution $u_{\text{glob},k}$ is composed by "patching together" the subdomain approximate solutions $\{u_{s,k}\}$, using the following rule:

- For every non-overlapping point, i.e., a point that belongs to only one subdomain, the global solution attains the same value as that inside the host subdomain.
- For every overlapping point, let us denote by n_{total} the total number of host subdomains that own this point. Let also n_{interior} denote the number of subdomains, among those n_{total} host subdomains, which do not have the point lying on their internal boundaries. (The overlapping subdomains must satisfy the requirement $n_{\mathrm{interior}} \geq 1$.) Then, the average of the n_{interior} local values becomes the global solution on the point. The other $n_{\mathrm{total}} - n_{\mathrm{interior}}$ local values are not used, because the point lies on the internal boundary there. Finally, the obtained global solution is duplicated in each of the n_{total} host subdomains. For the $n_{\mathrm{total}} - n_{\mathrm{interior}}$ host subdomains, which have the point lying on their internal boundary, the obtained global solution value will be used as the artificial Dirichlet condition during the next Schwarz iteration.

The Communication Procedure

To compose the global approximate solution and update the artificial Dirichlet conditions, as described by the above rule, we need to carry out a communication procedure among the neighboring subdomains at the end of each Schwarz iteration. During this communication procedure, each pair of neighboring subdomains exchanges an array of values associated with their shared overlapping points. Convergence of $\boldsymbol{x}_{s,k}$ toward the correct solution $\boldsymbol{x}|_{\Omega_s}$ implies that the difference between the subdomain solutions in an overlap zone will eventually disappear.

Let us describe the communication procedure by looking at the simple case of two-dimensional uniform subdomain meshes. If the number of overlapping cell layers is larger than one, we need to also consider the four "corner neighbors", in addition to the four "side neighbors", see Figure 9.1. In this case, some overlapping points are interior points in more than one subdomain (i.e., $n_{\mathrm{interior}} > 1$), so averaging n_{interior} values over each such point should be done, as described in the preceding text. For the convenience of implementation, the value on each overlapping interior point is scaled by a factor of $1/n_{\mathrm{interior}}$ before being extracted into the outgoing messages, whereas the value on each internal boundary point is simply set to zero. (Note that an internal boundary point may be provided with multiple values coming from the interior of $n_{\mathrm{interior}} > 1$ subdomains.) The benefit of such an implementation is that multiplications are only used when preparing the outgoing messages before communication, and after communication, values of the incoming messages can be directly added upon the local scaled entries (see the third task to be described below).

For a two-dimensional array `x` of dimension `[0:loc_nx+1]x[0:loc_ny+1]`, i.e., indices are from `0` to `loc_nx` or `loc_ny`, the pre-communication scaling task can be done as follows (using the efficient in-place modification of the array entries through the `*=` operator):

```
x[1:w0,:] *= 0.5   # overlapping interior points
x[loc_nx-w0+1:loc_nx,:] *= 0.5
x[:,1:w1] *= 0.5
```

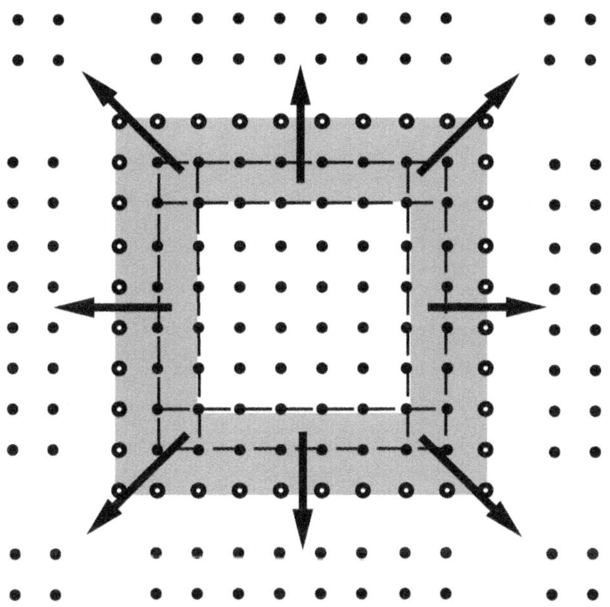

Fig. 9.1. One example two-dimensional rectangular subdomain in an overlapping setting. The shaded area denotes the overlap zone, which has two layers of overlapping cells in this example. Data exchange involves (up to) eight neighbors, and the origin of each arrow indicates the source of an outgoing message.

```
x[:,loc_ny-w1+1:loc_ny] *= 0.5
x[0,:] = 0.0          # internal boundary points
x[loc_nx,:] = 0.0
x[:,0] = 0.0
x[:,loc_ny] = 0.0
```

We note that `w0` denotes the number of overlapping cell layers in the x-direction, and `w1` is for the y-direction. The overlapping interior points in the four corners of the overlap zone are scaled twice by the factor 0.5, resulting in the desired scaling of 0.25 (since $n_{\text{interior}} = 4$). The reason to multiply all the internal boundary values by zero is because these values are determined entirely by the neighboring subdomains.

The second task in the communication procedure is to exchange data with the neighbors. First, we "cut out" eight portions of the `x` array, where the overlapping interior values are already scaled with respect to n_{interior}, to form outgoing communication data to the eight neighbors:

- `x[1:w0+1,:]` to the negative-x side neighbor,
- `x[loc_nx-w0:loc_nx,:]` to the positive-x side neighbor,
- `x[:,1:w1+1]` to the negative-y side neighbor,
- `x[:,loc_ny-w1:loc_ny]` to the positive-y side neighbor,
- `x[1:w0+1,1:w1+1]` to the lower-left corner neighbor,
- `x[1:w0+1,loc_ny-w1:loc_ny]` to the upper-left corner neighbor,

- `x[loc_nx-w0:loc_nx,1:w1+1]` to the lower-right corner neighbor, and
- `x[loc_nx-w0:loc_nx,loc_ny-w1:loc_ny]` to the upper-right corner neighbor.

Then, we receive incoming data from the eight neighboring subdomains and store them in eight internal data buffers.

The third and final task of the communication procedure is to add the incoming data values, which are now stored in the eight internal buffers, on top of the corresponding entries in `x`. Roughly speaking, the overlapping entries in the corners add values from three neighbors on top of their own scaled values, the remaining overlapping entries add values from one neighbor. For the details we refer to the `add_incoming_data` method in Section 9.4.2, which is a generic Python implementation of the communication procedure.

We remark that for the case of only one overlapping cell layer, the above communication procedure becomes the same as that of the array-level parallelization approach from Section 9.3.2.

Parallelization

In the Schwarz-type parallelization approach, we aim to re-use an entire serial PDE solver as a "black-box" for solving (9.6). Here, we assume that the existing serial PDE solver is flexible enough to work on any prescribed solution domain, on which it builds data arrays of appropriate size and carries out discretizations and computations. The only new feature is that the physical boundary condition valid on $\partial\Omega$ does not apply on the internal boundary of Ω_s, where the original boundary condition needs to be replaced by an artificial Dirichlet condition. So a slight modification/extension of the serial code may be needed with respect to boundary condition enforcement.

A simple Python program can be written to administer the Schwarz iterations, where the work in each iteration consists of calling the serial computational module(s) and invoking the communication procedure. Section 9.5.2 will show a concrete case of the Schwarz-type parallelization approach. Compared with the array-level parallelization approach, we can say that the Schwarz-type approach parallelizes serial PDE solvers at a higher abstraction level.

The difference between the array-level parallelization approach and the Schwarz-type approach is that the former requires a detailed knowledge of an existing serial PDE solver on the level of loops and arrays. The latter approach, on the other hand, promotes the re-use of a serial solver as a whole, possibly after a slight code modification/extension. Detailed knowledge of every low-level loop or array is thus not mandatory. This is particularly convenient for treating very old codes in Fortran 77. However, the main disadvantage of the Schwarz-type approach is that there is no guarantee for the mathematical strategy behind the parallelization to work for any type of PDE.

9.4 Python Software for Parallelization

This section will first explain how data communication can be done in Python. Then, we describe the implementation of a Python class hierarchy containing generic methods, which can ease the programming of parallel Python PDE solvers by re-using serial PDE code.

9.4.1 Message Passing in Python

The *message-passing* model will be adopted for inter-processor communication in this chapter. In the context of parallel PDE solvers, a message is typically an array of numerical values. This programming model, particularly in the form of using the MPI standard [11, 18], is applicable on all types of modern parallel architectures. MPI-based parallel programs are also known to have good performance. There exist several Python packages providing MPI wrappers, the most important being `pypar` [22], `pyMPI` [21], and `Scientific.MPI` [23] (part of the ScientificPython package). The `pypar` package concentrates only on an important subset of the MPI library, offering a simple syntax and sufficiently good performance. On the other hand, the `pyMPI` package implements a much larger collection of the MPI routines and has better flexibility. For example, MPI routines can be run interactively via `pyMPI`, which is very convenient for debugging. All the three packages are implemented in C.

Syntax and Examples

To give the reader a feeling of how easily MPI routines can be invoked via `pypar`, we present in the following a simple example of using the two most important methods in `pypar`, namely `pypar.send` and `pypar.receive`. In this example, a NumPy array is relayed as a message between the processors virtually connected in a loop. Each processor receives an array from its "upwind" neighbor and passes it to the "downwind" neighbor.

```
import pypar
myid = pypar.rank()              # ID of a processor
numprocs = pypar.size()          # total number of processors
msg_out = zeros(100, Float)      # NumPy array, communicated

if myid == 0:
    pypar.send (msg_out, destination=1)
    msg_in = pypar.receive(numprocs-1)
else:
    msg_in = pypar.receive(myid-1)
    pypar.send (msg_out, destination=(myid+1)%numprocs)

pypar.finalize()                 # finish using pypar
```

In comparison with an equivalent C/MPI implementation, the syntax of the `pypar` implementation is greatly simplified. The reason is that most of the arguments to the `send` and `receive` methods in `pypar` are optional and have well-chosen default values. Of particular interest to parallel numerical applications, the above example demonstrates that a NumPy array object can be used directly in the send and receive methods.

To invoke the most efficient version of the send and receive commands, which avoid internal safety checks and transformations between arrays and strings of characters, we must assign the optional argument bypass with value True. That is,

```
pypar.send (msg_out, destination=to, bypass=True)
msg_in = pypar.receive (from, buffer=msg_in_buffer,
                        bypass=True)
```

We also remark that using bypass=True in the pypar.receive method must be accompanied by specifying a message buffer, i.e., msg_in_buffer in the above example. The message buffer is assumed to be an allocated one-dimensional Python array of appropriate length. In this way, we can avoid the situation that pypar creates a new internal array object every time the receive method is invoked.

The communication methods in the pyMPI package also have a similarly user-friendly syntax. The send and recv methods of the pyMPI package are very versatile in the sense that any Python objects can be sent and received. However, an intermediate character array is always used internally to hold a message inside a pyMPI communication routine. Performance is therefore not a strong feature of pyMPI. Scientific.MPI works similarly as pypar but requires more effort with installation. Consequently, we will only use the MPI wrappers of pypar in our numerical experiments in Section 9.5.

Latency and Bandwidth

Regarding the actual performance of pypar, in terms of latency and bandwidth, we have run two ping-pong test programs, namely ctiming.c and pytiming, which are provided in the pypar distribution. Both the pure C test program and the pure Python-pypar test program measure the time usage of exchanging a series of messages of different sizes between two processors. Based on the series of measurements, the actual values of latency and bandwidth are estimated using a least squares strategy. We refer to Table 9.1 for the estimates obtained on a Linux cluster with Pentium III 1GHz processors, inter-connected through a switch and a fast Ethernet based network, which has a theoretical peak bandwidth of 100 Mbit/s. The version of pypar was 1.9.1. We can observe from Table 9.1 that there is no difference between C and pypar with respect to the actual bandwidth, which is quite close to the theoretical peak value. Regarding the latency, it is evident that the extra Python layer of pypar results in larger overhead.

9.4.2 Implementing a Python Class Hierarchy

To realize the two parallelization approaches outlined in Sections 9.3.2 and 9.3.3, we will develop in the remaining text of this section a hierarchy of Python classes containing the needed functionality, i.e., mesh partitioning, internal data structure

Table 9.1. Comparison between C-version MPI and `pypar`-layered MPI on a Linux-cluster, with respect to the latency and bandwidth.

	Latency	Bandwidth
C-version MPI	133×10^{-6} s	88.176 Mbit/s
`pypar`-layered MPI	225×10^{-6} s	88.064 Mbit/s

preparation for communication, and updating internal boundaries and overlap zones. Our attention is restricted to box-shaped subdomains, for which the efficient array-slicing functionality (see e.g. [5]) can be extensively used in the preparation of outgoing messages and in the extraction of incoming messages. A pure Python implementation of the communication operations will thus have sufficient efficiency for such cases. However, it should be noted that in the case of unstructured subdomain meshes, indirect indexing of the data array entries in an unstructured pattern will become inevitable. A mixed-language implementation of the communication operations, which still have a Python "appearance", will thus be needed for the sake of efficiency.

Class BoxPartitioner

Our objective is to implement a Python class hierarchy, which provides a unified interface to the generic methods that are needed in programming parallel PDE solvers on the basis of box-shaped subdomain meshes. The name of the base class is `BoxPartitioner`, which has the following three major methods:

1. `prepare_communication`: a method that partitions a global uniform mesh into a set of subdomain meshes with desired amount of overlap. In addition, the internal data structure needed in subsequent communication operations is also built up. This method is typically called right after an instance of a chosen subclass of `BoxPartitioner` is created.
2. `update_internal_boundaries`: a method that lets each subdomain communicate with its neighbors and update those points lying on the internal boundaries.
3. `update_overlap_regions`: a method that lets each subdomain communicate with its neighbors and update all the points lying in the overlap zones, useful for the Schwarz-type parallelization described in Section 9.3.3. This method assumes that the number of overlapping cell layers is larger than one in at least one space direction, otherwise one can use `update_internal_boundaries` instead.

Note that the reason for having a separate `update_internal_boundaries` method, in addition to `update_overlap_regions`, is that it may be desirable to communicate and update *only* the internal boundaries when the number of overlapping cell layers is larger than one.

Subclasses of BoxPartitioner

Three subclasses have been developed on the basis of `BoxPartitioner`, i.e., `BoxPartitioner1D`, `BoxPartitioner2D`, and `BoxPartitioner3D`. These first-level subclasses are designed to handle dimension-specific operations, e.g., extending the base implementation of the `prepare_communication` method in class `BoxPartitioner`. However, all the MPI-related operations are deliberately left out in the three first-level subclasses. This is because we want to introduce another level of subclasses in which the concrete message passing operations (using e.g. `pypar` or `pyMPI`) are programmed. For example, `PyMPIBoxPartitioner2D` and `PyParBoxPartitioner2D` are two subclasses of `BoxPartitioner2D`. A convenient effect of such a design is that a parallel Python PDE solver can freely switch between different MPI wrapper modules, e.g., by simply replacing an object of type `PyMPIBoxPartitioner2D` with a `PyParBoxPartitioner2D` object.

The constructor of every class in the `BoxPartitioner` hierarchy has the following syntax:

```
def __init__(self, my_id, num_procs,
                  global_num_cells=[], num_parts=[],
                  num_overlaps=[]):
```

This is for specifying, respectively, the subdomain ID (between 0 and $P-1$), the total number of subdomains P, the numbers of cells in all the space directions of the global uniform mesh, the numbers of partitions in all the space directions, and the numbers of overlapping cell layers in all the space directions. The set of information will be used later by the `prepare_communication` method.

Class PyParBoxPartitioner2D

To show a concrete example of implementing PDE-related communication operations in Python, let us take a brief look at class `PyParBoxPartitioner2D`.

```
class PyParBoxPartitioner2D(BoxPartitioner2D):
    def __init__(self, my_id=-1, num_procs=-1,
                      global_num_cells=[], num_parts=[],
                      num_overlaps=[1,1]):
        BoxPartitioner.__init__(self, my_id, num_procs,
                                      global_num_cells, num_parts,
                                      num_overlaps)

    def update_internal_boundaries (self, data_array):
        # a method for updating the internal boundaries
        # implementation is skipped

    def update_overlap_regions (self, data_array):

        # call 'update_internal_boundary' when the
        # overlap is insufficient
        if self.num_overlaps[0]<=1 and self.num_overlaps[1]<=1:
            self.update_internal_boundary (data_array)
            return

        self.scale_outgoing_data (data_array)
        self.exchange_overlap_data (data_array)
        self.add_incoming_data (data_array)
```

Note that we have omitted the implementation of `update_internal_boundaries` in the above simplified class definition of `PyParBoxPartitioner2D`. Instead, we will concentrate on the `update_overlap_regions` method consisting of three tasks (see also Section 9.3.3):

1. Scale portions of the target data array before forming the outgoing messages.
2. Exchange messages between each pair of neighbors.
3. Extract the content of each incoming message and add it on top of the appropriate locations inside the target data array.

In the following, let us look at the implementation of each of the three tasks in detail.

```
def scale_outgoing_data (self, data_array):

    # number of local cells in the x- and y-directions:
    loc_nx = self.subd_hi_ix[0]-self.subd_lo_ix[0]
    loc_ny = self.subd_hi_ix[1]-self.subd_lo_ix[1]

    # IDs of the four main neighbors:
    lower_x_neigh = self.lower_neighbors[0]
    upper_x_neigh = self.upper_neighbors[0]
    lower_y_neigh = self.lower_neighbors[1]
    upper_y_neigh = self.upper_neighbors[1]

    # width of the overlapping zones:
    w0 = self.num_overlaps[0]
    w1 = self.num_overlaps[1]

    # in case of a left neighbor (in x-dir):
    if lower_x_neigh>=0:
        data_array[0,:] = 0.0;
        if w0>1:
            data_array[1:w0,:] *= 0.5

    # in case of a right neighbor (in x-dir):
    if upper_x_neigh>=0:
        data_array[loc_nx,:] = 0.0;
        if w0>1:
            data_array[loc_nx-w0+1:loc_nx,:] *= 0.5

    # in case of a lower neighbor (in y-dir):
    if lower_y_neigh>=0:
        data_array[:,0] = 0.0;
        if w1>1:
            data_array[:,1:w1] *= 0.5

    # in case of a upper neighbor (in y-dir):
    if upper_y_neigh>=0:
        data_array[:,loc_ny] = 0.0;
        if w1>1:
            data_array[:,loc_ny-w1+1:loc_ny] *= 0.5
```

Note that the IDs of the four side neighbors (such as `self.lower_neighbors[0]`) have been extensively used in the above `scale_outgoing_data` method. These neighbor IDs are already computed inside the `prepare_communication` method belonging to the base class `BoxPartitioner`. The absence of a neighbor is indicated by a negative integer value. Note also that the internal arrays `subd_hi_ix` and `subd_lo_ix` are computed inside the `prepare_communication` method. These two arrays together determine where inside a virtual global array one local array should be mapped to. Moreover, the two integers `w0` and `w1` contain the number of overlapping cell layers in the x- and y-direction, respectively.

```
def exchange_overlap_data (self, data_array):

    # number of local cells in the x- and y-directions:
    loc_nx = self.subd_hi_ix[0]-self.subd_lo_ix[0]
    loc_ny = self.subd_hi_ix[1]-self.subd_lo_ix[1]

    # IDs of the four side neighbors:
    lower_x_neigh = self.lower_neighbors[0]
    upper_x_neigh = self.upper_neighbors[0]
    lower_y_neigh = self.lower_neighbors[1]
    upper_y_neigh = self.upper_neighbors[1]

    # width of the overlapping zones:
    w0 = self.num_overlaps[0]
    w1 = self.num_overlaps[1]

    if w0>=1 and lower_x_neigh>=0:
        #send message to left neighbor
        pypar.send (data_array[1:w0+1,:], lower_x_neigh,
                    bypass=True)
        #send message to lower left corner neighbor
        if w1>=1 and lower_y_neigh>=0:
            pypar.send (data_array[1:w0+1,1:w1+1],
                        lower_y_neigh-1, bypass=True)

    if w0>=1 and lower_x_neigh>=0:
        #receive message from left neighbor
        self.buffer1 = pypar.receive(lower_x_neigh,
                                     buffer=self.buffer1,
                                     bypass=True)
        if w1>=1 and lower_y_neigh>=0:
            #receive message from lower left corner
            self.buffer5 = pypar.receive (lower_y_neigh-1,
                                          buffer=self.buffer5,
                                              bypass=True)

    # the remaining part of the method is skipped
```

It should be observed that the `scale_outgoing_data` method has already scaled the respective portions of the target data array, so the `exchange_overlap_data` method can directly use array slicing to form the eight outgoing messages. For the incoming messages, eight internal buffers (such as `self.overlap_buffer1`) are heavily used. In the class `BoxPartitioner2D`, these eight internal buffers are already allocated inside the `prepare_communication` method. We also note that the option `bypass=True` is used in every `pypar.send` and `pypar.receive` call for the sake of efficiency.

```
def self.add_incoming_data (self, data_array)

    # number of local cells in the x- and y-directions:
    loc_nx = self.subd_hi_ix[0]-self.subd_lo_ix[0]
    loc_ny = self.subd_hi_ix[1]-self.subd_lo_ix[1]

    # IDs of the four main neighbors:
    lower_x_neigh = self.lower_neighbors[0]
    upper_x_neigh = self.upper_neighbors[0]
    lower_y_neigh = self.lower_neighbors[1]
    upper_y_neigh = self.upper_neighbors[1]

    # width of the overlapping zones:
    w0 = self.num_overlaps[0]
    w1 = self.num_overlaps[1]

    if w0>=1 and lower_x_neigh>=0:
        # contribution from the left neighbor
        data_array[0:w0,:] +=
            reshape(self.buffer1,[w0,loc_ny+1])

        # contribution from the lower left corner neighbor
        if w1>=1 and lower_y_neigh>=0:
            data_array[0:w0,0:w1] +=
                reshape(self.buffer5,[w0,w1])

    # the remaining part of the method is skipped
```

The above `add_incoming_data` method carries out the last task within the communication method `update_overlap_regions`. We note that all the internal data buffers are one-dimensional NumPy arrays. Therefore, these one-dimensional arrays must be re-ordered into a suitable two-dimensional array before being added on top of the appropriate locations of the target data array. This re-ordering operation is conveniently achieved by calling the `reshape` functionality of the NumPy module.

9.5 Test Cases and Numerical Experiments

This section will present two cases of parallelizing serial PDE solvers, one using an explicit numerical scheme for a C code and the other using an implicit scheme for a very old Fortran 77 code. The purpose is to demonstrate how the generic parallelization software from Section 9.4 can be applied. Roughly speaking, the parallelization work consists of the following tasks:

1. slightly modifying the computational modules of the serial code to have the possibility of accepting assigned local array sizes and, for the Schwarz-type parallelization approach, adopting an artificial Dirichlet boundary condition on the internal boundary,
2. implementing a light-weight wrapper code around each serial computational module, such that it becomes callable from Python,
3. writing a simple Python program to invoke the computational modules through the wrapper code and drive the parallel computations,

4. creating an object of `PyParBoxPartitioner2D` or `PyParBoxPartitioner3D` to carry out the domain partitioning task and prepare the internal data structure for the subsequent communication, and
5. inserting communication calls such as `update_internal_boundaries` and/or `update_overlap_regions` at appropriate locations of the Python program.

Although parallelizing PDE solvers can also be achieved using the traditional programming languages, the use of Python promotes a more flexible and user-friendly programming style. In addition, Python has a superior capability of easily interfacing existing serial code segments written in different programming languages. Our vision is to use Python to glue different code segments into a parallel PDE solver.

The upcoming numerical experiments of the parallel Python solvers will be performed on the Linux cluster described in Section 9.4.1.

9.5.1 Parallelizing an Explicit Wave Equation Solver

As the first test case, we consider the following linear wave equation with a source term:

$$\frac{\partial^2 u(\boldsymbol{x},t)}{\partial t^2} = c^2\nabla^2 u(\boldsymbol{x},t) + f(\boldsymbol{x},t) \quad \text{in } \Omega, \tag{9.7}$$

$$u(\boldsymbol{x},t) = g(\boldsymbol{x},t) \quad \text{on } \partial\Omega, \tag{9.8}$$

where c is a constant representing the wave speed, and the coordinates $\boldsymbol{x}$ are in either one, two, or three space dimensions. The above mathematical model (9.7)-(9.8) can have initial conditions of the form:

$$\frac{\partial u(\boldsymbol{x},0)}{\partial t} = 0 \quad \text{and} \quad u(\boldsymbol{x},0) = I(\boldsymbol{x}). \tag{9.9}$$

We consider here a scheme of the explicit type, which in three dimensions translates into the following time-stepping scheme:

$$\begin{aligned} u_{i,j,k}^{l+1} = & -u_{i,j,k}^{l-1} + 2u_{i,j,k}^{l} \\ & + c^2\frac{\Delta t^2}{\Delta x^2}\left(u_{i-1,j,k}^{l} - 2u_{i,j,k}^{l} + u_{i+1,j,k}^{l}\right) \\ & + c^2\frac{\Delta t^2}{\Delta y^2}\left(u_{i,j-1,k}^{l} - 2u_{i,j,k}^{l} + u_{i,j-1,k}^{l}\right) \\ & + c^2\frac{\Delta t^2}{\Delta z^2}\left(u_{i,j,k-1}^{l} - 2u_{i,j,k}^{l} + u_{i,j,k+1}^{l}\right) \\ & + \Delta t^2 f(x_i, y_j, z_k, l\Delta t). \end{aligned} \tag{9.10}$$

Here, we have assumed that the superscript l indicates the time level and the subscripts i, j, k refer to a uniform spatial computational mesh with constant cell lengths Δx, Δy, and Δz.

The Serial Wave Solver

A serial solver has already been programmed in C, where the three-dimensional computational module works at each time level as follows:

```
pos = offset0+offset1; /* skip lower boundary layer (in x)*/
for (i=1; i<nx; i++) {
  for (j=1; j<ny; j++) {
    for (k=1; k<nz; k++) {
      ++pos;
      u[pos] = -um2[pos] + 2*um[pos] +
       Cx2*(um[pos-offset0] - 2*um[pos] + um[pos+offset0]) +
       Cy2*(um[pos-offset1] - 2*um[pos] + um[pos+offset1]) +
       Cz2*(um[pos-1] - 2*um[pos] + um[pos+1]) +
       dt2*source3D(x[i], y[j], z[k], t_old);
    }
    pos += 2;          /* skip the two boundary layers (in z)*/
  }
  pos += 2*offset1; /* skip the two boundary layers (in y)*/
}
```

Here, the arrays u, um, and um2 refer to u^{l+1}, u^l, u^{l-1}, respectively. For the purpose of computational efficiency, each conceptually three-dimensional array actually uses the underlying data structure of a long one-dimensional array. This underlying data structure is thus utilized in the above three-level nested for-loop, where we have offset0=(ny+1)*(nz+1) and offset1=nz+1. The variables Cx2, Cy2, Cz2, dt2 contain the constant values $c^2\Delta t^2/\Delta x^2$, $c^2\Delta t^2/\Delta y^2$, $c^2\Delta t^2/\Delta z^2$, and Δt^2, respectively.

Parallelization

Regarding the parallelization, the array-level approach presented in Section 9.3.2 applies directly to this type of explicit finite difference schemes. The main modifications of the serial code involve (1) the global computational mesh is divided into a $P = N_x \times N_y$ or $P = N_x \times N_y \times N_z$ lattice, (2) each processor only constructs its local array objects including ghost points on the internal boundary, and (3) during each time step, a communication procedure for updating the internal boundary values is carried out after the serial computational module is invoked.

We show in the following the main content of a parallel wave solver implemented in Python:

```
from BoxPartitioner import *
# read 'gnum_cells' & 'parts' from command line ...
partitioner=PyParBoxPartitioner3D(my_id=my_id,
                              num_procs=num_procs,
                              global_num_cells=gnum_cells,
                              num_parts=parts,
                              num_overlaps=[1,1,1])
partitioner.prepare_communication ()
loc_nx,loc_ny,loc_nz = partitioner.get_num_loc_cells ()
# create the subdomain data arrays u, um, um2 ...
# enforce the initial conditions (details skipped)
```

Table 9.2. A comparison of the two-dimensional performance between a mixed Python-C implementation and and pure C implementation. The global mesh has 2000×2000 cells (i.e., 2001×2001 points) and there are 5656 time steps.

	Mixed Python-C		Pure C	
P	Wall-time	Speedup	Wall-time	Speedup
1	2137.18	N/A	1835.38	N/A
2	1116.41	1.91	941.59	1.95
4	649.06	3.29	566.34	3.24
8	371.98	5.75	327.98	5.60
12	236.41	9.04	227.55	8.07
16	193.83	11.03	175.18	10.48

```
import cloops
t = 0.0
while t <= tstop:
    t_old = t;   t += dt
    u = cloops.update_interior_pts3D(u,um,um2,x,y,z,
                                     Cx2,Cy2,Cz2,
                                     dt2,t_old)
    partitioner.update_internal_boundaries (u)
    # enforce boundary condition where needed ...
    tmp = um2; um2 = um; um = u; u = tmp;
```

Note that the `get_num_loc_cells` method of class `BoxPartitioner` returns the number of local cells for each subdomain. The `cloops.update_interior_pts3D` call invokes the serial three-dimensional computational module written in C. We also assume that the C computational module is placed in a file named `cloops.c`, which also contains the wrapper code needed for accessing the Python data structures in C. For more details we refer to [1].

Large-Scale Simulations

A two-dimensional 2000×2000 mesh and a three-dimensional $200 \times 200 \times 200$ mesh are chosen for testing the parallel efficiency of the mixed Python-C wave equation solver. For each global mesh, the value of P is varied between 1 and 16, and parallel simulations of the model problem (9.7)-(9.9) are run on P processors. Tables 9.2 and 9.3 report the wall-clock time measurements (in seconds) of the entire `while`-loop, i.e., the time-stepping part of solving the wave equation. We remark that the number of time steps is determined by the global mesh size, independent of P.

It can be observed from both tables that the mixed Python-C implementation is of approximately the same speed as the pure C implementation. The speed-up results of the two-dimensional mixed Python-C implementation scale slightly better than the pure C implementation. There are at least two reasons for this somewhat unexpected behavior. First, the Python-C implementation has a better computation-communication ratio, due to the overhead of repeatedly invoking a C function and executing the `while`-loop in Python. We note that this type of overhead is also present

Table 9.3. A comparison of performance between a mixed Python-C implementation and a pure C implementation for simulations of a three-dimensional wave equation. The global mesh has $200 \times 200 \times 200$ cells and there are 692 time steps.

	Mixed Python-C		Pure C	
P	Wall-time	Speedup	Wall-time	Speedup
1	735.89	N/A	746.96	N/A
2	426.77	1.72	441.51	1.69
4	259.84	2.83	261.39	2.86
8	146.96	5.01	144.27	5.18
12	112.01	6.57	109.27	6.84
16	94.20	7.81	89.33	8.36

when $P = 1$. Regarding the overhead associated only with communication, we remark that it consists of three parts: preparing outgoing messages, message exchange, and extracting incoming messages. The larger latency value of `pypar`-MPI than that of C-MPI thus concerns only the second part, therefore it may have an almost invisible impact on the overall performance. In other words, care should be taken to implement the message preparation and extraction work efficiently, and the index slicing functionality of Python arrays clearly seems to have sufficient efficiency for these purposes. Second, the better speed-up results of the mixed Python-C implementation may sometimes be attributed to the cache effects. To understand this, we have to realize that any mixed Python-C implementation always uses (slightly) more memory than its pure C counterpart. Consequently, the mixed Python-C implementation has a better chance of letting bad cache use "spoil" its serial ($P = 1$) performance, and it will thus in some special cases scale better when $P > 1$.

We must remark that the relatively poor speed-up results in Table 9.3, for both pure C and mixed Python-C implementations, are due to the relatively small computation-communication ratio in these very simple three-dimensional simulations, plus the use of blocking MPI send/receive routines [11]. For $P = 1$ in particular, the better speed of the mixed Python-C implementation is due to a fast one-dimensional indexing scheme of the data in the `cloop.update_interior_pts3D` function (which is migrated to C), as opposed to the relatively expansive triple indexing scheme (such as `u[i][j][k]`) in the pure C implementation. The relatively slow communication speed of the Linux cluster also considerably affects the scalability. Non-blocking communication routines can in principle be used for improving such simple parallel simulations. For example, in the pure C implementation, routines such as `MPI_Isend` and `MPI_Irecv` may help to overlap communication with computation and thus hide the overhead of communication. At the moment of this writing, the `pypar` package has not implemented such non-blocking communication routines, whereas `pyMPI` has non-blocking communication routines but with relatively slow performance. On the other hand, measurements in Table 9.3 are meant to provide an accurate idea of the size of the Python-induced *communication overhead*, so we have deliberately sticked to blocking communication routines in our pure C implementations.

9.5.2 Schwarz-Type Parallelization of a Boussinesq Solver

In the preceding case, an explicit finite difference algorithm constitutes the PDE solver. The parallelization is thus carried out at the array level, see Section 9.3.2. To demonstrate a case of parallelizing a PDE solver at a higher level, as discussed in Section 9.3.3, let us consider the parallelization of an implicit PDE solver implemented in a legacy Fortran 77 code. The original serial code consists of loop-based subroutines, which heavily involve low-level details of array indexing. It was developed 15 years ago without any paying attention to parallelization. Thus, this case study shows how the new Python class hierarchy `BoxPartitioner` can help to implement a Schwarz-type parallelization of old legacy codes.

The choice of the Schwarz-type approach for this case is motivated by two factors. First, the serial numerical scheme is not straightforward to parallelize using the array-level approach. Second, which is actually a more important motivation, the Schwarz-type parallelization approach alleviates the need of having to completely understand the low-level details of the internal loops and array indexing in the old-style Fortran code.

The Boussinesq Water Wave Equations

The mathematical model of the present case is the following system of Boussinesq water wave equations:

$$\frac{\partial \eta}{\partial t} + \nabla \cdot \boldsymbol{q} = 0, \qquad (9.11)$$

$$\frac{\partial \phi}{\partial t} + \frac{\alpha}{2}\nabla\phi\cdot\nabla\phi + \eta - \frac{\epsilon}{2}H\nabla\cdot\left(H\nabla\frac{\partial \phi}{\partial t}\right) + \frac{\epsilon}{6}H^2\nabla^2\frac{\partial \phi}{\partial t} = 0, \qquad (9.12)$$

which can be used to model weakly dispersive and weakly nonlinear water surface waves. The primary unknowns of the above PDEs are the water surface elevation $\eta(x, y, t)$ and the depth-averaged velocity potential $\phi(x, y, t)$. Equation (9.11) is often referred to as the continuity equation, where the flux function $\boldsymbol{q}$ is given by

$$\boldsymbol{q} = (H + \alpha\eta)\nabla\phi + \epsilon H\left(\frac{1}{6}\frac{\partial \eta}{\partial t} - \frac{1}{3}\nabla H\cdot\nabla\phi\right)\nabla H, \qquad (9.13)$$

and $H(x, y)$ denotes the water depth. Equation (9.12) is a variant of the Bernoulli (momentum) equation, where the constants α and ϵ control the degree of nonlinearity and dispersion. For more details we refer to [25].

For the present test case, we assume that the two-dimensional solution domain Ω is of a rectangular shape, i.e., $\Omega = [0, L_x] \times [0, L_y]$. On the boundary $\partial\Omega$, no-flux conditions are valid as

$$\boldsymbol{q}\cdot\boldsymbol{n} = 0 \quad \text{and} \quad \nabla\phi\cdot\boldsymbol{n} = 0, \qquad (9.14)$$

where $\boldsymbol{n}$ denotes as usual the outward unit normal on $\partial\Omega$. In addition, the Boussinesq equations are supplemented with initial conditions in the form of prescribed $\eta(x, y, 0)$ and $\phi(x, y, 0)$.

The Numerical Scheme

As a concrete case of applying the Schwarz-type parallelization approach, we consider an old serial legacy Fortran 77 code that uses a semi-implicit numerical scheme based on finite differences. Most of the parallelization effort involves using the F2PY tool to automatically generate wrapper code of the Fortran subroutines and calling methods of `PyParBoxPartitioner2D` for the needed communication operations. It should be noted that other implicit numerical schemes can also be parallelized in the same approach. We refer to [12, 6] for another example of an implicit finite element numerical scheme, which has been parallelized on the basis of additive Schwarz iterations.

The legacy code adopts a time stepping strategy for uniform discrete time levels $\ell\Delta t$, $\ell \geq 1$, and the temporal discretization of (9.11)-(9.12) is as follows:

$$\frac{\eta^\ell - \eta^{\ell-1}}{\Delta t} + \nabla\cdot\left(\left(H + \alpha\frac{\eta^{\ell-1}+\eta^\ell}{2}\right)\nabla\phi^{\ell-1} + \epsilon H\left(\frac{1}{6}\frac{\eta^\ell - \eta^{\ell-1}}{\Delta t} - \frac{1}{3}\nabla H\cdot\nabla\phi^{\ell-1}\right)\nabla H\right) = 0, \tag{9.15}$$

$$\frac{\phi^\ell - \phi^{\ell-1}}{\Delta t} + \frac{\alpha}{2}\nabla\phi^{\ell-1}\cdot\nabla\phi^{\ell-1} + \eta^\ell - \frac{\epsilon}{2}H\nabla\cdot\left(H\nabla\left(\frac{\phi^\ell - \phi^{\ell-1}}{\Delta t}\right)\right) + \frac{\epsilon}{6}H^2\nabla^2\left(\frac{\phi^\ell - \phi^{\ell-1}}{\Delta t}\right) = 0. \tag{9.16}$$

Moreover, by introducing an intermediate solution field FT (note that FT is a single entity, not F times T),

$$FT^\ell \equiv \frac{\phi^\ell - \phi^{\ell-1}}{\Delta t}, \quad \text{such that } FT^\ell_{i,j} \approx \frac{\partial\phi}{\partial t}(i\Delta x, j\Delta y, l\Delta t), \tag{9.17}$$

we can simplify the discretized Bernoulli equation (9.16) as

$$FT^\ell - \frac{\epsilon}{2}H\nabla\cdot(H\nabla FT^\ell) + \frac{\epsilon}{6}H^2\nabla^2 FT^\ell = -\frac{\alpha}{2}\nabla\phi^{\ell-1}\cdot\nabla\phi^{\ell-1} - \eta^\ell. \tag{9.18}$$

In other words, the computational work of the numerical scheme at time level ℓ consists of the following sub-steps:

1. Solve (9.15) with respect to η^ℓ using $\eta^{\ell-1}$ and $\phi^{\ell-1}$ as known quantities.
2. Solve (9.18) with respect to FT^ℓ using η^ℓ and $\phi^{\ell-1}$ as known quantities.
3. Update the ϕ solution by

$$\phi^\ell = \phi^{\ell-1} + \Delta t\, FT^\ell. \tag{9.19}$$

We note that the notation FT arises from the corresponding variable name used in the legacy Fortran code.

By a standard finite difference five-point stencil, the spatial discretization of (9.15)-(9.18) will give rise to two systems of linear equations:

$$A_\eta(\phi^{\ell-1})\eta^\ell = b_\eta(\eta^{\ell-1}, \phi^{\ell-1}), \qquad (9.20)$$

$$A_{FT}FT^\ell = b_{FT}(\eta^\ell, \phi^{\ell-1}). \qquad (9.21)$$

The η^ℓ vector contains the approximate solution of η at $(i\Delta x, j\Delta y, l\Delta t)$, while the FT^ℓ vector contains the approximate solution of $\partial\phi/\partial t$ at $t = \ell\Delta t$. The entries of matrix A_η depend on the latest ϕ approximation, whereas matrix A_{FT} remains unchanged throughout the entire time-stepping process. The right-hand side vectors b_η and b_{FT} depend on the latest η and ϕ approximations.

Parallelization

The existing serial Fortran 77 code uses the line-version of SSOR iterations to solve the linear systems (9.20) and (9.21). That is, all the unknowns on one mesh line are updated simultaneously, which requires solving a tridiagonal linear system per mesh line (in both x- and y-direction). Such a serial numerical scheme is hard to parallelize using the array-level approach (see Section 9.3.2). Therefore, we adopt the Schwarz-type parallelization approach from Section 9.3.3, which in this case means that the subdomain solver needed in (9.6) invokes one or a few subdomain line-version SSOR iterations embedded in the legacy Fortran 77 code.

A small extension of the Fortran 77 code is necessary because of the involvement of the artificial boundary condition (9.6) on each subdomain. However, this extension is not directly programmed into the old Fortran subroutines. Instead, two new "wrapper" subroutines, `iteration_continuity` and `iteration_bernoulli`, are written to handle the discretized continuity equation (9.15) and the discretized Bernoulli equation (9.18), respectively. These two wrapper subroutines use F2PY and are very light-weight in the sense that they mainly invoke the existing old Fortran subroutines as "black boxes". The only additional work in the wrapper subroutines is on testing whether some of the four side neighbors are absent. For each absent side neighbor, the original physical boundary conditions (9.14) must be enforced. Otherwise, the communication method `update_overlap_regions` takes care of enforcing the artificial Dirichlet boundary condition required in (9.6).

The resulting parallel Python program runs additive Schwarz iterations at a high level for solving both (9.20) and (9.21) at each time level. In a sense, the Schwarz framework wraps the entire legacy Fortran 77 code with a user-friendly Python interface. The main content of the Python program is as follows:

```
from BoxPartitioner import *
# read in 'gnum_cells','parts','overlaps' ...
partitioner=PyParBoxPartitioner2D(my_id=my_id,
                              num_procs=num_procs,
                              global_num_cells=gnum_cells,
                              num_parts=parts,
                              num_overlaps=overlaps)
partitioner.prepare_communication ()
loc_nx,loc_ny = partitioner.get_num_loc_cells ()
# create subdomain data arrays ...
# enforce initial conditions ...
```

```
lower_x_neigh = partitioner.lower_neighbors[0]
upper_x_neigh = partitioner.upper_neighbors[0]
lower_y_neigh = partitioner.lower_neighbors[1]
upper_y_neigh = partitioner.upper_neighbors[1]

import BQ_solver_wrapper as f77 # interface to legacy code

t = 0.0
while t <= tstop:
    t += dt

    # solve the continuity equation:
    dd_iter = 0
    not_converged = True
    nbit = 0

    while not_converged and dd_iter < max_dd_iters:
        dd_iter++
        Y_prev = Y.copy()  # remember old eta values
        Y, nbit = f77.iteration_continuity (F, Y, YW, H,
                                QY, WRK, dx, dy, dt, kit,
                                ik, gg, alpha, eps, nbit,
                                lower_x_neigh, upper_x_neigh,
                                lower_y_neigh, upper_y_neigh)
        # communication
        partitioner.update_overlap_regions (Y)
        not_converged = check_convergence (Y, Y_prev)

    # solve the Bernoulli equation:
    dd_iter = 0
    not_converged = True
    nbit = 0

    while not_converged and dd_iter < max_dd_iters:
        dd_iter++
        FT_prev = FT.copy()  # remember old FT values
        FT, nbit = f77.iteration_bernoulli (F, FT, Y, H,
                                QR,R,WRK, dx, dy, dt, ik,
                                alpha, eps, nbit,
                                lower_x_neigh, upper_x_neigh,
                                lower_y_neigh, upper_y_neigh)
        # communication
        partitioner.update_overlap_regions (FT)
        not_converged = check_convergence (FT, FT_prev)

    F += dt*FT  # update the phi field
```

The interface to the Fortran wrapper code is imported by the statement

```
import BQ_solver_wrapper as f77
```

The code above shows that a large number of NumPy arrays and other parameters are sent as input arguments to the two wrapper subroutines `iteration_continuity` and `iteration_bernoulli`. The arrays are actually required by the old Fortran subroutines, which are used as "black boxes". The large number of input arguments shows a clear disadvantage of old-fashion Fortran programming. In particular, the

Table 9.4. The wall-time measurements of the parallel Python Boussinesq solver. The global uniform mesh is 1000×1000 and the number of time steps is 40. The number of overlapping cell layers between neighboring subdomains is 8.

P	Partitioning	Wall-time	Speed-up
1	N/A	1810.69	N/A
2	1×2	847.53	2.14
4	2×2	483.11	3.75
6	2×3	332.91	5.44
8	2×4	269.85	6.71
12	3×4	187.61	9.65
16	2×8	118.53	15.28

arrays `Y`, `FT`, and `F` contain the approximate solutions of η, FT, and ϕ, respectively. The other arrays contain additional working data, and will not be explained here. The IDs of the four side neighbors, `lower_x_neigh`, `upper_x_neigh`, `lower_y_neigh`, and `upper_y_neigh`, are also sent in as the input arguments. These neighbor IDs are needed in the the wrapper subroutines for deciding whether to enforce the original physical boundary conditions on the four sides. We also note that the input/output argument `nbit` (needed by the existing old Fortran subroutines) is an integer accumulating the total number of line-version SSOR iterations used so far in each loop of the additive Schwarz iterations.

To test the convergence of the additive Schwarz iterations, we have coded a simple Python method `check_convergence`. This method compares the relative difference between $\boldsymbol{x}_{s,k}$ and $\boldsymbol{x}_{s,k-1}$. More precisely, convergence is considered achieved by the kth additive Schwarz iteration if

$$\max_s \frac{\|\boldsymbol{x}_{s,k} - \boldsymbol{x}_{s,k-1}\|}{\|\boldsymbol{x}_{s,k}\|} \leq \varepsilon. \tag{9.22}$$

Note that a collective communication is needed in `check_convergence`. In (9.22) ε denotes a prescribed convergence threshold value.

Measurements

We have chosen a test problem with $\alpha = \epsilon = 1$ and

$$L_x = L_y = 10, \;\; H = 1, \;\; \phi(x, y, 0) = 0, \;\; \eta(x, y, 0) = 0.008 \cos(3\pi x) \cos(4\pi y).$$

The global uniform mesh is chosen as 1000×1000, and the time step size is $\Delta t = 0.05$ with the end time equal to 2. The number of overlapping cell layers between the neighboring subdomains has been chosen as 8. One line-version SSOR iteration is used as the subdomain solver associated with solving (9.15), and three line-version SSOR iterations are used as the subdomain solver associated with solving (9.18). The convergence threshold value for testing the additive Schwarz iterations is chosen as $\varepsilon = 10^{-3}$ in (9.22).

We remark that the legacy Fortran code is very difficult to parallelize in a pure Fortran manner, i.e., using the array-level approach. Therefore, the performance of the Schwarz-type parallel Python-Fortran solver is not compared with a "reference" parallel Fortran solver, but only studied by examining its speed-up results in Table 9.4. It can be seen that Table 9.4 has considerably better speed-up results than Table 9.2. This is due to two reasons. First, the computation-communication ratio is larger when solving the Boussinesq equations, because one line-version SSOR iteration is more computation-intensive than one iteration of a seven-point stencil, which is used for the three-dimensional linear wave equation. Second, allocating a large number of data arrays makes the Boussinesq solver less cache-friendly than the linear wave solver for $P = 1$. This explains why there is a superlinear speed-up from $P = 1$ to $P = 2$ in Table 9.4. These two reasons thus have a combined effect for more favorable speed-up results, in spite of the considerable size of the overlap zone (8 layers).

9.6 Summary

It is well known that Python handles I/O, GUI, result archiving, visualization, report generation, and similar tasks more conveniently than the low-level languages like Fortran and C (and even C++). We have seen in this chapter the possibility of using high-level Python code for invoking inter-processor communications. We have also investigated the feasibility of parallelizing serial PDE solvers with the aid of Python, which is quite straightforward due to easy re-use of existing serial computational modules. Moreover, the flexible and structured style of Python programming helps to code the PDE-independent parallelization tasks as generic methods. As an example, an old serial legacy Fortran 77 code has been parallelized using Python with minor efforts. The performance of the resulting parallel Python solver depends on two things: (1) good serial performance which can be ensured by the use of NumPy arrays, vectorization, and mixed-language implementation, (2) high-performance message passing and low cost of constructing and extracting data messages, for which it is important to combine the `pypar` package with the functionality of array slicing and reshaping. Doing this right, our numerical results show that comparable parallel performances with respect to pure Fortran/C implementations can be obtained.

Acknowledgement

The authors would like to thank Prof. Geir Pedersen at the Department of Mathematics, University of Oslo, for providing access to the Fortran Boussinesq legacy code and guidance on the usage.

References

1. D. Ascher, P. F. Dubois, K. Hinsen, J. Hugunin, and T. Oliphant. Numerical Python. Technical report, Lawrence Livermore National Lab., CA, 2001. `http://www.pfdubois.com/numpy/numpy.pdf`.
2. D. L. Brown, W. D. Henshaw, and D. J. Quinlan. Overture: An object-oriented framework for solving partial differential equations. In Y. Ishikawa, R. R. Oldehoeft, J. V. W. Reynders, and M. Tholburn, editors, *Scientific Computing in Object-Oriented Parallel Environments*, Lecture Notes in Computer Science, vol 1343, pages 177–184. Springer, 1997.
3. A. M. Bruaset, X. Cai, H. P. Langtangen, and A. Tveito. Numerical solution of PDEs on parallel computers utilizing sequential simulators. In Y. Ishikawa, R. R. Oldehoeft, J. V. W. Reynders, and M. Tholburn, editors, *Scientific Computing in Object-Oriented Parallel Environments*, Lecture Notes in Computer Science, vol 1343, pages 161–168. Springer, 1997.
4. X. Cai and H. P. Langtangen. Developing parallel object-oriented simulation codes in Diffpack. In H. A. Mang, F. G. Rammerstorfer, and J. Eberhardsteiner, editors, *Proceedings of the Fifth World Congress on Computational Mechanics*, 2002.
5. X. Cai, H. P. Langtangen, and H. Moe. On the performance of the Python programming language for serial and parallel scientific computations. *Scientific Programming*, 13(1):31–56, 2005.
6. X. Cai, G. K. Pedersen, and H. P. Langtangen. A parallel multi-subdomain strategy for solving Boussinesq water wave equations. *Advances in Water Resources*, 28(3):215–233, 2005.
7. T. F. Chan and T. P. Mathew. Domain decomposition algorithms. In *Acta Numerica 1994*, pages 61–143. Cambridge University Press, 1994.
8. D. B. et al. Swig 1.3 Development Documentation, 2004. `http://www.swig.org/doc.html`.
9. F2PY software package. `http://cens.ioc.ee/projects/f2py2e`.
10. L. Formaggia, M. Sala, and F. Saleri. Domain decomposition techniques. In A. M. Bruaset and A. Tveito, editors, *Numerical Solution of Partial Differential Equations on Parallel Computers*, volume 51 of *Lecture Notes in Computational Science and Engineering*, pages 135–163. Springer-Verlag, 2005.
11. M. P. I. Forum. MPI: A message-passing interface standard. *Internat. J. Supercomputer Appl.*, 8:159–416, 1994.
12. S. Glimsdal, G. K. Pedersen, and H. P. Langtangen. An investigation of domain decomposition methods for one-dimensional dispersive long wave equations. *Advances in Water Resources*, 27(11):1111–1133, 2005.
13. C. Hughes and T. Hughes. *Parallel and Distributed Programming Using C++*. Addison Wesley, 2003.
14. H. P. Langtangen. *Python Scripting for Computational Science*. Texts in Computational Science and Engineering, vol 3. Springer, 2004.
15. H. P. Langtangen and X. Cai. A software framework for easy parallelization of PDE solvers. In C. B. Jensen, T. Kvamsdal, H. I. Andersson, B. Pettersen, A. Ecer, J. Periaux, N. Satofuka, and P. Fox, editors, *Parallel Computational Fluid Dynamics*. Elsevier Science, 2001.
16. Matlab code vectorization guide. `http://www.mathworks.com/support/tech-notes/1100/1109.html`.
17. Numerical Python software package. `http://sourceforge.net/projects/numpy`.

18. P. S. Pacheco. *Parallel Programming with MPI.* Morgan Kaufmann Publishers, 1997.
19. Parallel software in C and C++.
`http://www.mathtools.net/C_C__/Parallel/`.
20. R. Parsones and D. Quinlan. A++/P++ array classes for architecture independent finite difference computations. Technical report, Los Alamos National Lab., NM, 1994.
21. PyMPI software package. `http://sourceforge.net/projects/pympi`, 2004.
22. PyPar software package. `http://datamining.anu.edu.au/~ole/pypar`, 2004.
23. ScientificPython software package.
`http://starship.python.net/crew/hinsen`.
24. B. F. Smith, P. E. Bjørstad, and W. Gropp. *Domain Decomposition: Parallel Multilevel Methods for Elliptic Partial Differential Equations.* Cambridge University Press, 1996.
25. D. M. Wu and T. Y. Wu. Three-dimensional nonlinear long waves due to moving surface pressure. *Proc. 14th Symp. Naval Hydrodyn.*, pages 103–129, 1982.

10

Parallel PDE-Based Simulations Using the Common Component Architecture

Lois Curfman McInnes[1], Benjamin A. Allan[2], Robert Armstrong[2], Steven J. Benson[1], David E. Bernholdt[3], Tamara L. Dahlgren[4], Lori Freitag Diachin[4], Manojkumar Krishnan[5], James A. Kohl[3], J. Walter Larson[1], Sophia Lefantzi[6], Jarek Nieplocha[5], Boyana Norris[1], Steven G. Parker[7], Jaideep Ray[8], and Shujia Zhou[9]

[1] Mathematics and Computer Science Division, Argonne National Laboratory, Argonne, IL, USA
[mcinnes,benson,larson,norris]@mcs.anl.gov
[2] Scalable Computing R & D, Sandia National Laboratories (SNL), Livermore, CA, USA
[baallan,rob]@ca.sandia.gov
[3] Computer Science and Mathematics Division, Oak Ridge National Laboratory, Oak Ridge, TN, USA
[bernholdtde,kohlja]@ornl.gov
[4] Center for Applied Scientific Computing, Lawrence Livermore National Laboratory, Livermore, CA, USA
[dahlgren1,diachin2]@llnl.gov
[5] Computational Sciences and Mathematics, Pacific Northwest National Laboratory, Richland, WA, USA
[manojkumar.krishnan,jarek.nieplocha]@pnl.gov
[6] Reacting Flow Research, SNL, Livermore, CA, USA
slefant@ca.sandia.gov
[7] Scientific Computing and Imaging Institute, University of Utah, Salt Lake City, UT, USA
sparker@cs.utah.edu
[8] Advanced Software R & D, SNL, Livermore, CA, USA
jairay@ca.sandia.gov
[9] Northrop Grumman Corporation, Information Technology Sector, Chantilly, VA, USA
szhou@pop900.gsfc.nasa.gov

Summary. The complexity of parallel PDE-based simulations continues to increase as multimodel, multiphysics, and multi-institutional projects become widespread. A goal of component-based software engineering in such large-scale simulations is to help manage this complexity by enabling better interoperability among various codes that have been independently developed by different groups. The Common Component Architecture (CCA) Forum is defining a component architecture specification to address the challenges of high-performance scientific computing. In addition, several execution frameworks, supporting infrastructure, and general-purpose components are being developed. Furthermore, this group is collaborating with others in the high-performance computing community to design suites of domain-specific component interface specifications and underlying implementations.

This chapter discusses recent work on leveraging these CCA efforts in parallel PDE-based simulations involving accelerator design, climate modeling, combustion, and accidental fires and explosions. We explain how component technology helps to address the different challenges posed by each of these applications, and we highlight how component interfaces built on existing parallel toolkits facilitate the reuse of software for parallel mesh manipulation, discretization, linear algebra, integration, optimization, and parallel data redistribution. We also present performance data to demonstrate the suitability of this approach, and we discuss strategies for applying component technologies to both new and existing applications.

10.1 Introduction

The complexity of parallel simulations based on partial differential equations (PDEs) continues to increase as multimodel, multiphysics, multidisciplinary, and multi-institutional projects are becoming widespread. Coupling models and different types of science increases the complexity of the simulation codes. Collaboration across disciplines and institutions, while increasingly necessary, introduces new social intricacies into the software development process, such as different programming styles and different ways of thinking about problems. Added to these challenges, the software must cope with the multilevel memory hierarchies common to modern parallel computers where there may be three to five levels of data locality.

These challenges make it clear that the high-performance scientific computing community needs an approach to software development for parallel PDEs that facilitates managing such complexity while maintaining scalable and efficient parallel performance. Rather than being overwhelmed by the tedious details of parallel computing, computational scientists must be able to focus on the particular part of a simulation that is of primary interest to them (e.g., the physics of combustion) and employ well-tested and optimized code developed by experts in other facets of a simulation (e.g., parallel linear algebra and visualization). Traditional approaches, such as the widespread use of software libraries, have historically been valuable, but these approaches are being severely strained by this new complexity.

One goal of component-based software engineering (CBSE) is to enable interoperability among software modules that have been developed independently by different groups. CBSE treats applications as assemblies of software *components* that interact with each other only through well-defined *interfaces* within a particular execution environment, or *framework*. Components are a logical means of encapsulating knowledge from one scientific domain for use by those in others, thereby facilitating multidisciplinary interactions. The complexity of a given simulation is decomposed into bite-sized components that one or a few investigators can develop independently, thus enabling the collaboration of scores of researchers in the development of a single simulation. The glue that binds the components together is a set of common, agreed-upon interfaces. Multiple component implementations conforming to the same external interface standard should be interoperable, while providing flexibility to accommodate different aspects such as algorithms, performance characteristics, and coding styles. At the same time, the use of common interfaces facilitates the reuse

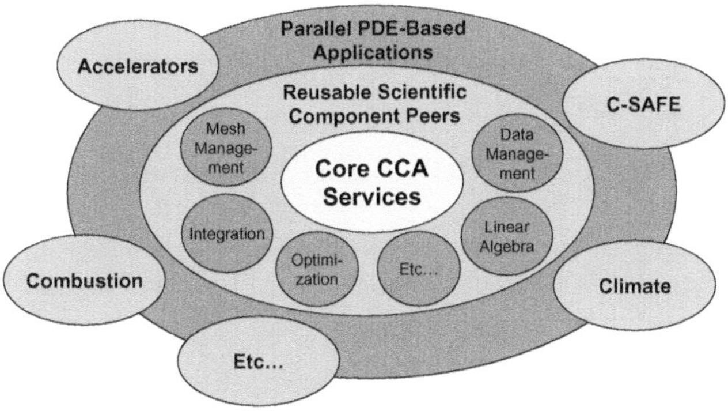

Fig. 10.1. Complete parallel PDE-based applications can be built by combining reusable scientific components with application-specific components; both can employ core CCA services to manage inter-component interactions.

of components across multiple applications. Even though details differ widely, many PDE-based simulations share the same overall software structure. Such applications could employ similar sets of components, which might conform to many of the same interfaces but differ in implementation details. This kind of software reuse enables the cross-pollination of both components and concepts across applications, projects, and problem domains.

The Common Component Architecture (CCA) [28, 19, 8] is designed specifically for the needs of parallel, scientific high-performance computing (HPC) in response to limitations in the general HPC domain of other, more widely used component approaches (see Section 10.3). The general-purpose design of the CCA is intended for use in a wide range of scientific domains for both PDE-based and non-PDE-based simulations.

As depicted in Figure 10.1, complete parallel PDE-based applications can be built in a CCA environment by combining various reusable scientific components with application-specific components. In keeping with the emphasis of this book, we explain (1) how component software can help manage the complexity of PDE-based simulations and (2) how the CCA, in particular, facilitates parallel scientific computations. We do this in the context of four motivating PDE-based application areas, which are introduced in Section 10.2. After presenting the basic concepts of the CCA in Section 10.3, we provide an overview of some reusable scientific components and explain how component interfaces built on existing parallel toolkits facilitate the reuse of software for parallel mesh manipulation, discretization, linear algebra, integration, optimization, and parallel data redistribution. Section 10.5 discusses strategies for applying component technologies to both new and existing applications, with an emphasis on approaches for the decomposition of PDE-based problems, including considerations for how to move from particular implementations to more general abstractions. Section 10.6 integrates these ideas through case studies that illustrate the application of component technologies and reusable components in the four motivating applications. Section 10.7 discusses conclusions and areas of future work.

10.2 Motivating Parallel PDE-Based Simulations

This section introduces four PDE-based application areas that motivate our work: accelerator design, climate modeling, combustion, and accidental fires and explosions.

10.2.1 Accelerator Modeling

Accelerators produce high-energy, high-speed beams of charged subatomic particles for research in high-energy and nuclear physics, synchrotron radiation research, medical therapies, and industrial applications. The design of next-generation accelerator facilities, such as the Positron-Electron Project (PEP)-II and Rare Isotope Accelerator (RIA), relies heavily on a suite of software tools that can be used to simulate many different accelerator experiments. Two of the codes used by accelerator scientists at the Stanford Linear Accelerator Center (SLAC) are Omega3P [117] and Tau3P [134]. Omega3P is an extensible, parallel, finite element-based code for the eigenmode modeling of large, complex three-dimensional electromagnetic structures in the frequency domain, while Tau3P provides solutions to electromagnetics problems in the time domain. Both codes make extensive use of unstructured mesh infrastructures to accommodate the complex geometries associated with accelerator models. In order to overcome barriers to computation and to improve functionality, both codes are being evaluated for possible extension.

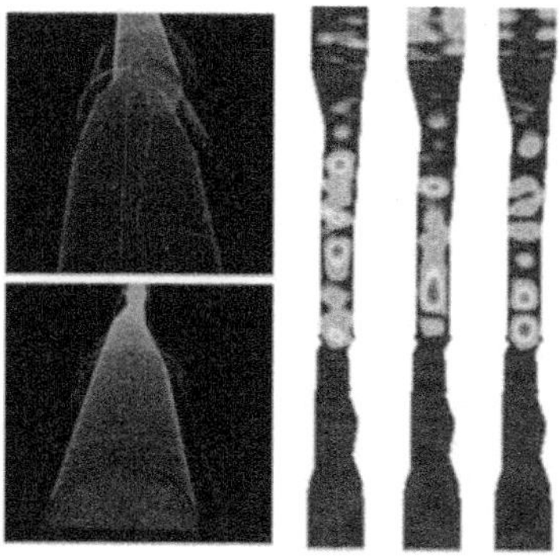

Fig. 10.2. State-of-the-art simulation tools are used to help design the next generation of accelerator facilities. *(Left)*: Mesh generated for the PEP-II interaction region using the CUBIT mesh generation package. Image courtesy of Tim Tautges of Sandia National Laboratories. *(Right)*: Excited fields computed using Tau3P. Image courtesy of the numerics team at SLAC. (For the color version, see Figure A.19 on page 476).

For Tau3P, different discretization strategies are being explored to address long-time instabilities on certain types of meshes. Tau3P is based on a modified Yee algorithm formulated on an unstructured grid and uses a discrete surface integral (DSI) method to solve Maxwell's equations. Since the DSI scheme is known to have potential instabilities on nonorthogonal meshes, scientists are using a time filtering technique that maintains stability in most cases, but at a significantly higher computational cost. Unfortunately, integrating new discretization techniques is costly;

and, because of resource constraints, several potentially useful methods cannot be investigated. A component-based approach that allows scientists to easily prototype different discretization and meshing strategies in a plug-and-play fashion would be useful in overcoming this obstacle.

For Omega3P, solutions are being explored that yield more accurate results without increasing the computational cost. That is, scientists are satisfied with the finite-element-based solver but cannot increase mesh resolution to reduce the large errors that occur in small regions of the computational domain. To overcome this barrier, SLAC scientists are working with researchers at Rensellaer Polytechnic Institute (RPI) to develop an adaptive mesh refinement (AMR) capability. Despite initially using a file-based information transfer mechanism, this effort has clearly demonstrated the advantage of AMR techniques to compute significantly more accurate solutions at a reduced computational cost. As described in Section 10.6.1, current efforts are centered on directly deploying these advanced capabilities in the Omega3P code by using a component approach. This approach has made the endeavor more tractable and has given scientists the flexibility of later experimenting with different underlying AMR infrastructures at little additional cost.

In order to facilitate the use of different discretization and meshing strategies, there is a need for a set of common interfaces that provide access to mesh and geometry information. A community effort to specify such interfaces is described in Section 10.4.1, and results of a performance study using a subset of those interfaces are discussed in Section 10.4.8.

10.2.2 Climate Modeling

Climate is the overall product of the mutual interaction of the Earth's atmosphere, oceans, biosphere, and cryosphere. These systems interact by exchanging energy, momentum, moisture, chemical fluxes, etc. The inherent nonlinearity of each subsystem's equations of evolution makes direct modeling of the *climate*—which is the set of statistical moments sampled over a large time scale—almost impossible. Instead, climate modeling is accomplished through integrations of coupled climate system models for extended periods, ranging from the century to millennial time scales, logging of model history output sampled at short time scales, and subsequent off-line analysis to compute climate statistics.

PDEs arise in many places in the climate system, most significantly in the dynamics of the atmosphere, ocean, and sea-ice. The ocean and atmosphere are both modeled as thin spherical shells of fluid in a rotating reference frame, using in each case a system of coupled PDEs governing mass, energy, and momentum conservation, called *the primitive equations*. Modern sea-ice models simulate the formation and melting of ice (the *thermodynamics* of the problem), how the ice pack is forced by surface winds and ocean currents, and how it behaves as a material (its *dynamics* and *rheology*). Schemes such as the elastic-viscous plastic (EVP) scheme [60] involve the solution of PDEs.

Climate modeling is a grand challenge high-performance computing application, requiring highly efficient and scalable algorithms capable of providing the high throughput needed for long-term integrations. To illustrate the high simulation costs, we consider the NASA finite-volume General Circulation Model. A 500-model-day simulation using this model, with a horizontal resolution of 0.5° latitude by 0.625° longitude and 32 vertical layers, takes a wall-clock day to run on a 1.25-GHz Compaq AlphaServer SC45 with 250 CPUs [83].

The requirements for coping with multiple, coupled physical processes as well the requirements for parallel computing make software development even more challenging. The traditional development process for these models was the creation of highly entangled applications that made little reuse of code and made the interchange of functional units difficult. In recent years, the climate/weather/ocean (CWO) community has embarked on an effort to increase modularity and interoperability, the main motivation being a desire to accelerate the development, testing, and validation cycle. This effort is positioning the community for the introduction of software component technology, and there is now an emerging community wide application framework, the Earth System Modeling Framework [67].

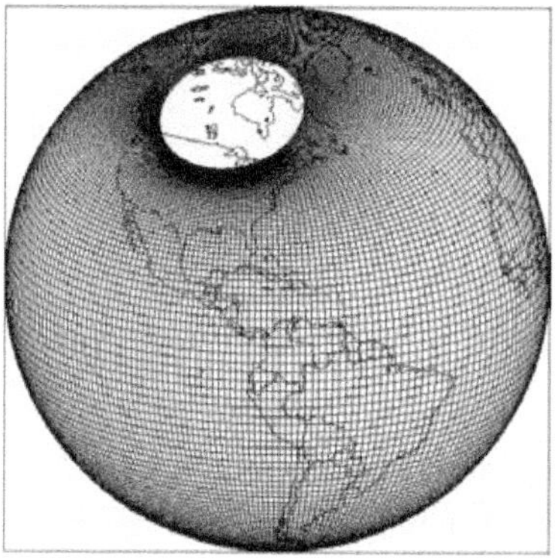

Fig. 10.3. Displaced pole grid on which the Parallel Ocean Program ocean model [101] solves its primitive equations. The polar region is displaced to lie over land, thereby minimizing the problems encountered at high latitudes by finite-difference schemes. Image courtesy of Phillip Jones and Richard Smith, Los Alamos National Laboratory.

In Section 10.4.7 we discuss the use of CCA components for climate model coupling. In Section 10.6.2 we describe the multiple software scales at which component technology is appropriate in climate system models. We briefly describe the ESMF and its relationship to the CCA, and we provide an example of CWO code refactoring to make it component friendly. We also describe a prototype component-based advection model that combines the interoperable component paradigms of the CCA and ESMF.

10.2.3 Combustion

The study of flames, experimentally and computationally, requires the resolution of a wide range of length and time scales arising from the interaction of chemistry, radiation, and transport (diffusive and convective). The complexity and expense involved in the experimental study of flames were recognized two decades ago, and the Combustion Research Facility [37] was created as a "user facility" whose equipment and expertise would be freely available to industry and academia. Today a similar challenge is being faced in the high-fidelity numerical simulations of flames [102]. Existing simulations employ a variety of numerical and parallel computing strategies to achieve an accurate resolution of physics and scales, with the unfortunate side effect of producing large, complex and ultimately unwieldy codes. Their lack of extensibility and difficulty of maintenance have been a serious impediment and were the prime motive for establishing in 2001 the Computational Facility for Reacting Flow Science (CFRFS) [90], a "simulation facility" where various numerical algorithms, physical and chemical models, meshing strategies, and domain partitioners may be tested in flame simulations.

In the CFRFS project, flames are solved by using the low Mach number form of the Navier-Stokes equation [130, 91], augmented by evolution equations for the various chemical species and an energy equation with a source term to incorporate the contribution from chemical reactions. The objective of the project is to simulate laboratory-sized flames with detailed chemistry, a problem that exhibits a wide spectrum of length and time scales. Block-structured adaptive meshes [17] are used to limit fine meshes only where (and when) required; operator-splitting [114, 68] is used to treat stiff chemical terms implicitly in time, while the convective and diffusive terms are advanced explicitly. In many cases, the stiff chemical system can be rendered nonstiff (without any appreciable loss of fidelity) by projection onto a lower-dimensional manifold. The identification of this manifold and the projection onto it are achieved by computational singular perturbation (CSP) [70, 75], a multiscale asymptotic method that holds the promise of significantly reducing the cost of solving the chemical system.

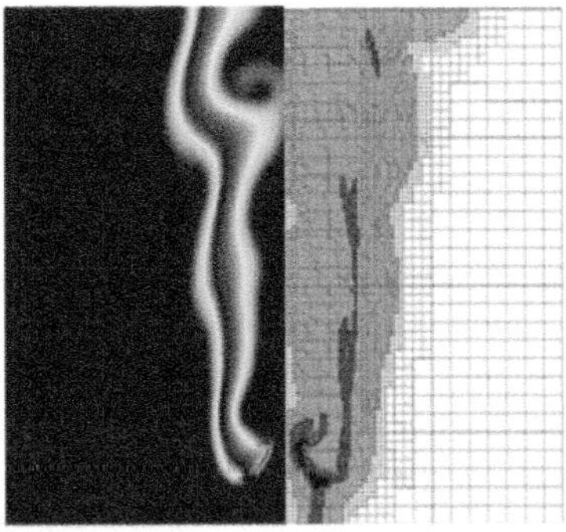

Fig. 10.4. A 10-cm-high pulsating methane-air jet flame, computed on an adaptive mesh. On the left is the temperature field with a black contour showing regions of high heat release rates. On the right is the adaptive mesh, in which regions corresponding to the jet shear layer are refined the most. (For the color version, see Figure A.20 on page 476).

Given the scope of the simulation facility, the requisite degree of flexibility and extensibility clearly could not be achieved without a large degree of modularization and without liberating the users (with widely varying levels of computational expertise) from the strait jacket imposed by global data-structures and models. Modularization was achieved by adopting a component-based architecture, and the multidimensional Fortran array was adopted as the basic unit of data exchange among scientific components. The simulation facility can thus be viewed as a toolkit of components, each embodying a certain numerical or physical functionality, mostly implemented in Fortran 77, with thin C++ "wrappers" around them.

In Section 10.4.2 we discuss SAMR components used in this application, and in Section 10.5 we detail the strategy we adopted to decompose mathematical and simulation requirements into modules, while preserving a close correspondence between the software components and identifiable physics in the governing equations. In Section 10.6.3 we demonstrate the payoffs of adopting such a *physics-based* approach.

10.2.4 Accidental Fires and Explosions

In 1997 the University of Utah created an alliance with the U.S. Department of Energy (DOE) Accelerated Strategic Computing Initiative (ASCI) to form the Center for the Simulation of Accidental Fires and Explosions (C-SAFE) [55]. C-SAFE focuses on providing state-of-the-art, science-based tools for the numerical simulation of accidental fires and explosions, especially within the context of handling and storing highly flammable materials. The primary objective of C-SAFE is to provide a software system in which fundamental chemistry and engineering physics are fully coupled with nonlinear solvers, optimization, computational steering, visualization, and experimental data verification, thereby integrating expertise from a wide variety of disciplines. Simulations using this system will help to better evaluate the risks and safety issues associated with fires and explosions in accidents involving both hydrocarbon and energetic materials. A typical C-SAFE problem is shown in Figure 10.5. Section 10.6.4 discusses the use of component concepts in this application and demonstrates scalable performance on a variety of parallel architectures.

Fig. 10.5. A typical C-SAFE problem involving hydrocarbon fires and explosions of energetic materials. This simulation involves fluid dynamics, structural mechanics, and chemical reactions in both the flame and the explosive. Accurate simulations of these events can lead to a better understanding of high-energy explosives, can help evaluate the design of shipping and storage containers for these materials, and can help officials determine a response to various accident scenarios. The fire image is courtesy of Schonbucher Institut for Technische Chemie I der Universitat Stuttgart, and the images of the container and explosion are courtesy of Eric Eddings of the University of Utah. (For the color version, see Figure A.21 on page 477).

10.3 High-Performance Components

High-performance components offer a means to deal with the ever-increasing complexity of scientific software, including the four applications introduced in Section 10.2. We first introduce general component concepts, discuss the Common Component Architecture (CCA), and then introduce two simple PDE-based examples to help illustrate CCA principles and components.

10.3.1 Component-Based Software Engineering

In addition to the advantages of component-based software engineering (CBSE; see, e.g., [118]) discussed in Section 10.1, component-based approaches offer additional benefits, including the following:

- *Plug-and-play assembly* improves productivity, especially when a significant number of components can be used without customization, and simplifies the evolution of applications to meet new requirements or address new problems.
- *Clear interfaces and boundaries* around components simplify the composition of multiple componentized libraries in ways that may be difficult or impossible with software libraries in their traditional forms. This approach also helps researchers to focus on the particular aspects of the problem corresponding to their interests and expertise.
- *Components enable adaptation* of applications in ways that traditional design cannot. For example, interface standards facilitate swapping of components to modify behavior or performance; such changes can even be made automatically without user intervention [94].

As implied above and in Section 10.1, CBSE can be thought of, in many respects, as an extension and refinement of the use of software libraries—a popular and effective approach in modern scientific computing. Components are also related to "domain-specific computational frameworks" or "application frameworks," which have become popular in recent years (e.g., Cactus [6], ESMF [67], and PRISM [53]). Typically, such environments provide deep computational support for applications in a given domain, and applications are constructed at a relatively high level. Many application frameworks even have a componentlike structure at the high level, allowing arbitrary code to be plugged in to the framework. Application frameworks are more constrained than general component environments because the ability to reuse components across scientific domains is quite limited, and the framework tends to embody assumptions about the workflow of the problem domain. General component models do not impose such constraints or assumptions and provide broader opportunities for reuse. Domain-specific frameworks can be constructed within general component environments by casting the domain-specific infrastructure and workflow as components.

Several component models have attained widespread use in mainstream computing, especially Enterprise JavaBeans [44, 109], Microsoft's COM/DCOM [26], and the Object Management Group's CORBA and the CORBA Component Model [95]. Despite its advantages, however, CBSE has found only limited adoption in the scientific computing community to date [63, 99, 84]. Unfortunately, the commodity component models tend to emphasize distributed computing while more or less ignoring parallel computing, impose significant performance overheads, or require significant changes to existing code to enable it to operate within the component environment. Additional concerns with many component models include support for programming languages important to scientific computing, such as Fortran; support for data types, such as complex numbers and arrays; and operating system support. The Common

Component Architecture has been developed in direct response to the need for a component environment targeted to the needs of high-performance scientific computing.

10.3.2 The Common Component Architecture

The Common Component Architecture [28] is the core of an extensive research and development program focused on understanding how best to utilize and implement component-based software engineering practices in the high-performance scientific computing area, and on developing the specifications and tools that will lead to a broad spectrum of CCA-based scientific applications. A comprehensive description of the CCA, including more detailed presentations of many aspects of the environment is available [19]; here we present a brief overview of the CCA environment, focusing on the aspects most relevant to parallel PDE-based simulations.

The specification of the Common Component Architecture [29] defines the rights, responsibilities, and relationships among the various elements of the model. Briefly, the elements of the CCA model are as follows:

- *Components* are units of software functionality that can be composed together to form applications. Components encapsulate much of the complexity of the software inside a black box and expose only well-defined interfaces.
- *Ports* are the abstract interfaces through which components interact. Specifically, CCA ports provide procedural interfaces that can be thought of as a class or an interface in object-oriented languages, or a collection of subroutines, or a module in a language such as Fortran 90. Components may provide ports, meaning that they implement the functionality expressed in a port (called *provides ports*), or they may use ports, meaning that they make calls on a port provided by another component (called *uses ports*). The notion of CCA ports is less restrictive than hardware ports: ports are not assumed to be persistent, e.g., available throughout an application's lifetime, and each port can have different access attributes, such as the number of simultaneous connections.
- *Frameworks* manage CCA components as they are assembled into applications and executed. The framework is responsible for connecting *uses* and *provides* ports without exposing the components' implementation details. The framework also provides a small set of standard services that are available to all components.

Several frameworks that implement the CCA specification and support various computing environments have been developed. Ccaffeine [3] and SCIRun2 [135], used by the applications in this chapter, focus on high-performance parallel computing, while XCAT [61, 52] primarily supports distributed computing applications; several other frameworks are being used as research tools.

The importance of efficient and scalable performance in scientific computing is reflected in both the design of the CCA specification and the features of the various framework implementations. The CCA's *uses/provides* design pattern allows components in the same process address space to be invoked directly, without intervention by the framework, and with data passed by reference if desired (also referred to as "direct connect," "in-process," or "co-located" components). In most CCA frameworks, this approach makes local method calls between components equivalent to C++ virtual function calls, an overhead of roughly 50 ns on a 500 MHz Pentium system (compared to 17 ns for a subroutine call in a non-object-oriented language such as C or Fortran) [21].

The CCA approach to parallelism. For parallel computing, the CCA has chosen not to specify a particular parallel programming model but rather to allow framework and application developers to use the programming models they prefer. This approach has several advantages, the most significant of which is that it allows component developers to use the model that best suits their needs, greatly facilitating the incorporation of existing parallel software into the CCA environment. Figure 10.6 shows schematically a typical configuration for a component-based parallel application in the Ccaffeine framework. For a single-program multiple-data (SPMD) application, each parallel process would be loaded with the same set of components, with their ports connected in the same way. Interactions *within a given parallel process* occur through normal CCA mechanisms, getting and releasing ports on other components and invoking methods on them. These would generally use the local direct connect approach mentioned above, to minimize the CCA-related overhead. Interactions within the parallel cohort of a given component are free to use the parallel programming model they prefer, for example MPI [89], PVM [48], or Global Arrays [92]. Different sets of components may even use different programming models, an approach that facilitates the assembly of applications from components derived from software developed for different programming models. This approach imposes no CCA-specific overhead on the application's parallel performance. Such mixing of programming models can occur for components that interact at relatively coarse grained levels with loose coupling (for example, two parts of a multi-model physics application, such as the climate models discussed in Section 10.6.2). In contrast, sets of relatively fine grain and tightly coupled components (for example, the mesh and discretization components shown in Figure 10.7) must employ compatible parallel programming models. Multiple-program multiple-data (MPMD) applications are also supported through a straightforward generalization of the SPMD model. It is also possible for a particular CCA framework implementation to provide its own parallel programming model, as is the case with the Uintah framework discussed in Section 10.6.4.

Language interoperability. A feature of many component models, including the CCA, is that components may be composed together to form applications regardless of the programming language in which they have been implemented. The CCA provides this capability through the Scientific Interface Definition Language (SIDL) [38], which component developers can employ to express component interfaces. SIDL works in conjunction with the Babel language interoperability tool [74, 38], which currently supports C, C++, Fortran 77, Fortran 90/95, and Python, with work under way on Java. SIDL files are processed by the Babel compiler, which generates the glue code necessary to enable the caller and callee to be in any supported language. The generated glue code handles the translation of argu-

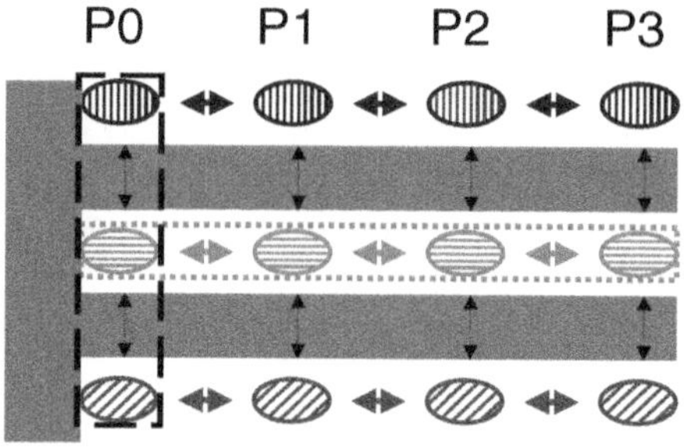

Fig. 10.6. A schematic representation of the CCA parallel programming environment in the single component/multiple data (SCMD) paradigm. Parallel processes, labeled P0,...,P3, are loaded with the same set of three components. Components in the same process (vertical dashed box) interact using standard CCA port-based mechanisms, while parallel components of the same type (horizontal dotted box) interact using their preferred parallel programming model.

ments and method calls between languages. Babel also provides an object-oriented (OO) model, which can be used even in non-OO languages such as C and Fortran. On the other hand, neither Babel nor the CCA requires that interfaces be strongly object-oriented; such design decisions are left to the component and interface designers.

The developers of Babel are also sensitive to concerns about performance. Where Babel must translate arguments for method calls (because of differing representations in the underlying languages), there will clearly be some performance penalty. Since most numerical types do not require translation, however, in many cases Babel can provide language interoperability with no additional performance cost [21]. In general, the best strategy is for designers and developers to be aware of translation costs, and take them into account when designing interfaces, so that wherever possible enough work is done within the methods so that the translation costs are amortized; see Section 10.4.8 for performance overhead studies.

Incorporating components. The CCA employs a minimalist design philosophy to simplify the task of incorporating existing software into the CCA environment. Generally, as discussed in Section 10.3.3, one needs to add to an existing software module just a single method that informs the framework which ports the component will provide for use by other components and which ports it expects to use from others. Within a component, calls to ports on other components may have slightly different syntax, and calls must be inserted to obtain and release the handle for the port. Experience has shown that componentization of existing software in the CCA environment is straightforward when starting from well-organized code [93, 5, 79]. Moreover, the componentization can be done incrementally, starting with a coarse-grained decomposition (possibly even an entire simulation, if the goal is coupled simulations) and successively refining the decomposition when opportunities arise to replace functionality with a better-performing component.

Common interfaces. Interfaces are clearly a key element of the CCA and of the general concept of component-based software engineering; they are central to the interoperability and reuse of components. We note that except for a very small number of interfaces in the CCA specification, typically associated with framework services, the CCA does *not* dictate "standard" interfaces – application and component developers are free to define and use whatever interfaces work best for their purposes.

However, we do strongly encourage groups of domain experts to work together to develop interfaces that can be used across a variety of components and applications. Numerous such efforts are under way, including mesh management, linear algebra, and parallel data redistribution, all of which are related to the applications described in this chapter and are discussed in Section 10.4. Anyone interested in these efforts, or in launching other standardization efforts, is encouraged to contact the authors.

10.3.3 Simple PDE Examples

We next introduce two simple PDE examples to help illustrate CCA principles and components. While we have deliberately chosen these examples to be relatively simple and thus straightforward to explain, they incorporate numerical kernels and phases of solution that commonly arise in the more complicated scientific applications that motivate our work, as introduced in Section 10.2.

Steady-State PDE Example

The first example is Laplace's equation on a two-dimensional rectangular domain: $\nabla^2\phi(x,y) = 0,\ x \in [0,1],\ y \in [0,1]$, with $\phi(0,y) = 0$, $\phi(1,y) = \sin(2\pi y)$, and $\frac{\partial\phi}{\partial y}(x,0) = \frac{\partial\phi}{\partial y}(x,1) = 0$. This system can be discretized by using a number of different methods, including finite difference, finite element, and finite volume techniques on either a structured or an unstructured mesh. This example has characteristics of the large, sparse linear systems that are at the heart of many scientific simulations, yet it is sufficiently compact to enable the demonstration of CCA concepts and code.

The composition of this CCA application is shown by a component wiring diagram in the upper portion of Figure 10.7; the graphical interface of the Ccaffeine [3] framework enables similar displays of component interactions. This example employs components (as represented by large gray boxes) for unstructured mesh management, discretization, and linear solution, which are further discussed in Section 10.4, as well as an application-specific driver component, which is discussed below. The lines in the diagram between components represent connections between *uses* and *provides* ports, which are denoted by rectangular boxes that are white and checkered, respectively. For example, the discretization component's "Mesh" *uses* port is connected to the unstructured mesh component's "Mesh" *provides* port, so that the discretization component can invoke the mesh interface methods that the mesh component has implemented. The special *GoPort* (named "Go" in this application driver) starts the execution of the application.

The application scientist's perspective. The application-specific driver component plays the role of a user-defined `main` program in traditional library-based applications. CCA frameworks do not require that an application contain a definition of a `main` subroutine. In fact, in many cases, `main` is not defined by the user; instead, a definition in the framework is used. In that case, a driver component partially fulfills the role of coordinating some of the application's components; the actual instantiation and port connections can be part of the driver as well, or these tasks can be

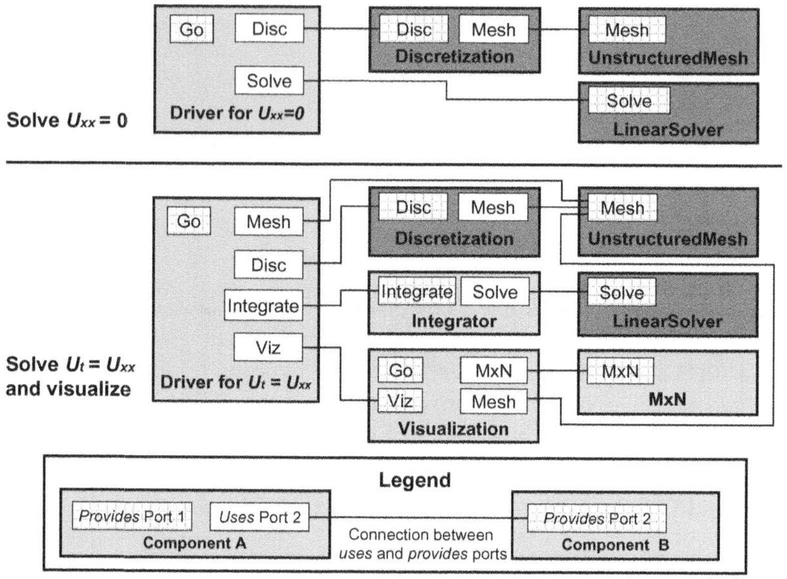

Fig. 10.7. Two component wiring diagrams for (*top*) a steady-state PDE example and (*bottom*) a time-dependent PDE example demonstrate the reuse of components for mesh management, discretization, and linear solvers in two different applications.

accomplished via a user-defined script or through a graphical user interface. A CCA framework can support multiple levels of user control over component instantiation and connection; here we present only one of the higher levels, where the user takes advantage of a framework-supplied `main` program, as well as framework-specific concise mechanisms for application composition. In this example, the application could be composed by using a graphical user interface, such as that provided with the Ccaffeine [3] framework, by selecting and dragging component classes to instantiate them, and then clicking on pairs of corresponding ports to establish connections. Alternatively, the application could be composed with a user-defined script.

In addition to writing a driver component, typical application scientists would also write custom components for the other parts of the simulation that are of direct interest to their research, for example the discretization of a PDE model (see Section 10.6.3 for a discussion of the approach used by combustion researchers). These application-specific components can then be used in conjunction with external component-based libraries for other facets of the simulation, for example, unstructured mesh management (see Section 10.4.1) and linear solvers (see Section 10.4.4). As discussed in detail in Section 10.4.1, if multiple component implementations of a given functionality adhere to common port specifications, then different implementations, which have been independently developed by different groups, can be seamlessly substituted at runtime to facilitate experimentation with a variety of algorithms and data structures.

```
package laplace version 1.0 {
  class Driver implements gov.cca.Component,
                          gov.cca.ports.GoPort
  {
    // The only method required to be a CCA component.
    void setServices(in gov.cca.Services services);
    // The GoPort method that returns 0 if successful.
    int go();
  }
}
```

Fig. 10.8. SIDL definition of the driver component for the steady-state PDE example.

A closer look at the application-specific driver component. Figure 10.8 shows the SIDL definition of the driver component for the solution of the steady-state PDE example in Figure 10.7. As discussed in Section 10.3.2, the use of SIDL for the component interface enables the component to interact easily with other components that may be written in a variety of programming languages. The `Driver` SIDL class must implement the `setServices` and `go` methods, which are part of the `gov.cca.Component` and `gov.cca.ports.GoPort` interfaces, respectively [29]. For this example, we used Babel to generate a C++ implementation skeleton, to which we then added the application-specific implementation details, portions of which are discussed next.

Figure 10.9 shows the implementation of the `setServices` method, which is generally used by components to save a reference to the framework `Services` object and to register *provides* and *uses* ports with the framework. The user-defined data member `frameworkServices` in the `Driver_impl` class stores the reference to the services object, which can be used subsequently to obtain and release ports from the framework and for other services. To provide the "Go" port, the driver component's `self` data member (a Babel-generated reference similar to the `this` pointer in C++) is first cast as a `gov::cca::Port` in the assignment of `self` to `port`; then the `addProvidesPort` services method is used to register the *provides* port of type `gov.cca.ports.GoPort` with the framework, giving it the name "Go". A `gov.cca.TypeMap` object, `tm`, is created and passed to each call that registers *provides* and *uses* ports; in larger applications, these name-value-type dictionaries can be used for storing problem and other application-specific parameters.

Figure 10.10 shows an abbreviated version of the `go` method implementation for the simple steady-state PDE example. First, we obtain a reference to the discretization port "Disc" from the framework services object, `frameworkServices`. Note that in Babel-generated C++ code, the casting of the `gov::cca::Port` object returned by `getPort` to type `disc::Discretization` is performed automatically. The discretization component uses finite elements to assemble the linear system in the implementation of the `createFESystem` method, which includes information exchange with the unstructured mesh component through the "Mesh" port. The linear system is then solved by invoking the `apply` method on the linear solver component. Finally, all ports obtained in the `go` method are released via the `releasePort` framework services method.

The linear algebra interfaces in this example are based on the TOPS solver interfaces [111] (also see Section 10.4.4). The matrix and vector objects in this example are not components themselves, but are created as regular objects by the driver component and then modified and used in the mesh, discretization, and solver components. While such linear algebra objects could be implemented as components themselves, we chose to use a slightly more lightweight approach (avoiding one layer of abstraction) because they have relatively fine-grain interfaces, e.g., setting individual vector and matrix elements. In contrast, the solver component, for example, provides a port whose methods perform enough computation to make the overhead of port-based method invocation negligible (see, e.g., [93]).

Time-Dependent PDE Example

The second PDE that we consider is the heat equation, given by $\frac{\partial \phi}{\partial t} = \nabla^2 \phi(x, y, t)$, $x \in [0, 1]$, $y \in [0, 1]$, with $\phi(0, y, t) = 0$, $\phi(1, y, t) = \frac{1}{2} \sin(2\pi y) \cos(t/2)$, $\frac{\partial \phi}{\partial y}(x, 0, t) = \frac{\partial \phi}{\partial y}(x, 1, t) = 0$. The initial condition is $\phi(x, y, 0) = \sin(\frac{1}{2}\pi x) \sin(2\pi y)$. As shown by the component wiring diagram in the lower portion of Figure 10.7, this application reuses the unstructured mesh, discretization, and linear algebra components employed by the steady-state PDE example and introduces a time integration component as well as components for parallel data redistribution and visualization. These reusable scientific components are discussed in further detail in Section 10.4.

Another component-based solution of the heat equation, but on a *structured mesh*, can be found at [106]. This approach employs different discretization and mesh components from those discussed above but reuses the same integrator. This CCA example is freely downloadable from [106], including scripts for running the code.

```
void laplace::Driver_impl::setServices (
  /*in*/ ::gov::cca::Services services )
throw ( ::gov::cca::CCAException)
{
  // frameworkServices is a programmer-defined private data member of
  // the Driver_impl class, declared as
  //     ::gov::cca::Services frameworkServices
  // in the Babel-generated laplace_Driver_impl.hh file
  frameworkServices = services;

  // Provide a Go port; the following statement performs an implicit cast
  gov::cca::Port port = self;
  gov::cca::TypeMap tm = frameworkServices.createTypeMap();
  frameworkServices.addProvidesPort(port, "Go", "gov.cca.ports.GoPort",tm);

  // Use Discretization and Solver ports
  frameworkServices.registerUsesPort("Disc", "disc.Discretization",tm);
  frameworkServices.registerUsesPort("Solver", "solvers.LinearSolver",tm);
}
```

Fig. 10.9. Laplace application driver code fragment showing the C++ implementation of the `setServices` method of the `gov.cca.Component` interface.

More detailed CCA tutorial materials, including additional sample component codes as well as the Ccaffeine framework and Babel language interoperability tool, are available via `http://www.cca-forum.org/tutorials`. We recommend this site as a starting point for individuals who are considering the use of CCA tools and components.

These examples illustrate one of the ways that components can participate in a scientific application. In larger applications, such as those introduced in Section 10.2, different components are typically developed by different teams, often at different sites and times. Some of these components are thin wrappers over existing numerical libraries, while others are implemented from scratch to perform some application-specific computation, such as the discretization components in Figure 10.7. The CCA component model, like other component models, provides a specification and tools that facilitate the development of complex, multi-project, multi-institutional software. In addition to helping manage software development complexity, the simple port abstraction (1) enables the definition of explicit interaction points between parts

```
int32_t laplace::Driver_impl::go() throw () {

  disc::Discretization discPort;
  solvers::Solver linearSolverPort;
  try {
    // Get the discretization port.
    discPort = frameworkServices.getPort("Disc");

    // The layout object of type solvers::Layout_Rn is a data member
    // of the Driver_impl class describing how vector and matrix
    // data is laid out across processors; it also provides a factory
    // interface for creating parallel vectors and matrices.

    // Create the matrix, A, and right-hand-side vector, b
    solvers::Vector_Rn b = layout.createGlobalVectors(1)[0];
    solvers::Matrix_Rn A = layout.createOperator(layout);

    // Assemble A and b to define the linear system, Ax=b
    discPort.createFESystem(A, b);

    // Get the solver port
    linearSolverPort = frameworkServices.getPort("Solver");

    // Create the solution vector, x
    solvers::Vector_Rn x = layout.createGlobalVectors(1)[0];

    // Initialize and solve the linear system
    linearSolverPort.setOperator(A);
    linearSolverPort.apply(b, x);

    // Release ports
    frameworkServices.releasePort("Disc");
    frameworkServices.releasePort("Solver");
    return 0;
  } catch ( gov::cca::CCAException& e) { return -1; }
}
```

Fig. 10.10. Laplace application driver code fragment showing the C++ implementation of the go method from the `gov.cca.ports.GoPort` interface. Exceptions are converted to the function return code specified in Figure 10.8 with the try/catch mechanism.

of an application; (2) facilitates the use of thoroughly tuned external components implemented by experts; and (3) allows individual components to be developed, maintained, and extended independently, with minimal impact on the remainder of the application.

10.4 Reusable Scientific Components

Various scientific simulations often have similar mathematics and physics, but currently most are written in a stovepipe fashion by a small group of programmers with minimal code reuse. As demonstrated in part by Figure 10.7, a key advantage of component-oriented design is software reuse. Components affect reuse in two ways: (1) because an exported port interface is simpler to use than the underlying software, a component should be easier to import into a simulation than to rewrite from scratch; and (2) because *common* interfaces for particular functionalities can be employed by many component implementations, different implementations can be easily substituted for one another to enhance performance for a target machine, data layout, or parameter set. In Sections 10.4.1 through 10.4.7 we detail these two facets of reusability in terms of several current efforts to develop component implementations and domain-specific groups devoted to defining common interfaces for various numerical and parallel computing capabilities. In Section 10.4.8 we demonstrate that the overhead associated with CCA components is negligible when appropriate levels of abstraction are employed.

Component implementations can directly include the code for core numerical and parallel computing capabilities, and indeed new projects that start from scratch typically do so. However, many of the component implementations discussed in this section employ the alternative approach of providing thin wrappers layered on top of existing libraries, thereby offering optional new interfaces that make these independently developed packages easier to use in combination with one another in diverse projects. Section 10.5 discusses some of the issues that we have found useful to consider when building these component interfaces. The web site `http://www.cca-forum.org` has current information on the availability of these components as well as others.

10.4.1 Unstructured Mesh Management

Unstructured meshes are employed in many PDE-based models, including the accelerator application introduced in Section 10.2.1 and the simple examples discussed in Section 10.3.3. The Terascale Simulation Tools and Technologies (TSTT) Center [121], established in 2001, is developing common interface abstractions for managing mesh, geometry, and field data for the numerical solution of PDEs. As shown in Figure 10.11, the common TSTT mesh interface facilitates experimentation with different mesh management infrastructures by alleviating the need for scientists to write separate code to manage the interactions between an application and different meshing tools.

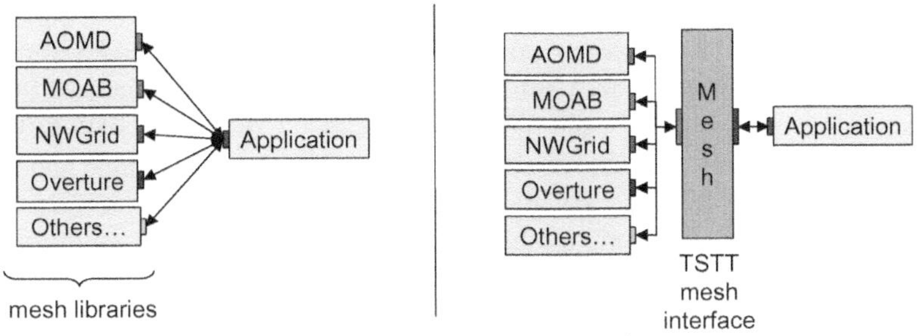

Fig. 10.11. *(Left)*: The current interface situation connecting an application to m mesh management systems through m different interfaces. Because developing these connections is often labor-intensive for application scientists, experimentation with various mesh systems is severely inhibited. *(Right)*: The desired interface situation, in which many mesh systems that provide similar functionality are compliant with a single common interface. Scientists can more easily explore the use of different systems without needing to commit to a particular solution strategy that could prematurely lock their application into a specific mesh management system.

TSTT Mesh Interfaces

The TSTT interfaces include *mesh data*, which provides the geometric and topological information associated with the discrete representation of a computational domain; *geometric data*, which provides a high-level description of the domain boundaries, for example, a CAD model; and *field data*, which provides the time-dependent physics variables associated with application solutions. The TSTT data model covers a broad spectrum of mesh types and functionalities, ranging from a nonoverlapping, connected set of entities (e.g., a finite element mesh) to a collection of such meshes that may or may not overlap to cover the computational domain. To date, TSTT efforts have focused on the development of mesh query and modification interfaces at varying levels of granularity. The basic building blocks for the TSTT interfaces are *mesh entities*, for example, vertices, edges, faces, and regions, and *entity sets*, which are arbitrary groupings of mesh entities that can be related hierarchically or by subsets. Functions have been defined that allow the user to access mesh entities using arrays or iterators, attach user-defined data through the use of tags, and manipulate entity sets using Boolean set operations.

The TSTT mesh interface is divided into several ports. The core interface provides basic functionality for loading and saving the mesh, obtaining global information such as the number of entities of a given type and topology, and accessing vertex coordinate information and adjacency information using both primitive arrays and opaque entity handle arrays. Additional ports are defined that provide single entity iterators, workset iterators, and mesh modification functionality. Figure 10.12 shows an example C-code client that uses the TSTT interface. The `mesh` variable represents a pointer to an object of type `TSTT_mesh`, which can be either an object created via a call to a Babel-generated constructor or a *uses* port provided by a mesh component instantiated in a CCA framework (for example, see Figure 10.7).

The mesh data is loaded from a file whose name is specified via a string. Because many of the TSTT functions work on both the full mesh and on subsets of the mesh, the user must first obtain the root entity set using `getRootSet` to access the vertex and face information. In this example, the user asks for the handles associated with the triangular elements with the `getEntities` call. Once the triangle handles have been obtained, the user can access the adjacent vertices and their coordinate information either one handle at a time as shown in the example, or using a single function call that returns the adjacency information for all of the handles simultaneously.

```
#include "TSTT.h"

/* ... */
  void   *root_entity_set;
  void   **tri_handles, **adj_vtx_handles;
  int    i, num_tri, num_vtx, coords_size;
  double *coords;

  /* Load the data into the previously created mesh object */
  TSTT_mesh_load(mesh,''mesh.file'');

  /* Obtain a handle to the root entity set */
  root_entity_set = TSTT_mesh_getRootSet(mesh);

  /* Obtain handles to the triangular elements in the mesh */
  TSTT_mesh_getEntities(mesh,root_entity_set, TSTT_EntityType_FACE,
                        TSTT_EntityTopology_TRIANGLE, &tri_handles,
                        &num_tri);

  /* For each triangle, obtain the corner vertices and their
     coordinates */
  for (i=0;i<num_tri;i++) {
    TSTT_mesh_getEntAdj(mesh,tri_handle[i],TSTT_EntityType_VERTEX,
                        &adj_vtx_handles,&num_vtx);
    TSTT_mesh_getVtxArrCoords(mesh,adj_vtx_handles,num_vtx,
                        TSTT_StorageOrder_BLOCKED,&coords,&coords_size);
  }
/* ... */
```

Fig. 10.12. An example code fragment showing the use of the TSTT interface in C to load a mesh and retrieve the triangular faces and their corner vertex coordinates.

More information on the mesh interfaces and the TSTT Center can be found in [121]. Preliminary results of a performance study of the use of a subset of the mesh interfaces can be found in Section 10.4.8.

TSTT Mesh Component Implementations

Implementations of the TSTT mesh interfaces are under way at several institutions. The ports provided by these mesh components include the "Mesh" port, which gives basic access to mesh entities through primitive arrays and opaque entity handles. In addition to global arrays, entities can be accessed individually through iterators in the "Entity" port or in groups of a user-defined size in the "Arr" port. These components also support the the `Tag` interface, which is a generic capability that allows the user to add, set, remove, and change tag information associated with the mesh or individual entities; the `Set` interface, which allows the creation, deletion, and definition of relations among entity sets; and the `Modify` interface, which allows the creation and deletion of mesh entities as well as the modification of vertex coordinate locations.

For the performance studies presented in Section 10.4.8, we use a simple implementation of the interface that supports two-dimensional simplicial meshes. The mesh software is written in C and uses linked lists to store the element and vertex data. This component has been used primarily as a vehicle for demonstrating the benefits of the component approach to mesh management and for evaluating the performance costs associated therein [93].

10.4.2 Block-Structured Adaptive Mesh Refinement

Block-structured adaptive mesh refinement (SAMR) is a domain discretization technique that seeks to concentrate resolution where required, while leaving the bulk of the domain sparsely meshed. Regions requiring resolution are identified, collated into rectangular patches (or boxes), and then resolved by a finer grid. Note that the fine grid is *not* embedded in the coarser one; rather, distinct meshes of different resolutions are maintained for the same region in space. Data, in each of these boxes, is often stored as multidimensional Fortran 77 arrays in a blocked format; that is, the same variable (e.g., temperature) for all the grid points are stored contiguously, followed by the next variable. This approach allows operations involving a spatial operator (e.g., interpolations, ghost cell updates across processors) to be written for one variable and reused for others while exploiting cache locality. This approach also allows scientific operations on these boxes to be performed by using legacy codes. The collection of boxes that constitute the discretized domain on a CPU are usually managed by using an object-oriented approach.

SAMR is used by `GraceComponent` [79], a CCA component based on the GrACE library [97], as well as by Chombo [35], a block-structured mesh infrastructure with similar functionality developed by the APDEC [34] group. While each has a very different object-oriented approach (and interface) to managing the collection of boxes, individual boxes are represented in a very similar manner, and the data is stored identically. This fundamental similarity enables a simple, if slightly cumbersome, approach to interoperability.

Briefly, the data pointer of each box is cached in a separate component, along with a small amount of metadata (size of array, position of the box in space, etc.), and keyed to an opaque handle (an integer, in practice). These handles can be exchanged among components, and the entire collection of patches can be recreated by retrieving them from the cache. The interoperability interface is easy to understand and implement; however, the frequent remaking of the box container imposes some overhead, though not excessively so, since array data is not copied. Because the main purpose of this AMR interoperability is to exploit specialized solvers and input/output routines in various packages, metadata overhead is not expected to be significant.

Further, this approach is not a preferred means of interoperability on an individual-box basis unless the box is large or the operation very intensive. A prototype implementation of this exchange is being used to exploit Chombo's elliptic solvers in the CFRFS combustion application introduced in Section 10.2.3. Section 10.6.3 includes further information about the use of SAMR components in this application.

10.4.3 Parallel Data Management

The effective management of parallel data is a key facet of many PDE-based simulations. The `GlobalArray` component, based on the Global Array library [92], includes a set of capabilities for managing distributed dense multidimensional arrays that can be used to represent multidimensional meshes. In addition to a rich set of operations on arrays, the user can create ghost cells with a specified width around each of the mesh sections assigned to a processor. Once an update operation is complete, the local data on each processor contains the locally held *visible* data plus data from the neighboring elements of the global array, which has been used to fill in the ghost cells. Two types of update operations are provided to facilitate data transfer from neighboring processors to the ghost regions: a collective update of all ghost cells by assuming periodic, or wraparound, boundary conditions and another nonblocking and noncollective operation for updating ghost cells along the specified dimension and direction with the option to include or skip corner ghost cell updates. The first of these two operations was optimized to avoid redundant communication involving corner points, whereas the second was designed to enable overlapping communication involved in updating ghost cells with computations [96].

Unstructured meshes are typically stored in a compressed sparse matrix form, in which the arrays that represent the data structures are one-dimensional. Computations on such unstructured meshes often lead to irregular data access and communication patterns. The `GlobalArray` component provides a set of operations that can be used to implement and manage distributed sparse arrays (see [24, 30]). Modeled after similar functions in the CMSSL library of the Thinking Machines CM-2/5, these operations have been used to implement the NWPhys/NWGrid [120] adaptive mesh refinement code. Additional `GlobalArray` numerical capabilities have been employed with the optimization solvers discussed in Section 10.4.6 [13, 65].

10.4.4 Linear Algebra

High-performance linear algebra operations are key computational kernels in many PDE-based applications. For example, vector and matrix manipulations, along with linear solvers, are needed in each of the motivating applications introduced in Sections 10.2 and 10.3.3, as well as in the integration and optimization components discussed in Sections 10.4.5 and 10.4.6, respectively.

Linear Algebra Interfaces

Linear algebra has been an area of active interface development in recent years. Abstract interfaces were defined in the process of implementing numerical linear algebra libraries, such as the Hilbert Class Library [51], the Template Numerical Toolkit [103], the Matrix Template Library [85], PLAPACK [7], uBLAS (part of the BOOST collection) [25], BLITZ++ [126], and the Linear System Analyzer [27]. Many of these packages were inspired by or evolved from legacy linear algebra software, such as BLAS and LAPACK. This approach allowed the flexibility of object-oriented design to be combined with the high performance of optimized library codes. In some cases, such as BLITZ++, the goal is to extract high performance even if computationally intensive operations are implemented by using high-level language features; in that case, the library assumes a burden similar to that of a compiler in order to ensure that array operations are performed in a way that exploits temporal and spatial data locality.

Starting in 1997, the Equation Solvers Interface (ESI) working group [31] focused on developing interfaces targeted at the needs of large-scale DOE ASCI program computations, but with the goal of more general use and acceptance. The ESI includes interfaces for linear equation solvers, as well as support for linear algebra entities such as index sets, vectors, and matrices.

More recently, the Terascale Optimal PDE Simulation (TOPS) Center [66], whose mission is to develop a set of compatible toolkits of open-source, optimal complexity solvers for nonlinear partial differential equations, has produced a prototype set of linear algebra interfaces expressed in SIDL [111]. This language-independent specification enables a wide variety of new underlying implementations as well as access to existing libraries. Special care has been taken to separate functionality from the details of accessing the underlying data representation. The result is a hierarchy of interfaces that can be used in a variety of ways depending on the needs of particular applications. See section 10.3.3 for some simple examples and code using linear algebra components based on the TOPS interfaces.

Linear Algebra Component Implementations

As discussed in detail in [93], an early CCA linear solver port was based on ESI [31] and was implemented by two components based on Trilinos [56] and PETSc [11, 10]. The creation of basic linear algebra objects (e.g., vectors and matrices) was implemented as an abstract factory, with specific factory implementations based on Trilinos and PETSc provided as components. The factory and linear solver components were successfully reused in several unrelated component-based applications. Linear algebra ports and components based on TOPS [66] interfaces are currently under development. One of the most significant advancements since the original simple linear solver ports and components were developed is the use of SIDL for interface definition, alleviating implementation language restrictions. By contrast, the ESI-based ports and components used C++, making the incorporation in non-C or C++ applications more difficult. The most recent linear solver component implementations, such as those shown in the examples in Section 10.3.3, are based on the language-independent TOPS interfaces [111].

10.4.5 Integration

Ordinary differential equations (ODEs) are solved routinely when modeling with PDEs. Often the solution is needed only locally, for example, when integrating stiff nonlinear chemical reaction rates at a point in space over a short time span. At other times we need a large parallel ODE solution to a method-of-lines problem.

By wrapping the CVODE [33] library, we have created the `CVODEComponent` [112, 79] for the solution of local ODEs in operator-splitting schemes, such as in the combustion modeling application introduced in Section 10.2.3. Like the library it wraps, `CVODEComponent` can be used to solve ODEs by using variable-order Adams-Bashforth or backward-difference formula techniques. The library provides its own linear solvers and requires the application to provide data using a particular vector representation. The application can provide a sparse, banded, or dense gradient or can request CVODE to construct a finite difference approximation of the gradient if needed.

For parallel ODE solutions we have refactored the LSODE [104, 58] library to allow the application to provide abstract linear solvers and vectors. The parallelism of the ODE is hidden from the integration code by the vector and solver implementations. The resulting `IntegratorLSODE` and related components are described in [5]. These components have been coupled with PETSc-based linear algebra components described in Section 10.4.4 to solve finite element models [93].

10.4.6 Optimization

The solution to boundary value problems and other PDEs often can be represented as a function $u \in U$ such that $J(u) = \inf_{v \in U} J(v)$. In this formulation, U is a set of admissible functions, and $J : U \rightarrow \Re$ is a functional representing the total energy associated with an element in U. This formulation of a PDE is often preferred for nonlinear PDEs with more than one solution. While each solution satisfies the first-order optimality conditions of the corresponding minimization problem, the solution that minimizes the energy functional is often more stable and of greater interest.

The minimization approach also enables inequality constraints to be incorporated in the model. Obstacle problems, for example, have a free boundary that can be modeled by using variational inequalities. Efficient algorithms with rigorous proof of convergence can be applied to minimization problems with inequality constraints [14, 16]. Even PDEs whose corresponding minimization problem is unconstrained or equality constrained can benefit from optimization solvers [88].

Optimization components [110] based on the TAO library [15] encapsulate the algorithmic details of various solvers. These details include line searches, trust regions, and quadratic approximations. The components interact with the application model, which computes the energy function, derivative information, and constraints [65]. The optimization solvers achieve scalable and efficient parallel performance by leveraging general-purpose linear solvers and specialized preconditioners available in external components, including the data management components discussed in Section 10.4.3 and the linear algebra components described in Section 10.4.4 [13, 65, 93].

10.4.7 Parallel Data Redistribution

As discussed in Section 10.2, scientific simulations are increasingly composed of multiple distinct physics models that work together to provide a more accurate overall system model or to otherwise enhance fidelity by replacing static boundaries with dynamically computed data values from a live companion simulation. Often each of these models constitutes its own independent code that must be integrated, or *coupled*, with the other models' codes to form a unified simulation. This coupling is usually performed by sharing or exchanging certain common or relevant data fields among the individual models, for example, heat and moisture fluxes in a coupled climate simulation. Because most high-performance scientific simulations require parallel algorithms, the coupling of data fields among these parallel codes raises a number of challenges. Even for the *same* basic data field, each distinct model often applies a unique distributed data decomposition to optimize data access patterns in its localized portion of a parallel algorithm. Further, each model can use a different number of parallel processors, requiring complex mappings between disparate parallel resource topologies (hence the characterization of this mapping as the "MxN" problem – transferring data from "M" parallel processors to another set of "N" processors, where M and N are not in general equal). These mappings require both an understanding of the distributed data decompositions for each distinct model, to construct a "communication schedule" of the elements between the source and destination data fields, as well as special synchronization handling to ensure that data consistency is maintained in any MxN exchanges.

Worse yet, each model may compute using a different time step or may store data elements on a unique mesh using wholly different coordinate systems or axes. This situation necessitates the use of complex spatial and temporal interpolation schemes, and often the preservation of key energy or flux conservation laws. Such complications further exacerbate the already complex infrastructure necessary for coupling disparate data arrays, and require the incorporation of a diverse set of interpolation schemes that are often chosen as a part of the system's scientific requirements, or simply to ensure backwards compatibility with legacy code. Some software packages capable of addressing these issues exist, notably the Mesh-based parallel Code Coupling Interface (MpCCI) [2, 1] and the Model Coupling Toolkit (MCT) [72, 71]. The details of interpolation schemes and their inclusion in MxN infrastructure are beyond the scope of the current work. Yet this important follow-on research will commence upon satisfactory completion of the fundamental generalized MxN data exchange technology. In the meantime, such data interpolation must be handled manually and separately by each distinct code or by an intermediary piece of coupling software.

Parallel Data Redistribution Interfaces

The CCA project is developing generalized interfaces for specifying and controlling MxN parallel data redistribution operations [22]. These interfaces and their accompanying prototype implementations address synchronization and data movement issues in parallel component-based model coupling. Initial efforts have focused on independently defining the local data allocation and decomposition information within a given parallel component and then applying these details to automatically generate efficient mappings, or communication schedules, for executing MxN transfers. These evolving interfaces are sufficiently flexible and high-level to minimize the instrumentation cost for legacy codes, and to enable nearly *transparent* operation by most of the coupling participants.

Parallel Data Redistribution Component Implementations

A variety of general-purpose prototype MxN parallel data redistribution components have been developed by using existing technology. Initial prototypes, which were loosely based on the CUMULVS [49, 69] and PAWS [64, 12] systems, provided a proof of concept to verify the usefulness of the MxN interface specification and to assist in evolving this specification toward a stable and flexible standard. The `CumulvsMxN` component continues to be extended to cover a wider range of data objects, including structured and unstructured dense meshes and particle-based decompositions. The underlying messaging substrates for MxN data transfers are also being generalized to improve their applicability to common scientific codes. Additional MxN component solutions are being developed based on related tools such as Meta-Chaos [42, 105] and ECho [43, 133].

Special-purpose MxN components [73] for use with climate modeling simulations have been built using the Model Coupling Toolkit (MCT) [72, 71] (see Section 10.6.2). These prototype coupler components provide crucial scientific features beyond fundamental parallel data transfer and redistribution, including spatial interpolation, time averaging and accumulation of data, merging of data from multiple components for use in another component, and global spatial integrals for enforcing conservation laws [22]. Currently MCT is being employed to couple the atmosphere, ocean, sea-ice, and land components in the Community Climate System Model [54] and to implement the coupling interface for the Weather Research and Forecasting Model [129]. Similar coupling capabilities for climate modeling are also being explored as part of the Earth System Modeling Framework (ESMF) [67] effort, with specially tailored climate-specific interfaces and capabilities.

10.4.8 Performance Overhead Studies

Object-oriented programming in general and components in particular adopt a strict distinction between interfaces and implementations of functionality. The following subsections demonstrate that the interface-implementation separation results in negligible overhead when appropriate levels of abstraction are employed.

CCA Components

To quantify the overhead associated with CCA components, we solved an ODE, $Y_t = F(Y, t)$, where Y is a 4-tuple, implicitly in time using the `CVODEComponent` integration component introduced in Section 10.4.5. The numerical scheme discretizes and linearizes the problem into a system of the form $Ax = b$, which is solved iteratively. A is derived from the Jacobian of the system, $\partial F/\partial Y$, which is calculated numerically by evaluating F repeatedly. Each F invocation consists of one log evaluation and 40 exponentials and corresponds to the simplest detailed chemical mechanism described in Section 10.6.3. A three-component assembly was created (the `Driver`, `Integrator` and `F`), and the `F` evaluations were timed. These were then compared with timings from a non-component version of the code. In order to reduce instrumentation-induced inaccuracies, the problem was repeated multiple times and the total time measured. These steps were taken to ensure that the invocation overhead was exercised repeatedly and the time being measured was significant. The code for F was written in C++ and compiled using `g++ -O2` using egcs version 2.91 compilers on a RedHat 6.2 Intel/Linux platform.

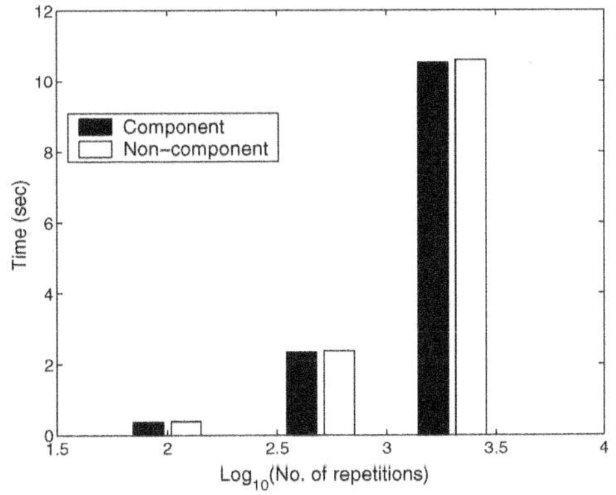

Fig. 10.13. Timings (in seconds) for component and non-component versions of an ODE application. The differences clearly are insignificant. The x-axis shows the number of times the same problem was solved to increase the execution time between instrumentation invocations.

Figure 10.13 plots the solution time using the C++ component and the non-component versions. The differences are clearly insignificant and can be attributed to system noise. For this particular case, the function invocation overheads were small compared to the function execution time. Likewise, negligible overhead for both C++ and SIDL variants of optimization components has been shown [13, 93]. The actual overhead of the virtual pointer lookup has been estimated to be of the order of a few hundred nanoseconds (around 150 ns on a 500 MHz Intel processor) [21]. Thus, the overhead introduced by componentization is expected to be insignificant unless the functions are exceptionally lightweight, such as pointwise data accessor methods.

TSTT Mesh Interfaces

We next evaluate the performance ramifications of using a component model for the finer-grained activity of accessing core mesh data structures, where we used the simple mesh management component described in Section 10.4.1. Since the granularity

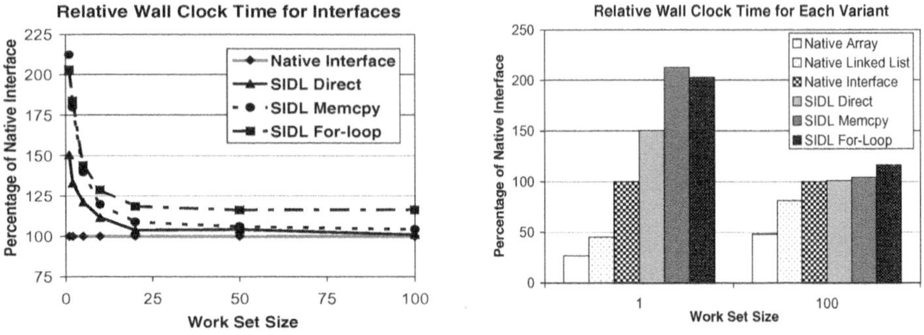

Fig. 10.14. Average wall clock time for traversing all mesh elements by work set size relative to Native Interface. *(Left)*: A comparison of the four TSTT interface approaches for work set sizes 1 through 100 entities. *(Right)*: A comparison of the six variants for work sets of size 1 and 100 entities only.

of access is a major concern, initial experiments focused on the traversal of mesh entities using work set iterators. These iterators allow the user to access mesh entities in blocks of a user-defined size, N. That is, for each call to the iterator, N entity handles are returned in a SIDL array, and it is expected that as N increases, the overhead associated with the function call will be amortized. For comparison purposes, experiments were also performed using native data structures to quantify the base costs.

Figure 10.14 shows the relative costs of obtaining entity handles from the mesh using both native data structures and interfaces. In the experiments, six different mechanisms were used for data access. The native variants consist of timing array (*Native Array*), linked list (*Native Linked List*), and language-specific TSTT interface (*Native Interface*) versions. In order to test the performance of the language interoperability layer created by SIDL, three variations of managing the conversion between native and SIDL arrays were developed. The first two, referred to as *SIDL Direct* and *SIDL Memcpy*, take advantage of the fact that the language interoperability layer and native implementation are both written in C. The former allows the underlying implementation to directly manage the SIDL array contents, while the latter is able to use the `memcpy` routine. The general-purpose variant, called *SIDL For-loop*, individually copies the pointers from the native into the SIDL array. Babel 0.9 was used to generate the interoperability layer for the underlying mesh implementation. The results were obtained on a dedicated Linux workstation with a 1.7 GHz Intel Pentium processor and 1 GB RD RAM using three meshes that ranged in size from 13,000 to 52,000 elements and work set sizes from 1 to 2,000 elements. The codes were compiled *without* optimization, and the timing data was measured in microseconds. Because of the consistency in results across the three different mesh sizes, the average value of all runs on the meshes is reported for each work set size.

The left-hand side of Figure 10.14 reports the percentage increase for the SIDL-based accesses compared to the baseline native interface access for increasing work set sizes. As expected, the additional function call and array conversion overhead of the SIDL interoperability layer is most noticeable when accessing entities using a work set size of 1 and ranges from 50% more expensive than the native interface for the *SIDL Direct* to 112% more expensive for the *SIDL Memcpy* variant. From work set size 2 on, the *SIDL For-loop* variant is the worst performing. For all cases, the *SIDL Direct* gives the best performance. As the work set size increases to 20 entities and beyond, the SIDL-related overhead decreases to 3.7% more than the native interface in the direct case and approximately 18.4% more in the for-loop case.

To gauge the overhead associated with functional interfaces compared to accessing the data directly using native data structures, the costs of all six access mechanisms are shown on the right-hand side of Figure 10.14. The results for the interface-based versions are displayed for work set sizes of 1 and 100 entities. As expected, the array traversal is the fastest and is 40% faster than traversing a linked list. Going through the native C interface is about 2.2 and 1.2 times slower than using the linked lists directly for work set sizes 1 and 100, respectively. For work sets of size 1 and 100, the *SIDL Direct* method gives the best SIDL performance and is 3.3 and 1.2 times slower, respectively, than using linked lists. For work sets of size 1, the memcpy SIDL variant is 4.7 times slower (versus 4.5 times slower for the for-loop version). This position is reversed for work sets of size 100, where the for-loop version is 44% slower (versus 29% for the memcpy version).

These experiments demonstrate that the granularity of access is critical in determining the performance penalty involved in restructuring an application to use an interface-based implementation. However, the granularity does not need to be very large. In fact, our experiments show that work sets of size 20 were sufficient to amortize the function call overhead. We also found that the additional overhead associated with transitioning from a native to language interoperable version of the interfaces can be negligible for suitable work set sizes.

10.5 Componentization Strategies

Next we examine strategies for developing software components for parallel PDE-based applications, including projects that incorporate legacy code as well as completely new undertakings. These general considerations have been employed when developing the reusable components discussed in Section 10.4 as well as throughout the application case studies presented in Section 10.6. Useful component designs may be coarse-grained (handling a large subset of the overall simulation task per component) or fine-grained. Similarly, an interface (port) between components may be a simple interface with just a few functions having a few simple arguments or a complex interface having all the functions of a library such as MPI.

The first step in defining components and ports for PDE-based simulations is considering in detail the granularity and application decomposition of the desired software. The second step is evaluating the impact on implementers and users of the chosen component and interface designs. This step may involve implementation and testing. The third step is iterating the design and implementation steps until a sufficient subset of the implementers and users is content with the resulting components and interfaces. The finest details of CCA component software design [5] and construction [19] are beyond our scope.

10.5.1 Granularity

How much functionality is inside each component? What information appears in public interfaces? In what format does the information appear? The answers to these questions determine the granularity of any component. We may want fine or coarse granularity or a combination.

Very coarse-grained designs. At the coarsest granularity, an entire PDE simulation running on a parallel machine can be treated as a single component. In most cases of this scenario, the component simply uses input files and provides output files. This approach permits the integration of separate programs by applying data file transformations. The overhead of transferring data using intermediate files may be reduced by instead transporting data directly from one application to another in any of several ways, but the essential aspect of copying data from one simulation stage to the next remains (see Section 10.4.7). The cost of moving large data sets at each iteration of an algorithm to and from file systems in large parallel machines can be prohibitive. Nonetheless, the CCA specification allows components to interact by file exchange. Software integration through files (see, e.g., Section 10.6.1) can be a useful starting point for the evolutionary development of a coupled simulation. Many tools have been built on this style of componentization, but scaling up to very large, multidomain, multiphysics problems is often inefficient or even impossible.

Finer-grained designs. For the remainder of this section we are concerned with building an application from component libraries that will result in executing a single job. Nearly every PDE-based application has one or more custom drivers that manage a suite of more general libraries handling different areas of concern, such as mesh definition, simulation data, numerical algorithms, and auxiliary services, such as file input and output, message passing, performance profiling, and visualization. Building component software for PDEs means teasing apart these many areas of concern into separate implementations and defining suitable functional interfaces for exchanging data among these implementations. Once clearly separated, the substitution and testing of an alternate implementation in any individual area is much easier.

10.5.2 Application Decomposition and Interface Design

Making components for a new or existing PDE-based simulation will be straightforward or difficult depending on how well the code is already decomposed into public interfaces and private implementations. We have found that the answers to the following technical questions about a PDE software system can aid in creating a good first approximation to an equivalent component design. Like all good software design, an iterative implementation and evaluation process is required to arrive at a final design.

- What are the equations to be solved, the space and time domains in which these equations apply, and the kinds of boundary and initial conditions that apply at the edges of these domains?
- What are appropriate meshing and spatial discretization techniques and an efficient machine representation of the resulting mesh?
- How is data stored for the simulation variables, and how are relationships of the stored data to various mesh entities (points, edges, faces, cells) represented?
- What are appropriate algorithms for solving the equations?
- What are appropriate implementations of the algorithms, given the algorithms and data structures selected? Can these implementations be factored into multiple independent levels, such as vector and matrix operations, linear solvers, nonlinear solvers, and time marching solvers?
- How can we identify, analyze, and store interesting data among the computed results?

Usually many mathematical and software design solutions exist for each of these questions, and of course they are interconnected. Our fundamental goal in component design for an application involving large-scale or multidomain PDEs is to separate these areas of concern into various software modules without reducing overall application performance. Many packages used for solving PDEs are already reasonably well decomposed into subsystems with clear (though often rather large) interfaces that do not impede performance; extending these packages to support component programming is usually simple unless they rely on global variables being set by the end user.

The performance requirement directly impacts the interfaces *between* components. Arrays and other large data structures that are operated on by many components must be exchanged by passing pointers or other lightweight handles. This requirement in turn necessitates specific public interface commitments to array and structure layout, which depend on the type of PDE and the numerical algorithms being used. CCA componentization allows similar packages to work together, but it does not provide a magic bullet solution to integrating packages based on fundamentally different assumptions and requirements. For example, a mesh and data component representing SAMR data as three-dimensional dense arrays [79, 77] is simply inappropriate for use in adaptive finite-element algorithms that call for data trees. When constructing new components, it is useful for interoperability to design them to be as general as reasonably possible with respect to details of the data structures they can accept. For example, we suggest dealing with arrays specified by strides through memory for each dimension rather than restricting a code to either row-major or column-major layout.

Some preprocessing or postprocessing components may demand flexibility in accessing many different kinds of mesh or data subsets. In this situation public interface definitions that require a function call to access data for each individual node or other mesh entity may be useful. However, as seen in testing the TSTT mesh interfaces (Section 10.4.8), single entity access function overhead slows inner loops, so that care should be taken when deciding to use such interfaces.

Success is often in the eye of the beholder when choosing a software decomposition and when naming functions within individual interfaces. Compact, highly reusable objects, such as vectors and other simple objects in linear algebra, may appear at first to be tempting to convert into components or ports. Instead, in many CCA applications, experience has shown that these objects are best handled as function arguments. Many high-performance, component-based implementations for PDEs have been reported. Most feature a combination of a mesh definition component, a data management component, and various numerical algorithm components [78, 73, 135, 93, 4, 19].

10.5.3 Evaluating a Component System Design

Each reusable component provides a set of public interfaces and may also use interfaces implemented by other components. Armed with our technical answers about the PDE software system that we will wrap, refactor, or create from scratch, we must ask questions about how we expect the components to be used and implemented. Is some aspect of the design too complex to be useful to the target users? Have we introduced an interface in such a way that we lose required efficiency or capability?

Component granularity checks. An application is formed by connecting a set of components and their ports together, and the more components involved in an application, the more complex it is to define. Too many components *may* be an indication of an overly fine decomposition. A good test is to consider the replacement of each component individually with an alternative implementation. If this would necessitate changing multiple components at once, then that group may be a candidate for merging into a single component. As a rule of thumb, we have found that if a *simple* test application, such as those discussed in Section 10.3.3, requires more than seven components, then the decomposition may be worth revisiting.

Similarly, a component providing or using too many ports may be an indication that it contains too much functionality to be manageable and should be decomposed further. Multiple ports may offer alternative interfaces to the same functionality, but many unrelated ports may signal a candidate for further study. A detailed case study of component designs for ODE integrations is given in [5].

User experience. Empirical evidence determined by CFRFS combustion researchers, whose work is introduced in Section 10.2.3, indicates that the final granularity of a component-based application code is arrived at iteratively. One starts with a coarse-grained decomposition, which in their case was dictated by the nature of their time-integration scheme, and moves to progressively more refined and fine-grained designs. For the CFRFS researchers, the more refined decompositions were guided by physics as represented as mathematical operators and terms in the PDE system being solved. The finest granularity was achieved at the level of physical/-chemical models. For example, in their flame simulation software, the transport system formed an explicit-integration system. In the second level of refinement, transport was separated into convection and diffusion, which occurs as separate terms in their governing equations. In the final decomposition, diffusion was separated into a mathematical component that implemented the discretized diffusion operator, while the functionality of calculating diffusion coefficients (required to calculate the diffusion term and used in the discretized diffusion operator) was separated as a specialized component [79]. Such a decomposition enables the testing of various diffusion coefficient models and discretizations by simply replacing the relevant component. Since these components implement the same ports, this activity is literally plug-and-play.

Port complexity checks. Many ports for PDE computations are as simple as a function triplet (setup, compute, finish) with a few arguments to each function (see, for example, the climate component discussion in Section 10.6.2). Port interface design complexity can be measured in terms of the number of functions per port and the number of arguments per function. If both these numbers are very high, the port may be difficult to use and need further decomposition. Should some of the functions instead be placed in a component configuration port that is separate from the main computation function port? Is a subset of the functions in the port unique to a specific implementation, making the port unlikely to be useful in any other component? Are there several subsets of functions in the port such that using one subset implies ignoring other subsets? Conversely, there may be so few functions in a port that it is always connected in parallel with another port. In this case the two ports may be combined.

10.5.4 Adjusting Complexity and Granularity

Of course a degree of complexity is often unavoidable in the assembly of modern applications. The CCA specification provides for *containers*, which can encapsulate assemblies of components to further manage complexity [20]. This capability allows a component set to appear as a single component in an application, so that the granularity of components and complexity of interfaces can be revised for new audiences without major re-implementation. The container may expose a simplified port with fewer or simpler functions (and probably more built-in assumptions) than a similar complex port appearing on one of its internally managed components. Some or all of the ports not connected internally may be forwarded directly to the exterior of the container for use at the application level.

10.6 Case Studies: Tying Everything Together

The fundamental contribution of component technology to the four parallel PDE-based applications introduced in Section 10.2 and discussed in detail in this section is the enforcement of modularity. This enforcement is *not* the consequence of programming discipline; rather, it is a fundamental property of the component paradigm itself. The advantages observed are those naturally flowing from modularization:

1. *Maintainability*: Components divide complexity into manageable chunks, so that errors and substandard implementations are localized and easily identifiable, and consequently may be quickly repaired. Further, the consequences of careless design of one component often stop at its boundary.

2. *Extensibility*: Modularization limits the amount of detail that one has to learn before beginning to contribute to a component-based application. This simplifies and accelerates the process of using and contributing to an external piece of software and thus makes it accessible to a wider community.
3. *Consistency*: Even though the component designs for the various projects have been agreed to rather informally within small communities, using the component-based architecture ensures through compile-time checking that object-oriented, public interfaces (ports) are used *consistently* everywhere. This approach eliminates errors often associated with older styles of interface definition such as header files with global variables and C macros that may depend on compiler flags or potentially conflicting Fortran common block declarations in multiple source files. For example, in the CFRFS project introduced in Section 10.2.3, interfaces and the overall design could be (and was) changed often and without much formal review. However, each port change *had to be* propagated through all the components dependent on that port interface in order for the software to compile and link correctly. This requirement enforced uniformity and consistency in the design. As the CFRFS toolkit grew, major interface changes became more time-consuming, compelling the designers to design with care and completeness in the first place, a good software engineering practice under any conditions.

We now discuss how component technology has been applied in these four scientific applications, each of which faces different challenges and is at a different stage of incorporating component concepts. We begin in Section 10.6.1 by discussing the accelerator project, introduced in Section 10.2.1, which is currently at an early phase of exploring a component philosophy for mesh infrastructure to facilitate experimentation with different meshing technologies, as introduced in Section 10.4.1. Section 10.6.2 explains how climate researchers have decomposed their models using the general principles introduced in Section 10.5 to develop next-generation prototype applications that handle model coupling issues, which were introduced in Sections 10.2.2 and 10.4.7. Section 10.6.3 highlights how the plug-and-play nature of CCA components enables combustion scientists to easily explore different choices in algorithms and data structures and thereby achieve their scientific goals, introduced in Section 10.2.3. The final application, discussed in Section 10.6.4, explains how CCA components help to harness the complexity of interdisciplinary simulations involving accidental fires and explosions, as introduced in Section 10.2.4. Here components allow diverse researchers to work together without being in lock step, so that a large, multiphysics application can achieve efficient and scalable performance on a wide range of parallel architectures.

10.6.1 Accelerator Modeling

As introduced in Section 10.2.1, ongoing collaborations among scientists in the TSTT Center and SLAC have resulted in a number of improvements to the mesh generation tools and software infrastructure used for accelerator modeling. Although the TSTT mesh interfaces are not yet mature enough for direct use in an application code, a component philosophy is being employed to insert TSTT adaptive mesh capabilities into the finite element-based frequency domain code, Omega3P.

Initially, the goal was to demonstrate the benefits of adaptive mesh refinement without changing a line of code in the core of Omega3P. This goal was accomplished by cleanly dividing the responsibilities of the different pieces of software and iteratively processing the mesh until convergence was achieved. In particular, TSTT tools developed at RPI handled error estimation and adaptive refinement of the mesh, while Omega3P computed the solution fields. Information was exchanged between Omega3P and the TSTT meshing tools by using a file-based mechanism in which the current mesh and solution fields were written to a file that was then read by the RPI tools.

Although the performance of a file-based mechanism for information transfer between adaptive refinement steps is clearly not ideal, it proved to be an excellent starting point because it allowed a very quick demonstration of the potential benefits of adaptive mesh refinement for the Omega3P code. As mentioned in Section 10.2.1, a high degree of accuracy is required in the frequency domain results. For one commonly used test geometry, the Trispal geometry, the results from the adaptive refinement loop were more accurate than those from prior simulations and provided the best match to experimental results at a fraction of the computational cost [47]. Furthermore, this work has showcased the benefits of modularizing various aspects of the simulation process by allowing SLAC researchers to quickly use AMR technologies without requiring a wholesale change to their code. Based on the success of this demonstration, work is now proceeding to insert the adaptive refinement loop directly into Omega3P using the TSTT interface philosophy and the underlying implementations at RPI.

10.6.2 Climate Modeling

In Section 10.2.2 we described how the climate system's complexity leads to software complexity in climate system models. Here we discuss in greater detail the practices of the climate/weather/ocean (CWO) community; for brevity, our scope is restricted to atmospheric global climate models. We discuss the refactoring of CWO software to make it more component friendly and to alleviate complexity. We describe the CWO community's effort to create its own component specification (ESMF). We also present a prototype component-based atmospheric advection model, which uses both the ESMF and CCA paradigms. See [73] for further information on these topics.

Model Decomposition

As mentioned in Section 10.2.2, the fundamental equations for atmosphere and ocean dynamics are called the primitive equations. Their solvers are normally structured in two parts. The first part, which solves the primitive equations, is called the *dynamics*, *dynamical core*, or *dycore*. The second part, which models source and sink terms

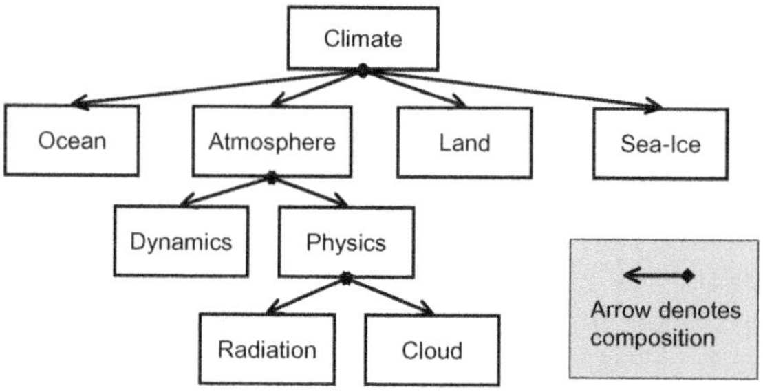

Fig. 10.15. Diagram of climate models decomposed in terms of components.

resulting from length scales shorter than those used in the dynamical core's discretization, is called the *physics*. Examples of parameterized physical processes in the atmosphere include convection, cloud formation, and radiative transfer. In principle, one could use the same solver infrastructure for both atmosphere and ocean dynamics, but this approach is rarely used because of differences in model details.

Climate models are a natural application for component technology. Figure 10.15 illustrates one of the ways for the component decomposition at several levels. The highest level integrates the major subsystems of the earth's climate (ocean, atmosphere, sea-ice, and land-surface), which are each a component. Within the atmosphere, we see a component decomposition of the major parts of the model—the dynamics and the physics. Within the physics parameterization package, each subgridscale process can also be packaged as a component.

The plug-and-play capabilities of component-based design, as introduced in Section 10.3, can aid researchers in exploring trade-offs among various choices in meshing, discretization, and numerical algorithms. As an example we consider atmospheric global climate models, in which various approaches, each with different advantages and disadvantages, can be used for solving the primitive equations. The main solvers are finite-difference [41] and spectral methods [50]; semi-Lagrangian and finite element techniques are also sometimes used. Various solvers have been developed for each approach, including the Aries dycore [115], the GFDL and NCAR spectral dycores [50, 123], and the Lin-Rood finite-volume dycore [83]. An additional challenge is handling the physical mesh definition; choices for the horizontal direction include logically Cartesian latitude-longitude, geodesic [36], and cubed-sphere [86]. Various choices for the vertical coordinate include pressure, sigma-pressure, isentropic, or a combination of these.

A primary challenge in developing atmospheric models is achieving scalable performance that is portable across a range of parallel architectures. For example, parallel domain decompositions of atmospheric dycores are usually one- or two-dimensional in the horizontal direction. Most codes use MPI or an MPI/OpenMP hybrid scheme for parallelization because it provides solid performance and portability. Component-based design helps to separate issues of parallelism from portions of code that handle physics and mathematics, thereby facilitating experimentation with different parallel computing strategies. Another challenge is language interoperability. The modernization path for most climate models has been a migration from Fortran 77 to Fortran 90, combined with some refactoring to increase modularity. Few of these models are implemented in C or C++ [119]. As discussed in Section 10.3, the programming language gap between applications and numerical libraries is an issue that component technology can help to bridge.

In response to a CWO component initiative, scientists have been refactoring their application codes. Initially, this activity was undertaken simply to provide better modularity and sharing of software among teams. A good example is the refactoring of the Community Atmosphere Model (CAM) to split the previously entangled physics and dynamics portions of code. This entanglement made the change of one dycore for another an arduous process. Now that the physics and dynamics have been split, CAM has three dycores: the spectral dycore with semi-Lagrangian moisture transport, the Lin-Rood finite-volume dycore, and the Williamson-Rasch semi-Lagrangian scheme. This refactoring will ease the integration and testing of the newly developed NCAR spectral element dycore. An effort is also under way to repackage these dycores as ESMF components (see below), which will be the first introduction of component technology to CAM.

Model Coupling

The primitive equations are a boundary-value problem. For the atmosphere, the lateral boundary values are periodic, the top of the atmosphere's boundary condition is specified, and the boundary conditions at the Earth's surface are provided by the ocean, sea-ice, and land-surface components. This mutual interaction between multiple subsystems requires MxN parallel data transfers, such as those described in Section 10.4.7. This need for boundary data also poses the problem of how to schedule and execute the system's atmosphere, ocean, sea-ice, and land-surface components to maximizes throughput. There are two basic scheduling strategies. The first is a sequential event-loop strategy, in which the components run in turn on the same pool of processors (e.g., the Parallel Climate Model (PCM) [23]). The second strategy is concurrent component execution, in which each model executes independently on its own pool of processors (e.g., CCSM). Componentization of the land, atmosphere, ocean, and sea-ice models will increase overall flexibility in scheduling the execution of a climate model's constituents and thereby facilitate aggressive experimentation in creating previously unimplemented climate system models.

ESMF and the CCA

The great potential that components offer for enabling new science has inspired the CWO community embrace component technology. Of particular note is the NASA-funded interagency project to develop the Earth System Modeling Framework [67, 57]. The ESMF comprises a *superstructure* and an *infrastructure*. The superstructure provides the component specification and the overall component interfaces used in coupling. The infrastructure includes commonly needed low-level utilities for error handling, input/output, timing, and overall time management. The infrastructure also provides a common data model for coordinate grids, physical meshes, and layout of field data, as well as services for halo updates, parallel data transfer, and intergrid interpolation, much like the facilities described in Section 10.4.7. One distinguishing feature of ESMF components is that they have three methods: `Initialize`, `Run`, and `Finalize`. The ESMF supplies its coupling data in the form of the `ESMF_State` datatype.

ESMF developers have collaborated with the CCA to ensure framework interoperability, so that ESMF components may run in a CCA-compliant framework and vice versa. This effort will provide scientists application-specific ESMF services in composing climate components, while also enabling the use of CCA numerical components, such as those described in Section 10.4.

A Prototype Component-Based Climate Application

A prototype coupled atmosphere-ocean application, which employs both the CCA and ESMF component paradigms, has been developed as proof-of-concept application for CWO codes [137, 136]. The application combines the CCA component registration infrastructure and *uses-provides* interaction model introduced in Section 10.3 with the ESMF's component method specification (i.e., `Initialize`, `Run`, `Finalize`) and data model. (i.e., `ESMF_State`). This application includes a component common to the atmosphere and ocean dycores, namely, two-dimensional advection of a quantity Ψ by the horizontal velocity field (u, v):

$$\frac{\partial \Psi}{\partial t} + u\frac{\partial \Psi}{\partial x} + v\frac{\partial \Psi}{\partial y} = S,$$

where $\Psi(x, y, t)$ is the advected quantity, and $S(x, y, t)$ is the sum of all sources and sinks. The x-y spatial grid is rectangular, and the discretization method is a finite-difference scheme. Here we consider three finite difference variants, which are each forward in time and either forward-, central-, or backward-difference in space.

Following the CCA component specification, we created an advection component with a solver port definition for a finite-difference scheme. The advection equation can be solved by using forward-, central-, or backward-differencing in space. We employ a *proxy* design pattern [46] to allow the atmospheric model component the choice of one of these default solvers or a user-designated scheme. We also use CCA technology to enable the user to specify run-time parameters such as advection speed. The ability to easily swap in different implementations in this component-based advection application has proven useful in exploring differences in the accuracy and computational complexity of the various numerical methods.

10.6.3 Combustion

The objective of the CFRFS [90] project introduced in Section 10.2.3 is the creation of a component-based toolkit for simulating laboratory-sized (0.1^3 m) flames with detailed chemistry. Such flames contain tens of species, hundreds of reactions, spatial structures 10^{-4} meters in size, and timescales ranging from 10^{-9} seconds (chemical processes) to 10^{-1} seconds (convective processes). The low Mach Navier-Stokes equation [130, 91], and the equations for species' evolution comprise a set of coupled PDEs of the form

$$\frac{\partial \mathbf{\Phi}}{\partial t} = \mathbf{F}(\mathbf{\Phi}, \nabla \mathbf{\Phi}, \nabla^2 \mathbf{\Phi}, ...) + \mathbf{G}(\mathbf{\Phi}),$$

where $\mathbf{\Phi}$ consists of flow quantities such as density and temperature. The equation is discretized on rectangular meshes and solved in rectangular domains. For these systems $\mathbf{G}$ involves the variables only at a given mesh point, while $\mathbf{F}$, which involves spatial derivatives (computed by using finite-difference or finite-volume schemes), depends on the mesh point and its close neighbors. $\mathbf{G}$ is stiff, so that the ratio of the largest and the smallest eigenvalues of $\partial \mathbf{G}/\partial \mathbf{\Phi}$ is large, while $\mathbf{F}$ is non-stiff. Operator splitting [113, 114] is employed to evolve the stiff ($\mathbf{G}$) and nonstiff ($\mathbf{F}$) terms in a decoupled manner by following a $\mathbf{GFG}$ sequence, thus letting the stiff operator be the last in the time step, in order to achieve higher accuracy in the data reported at the end of the time step. A backward-difference formulation [33] and a Runge-Kutta-Chebyshev integrator [9] are used for the stiff and nonstiff problem, respectively. The solution vector $\mathbf{\Phi}$ exhibits steep spatial variations in scattered, time-evolving regions of the domain. Block-structured adaptive mesh refinement (SAMR) [18] and time refinement [17] are used to track and resolve these regions.

The CFRFS team used CCA-compliant component technology to explore the use of high-order spatial discretizations in a SAMR setting [76, 78, 107] for the first time and to perform scalability studies of reacting flow problems on SAMR meshes.

High-Order Spatial Discretizations and Block SAMR

PDEs can be discretized by a variety of methods [100]. Finite differences and volumes are popular for solving fluid flows. Typically, second-order spatial discretizations are used, although high-order spatial discretizations on *single-level* structured or unstructured meshes are becoming common [81, 127, 128, 82, 32, 62]. The CFRFS team explored the use of high-order (> 2) schemes in *multilevel* block-structured adaptive meshes [76, 78]. Multilevel, block-structured meshes are a conceptually elegant way of achieving resolution in simple domains. One starts with a coarse, structured, logically rectangular mesh. Regions requiring resolution are identified, collated into patches, and overlaid with a rectangular mesh of higher density. This high-density patch is not embedded; rather, it is preserved separately as a fine patch. This process is carried out recursively, leading to a hierarchy of patches, that is, a multilevel grid hierarchy [17]. In this way a given point in space is resolved at different resolutions simultaneously, by different levels of the grid hierarchy.

Deep grid hierarchies pose significant load-balancing problems. High-order spatial discretizations present a simple solution because they may provide an acceptable degree of accuracy on relatively coarse meshes (i.e., with relatively shallow grid hierarchies). Incorporating high-order schemes in a SAMR setting is nontrivial, however, as the software infrastructure associated with parallel SAMR codes is very complex. Indeed, the mathematical complexities of high-order schemes have

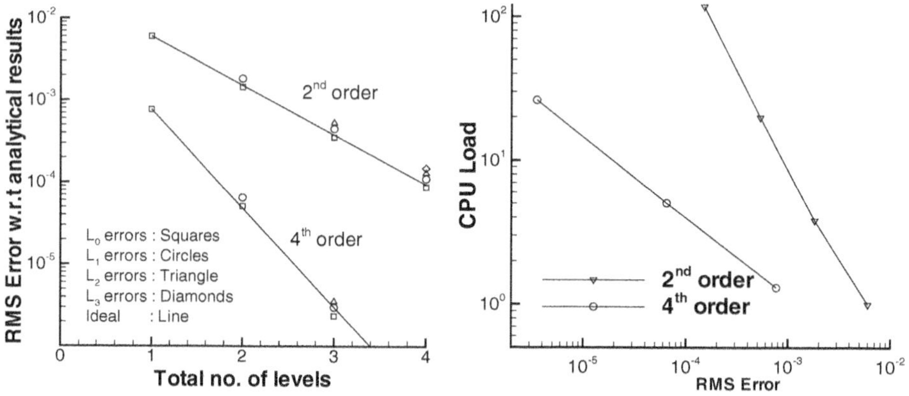

Fig. 10.16. *(Left)*: The root mean squared (RMS) error on the individual levels, as the simulation is run on a 1-, 2-, 3- and 4-level grid hierarchy. *(Right)*: The computational load versus RMS error for the second- and fourth-order approaches. Results have been normalized by the computational load of a second-order, 1-level grid hierarchy run.

restricted their use to relatively simple problems. The component design established a clear distinction between the various domains of expertise and a means of incorporating the contributions of diverse contributors without imposing a programming and data-structural straitjacket. Most contributions were written by experts in Fortran 77 and then componentized.

These components were used to simulate PDEs on multilevel meshes, with factor-of-two refinements between levels. The left-hand side of Figure 10.16 shows the recovery of the theoretical convergence rate as the effective resolution was increased by increasing the number of levels. The base grid has 100 cells in the [0,1] domain. Both second- and fourth-order discretizations were used. High-order schemes were found to be more economical than second-order schemes because they required sparser meshes to achieve a given level of accuracy. Further, higher-order schemes become progressively more economical vis-a-vis second-order approaches as the error tolerances become stringent. This behavior is evident in the right-hand side of Figure 10.16, which plots the computational loads (in terms of floating point operations count) normalized by the one-level grid hierarchy load (1,208,728,001 operations).

Strong-Scalability Analysis of a SAMR Reacting Flow Code

SAMR scalability studies are rare [131, 132] and usually specific to the applications being tested, that is, specific to the implemented algorithm, and hence are difficult to analyze and interpret. To explore the behavior of the parallel CCA-based block-SAMR toolkit and to identify the scalability bottlenecks, the CFRFS team performed a *strong scaling* study (i.e., the global problem size was kept constant while the number of processors increased linearly) for a two-dimensional reaction-diffusion problem with detailed hydrogen-air chemistry using three levels of refinement with

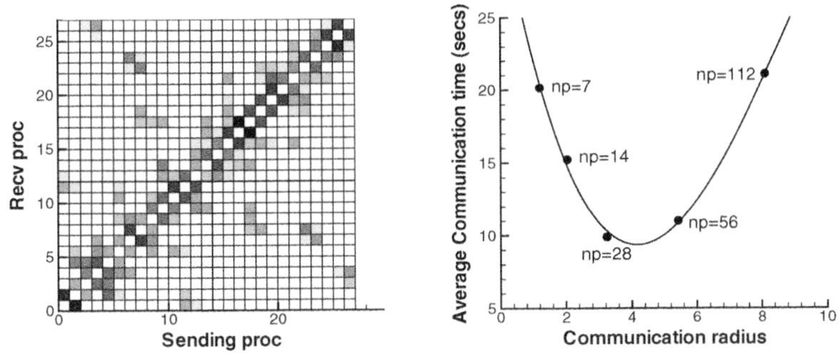

Fig. 10.17. *(Left)*: Communication patterns for 28 processors at timestep 40. *(Right)*: Communication costs as a function of the communication radius at timestep 40. (For the color version, see Figure A.22 on page 477).

a refinement factor of two [80]. The initial condition was a random distribution of temperature kernels in a stoichiometric hydrogen-air mixture. For this experiment, the parallel virtual machine expanded by a factor of two, starting with 7 processors and reaching 112 processors. Time and messaging volumes were measured by connecting the TAU [124] performance analysis component to the CFRFS component code assembly.

Results indicated that the overall scalability of the adaptive algorithm is highly dependent on the scalability of the intermediate time steps [80]. There were "scalable" and "nonscalable" time steps, depending on the quality of the domain decomposition. The nonscalable time steps were a consequence of synchronization times, where many processors idled because of severely uneven computational load partitioning. The scalable time steps showed good load-balance, but their communication times increased as the number of processors increased. The left-hand side of Figure 10.17 shows the communication map for a 28-processor run. While the bulk of the communication is with the nearest neighbors, there are a significant number of outliers. The remoteness of these outliers was characterized by an average communication radius r. The right-hand side of Figure 10.17 shows that the average communication time per processor increases with r (after $r \sim 4$), a counterintuitive result, as increasing r indicates more processors and smaller per-processor problem sizes. The explanation lies in the network topology. The scaling study was performed on a cluster with Myrinet, which has a Clos-network topology. Eight nodes are connected to a switch; a cascade of 16-port switches ensures full connectivity, though at increasing levels of indirection. As $r \longrightarrow 8$, increasing fractions of the total communication occur over the cascade, as opposed to in-switch communication. This situation results in message contentions and collisions and hence slower transfer speeds and larger communication costs.

10.6.4 Accidental Fires and Explosions

The simulation environment for the Center for the Simulation of Accidental Fires and Explosions (C-SAFE) [55], introduced in Section 10.2.4, is the Uintah Computational Framework (UCF) [40], which is a set of software components and libraries that facilitate the parallel simulation of PDEs on structured adaptive mesh refinement (SAMR) grids. The UCF is implemented in the context of the CCA-based SCIRun2 framework [135], which supports a wide range of computational and visualization applications. One of the challenges of creating component-based PDE software is achieving scalability, which is a global application property, through components that, by definition, make local decisions.

Managing Parallelism via Taskgraphs

To address this challenge of managing parallelism in multidisciplinary applications, the UCF employs a nontraditional approach. Instead of using explicit MPI calls throughout each component of the program, applications are cast in terms of a *taskgraph*, which describes the data dependencies among various steps of the problem.

Computations are expressed as directed acyclic graphs of *tasks*, each of which produces some output and consumes some input, which is in turn the output of some previous task. These inputs and outputs are specified for each patch in a structured AMR grid. Associated with each task is a method that performs the actual computation. This representation has many advantages, including efficient fine-grained coupling of multiphysics components, flexible load balancing mechanisms, and a separation of application and parallelism concerns. Moreover, UCF data structures are compatible with Fortran arrays, so that application writers can use Fortran subroutines to provide numerical kernels on each patch.

Each execution of a taskgraph integrates a single timestep, or a single nonlinear iteration, or some other coarse algorithmic step. Tasks communicate with each other through an entity called the *DataWarehouse*. The DataWarehouse is accessed through a simple name-based dictionary mechanism, and it provides each task with the illusion that all memory is global. If the tasks correctly describe their data dependencies, then the data stored in the DataWarehouse will match the data (variable and region of space) needed by the task. In other words, the DataWarehouse is an abstraction of a global single-assignment memory, with automatic data lifetime management and storage reclamation. Values stored in the DataWarehouse are typically array-structured. Communication is scheduled by a local algorithm that approximates the true globally optimal communication schedule. Because of the flexibility of single-assignment semantics, the UCF is free to execute tasks close to data or move data to minimize future communication.

The UCF storage abstraction is sufficiently high level that it can be efficiently mapped onto both message-passing and shared-memory communication mechanisms. Threads sharing a memory can access their input data directly; single-assignment dataflow semantics eliminate the need for any locking of values. Threads running in disjoint address spaces communicate by a message-passing protocol, and

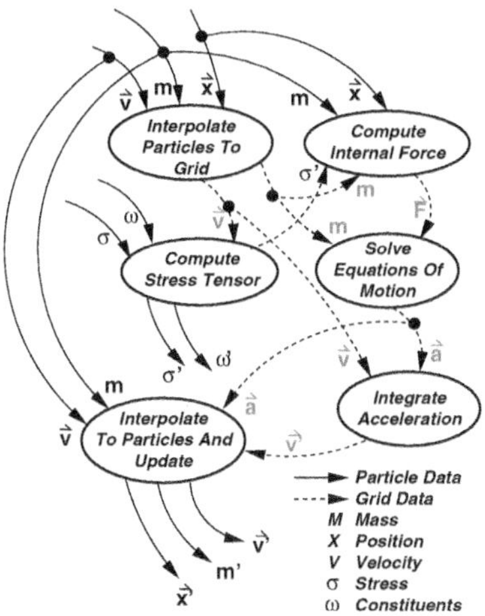

Fig. 10.18. An example UCF taskgraph, depicting a portion of the material point method (MPM) algorithm used to simulate solid materials in C-SAFE scenarios.

the UCF is free to optimize such communication by message aggregation. Tasks need not be aware of the transports used to deliver their inputs, and thus UCF has complete flexibility in control and data placement to optimize communication both between address spaces or within a single shared-memory symmetric multiprocessing node. Latency in requesting data from the DataWarehouse is not an issue; the correct data is deposited into the DataWarehouse before each task is executed.

Consider the taskgraph in Figure 10.18. Ovals represent tasks, each of which is a simple array-based subroutine. Edges represent named values stored by the UCF. Solid edges have values defined at each material point, and dashed edges have values defined at each grid vertex. Variables denoted with a prime (') have been updated during the time step. The figure shows a portion of the Uintah material point method (MPM) [116] taskgraph concerned with advancing Newtonian material point motion on one patch for a single time step.

The idea of the dataflow graph as an organizing structure for execution is well known. The SMARTS [125] dataflow engine that underlies the POOMA [108] toolkit shares goals and philosophy with the UCF. Sisal compilers [45] used dataflow concepts at a much finer granularity to structure code generation and execution. Dataflow is a simple, natural, and efficient way of exposing parallelism and managing computation and is an intuitive way of reasoning about parallelism. What distinguishes implementations of dataflow ideas is that each caters to a particular higher-level presentation. SMARTS is tailored to POOMA's C++ implementation and stylistic template-based presentation. The UCF supports a presentation catering to C++ and Fortran-based mixed particle/grid algorithms on structured adaptive meshes, and the primary algorithms of importance to C-SAFE are the MPM and Eulerian computational fluid dynamics algorithms.

This dataflow-based representation of parallel computation fits well with the structured AMR grids and with the nature of the computations that C-SAFE performs. In particular, we used this approach in order to accommodate multiphysics integration, load-balancing, and mixed thread/MPI programming. A more detailed discussion of these advantages (and disadvantages) can be found in [98].

The most important advantage for a large interdisciplinary project such as C-SAFE is that the taskgraph facilitates the separate development of simulation components and allows pieces of the simulation to evolve independently. Because C-SAFE is a research project, we need to accommodate the fact that most of the software is still under development. The component-based architecture allows pieces of the system to be implemented in a basic form at first and then to evolve as the technologies mature. Most importantly, the UCF allows the aspects of parallelism (schedulers, load-balancers, parallel input/output, and so forth) to evolve independently of the simulation components. This approach allows the computer science effort to focus on these problems without waiting for the completion of the scientific applications or vice-versa.

Components Involved

Figure 10.19 shows the main components involved in a typical C-SAFE simulation. The simulation controller component, which is in charge of the simulation, manages restart files if necessary and controls the integration through time. First, it reads the specification of the problem from an XML input file. After setting up the initial grid, it passes the description to the simulation component, which can implement various algorithms, including one of two different CFD algorithms, the MPM algorithm, or a coupled MPM-CFD algorithm. The simulation component defines a set of tasks for the scheduler. In addition, a data-archiver component describes a set of output tasks to the scheduler. These tasks save a specified set of variables to disk. Once all tasks are known to the scheduler, the load-balancer component uses the machine configuration to assign tasks to processing resources. The scheduler uses MPI for communication and then executes callbacks to the simulation or data-archiver components to perform the actual work. This process continues until the taskgraph has been fully executed. The execution process is then repeated to integrate further time steps.

Each of these components runs concurrently on each processor. The components communicate with their counterparts on other processors using MPI. However, the scheduler is typically the only component that needs to communicate with other processors. Figure 10.20 demonstrates that the resulting system scales well on various parallel architectures. Delegating responsibility for parallelism to the scheduler component allows complex multiphysics applications to utilize processor resources efficiently and reduces the programming burden for applications that require complex communication patterns to achieve good scalability.

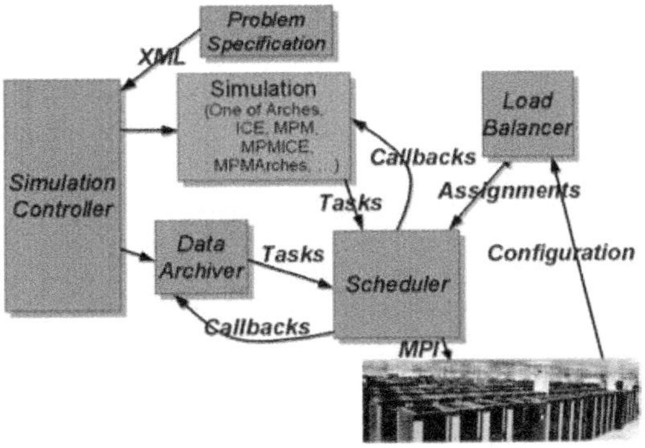

Fig. 10.19. UCF simulation components. The simulation describes tasks to a scheduling component, which are assigned to processing resources by a load-balancer component. Callbacks are made into the simulation component to perform the computation. Checkpointing and data input/output are performed automatically by the data-archiver component.

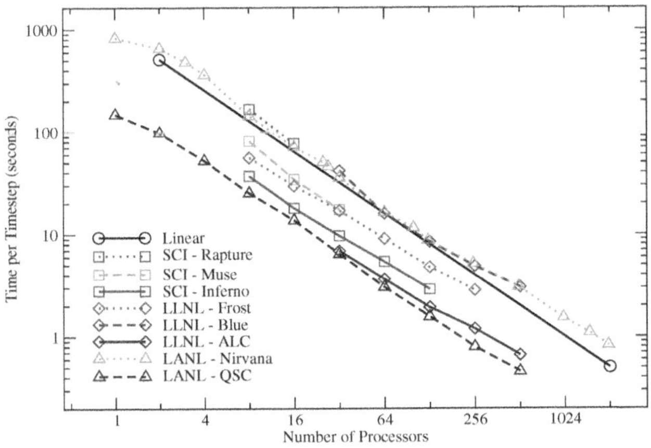

Fig. 10.20. Performance of the UCF MPM simulation on various architectures during the development of the simulation. Rapture: 32-processor 250 MHz Origin 2000, Muse: 64-processor 600 MHz Origin 3000, Inferno: 256-processor 2.4 GHz Pentium 4 Linux cluster, Frost: 1024-processor IBM SP Power 3, Blue: 3392-processor IBM SP PowerPC 604e, ALC: 1024-processor 2.4 GHz Pentium 4 Linux cluster, Nirvana: 2048-processor 250 MHz Origin 2000 at LANL, QSC: 256-processor 1.25 GHz Alpha. Data courtesy of Randy Jones, Jim Guilkey, and Todd Harman of the University of Utah.

10.7 Conclusions and Future Work

All component-based approaches to software seek to divide the inherent complexity of large-scale applications into sizes that human beings can deal with individually, so that more complex applications can be constructed from these manageable units. Parallel PDE-based applications are unique in this context only to the extent that they tend to be extremely complex and thus can profoundly benefit from component concepts. The CCA contributes a component model for high-performance scientific computing that may be of particular interest to investigators conducting simulations of detailed or comprehensive physical phenomena. Compliance with the CCA specification has enabled the scientific application teams featured in this chapter to perform several tasks more easily:

- Create sets of reusable, easy-to-maintain, and scalable components, each of which expresses a unique physical or numerical functionality [77, 79, 98, 73, 137]
- Use legacy code (originally written in Fortran or C) in the CCA environment without major code rewrites [77, 73]
- Easily test different physics and numerics modules with well-defined interfaces in a plug-and-play mode [79]
- Manage the evolution of complex scientific applications, by separating the disparate concerns of physics, numerical, and computer science issues [40, 98]
- Obtain good parallel performance [79, 39, 87] with negligible CCA overhead [21]

There is no point at which we envision the CCA as a component model will be finished. The CCA continues to respond to implementers' concerns, feature requests, and unforeseen conflicts created by CCA-specified mechanisms or the lack thereof. In addition, the CCA Forum is extending the prototype work of Section 10.4 and assembling a critical mass of components from which parallel simulations can be prototyped and evolved into meaningful simulations. Component concepts also provide unprecedented opportunities for automation. Recent work on *computational quality of service* [59, 94] allows component parameters and component configurations to be rearranged dynamically, thereby enabling the automatic selection and configuration of components to suit the computational conditions imposed by a simulation and its operating environment. The CCA Forum aims to enable next-generation high-performance scientific simulations by providing a means for tens or even hundreds of researchers to contribute to a single application as well as by developing the infrastructure to automate its construction and execution.

Acknowledgments

The CCA has been under development since 1998 by the CCA Forum and represents the contributions of many people, all of whom we gratefully acknowledge. We also thank our collaborators outside the CCA Forum, especially the domain scientists who have contributed to the four applications discussed in this chapter and the early adopters of the CCA, for the important contributions they have made both to our understanding of CBSE in the high-performance scientific computing context and to making the CCA a practical and usable environment. We thank Barry Smith for developing a PETSc-based implementation of the TOPS solver interfaces discussed in Section 10.4.4. In addition, we thank Are Magnus Bruaset, Jeff Keasler, Barry

Smith, and the anonymous reviewers of this chapter for valuable feedback that has enabled us to improve our presentation.

This work has been supported in part by the U. S. Department of Energy's Scientific Discovery through Advanced Computing (SciDAC) [122] initiative, through the Center for Component Technology for Terascale Simulation Software, of which Argonne, Lawrence Livermore, Los Alamos, Oak Ridge, Pacific Northwest, and Sandia National Laboratories, Indiana University, and the University of Utah are members. Members of the SciDAC Computational Facility for Reacting Flow Research have also contributed to this paper.

Research at Argonne National Laboratory was supported in part by the Mathematical, Information, and Computational Sciences Division subprogram of the Office of Advanced Scientific Computing Research, Office of Science, U.S. Department of Energy, under Contract W-31-109-ENG-38.

Some of this work was performed under the auspices of the U.S. Department of Energy by University of California Lawrence Livermore National Laboratory under contract No. W-7405-Eng-48.

Oak Ridge National Laboratory is managed by UT-Battelle, LLC for the US Dept. of Energy under contract DE-AC-05-00OR22725.

This research was performed in part using the Molecular Science Computing Facility (MSCF) in the William R. Wiley Environmental Laboratory at the Pacific Northwest National Laboratory (PNNL). The MSCF is funded by the Office of Biological and Environmental Research in the U.S. Department of Energy. PNNL is operated by Battelle for the U.S. Department of Energy under contract DE-AC06-76RLO 1830.

Research at the University of Utah is also sponsored by the National Science Foundation under contract ACI0113829, and the DOE ASC Program.

Some of the work in this paper was carried out by Northrop Grumman with funding provided by NASA's Computation Technologies (CT) Project, part of the Earth Science Technology Office (ESTO), under a contract with the National Aeronautics and Space Administration.

References

1. R. Ahrem, P. Post, B. Steckel, and K. Wolf. MpCCI: A tool for coupling CFD with other disciplines. In *Proceedings of the Fifth World Conference in Applied Fluid Dynamics, CFD-Efficiency and Economic Benefit in Manufacturing*, 2001.
2. R. Ahrem, P. Post, and K. Wolf. A communication library to couple simulation codes on distributed systems for multi-physics computations. In E. D'Hollander, G. Joubert, F. Peters, and H. Sips, editors, *Parallel Computing: Fundamentals and Applications, Proceedings of the International Conference ParCO 99*, pages 47–55. Imperial College Press, 1999.
3. B. Allan, R. Armstrong, S. Lefantzi, J. Ray, E. Walsh, and P. Wolfe. Ccaffeine – a CCA component framework for parallel computing. `http://www.cca-forum.org/ccafe/`, 2005.

4. B. A. Allan, R. C. Armstrong, A. P. Wolfe, J. Ray, D. E. Bernholdt, and J. A. Kohl. The CCA core specification in a distributed memory SPMD framework. *Concurrency and Computation: Practice and Experience*, 14(5):1–23, 2002.
5. B. A. Allan, S. Lefantzi, and J. Ray. ODEPACK++: Refactoring the LSODE Fortran library for use in the CCA high performance component software architecture. In *Proceedings of the 9th International Workshop on High-Level Parallel Programming Models and Supportive Environments (HIPS 2004)*, Santa Fe, NM, April 2004. IEEE Press. see `http://www.cca-forum.org/~baallan/odepp`.
6. G. Allen, W. Benger, T. Goodale, H. Hege, G. Lanfermann, A. Merzky, T. Radke, E. Seidel, and J. Shalf. The Cactus code: A problem solving environment for the Grid. In *High Performance Distributed Computing (HPDC)*, pages 253–260. IEEE Computer Society, 2000.
7. P. Alpatov, G. Baker, C. Edwards, J. Gunnels, G. Morrow, J. Overfelt, R. van de Geijn, and Y.-J. J. Wu. PLAPACK: Parallel linear algebra package - design overview. In *Proceedings of SC97*, 1997.
8. R. Armstrong, D. Gannon, A. Geist, K. Keahey, S. Kohn, L. McInnes, S. Parker, and B. Smolinski. Toward a Common Component Architecture for high-performance scientific computing. In *Proceedings of the Eighth IEEE International Symposium on High Performance Distributed Computing*, 1999.
9. L. F. S. B. P. Sommeijer and J. G. Verwer. RKC: an explicit solver for parabolic PDEs. *J. Comp. Appl. Math.*, 88:315–326, 1998.
10. S. Balay, K. Buschelman, W. Gropp, D. Kaushik, M. Knepley, L. McInnes, B. F. Smith, and H. Zhang. PETSc users manual. Technical Report ANL-95/11 - Revision 2.2.1, Argonne National Laboratory, 2004. `http://www.mcs.anl.gov/petsc`.
11. S. Balay, W. D. Gropp, L. C. McInnes, and B. F. Smith. Efficient management of parallelism in object oriented numerical software libraries. In E. Arge, A. M. Bruaset, and H. P. Langtangen, editors, *Modern Software Tools in Scientific Computing*, pages 163–202. Birkhauser Press, 1997.
12. P. Beckman, P. Fasel, W. Humphrey, and S. Mniszewski. Efficient coupling of parallel applications using PAWS. In *Proceedings of the 7th IEEE International Symposium on High Performance Distributed Computation*, July 1998.
13. S. Benson, M. Krishnan, L. McInnes, J. Nieplocha, and J. Sarich. Using the GA and TAO toolkits for solving large-scale optimization problems on parallel computers. Technical Report ANL/MCS-P1084-0903, Argonne National Laboratory, September 2003.
14. S. Benson, L. C. McInnes, and J. Moré. A case study in the performance and scalability of optimization algorithms. *ACM Transactions on Mathematical Software*, 27:361–376, 2001.
15. S. Benson, L. C. McInnes, J. Moré, and J. Sarich. TAO users manual. Technical Report ANL/MCS-TM-242 - Revision 1.7, Argonne National Laboratory, 2004. `http://www.mcs.anl.gov/tao/`.
16. S. Benson and J. Moré. A limited-memory variable-metric algorithm for bound-constrained minimization. Technical Report ANL/MCS-P909-0901, Mathematics and Computer Science Division, Argonne National Laboratory, 2001.
17. M. Berger and P. Colella. Local adaptive mesh refinement for shock hydrodynamics. *J. Comp. Phys.*, 82:64–84, 1989.
18. M. J. Berger and J. Oliger. Adaptive mesh refinement for hyperbolic partial differential equations. *J. Comp. Phys.*, 53:484–523, 1984.
19. D. E. Bernholdt, B. A. Allan, R. Armstrong, F. Bertrand, K. Chiu, T. L. Dahlgren, K. Damevski, W. R. Elwasif, T. G. W. Epperly, M. Govindaraju, D. S. Katz, J. A. Kohl,

M. Krishnan, G. Kumfert, J. W. Larson, S. Lefantzi, M. J. Lewis, A. D. Malony, L. C. McInnes, J. Nieplocha, B. Norris, S. G. Parker, J. Ray, S. Shende, T. L. Windus, and S. Zhou. A component architecture for high-performance scientific computing. *Intl. J. High-Perf. Computing Appl.*, 2005. Submitted to ACTS Collection special issue, in press.

20. D. E. Bernholdt, R. C. Armstrong, and B. A. Allan. Managing complexity in modern high end scientific computing through component-based software engineering. In *Proceedings. of the HPCA Workshop on Productivity and Performance in High-End Computing (P-PHEC 2004), Madrid, Spain.* IEEE Computer Society, 2004.
21. D. E. Bernholdt, W. R. Elwasif, J. A. Kohl, and T. G. W. Epperly. A component architecture for high-performance computing. In *Proceedings of the Workshop on Performance Optimization via High-Level Languages and Libraries (POHLL-02)*, 2002.
22. F. Bertrand, R. Bramley, K. Damevski, J. Kohl, J. Larson, and A. Sussman. MxN interactions in parallel component architectures. Technical Report TR604, Department of Computer Science, Indiana University, Bloomington, 2004.
23. T. Bettge, A. Craig, R. James, and V. Wayland. The DOE Parallel Climate Model (PCM): The computational highway and backroads. In V. N. Alexandrov, J. J. Dongarra, B. A. Juliano, R. S. Renner, and C. J. K. Tan, editors, *Proceedings of the International Conference on Computational Science (ICCS) 2001*, volume 2073 of *Lecture Notes in Computer Science*, pages 148–156, Berlin, 2001. Springer-Verlag.
24. G. E. Blelloch, M. A. Heroux, and M. Zagha. Segmented operations for sparse matrix computation on vector multiprocessor. Technical Report CMU-CS-93-173, Carnegie Mellon University, 1993.
25. Boost. `http://www.boost.org`, 2005.
26. D. Box. *Essential COM*. Addison-Wesley, December 1997.
27. R. Bramley, D. Gannon, T. Stuckey, J. Vilacis, E. Akman, J. Balasubramanian, F. Berg, S. Diwan, and M. Govindaraju. The linear system analyzer. In *Enabling Technologies for Computational Science*, Kluwer, 2000.
28. CCA Forum homepage. `http://www.cca-forum.org/`, 2005.
29. CCA Specification. `http://cca-forum.org/specification/`, 2005.
30. S. Chatterjee, G. E. Blelloch, and M. Zagha. Scan primitives for vector computers. In *Supercomputing 1990*, 1990.
31. R. Clay et al. ESI homepage. `http://www.terascale.net/esi`, 2001.
32. B. Cockburn and C.-W. Shu. The local discontinuous Galerkin method for time-dependent convection-diffusion systems. *SIAM J. Numer. Anal.*, 35(6):2440–2463, 1998.
33. S. D. Cohen and A. C. Hindmarsh. CVODE, a stiff/nonstiff ODE solver in C. *Computers in Physics*, 10(2):138–143, 1996.
34. P. Colella. An Algorithmic and Software Framework for Applied Partial Differential Equations Center (APDEC). `http://davis.lbl.gov/APDEC/`, 2005.
35. P. Colella et al. Chombo – Infrastructure for Adaptive Mesh Refinement. `http://seesar.lbl.gov/anag/chombo`, 2005.
36. Colorado State University. The CSU GCM (BUGS) homepage. `http://kiwi.atmos.colostate.edu/BUGS/`, 2005.
37. Combustion Research Facility. `http://www.ca.sandia.gov/CRF`, 2005.
38. T. Dahlgren, T. Epperly, and G. Kumfert. *Babel User's Guide.* CASC, Lawrence Livermore National Laboratory, version 0.9.0 edition, January 2004.
39. J. de St. Germain, A. Morris, S. Parker, A. Malony, and S. Shende. Integrating performance analysis in the Uintah software development cycle. In *The Fourth International Symposium on HighPerformance Computing (ISHPC-IV)*, pages 190–206, May 15-17 2002.

40. J. D. de St. Germain, J. McCorquodale, S. G. Parker, and C. R. Johnson. Uintah: A massively parallel prolem solving environment. In *Proceedings of the Ninth IEEE International Symposium on High Performance and Distributed Computing*, August 2000.
41. D. R. Durran. *Numerical Methods for Wave Equations in Geophysical Fluid Dynamics*. Springer, 1999.
42. G. Edjlali, A. Sussman, and J. Saltz. Interoperability of data-parallel runtime libraries. In *International Parallel Processing Symposium*, Geneva, Switzerland, April 1997. IEEE Computer Society Press.
43. G. Eisenhauer, F. Bustamante, and K. Schwan. Event services for high performance systems. *Cluster Computing: The Journal of Networks, Software Tools, and Applications*, 3(3), 2001.
44. R. Englander. *Developing Java Beans*. O'Reilly and Associates, June 1997.
45. J. T. Feo, D. C. Cann, and R. R. Oldehoeft. A report on the Sisal language project. *Journal of Parallel and Distributed Computing*, 10(4):349–366, 1990.
46. E. Gamma et al. *Design Patterns: Elements of Reusable Object-Oriented Software*. Addison-Wesley, 1994.
47. L. Ge, L. Lee, L. Zenghai, C. Ng, K. Ko, Y. Luo, and M. Shephard. Adaptive mesh refinement for high accuracy wall loss determination in accelerating cavity design. In *IEEE Conf. on Electromagnetic Field Computations*, June 2004.
48. G. A. Geist, A. Beguelin, J. Dongarra, W. Jiang, R. Manchek, and V. Sunderam. *PVM: Parallel Virtual Machine, A User's Guide and Tutorial for Networked Parallel Computing*. MIT Press, Cambridge, MA, 1994.
49. G. A. Geist, J. A. Kohl, and P. M. Papadopoulos. CUMULVS: Providing fault tolerance, visualization and steering of parallel applications. *Intl. J. High-Perf. Computing Appl.*, 11(3):224–236, 1997.
50. GFDL Flexible Modeling System. `http://www.gfdl.noaa.gov/fms`, 2004.
51. M. S. Gockenbach, M. J. Petro, and W. W. Symes. C++ classes for linking optimization with complex simulations. *ACM Transactions on Mathematical Software*, 25(2):191–212, 1999.
52. M. Govindaraju, S. Krishnan, K. Chiu, A. Slominski, D. Gannon, and R. Bramley. Merging the CCA component model with the OGSI framework. In *3rd IEEE/ACM International Symposium on Cluster Computing and the Grid*, 12–15 May 2003.
53. E. Guilyardi, R. G. Budich, and S. Valcke. PRISM and ENES: European approaches to Earth System Modelling. In *Proceedings of Realizing TeraComputing - Tenth Workshop on the Use of High Performance Computing in Meteorology*, November 2002.
54. L. Harper and B. Kauffman. Community Climate System Model. `http://www.ccsm.ucar.edu/`, 2005.
55. T. C. Henderson, P. A. McMurtry, P. J. Smith, G. A. Voth, C. A. Wight, and D. W. Pershing. Simulating accidental fires and explosions. *Comp. Sci. Eng.*, 2:64–76, 1994.
56. M. A. Heroux and J. M. Willenbring. Trilinos Users Guide. Technical Report SAND2003-2952, Sandia National Laboratories, 2003. `http://software.sandia.gov/Trilinos`.
57. C. Hill et al. The architecture of the earth system modeling framework. *Computing in Science and Engineering*, 6(1):18–28, 2003.
58. A. C. Hindmarsh. ODEPACK, a systematized collection of ODE solvers. *Scientific Computing*, 1993.
59. P. Hovland, K. Keahey, L. C. McInnes, B. Norris, L. F. Diachin, and P. Raghavan. A quality of service approach for high-performance numerical components. In *Proceedings of Workshop on QoS in Component-Based Software Engineering, Software Technologies Conference*, Toulouse, France, 20 June 2003.

60. E. C. Hunke and J. K. Dukowicz. An elastic-viscous-plastic model for sea ice dynamics. *J. Phys. Oc.*, 27:1849–1867, 1997.
61. Indiana University. XCAT homepage. `http://www.extreme.indiana.edu/xcat/`, 2005.
62. G. E. Karniadakis and S. J. Sherwin. *Spectral/Hp Element Methods for CFD*. Numerical Mathematics and Scientific Computation. Oxford University Press, 1999.
63. K. Keahey, P. Beckman, and J. Ahrens. Ligature: Component architecture for high performance applications. *Intl. J. High-Perf. Computing Appl.*, 14(4):347–356, Winter 2000.
64. K. Keahey, P. Fasel, and S. Mniszewski. PAWS: Collective interactions and data transfers. In *Proceedings of the High Performance Distributed Computing Conference*, San Francisco, CA, August 2001.
65. J. P. Kenny, S. J. Benson, Y. Alexeev, J. Sarich, C. L. Janssen, L. C. McInnes, M. Krishnan, J. Nieplocha, E. Jurrus, C. Fahlstrom, and T. L. Windus. Component-based integration of chemistry and optimization software. *J. of Computational Chemistry*, 25(14):1717–1725, 2004.
66. D. Keyes. Terascale Optimal PDE Simulations (TOPS) Center. `http://tops-scidac.org/`, 2005.
67. T. Killeen, J. Marshall, and A. da Silva. Earth System Modeling Framework. `http://www.esmf.ucar.edu`, 2005.
68. O. Knio, H. Najm, and P. Wyckoff. A semi-implicit numerical scheme for reacting flow. II. stiff, operator-split formulation. *J. Comp. Phys.*, 154:428–467, 1999.
69. J. A. Kohl and P. M. Papadopoulos. A library for visualization and steering of distributed simulations using PVM and AVS. In *High Performance Computing Symposium*, Montreal, CA, July 1995.
70. S. H. Lam and D. A. Goussis. The CSP method of simplifying kinetics. *International Journal of Chemical Kinetics*, 26:461–486, 1994.
71. J. Larson, R. Jacob, and E. Ong. The Model Coupling Toolkit: A new Fortran90 toolkit for building multi-physics parallel coupled models. Technical Report ANL/MCS-P1208-1204, Argonne National Laboratory, 2004. Submitted to Int. J. High Perf. Comp. App. See also `http://www.mcs.anl.gov/mct/`.
72. J. W. Larson, R. L. Jacob, I. T. Foster, and J. Guo. The Model Coupling Toolkit. In V. N. Alexandrov, J. J. Dongarra, B. A. Juliano, R. S. Renner, and C. J. K. Tan, editors, *Proceedings of the International Conference on Computational Science (ICCS) 2001*, volume 2073 of *Lecture Notes in Computer Science*, pages 185–194, Berlin, 2001. Springer-Verlag.
73. J. W. Larson, B. Norris, E. T. Ong, D. E. Bernholdt, J. B. Drake, W. R. Elwasif, M. W. Ham, C. E. Rasmussen, G. Kumfert, D. S. Katz, S. Zhou, C. DeLuca, and N. S. Collins. Components, the Common Component Architecture, and the climate/weather/ocean community. In *84th American Meteorological Society Annual Meeting*, Seattle, Washington, 11–15 January 2004. American Meteorological Society.
74. Lawrence Livermore National Laboratory. Babel. `http://www.llnl.gov/CASC/components/babel.html`, 2005.
75. J. C. Lee, H. N. Najm, M. Valorani, and D. A. Goussis. Using computational singular perturbation to analyze large scale reactive flows. In *Proceedings of the Fall Meeting of the Western States Section of the The Combustion Institute*, Los Angeles, California, October 2003. Distributed via CD-ROM.
76. S. Lefantzi, C. Kennedy, J. Ray, and H. Najm. A study of the effect of higher order spatial discretizations in SAMR (Structured Adaptive Mesh Refinement) simulations.

In *Proceedings of the Fall Meeting of the Western States Section of the The Combustion Institute*, Los Angeles, California, October 2003. Distributed via CD-ROM.
77. S. Lefantzi and J. Ray. A component-based scientific toolkit for reacting flows. In *Proceedings of the Second MIT Conference on Computational Fluid and Solid Mechanics, June 17-20, 2003, Cambridge, MA*, volume 2, pages 1401–1405. Elsevier, 2003.
78. S. Lefantzi, J. Ray, C. Kennedy, and H. Najm. A component-based toolkit for reacting flow with high order spatial discretizations on structured adaptively refined meshes. *Progress in Computational Fluid Dynamics: An International Journal*, 2004. To appear.
79. S. Lefantzi, J. Ray, and H. N. Najm. Using the Common Component Architecture to design high performance scientific simulation codes. In *Proceedings of the 17th International Parallel and Distributed Processing Symposium (IPDPS 2003), 22-26 April 2003, Nice, France*. IEEE Computer Society, 2003.
80. S. Lefantzi, J. Ray, and S. Shende. Strong scalability analysis and performance evaluation of a CCA-based hydrodynamic simulation on structured adaptively refined meshes. Poster in ACM/IEEE Conference on Supercomputing, November 2003, Phoenix, AZ.
81. S. Lele. Compact finite differnece schemes with spectral-like resolution. *J. Comp. Phys.*, 103:16–42, 1992.
82. Z. Lilek and M. Perić. A fourth-order finite volume method with collocated variable arrangement. *Computers & Fluids*, 24, 1995.
83. S. J. Lin et al. Global weather prediction and high-end computing at NASA. *Computing in Science and Engineering*, 6(1):29–35, 2003.
84. J. Lindemann, O. Dahlblom, and G. Sandberg. Using CORBA middleware in finite element software. In P. M. A. Sloot, C. J. K. Tan, J. J. Dongarra, , and A. G. Hoekstra, editors, *Proceedings of the 2nd International Conference on Computational Science*, Lecture Notes in Computer Science. Springer, 2002. To appear in *Future Generation Computer Systems* (2004).
85. A. Lumsdaine et al. Matrix Template Library. `http://www.osl.iu.edu/research/mtl`, 2005.
86. Massachussetts Institute of Technology. The MIT GCM homepage. `http://mitgcm.org/`, 2005.
87. J. McCorquodale, J. de St. Germain, S. Parker, and C. Johnson. The Uintah parallelism infrastructure: A performance evaluation on the SGI Origin 2000. In *High Performance Computing 2001*, Mar 2001.
88. J. J. Moré and S. J. Wright. *Optimization Software Guide*. SIAM Publications, Philadelphia, 1993.
89. MPI Forum. MPI: a message-passing interface standard. *International Journal of Supercomputer Applications and High Performance Computing*, 8(3/4):159–416, Fall-Winter 1994.
90. H. N. Najm et al. CFRFS homepage. `http://cfrfs.ca.sandia.gov/`, 2005.
91. H. N. Najm, R. W. Schefer, R. B. Milne, C. J. Mueller, K. D. Devine, and S. N. Kempka. Numerical and experimental investigation of vortical flow-flame interaction. SAND Report SAND98-8232, UC-1409, Sandia National Laboratories, Livermore, CA 94551-0969, February 1998.
92. J. Nieplocha, R. J. Harrison, and R. J. Littlefield. Global arrays: A non-uniform-memory-access programming model for high-performance computers. *J. Supercomputing*, 10(2):169, 1996.
93. B. Norris, S. Balay, S. Benson, L. Freitag, P. Hovland, L. McInnes, and B. Smith. Parallel components for PDEs and optimization: Some issues and experiences. *Parallel Computing*, 28(12):1811–1831, 2002.

94. B. Norris, J. Ray, R. Armstrong, L. C. McInnes, D. E. Bernholdt, W. R. Elwasif, A. D. Malony, and S. Shende. Computational quality of service for scientific components. In *Proc. of International Symposium on Component-Based Software Engineering (CBSE7), Edinburgh, Scotland*, 2004.
95. Object Management Group. CORBA component model. `http://www.omg.org/technology/documents/formal/components.htm`, 2002.
96. B. Palmer and J. Nieplocha. Efficient algorithms for ghost cell updates on two classes of MPP architectures. In *14th IASTED International Conference on Parallel and Distributed Computing and Systems*, Cambridge, MA, 2002.
97. M. Parashar et al. GrACE homepage. `http://www.caip.rutgers.edu/TASSL/Projects/GrACE/`, 2005.
98. S. G. Parker. A component-based architecture for parallel multi-physics PDE simulation. In *Proceedings of the International Conference on Computational Science-Part III*, pages 719–734. Springer-Verlag, 2002.
99. C. Pérez, T. Priol, and A. Ribes. A parallel CORBA component model for numerical code coupling. *Intl. J. High-Perf. Computing Appl.*, 17(4), Nov 2003.
100. R. Peyret and T. Taylor. *Computational Methods for Fluid Flow*, chapter 6. Springer Series in Computational Physics. Springer-Verlag, New York, 1983. Finite-Difference Solution of the Navier-Stokes Equations.
101. Phillip Jones. Parallel Ocean Program (POP) homepage. `http://climate.lanl.gov/Models/POP/`, 2004.
102. T. Poinsot, S. Candel, and A. Trouvé. Applications of direct numerical simulation to premixed turbulent combustion. *Progress in Energy and Combustion Science*, 21:531–576, 1995.
103. R. Pozo. Template Numerical Toolkit. `http://math.nist.gov/tnt`, 2004.
104. K. Radhakrishnan and A. C. Hindmarsh. Description and use of LSODE, the Livermore solver for ordinary differential equations. Technical Report UCRL-ID-113855, Lawrence Livermore National Laboratory, 1993.
105. M. Ranganathan, A. Acharya, G. Edjlali, A. Sussman, and J. Saltz. A runtime coupling of data-parallel programs. In *Proceedings of the 1996 International Conference on Supercomputing*, Philadelphia, PA, May 1996.
106. J. Ray, B. A. Allan, R. Armstrong, and J. Kohl. Structured mesh demo for supercomputing 2004. `http://www.cca-forum.org/~jaray/SC04/sc04.html`, 2004.
107. J. Ray, C. Kennedy, S. Lefantzi, and H. Najm. High-order spatial discretizations and extended stability methods for reacting flows on structured adaptively refined meshes. In *Proceedings of the Third Joint Meeting of the U.S. Sections of The Combustion Institute, March 16-19, 2003, Chicago, Illinois.*, 2003. Distributed via CD-ROM.
108. J. V. W. Reynders, J. C. Cummings, P. J. Hinker, M. Tholburn, M. S. S. Banerjee, S. Karmesin, S. Atlas, K. Keahey, and W. F. Humphrey. *POOMA: A FrameWork for Scientific Computing Applications on Parallel Architectures*, chapter 14. MIT Press, 1996.
109. E. Roman. *Mastering Enterprise JavaBeans*. O'Reilly and Associates, June 1997.
110. J. Sarich. A programmer's guide for providing CCA component interfaces to the Toolkit for Advanced Optimization. Technical Report ANL/MCS-TM-279, Argonne National Laboratory, 2004.
111. B. Smith et al. TOPS Solver Interface. `http://www-unix.mcs.anl.gov/scidac-tops/tops-solver-interface`, 2005.
112. K. Smith, J. Ray, and B. A. Allan. CVODE component user guidelines. Technical Report SAND2003-8276, Sandia National Laboratory, May 2003.

113. B. Sportisse. An analysis of operator splitting techniques in the stiff case. *J. Comp. Phys.*, 161:140–168, 2000.
114. G. Strang. On the construction and comparison of difference schemes. *SIAM J. Numer. Anal.*, 5(3):506–517, 1968.
115. M. J. Suarez and L. Takacs. Documentation of the Aries-GEOS dynamical core: Version 2. Technical Report TM-1995-104606, NASA, 1995.
116. D. Sulsky, Z. Chen, and H. L. Schreyer. A Particle Method for History Dependent Materials. *Comp . Methods Appl. Mech. Engrg*, 118, 1994.
117. Y. Sun, N. Folwell, Z. Li, and G. Golub. High precision accelerator cavity design using the parallel eigensolver Omega3P. In *Proc. of the 18th Annual Review of Progress in Applied Computational Electromagnetics ACES 2002*, Monterey, CA, 2002.
118. C. Szyperski. *Component Software: Beyond Object-Oriented Programming.* ACM Press, New York, 1999.
119. B. Talbot, S. Zhou, and G. Higgins. Software engineering support of the third round of scientific grand challenge investigations–earth system modeling software framework survey task4 report. Technical Report TM-2001-209992, NASA, 2001.
120. L. Trease, H.E.and Trease. NWGrid: A multi-dimensional, hybrid, unstructured, parallel mesh generation system. `http://www.emsl.pnl.gov/nwgrid`, 2005.
121. The Terascale Simulation Tools and Technologies (TSTT) Center. `http://www.tstt-scidac.org`, 2005.
122. U. S. Dept. of Energy. SciDAC Initiative homepage. `http://www.osti.gov/scidac/`, 2005.
123. University Corporation for Atmospheric Research. The Community Atmosphere Model (CAM) homepage. `http://www.ccsm.ucar.edu/models/atm-cam/`, 2005.
124. University of Oregon. TAU: Tuning and analysis utilities. `http://www.cs.uoregon.edu/research/tau`, 2005.
125. S. Vajracharya, S. Karmesin, P. Beckman, J. Crotinger, A. Malony, S. Shende, R. Oldehoeft, and S. Smith. Smarts: Exploiting temporal locality and parallelism through vertical execution. In *Proceedings of the 13th International Conference on Supercomputing (ICS 99)*, pages 302–310, Rhodes, Greece, 1999. ACM Press.
126. T. Veldhuizen et al. BLITZ++: Object-oriented scientific computing. `http://www.oonumerics.org/blitz`, 2005.
127. M. R. Visbal and D. V. Gaitonde. On the use of higher-order finite-difference schemes on curvilinear and deforming meshes. *J. Comp. Phys.*, 181:155–185, 2002.
128. Z. Wang and G. P. Huang. An essentially nonoscillatory high-order Padé-type (ENO-Padé) scheme. *J. Comp. Phys.*, 177:37–58, 2002.
129. Weather Research and Forecasting Model. `http://www.wrf-model.org/`, 2005.
130. F. Williams. *Combustion Theory.* Addison-Wesley, New York, 2nd edition, 1985.
131. A. Wissink, R. Hornung, S. Kohn, S. Smith, and N. Elliott. Large scale parallel structured AMR calculations using the SAMRAI framework. In *Proceedings of the SC01 Conf. High Perf. Network. and Comput*, Denver, CO, November 2001.
132. A. Wissink, D. Hysom, and R. Hornung. Enhancing scalability of parallel structured AMR calculations. In *Proceedings of the* 17th *ACM International Conference on Supercomputing (ICS03)*, pages 336–347, San Francisco, CA, June 2003.
133. M. Wolf, Z. Cai, W. Huang, and K. Schwan. Smart pointers: Personalized scientific data portals in your hand. In *Proceedings of Supercomputing 2002*, November 2002.
134. M. Wolf, A. Guetz, and C.-K. Ng. Modeling large accelerator structures with the parallel field solver Tau3P. In *Proc. of the 18th Annual Review of Progress in Applied Computational Electromagnetics ACES 2002*, Monterey, CA, 2002.

135. K. Zhang, K. Damevski, V. Venkatachalapathy, and S. Parker. SCIRun2: A CCA framework for high performance computing. In *Proceedings of the 9th International Workshop on High-Level Parallel Programming Models and Supportive Environments (HIPS 2004)*, Santa Fe, NM, April 2004. IEEE Press.
136. S. Zhou. Coupling earth system models: An ESMF-CCA prototype. `http://webserv.gsfc.nasa.gov/ESS/esmf_tasc`, 2003.
137. S. Zhou, A. da Silva, B. Womack, and G. Higgins. Prototyping the ESMF using DOE's CCA. In *NASA Earth Science Technology Conference 2003*, College Park, MD, 24–26 June 2003. `http://esto.nasa.gov/conferences/estc2003/papers/A4P3(Zhou).pdf`.

Part IV

Parallel Applications

11

Full-Scale Simulation of Cardiac Electrophysiology on Parallel Computers

Xing Cai and Glenn Terje Lines

Simula Research Laboratory, P.O. Box 134, NO-1325 Lysaker, Norway

Department of Informatics, University of Oslo, P.O. Box 1080, Blindern, NO-0316 Oslo, Norway

[xingca,glennli]@simula.no

Summary. In this chapter, we will present an advanced parallel electro-cardiac simulator, which employs anisotropic and inhomogeneous conductivities in realistic three-dimensional geometries for modeling both the heart and the torso. Since partial differential equations (PDEs) constitute the main part of the mathematical model, this chapter thus demonstrates a concrete example of solving PDEs on parallel computers. It will be shown that good overall parallel performance relies on at least two factors. First, the serial numerical strategy must find a parallel substitute that is scalable with respect to both convergence and work amount. Second, care must be taken to avoid unnecessary duplicated local computations while maintaining an acceptable level of load balance.

11.1 Introduction

The contraction of the heart is triggered by a signal wave that propagates through the muscle tissue. Between two heart beats, each muscle cell is in a resting phase and has a surplus of negative charge compared to the exterior of the cell. This is called a negative transmembrane potential. When the signal wave reaches a cell, the conductivity properties of the cell membrane are altered, and positive ions are able to enter the cell. This is the so-called *depolarization* of the cell. This depolarization causes the neighboring cells to also alter their membrane conductance, thus allowing ions to enter. In this way the signal is sustained and travels through the whole heart. The aggregated effect of all the cells being depolarized is measurable on the body surface as the peak of the ECG signal.

The currently most accurate mathematical model of the electrical activities of the heart tissue is the Bidomain model, which will be defined precisely in Section 11.2. It consists of two coupled PDEs, where the primary unknowns are the electrical potentials outside and inside of the cell. The difficulty with solving this mathematical model is that the current crossing the membrane, hereafter referred to as the ionic current, depends nonlinearly on the potentials and also on a large number of other entities, such as ionic concentrations and permeability of the membrane. These entities are typically modeled by a system of ordinary differential equations (ODEs). A complete mathematical model thus consists of the Bidomain PDE system coupled with an ODE system. For an example of computer simulations of cardiac electrophysiology, we refer to Figure 11.1 which shows two snapshots from a prototypical three-dimensional simulation done in [21].

The ODE system can be quite complex. For example, the model of Winslow et al. [38] involves 30 state variables. As more biological data become available, more detailed descriptions are incorporated into the models, and consequently the model complexity increases. There also exist simplified models with only a few state variables. For some applications these simple models might be sufficient, but they are typically valid only within a narrow range. The complex models are more realistic as they include many of the subsystems within the cell, which may be calcium stored inside the cell or the state of the different membrane proteins. These proteins allow current to pass through the membrane, forming so-called ionic channels. Since the complex models are more realistic, they have a wider area of application. For example, one can investigate the effect of a drug that blocks an ionic canal by simply reducing the current going through this channel in a complex model. There is, in general, a complex interplay between different the ionic channels, and it is not easy to predict how the blocking of one type of channel affects the cell as a whole, or indeed the performance of the entire heart. Simulation of such systems will therefore give us an insight that is not easily obtainable in any other way.

The time scale of the biochemical process in a cell is very small compared with the duration of a heart beat. The depolarization of a cell takes about 1ms, in comparison with about one second for a heart beat. From a numerical point of view, another challenge is that the activation signal wave front is very narrow, about 1mm compared with 10cm for the whole heart. A commonly reported value for the required spatial resolution is 0.2mm, which is necessary for rendering the narrow wave front with sufficient accuracy; see e.g. [3, 35]. Using such a mesh point density over the entire heart volume of an adult means that about 40 million mesh points are required; see [20]. It is obvious that the computational challenges are formidable for such a numerical problem.

To deal with the computational burden, one often resort to studying simplified mathematical models. One option is to use the Monodomain model, which is a scalar PDE and can be derived as a special case of the Bidomain model. Another option is to use simpler models for the ionic current. These simplifications together can reduce the computation time by at least a factor of 10. This time saving factor is due to a combination of at least four reasons: a) the number of unknowns in the Bidomain model is twice of that in the Monodomain model, b) the resulting linear system of the Bidomain model (from applying a linearization and spatial discretizations) requires considerably more iterations of an iterative solver than the Monodomain model, especially in the absence of a powerful preconditioner (see e.g. [29]), c) the computational cost of one matrix-vector product associated with the Bidomain model is approximately four times of that with the Monodomain model, and d) a simpler ionic current model results in an ODE system that has fewer equations and is easier to

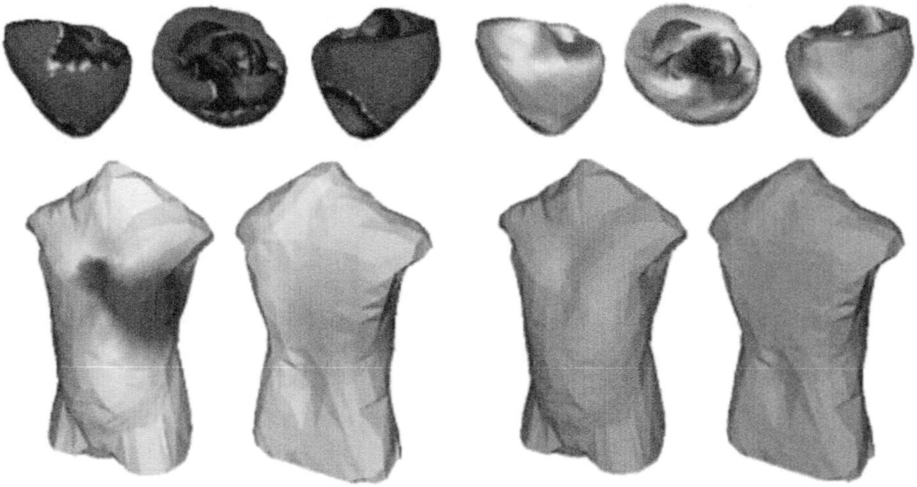

Fig. 11.1. Snapshots from two time levels of a simulation of the electrical field in the human heart and torso. At each time level, the electrical potential distribution on the heart surface is shown at three different angles, while the distribution on the torso surface is shown at two different angles. (For the color version, see Figure A.23 on page 478).

solve. To avoid the need for 40 million mesh points one can also consider a volume smaller than the entire heart, or, alternatively, use a coarser mesh resolution. Both these approaches are popular since they make it easier to fit the whole problem onto a serial computer, but at the expense of less realistic results.

Performing full-scale simulations of the entire heart is only possible by using parallel computers. We can argue for this observation roughly as follows. Assuming 40×10^6 mesh points in the heart domain and using the Winslow ODE model, the total number of degrees of freedom amounts to $32 \times 40 \times 10^6$, where for each mesh point there are two PDE degrees of freedom and 30 ODE degrees of freedom. At any given time level, storing the values of all the degrees of freedom will require approximately 10GB memory using double precisions. The time stepping scheme typically needs to store the values for two consecutive time levels. Moreover, the data structure needed for storing the computational meshes and the matrix for the discretized PDEs will require considerably more memory. Clearly, such a huge demand of memory can not be met by any serial computer, in addition to the equally huge demand of simulation time.

Parallel electro-cardiac simulation is still a relatively new research topic. Publications on this particular topic have been sparse. This is due to a combination of the complicated mathematical model, the necessity of an advanced numerical strategy, and the difficulty with writing a high-performance parallel implementation. Among related works, we can mention [28], [27], [26], [24], [35], [12], and [37]. In [28], a three-dimensional finite difference model with a parallel implementation is presented for solving the Bidomain equations with inhomogeneous and anisotropic conductivities. Four different parallelization schemes are examined in [27] for the simplified Monodomain model in two dimensions. A modular simulation system for the Bidomain equations is presented in [26] and speed-up results obtained on regular three-dimensional meshes are reported. Recently, a fully implicit parallel algorithm for the two-dimensional Bidomain equations is proposed in [24], adopting the advanced Newton-Krylov-Schwarz strategy but associated with a very simple cell model. Moreover, a computer three-dimensional heart model employing the Monodomain equation and a modified Luo-Rudy membrane model is reported in [35], where finite differences with explicit time stepping are used. In [12], both the monodomain and bidomain models are discussed, where the temporal discretization is of the semi-implicit type and the resulting bidomain stiffness matrix has a 2×2-block structure, in the same fashion as our numerical strategy to be presented in Section 11.3. However, the two PDEs are both of the reaction-diffusion type in the bidomain approach of [12], and one-level domain decomposition preconditioning techniques, which use a relatively simple subdomain solver, are reported for three-dimensional structured computational meshes. Finally, [37] reports some numerical experiences associated with a parallel multigrid preconditioner for a two-dimensional bidomain model based on rectangular meshes. The temporal discretization in [37] is done in an explicit fashion, such that the two PDEs are solved separately.

To the topic of parallel electro-cardiac simulations, the contributions from our earlier papers [21, 7, 6] and the present chapter are twofold. First, we have adopted realistic three-dimensional geometries and an implicit time-stepping scheme in parallel full-scale simulations, instead of simple rectangular domains. Our mathematical model (see Section 11.2) has also incorporated the torso domain, so that the ECG signal can be computed on the body surface, at the same time when the Bidomain equations are solved in the heart domain. Regarding our earlier papers [21] and [7], no parallel preconditioners were adopted in [21], whereas the numerical strategy used in [7] did not solve the two PDEs of the Bidomain model simultaneously. As a second contribution to parallel electro-cardiac simulations, we have devised a fast-convergent parallel block preconditioner, on the basis of an order-optimal serial block preconditioner that was derived in [31] and further analyzed in [23]. The parallel block preconditioner was first proposed in [6], and the present chapter contains, among other things, continued work on the parallelization. Using several performance enhancing techniques, which will be discussed in Section 11.5, our advanced parallel simulator is shown in Section 11.6 to obtain decent parallel performance even on a Linux cluster.

The focus of the present chapter is on a parallel full-scale electro-cardiac simulator, which employs anisotropic and inhomogeneous conductivity properties in realistic three-dimensional geometries modeling both the heart and the torso. Another advanced feature is that the two PDEs, which constitute the Bidomain system in the PDE part of the mathematical model, are solved together as a coupled 2×2 block system of linear equations during each time step. The 2×2 block linear system arises from a linearization based on an operator splitting technique, together with using an implicit time-stepping scheme associated with finite element discretizations;

see Section 11.3. We will take the starting point in an order-optimal serial block preconditioner (see [31, 23]) for the 2×2 block system. The overall parallelization strategy is based on dividing a global solution domain into subdomains. The main challenge arises from devising and implementing a parallel substitute of the serial block preconditioner; see Section 11.2, which is essential for the PDE part to obtain a rapid convergence of a Krylov iterative linear solver. To achieve scalability with respect to the convergence speed, we build our parallel preconditioner on the basis of *additive Schwarz iterations*; see [9, 30, 39, 11] Such iterations use a "divide-and-conquer" strategy for partitioning the work load and thus are compatible with the overall subdomain-based parallel strategy. In fact, the following advanced numerical ingredients need to be incorporated into the parallel implementation:

1. a two-block diagonal system is chosen as the preconditioner and is solved during each preconditioning operation within a parallel conjugate gradient solver,
2. additive Schwarz iterations are used as the parallel solvers for each of the two diagonal blocks,
3. multigrid V-cycles are used as the subdomain solvers within each Schwarz iteration, and
4. global coarse grid corrections are also incorporated into the additive Schwarz iterations.

Using a combination of the above numerical techniques, we have been able to obtain an order-optimal parallel preconditioner for the 2×2 block linear system that arises from discretizing the Bidomain equations. More specifically, the number of parallel Krylov iterations needed for solving the 2×2 block system remains independent of both the number of degrees of freedom and the number of subdomains. Moreover, the total amount of work during each preconditioning operation remains approximately linearly proportional to the number of degrees of freedom.

As the foundation for the parallel implementation, we have used a specially designed mesh partitioning scheme, which partitions both the heart domain and the torso domain in a communication-efficient manner. Moreover, each subdomain is equipped with a hierarchy of adaptively refined subdomain meshes, on which multigrid V-cycles can be run. Due to the mathematical requirement of overlap by the additive Schwarz iterations, we have also taken care to avoid unnecessary duplicated local computations. Consequently, satisfactory speed-up results have been obtained for the resulting parallel electro-cardiac simulator; see Section 11.6.

The remainder of the chapter is organized as follows. First, Section 11.2 presents the mathematical model with an emphasis on the involved PDEs. Then, Section 11.3 explains the overall numerical strategy, which is semi-implicit and is based on an operator splitting technique. The specially designed block preconditioner, in both its serial and parallel versions, is also explained. Afterwards, Section 11.4 focuses on the parallelization of the serial numerical strategy, where implementing the parallel block preconditioner is the main theme. To obtain a satisfactory parallel performance, Section 11.5 discusses the issue of reducing the overhead in parallel simulations. Finally, Section 11.6 reports some detailed measurements of the full-scale parallel electro-cardiac simulator.

11.2 The Mathematical Model

The *Bidomain equations* constitute a well-established mathematical model for describing the electrical activities in the heart; see [36] and also the references given in [20]. Let H denote the domain of the heart. We consider the Bidomain equations of the following form (11.1)-(11.2), involving two primary unknown functions in H, i.e., the transmembrane potential v and the extracellular potential u_e.

$$\chi C_{\mathrm{m}} \frac{\partial v}{\partial t} + \chi I_{\mathrm{ion}}(v, \mathbf{s}) = \nabla \cdot (M_i \nabla v) + \nabla \cdot (M_i \nabla u_e) \text{ in } H, \tag{11.1}$$

$$0 = \nabla \cdot (M_i \nabla v) + \nabla \cdot ((M_i + M_e) \nabla u_e) \text{ in } H. \tag{11.2}$$

In (11.1), the nonlinear function $I_{\mathrm{ion}}(v, \mathbf{s})$ represents the ionic current, where $\mathbf{s}$ denotes a vector of state variables describing the state of the cell membrane. There exist many models of I_{ion}, which together with different ways of modeling the interaction between the state variables give rise to different cell models. Almost all of the cell models rely on solving a system of ODEs to update the $\mathbf{s}$ vector; see e.g. [22, 38]. More specifically, an ODE system of the form

$$\frac{\partial \mathbf{s}}{\partial t} = \mathbf{F}(v, \mathbf{s}, t) \tag{11.3}$$

models the electrical behavior of the cardiac cells, at every point inside H. The more sophisticated the cell model, the larger the number of ODEs are involved in (11.3). Moreover, χ in (11.1) is a surface-to-volume scaling factor, C_{m} denotes the capacitance of the membrane, and M_i denotes the intracellular conductivity tensor. Similarly, M_e denotes the extracellular conductivity tensor in (11.2). Due to the fibrous structure of the heart muscle, the conductivity tensors are anisotropic. More specifically, the formulas for M_i and M_e are as follows:

$$M_{i,e} = \sigma^t_{i,e} I + (\sigma^l_{i,e} - \sigma^t_{i,e}) \mathbf{a}_l \mathbf{a}_l^T + (\sigma^n_{i,e} - \sigma^t_{i,e}) \mathbf{a}_n \mathbf{a}_n^T, \tag{11.4}$$

where the constants $\sigma^t_{i,e}$, $\sigma^l_{i,e}$, and $\sigma^n_{i,e}$ describe the different conductivity properties of the heart muscle in three orthogonal directions, I is the identity matrix, while the vectors $\mathbf{a}_l$ and $\mathbf{a}_n$ model the orientation of the muscle fibers and sheet layers, varying throughout the heart. The superscripts l, t, n denote, respectively, the direction along the fibers, the direction normal to the fibers but in the plane of the sheets, and the direction normal to the sheets. Examples of the $\sigma^t_{i,e}, \sigma^l_{i,e}, \sigma^n_{i,e}$ values are given in (11.23)-(11.23) in Section 11.6. For a more detailed description, we refer to [19, 32]. Figure 11.2 shows an example of the orientation of the muscle fibers and sheet layers in the heart, of which the $\mathbf{a}_l$ and $\mathbf{a}_n$ vectors have already been used in the different simulations of [21, 7, 33, 6]. This set of heart muscle data is provided by the Biomedical Engineering Group at the University of Auckland; see [13]. Inhomogeneity of the conductivity properties will arise when electrical signals propagate from the heart into the torso, between organs in the torso, and also when dead or diseased tissues need to be modelled in the heart.

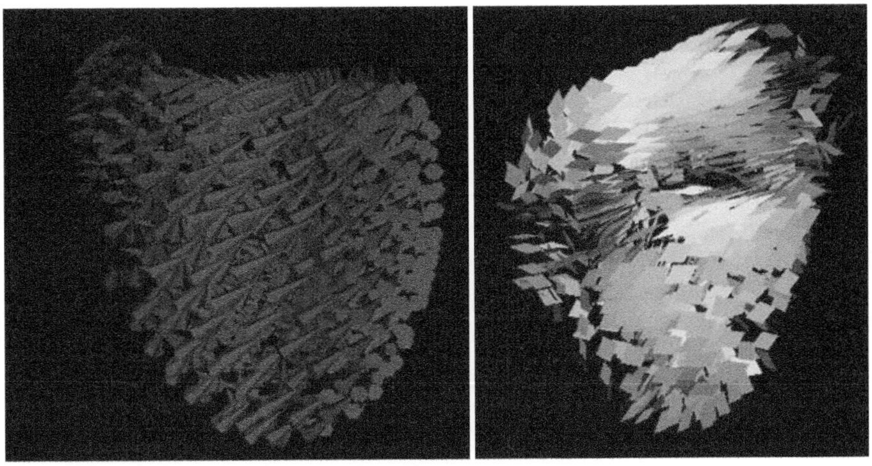

Fig. 11.2. The orientation of the muscle fibers (left) and sheet layers (right) in the heart. (For the color version, see Figure A.24 on page 478).

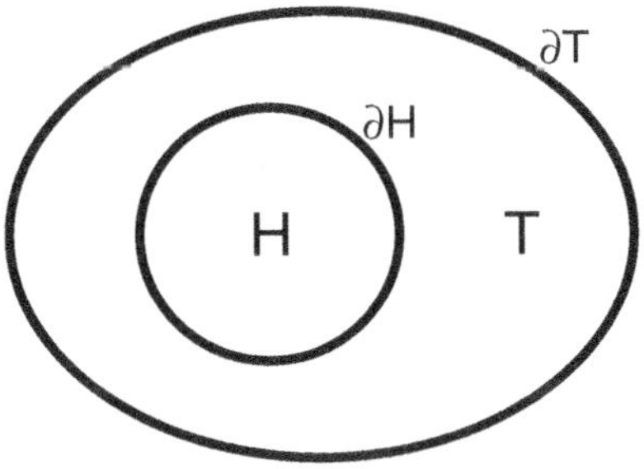

Fig. 11.3. A schematic 2D slice of the entire solution domain $\Omega = H \cup T$.

In order to compute the electrical potential on the body surface by the same simulation, we supplement the Bidomain equations (11.1)-(11.2) with the following elliptic PDE describing the propagation of the electrical signal in the torso T exterior to the heart:

$$\nabla \cdot (M_o \nabla u_o) = 0 \qquad \text{in } T, \tag{11.5}$$

where u_o is the electrical potential in the torso and M_o denotes the associated conductivity tensor. This supplementary PDE enables a direct comparison between ECG measurements and simulated results. We refer to [20] for a summary on this topic.

To summarize, our complete mathematical model consists of the ODE system (11.3) plus three PDEs (11.1)-(11.2), and (11.5), for which the combined solution domain $\Omega = H \cup T$ is depicted in Figure 11.3. As for the boundary conditions, we have

$$M_i \frac{\partial}{\partial n}(v + u_e) = 0, \quad u_e = u_o, \quad M_e \frac{\partial u_e}{\partial n} - M_o \frac{\partial u_o}{\partial n} = 0 \quad \text{on } \partial H, \quad (11.6)$$

$$M_o \frac{\partial u_o}{\partial n} = 0 \quad \text{on } \partial T, \qquad (11.7)$$

where $\partial/\partial n$ denotes the derivative in the normal direction on the boundary. Although the boundary conditions are not sufficient for uniquely determining u_e and u_o, they are sufficient for practical applications which are interested in the variations within the u_e and u_o solutions, rather than their absolute values.

In the temporal direction, the mathematical model is to be solved within a time domain with known initial values for v and $\mathbf{s}$.

11.3 The Numerical Strategy

11.3.1 The Time-Stepping Scheme

The starting point of the whole numerical strategy is to use the technique of *operator splitting* (see e.g. [38]) to decouple the nonlinear PDE (11.1) into two parts:

$$\frac{\partial v}{\partial t} = -\frac{1}{C_{\rm m}} I_{\rm ion}(v, \mathbf{s}), \qquad (11.8)$$

$$\chi C_{\rm m} \frac{\partial v}{\partial t} = \nabla \cdot (M_i \nabla v) + \nabla \cdot (M_i \nabla u_e), \qquad (11.9)$$

i.e., an ODE and a linear PDE. The ODE (11.8) is then joined with a chosen ODE system of form (11.3) to establish the following new system of ODEs:

$$\begin{cases} \dfrac{\partial v}{\partial t} = -\dfrac{1}{C_{\rm m}} I_{\rm ion}(v, \mathbf{s}), \\ \dfrac{\partial \mathbf{s}}{\partial t} = \mathbf{F}(v, \mathbf{s}, t), \end{cases} \qquad (11.10)$$

which needs to be solved during each time step. We remark that the choice of $I_{\rm ion}$ and $\mathbf{F}$ in the above ODE system determines the so-called cell model, for which a well-known example is the Winslow model [38].

In the temporal direction, the simulation time domain is divided into discrete time levels:

$$0 = t_0 < t_1 < t_2 \cdots,$$

where at each time level t_l, $l \geq 1$, the solutions from the previous time level, $v^{l-1}, u_e^{l-1}, u_o^{l-1}, \mathbf{s}^{l-1}$, are used as the starting values. Using a θ-rule, where $0 \leq \theta \leq 1$, we can construct a flexible time-stepping scheme whose work per time step consists of solving an ODE system twice, separated by the solution of a system of three PDEs. More specifically, the computational work for the time step $t_{l-1} \rightarrow t_l$ consists of the following sub-steps:

1. At every mesh point in H, an ODE system of form (11.10) is solved for $t \in (t_{l-1}, t_{l-1} + \theta(t_l - t_{l-1})]$, using the initial values v^{l-1} and $\mathbf{s}^{l-1}$. The solution results are an intermediate transmembrane potential solution $\tilde{v}^{l-1}$ and an updated $\tilde{\mathbf{s}}^{l-1}$ vector.
2. The three PDEs (11.9), (11.2), and (11.5) are solved simultaneously, where we remark that (11.9) is the remaining part of (11.1) after operator splitting. The temporal discretization, which uses a θ-rule, is of the following form:

$$\chi C_{\mathrm{m}} \frac{\hat{v}^l - \tilde{v}^{l-1}}{\Delta t} = (1-\theta)\left(\nabla \cdot (M_i \nabla \tilde{v}^{l-1}) + \nabla \cdot (M_i \nabla u_e^{l-1})\right)$$
$$+\theta\left(\nabla \cdot (M_i \nabla \hat{v}^l) + \nabla \cdot (M_i \nabla u_e^l)\right), \tag{11.11}$$
$$0 = \nabla \cdot (M_i \nabla \hat{v}^l) + \nabla \cdot ((M_i + M_e)\nabla u_e^l), \tag{11.12}$$
$$0 = \nabla \cdot (M_o \nabla u_o^l). \tag{11.13}$$

We remark that $\hat{v}^l, u_e^l, u_o^l$ are the unknowns to be found. If we combine the unknown values of u_e^l and u_o^l into one vector $\mathbf{u}^l$, the spatial discretization (using e.g. finite elements) will give rise to a 2×2 block linear system:

$$\begin{bmatrix} \chi C_{\mathrm{m}} \mathbf{I} + \theta \Delta t \mathbf{A}_{\mathrm{v}} & \theta \Delta t \hat{\mathbf{A}}_{\mathrm{v}} \\ \theta \Delta t \hat{\mathbf{A}}_{\mathrm{v}}^T & \theta \Delta t \mathbf{A}_{\mathrm{u}} \end{bmatrix} \begin{bmatrix} \hat{\mathbf{v}}^l \\ \mathbf{u}^l \end{bmatrix} \equiv \mathcal{A} \begin{bmatrix} \hat{\mathbf{v}}^l \\ \mathbf{u}^l \end{bmatrix} = \begin{bmatrix} \mathbf{b}^l \\ \mathbf{0} \end{bmatrix}. \tag{11.14}$$

In the above block linear system, $\mathbf{I}$ and $\mathbf{A}_{\mathrm{v}}$ represent the mass matrix and the stiffness matrix associated with M_i inside H, respectively. The $\mathbf{b}^l$ vector is computed on the basis of $\tilde{v}^{l-1}$ and u_e^{l-1}. The sparse matrix $\mathbf{A}_{\mathrm{u}}$ arises from discretizing (11.2) and (11.5) together. That is, we discretize a combined elliptic equation:

$$\nabla \cdot (M \nabla u) = 0 \quad \text{in } \Omega, \tag{11.15}$$

using $M = M_i + M_e$ in H and $M = M_o$ in T; see [31] for more details. The number of degrees of freedom in $\mathbf{u}^l$ equals the number of mesh points covering the combined domain $\Omega = H \cup T$. The $\hat{\mathbf{A}}_v$ matrix is the same as $\mathbf{A}_{\mathrm{v}}$, except that $\hat{\mathbf{A}}_v$ is padded with some additional zero columns to allow a multiplication of form $\hat{\mathbf{A}}_v \mathbf{u}$, whereas the $\hat{\mathbf{A}}_v^T$ matrix is the transpose of $\hat{\mathbf{A}}_v$.
3. At every mesh point in H, the ODE system (11.10) is solved again for $t \in (t_{l-1} + \theta(t_l - t_{l-1}), t_l]$, using the initial values $\hat{v}^l$ and $\tilde{\mathbf{s}}^{l-1}$. The computational results are stored in v^l and $\mathbf{s}^l$.

We refer to [31] for a detailed explanation of the above time-stepping scheme, and remark that using $\theta = 1/2$ in (11.11) gives rise to a second-order accurate temporal discretization. To handle the potentially stiff ODE system (11.10), due to steep propagation fronts, we adopt a third-order Runge-Kutta type scheme proposed in [34]. Although a second-order ODE solver is sufficient for ensuring the order of accuracy in the time direction, our experiences have indicated that the adopted third-order ODE solver is better at treating the situations of stiffness, so that fewer intermediate ODE steps are needed between two time levels. That is, the overall time spent by a third-order ODE solver is not necessarily larger than that of a lower order ODE solver.

Table 11.1. An example on the number of CG iterations (without preconditioner) needed for solving the 2×2 block system (11.14) of different sizes.

$N_H + N_\Omega$	I_{CG}
39,589	897
302,166	1999
1,552,283	4087

11.3.2 A Serial Block Preconditioner

By extending the proof given in [25] to also include the torso domain, we can deduce that the 2×2 block matrix $\mathcal{A}$ defined in (11.14) is *symmetric and positive semi-definite*. We remark that this property arises from the symmetry and positive definiteness of the conductivity tensors M_i, M_e, and M_o (consequently are the $\mathbf{A}_\mathrm{v}$ and $\mathbf{A}_\mathrm{u}$ matrices symmetric and positive semi-definite); see [25]. The existence of zero eigenvalues for $\mathcal{A}$ is due to the boundary conditions (11.6)-(11.7). Our earlier experiences have suggested that the conjugate gradient (CG) method is an appropriate choice for solving (11.14); see [31].

The next question is how fast can the CG method achieve convergence for (11.14)? To answer this question, we have conducted three experiments where we study the number of CG iterations, denoted by I_{CG}, which is needed to reduce the residual vector associated with the 2×2 block system (11.14) by a factor of 10^4 in the L_2-norm, compared with its initial value before the first CG iteration. We remark that a typical value of $\Delta t = 0.125$ms is chosen in building $\mathcal{A}$, and our experiences suggest that the size of Δt within the range of $[0.1, 0.5]$ms does not have an immediate effect on the number of CG iterations. In Table 11.1, N_H denotes the number of unknowns in the $\mathbf{v}$ vector, i.e., the number of mesh points in the three-dimensional heart domain, whereas N_Ω denotes the number of unknowns in the $\mathbf{u}$ vector, i.e., the number of mesh points in the entire torso domain including the heart. The different numbers of N_H and N_Ω are obtained from an adaptive mesh refinement process, which increases the mesh resolution mainly inside H and keeps the mesh resolution inside T mostly unchanged. (We note that all the numerical experiments reported in the present chapter are associated with realistic three-dimensional geometries.) It can be deduced that without preconditioning the number of CG iterations approximately doubles when the spacing between mesh points is halved. The convergence speed is determined by the elliptic operators. Consequently, the work amount of each CG iteration grows much faster than the number of degrees of freedom, when no preconditioner is used. Although the values of I_{CG} in Table 11.1 are associated with one particular time step, numerical experiments have suggested that I_{CG} remains roughly constant throughout an entire elector-cardiac simulation, i.e., I_{CG} is not sensitive to the changing right-hand side vector in (11.14).

To overcome the huge numbers of CG iterations reported in Table 11.1, a preconditioner (see e.g. [2]) needs to be used inside the CG iterations. The standard choices of preconditioning (such as SSOR and RILU) all have the weakness that the number of CG iterations grows considerably with respect to the number of degrees of

Table 11.2. An example on the number of CG iterations (using the serial multigrid-block preconditioner) needed for solving the 2×2 block system (11.14).

$N_H + N_\Omega$	I_{CG}
302,166	15
1,552,283	16

freedom (although the growth is slower than that reported in Table 11.1). We would thus like to have an *order-optimal* preconditioner, which means that the number of CG iterations remains constant independent of the number of degrees of freedom. Moreover, the work executed in each operation of such an order-optimal preconditioner should be linearly proportional to the number of degrees of freedom.

A so-called block preconditioner based on multigrid algorithms was first proposed in [31]. More specifically, the following diagonal block matrix

$$\mathcal{D} = \begin{bmatrix} \chi C_{\mathrm{m}}\mathbf{I} + \theta \Delta t \mathbf{A}_{\mathrm{v}} & \mathbf{0} \\ \mathbf{0} & \theta \Delta t \mathbf{A}_{\mathrm{u}} \end{bmatrix} \tag{11.16}$$

is used as a suitable preconditioner for (11.14). In other words, the original 2×2 block system (11.14) is replaced with

$$\mathcal{D}^{-1}\mathcal{A}\begin{bmatrix} \hat{\mathbf{v}}^l \\ \mathbf{u}^l \end{bmatrix} = \mathcal{D}^{-1}\begin{bmatrix} \mathbf{b}^l \\ \mathbf{0} \end{bmatrix}. \tag{11.17}$$

This means that during each preconditioning operation, the two diagonal blocks in $\mathcal{D}$ need to be solved (approximately). It has recently been shown in [23] that the number of CG iterations, when using the block preconditioner $\mathcal{D}$, will remain independent of the number of degrees of freedom. The rapid convergence of this block preconditioner is due to the fact that $\mathcal{D}$ is spectrally equivalent to the 2×2 block matrix $\mathcal{A}$ in (11.14). To make $\mathcal{D}$ an order-optimal serial preconditioner, we have chosen to apply one multigrid V-cycle (see [14, 1, 15]) as the approximate solver for both the diagonal blocks in $\mathcal{D}$. This is because multigrid V-cycles are powerful linear solvers and the work amount of one multigrid V-cycle is linearly proportional to the number of degrees of freedom.

To demonstrate the actual effect of the block preconditioner $\mathcal{D}$ defined in (11.16) using multigrid V-cycles, we list in Table 11.2 the number of CG iterations when the block preconditioner is in use. In comparison with Table 11.1, we can see that I_{CG} is dramatically reduced and stays independent of the number of degrees of freedom. Here, we remark that $N_H + N_\Omega = 39,589$ in Table 11.1 is related to the coarsest possible meshes we have for modeling our realistic heart and torso geometries (see Figure 11.1). The measurement of I_{CG} associated with $N_H + N_\Omega = 39,589$ is absent in Table 11.2 because no coarser meshes can be found to form a mesh hierarchy needed by the multigrid V-cycles.

11.3.3 A Parallel Block Preconditioner

The objective of finding a parallel version of the block preconditioner $\mathcal{D}$ defined in (11.16) is accompanied with the request of maintaining the rapid convergence speed and the scalability in work amount. To achieve this, we replace the multigrid V-cycles, which are used in the serial preconditioner for approximately solving each of the diagonal blocks in $\mathcal{D}$, with *additive Schwarz iterations* (see [9, 30, 11]). This is mainly motivated by the inherent parallelism of these domain decomposition algorithms. Moreover, the simple algorithmic structure of the additive Schwarz iterations (e.g., no special interface solvers are needed) suits very well for a parallel implementation. In addition, additive Schwarz iterations are also known to have very good convergence behavior, similar to the multigrid algorithms.

Roughly speaking, the starting point of any additive Schwarz iteration is that a global solution domain Ω is divided into a set of *overlapping* subdomains $\{\Omega_s\}$. Each subdomain becomes an independent working unit, which mostly concentrates on local discretizations within Ω_s and solving local linear systems. In addition, the subdomains frequently collaborate with each other, in a form that neighboring subdomains exchange local solutions within overlapping zones. A loose synchronization of the work progress on the subdomains also has to be enforced.

The Mathematical Framework of Additive Schwarz Iterations

The mathematics behind the additive Schwarz iterations can be understood as follows. Suppose we want to solve a global linear system

$$\boldsymbol{A}\boldsymbol{x} = \boldsymbol{b}, \tag{11.18}$$

which arises from discretizing a PDE in a global domain Ω. Given a set of P subdomains $\{\Omega_s\}$, $1 \leq s \leq P$, such that $\Omega = \cup\Omega_s$ and there is a certain amount of overlap between neighboring subdomains, we locally discretize the PDE in every subdomain Ω_s. The result of the local discretization is

$$\boldsymbol{A}_s\boldsymbol{x}_s = \boldsymbol{b}_s(\boldsymbol{x}|_{\partial\Omega_s\backslash\partial\Omega}). \tag{11.19}$$

The above linear system namely arises from restricting the discretization of the target PDE within Ω_s. The only special treatment happens on the so-called *internal boundary* $\partial\Omega_s\backslash\partial\Omega$, i.e., the part of $\partial\Omega_s$ that does not coincide with the physical boundary $\partial\Omega$ of the global domain. We remark that a requirement of the overlapping zones says that any point lying on the internal boundary of a subdomain must also be an interior point in at least one of the neighboring subdomains. On the internal boundary, artificial Dirichlet conditions are repeatedly updated using new values of the subdomain solutions computed in the neighboring subdomains. The involvement of the artificial Dirichlet conditions is indicated by the notation $\boldsymbol{b}_s(\boldsymbol{x}|_{\partial\Omega_s\backslash\partial\Omega})$ in (11.19). On the remaining part of $\partial\Omega_s$, the original boundary conditions of the target PDE are valid as before.

For the artificial Dirichlet conditions to converge toward the correct values on the internal boundary, iterations need to be carried out. That is, we generate on each subdomain a series of approximate solutions $\boldsymbol{x}_s^0, \boldsymbol{x}_s^1, \boldsymbol{x}_s^2 \ldots$, which will hopefully converge toward the correct solution $\boldsymbol{x}_s = \boldsymbol{x}|_{\Omega_s}$. The kth additive Schwarz iteration is thus defined as

$$x_s^k = A_s^{-1} b_s(x^{k-1}|_{\partial\Omega_s \backslash \partial\Omega}), \quad x^k = \text{composition of all } x_s^k. \tag{11.20}$$

The symbol A_s^{-1} in (11.20) means an inverse of the subdomain matrix A_s, but an approximate subdomain solver that is sufficiently close to A_s^{-1} is also allowed. The right-hand side vector b_s needs to be updated with artificial Dirichlet conditions on the internal boundary, using solution of the previous Schwarz iteration provided by the neighboring subdomains.

Parallel Computing with Additive Schwarz Iterations

We note that the subdomain local solves in (11.20) can be carried out independently in each additive Schwarz iteration. This immediately gives rise to the possibility of parallel computing. At the end of the kth additive Schwarz iteration, the (logically existing) global approximate solution x^k is composed on the basis of the subdomain approximate solutions $\{x_s^k\}$. In particular, the following rule for composing a global solution, using the principle of *partition of unity*, should be used:

- An overlapping point refers to a point that lies inside a zone of overlap, i.e., the point belongs to at least two subdomains.
- For every non-overlapping point, i.e., a point that belongs to only one subdomain, the global solution attains the same value as that inside the host subdomain.
- For every overlapping point, let us denote by n_{total} the total number of host subdomains that own this point. Let also n_{interior} denote the number of subdomains, among those n_{total} host subdomains, which do not have the point lying on their internal boundaries. (The setup of the overlapping subdomains ensures $n_{\text{interior}} \geq 1$.) Then, the average of the n_{interior} local values becomes the global solution on the point. The other $n_{\text{total}} - n_{\text{interior}}$ local values are not used, because the point lies on the internal boundary there. (Take Figure 11.4 for instance, where there are four overlapping points shared between the two subdomains. For the two "middle" overlapping points we have $n_{\text{interior}} = n_{\text{total}} = 2$, whereas for the two "outer" overlapping points we have $n_{\text{interior}} = 1$ and $n_{\text{total}} = 2$. That is, on each of the two "middle" overlapping points, the two values from the two subdomains should be averaged, whereas on each of the two "outer" overlapping points, the value from the neighboring subdomain should replace the value on the host subdomain.) Finally, the obtained global solution is enforced in each of the n_{total} host subdomains. For the $n_{\text{total}} - n_{\text{interior}}$ host subdomains, which have the point lying on their internal boundary, the obtained global solution will be used as the artificial Dirichlet condition during the next Schwarz iteration.

To compose the global solution and update the artificial Dirichlet conditions, as described by the above rule, we need to carry out a procedure of communication among the neighboring subdomains at the end of each additive Schwarz iteration. During this procedure of communication, each pair of neighboring subdomains exchanges between each other an array of values that are associated with their shared overlapping points. It is clear that if each subdomain solution x_s^k converges toward

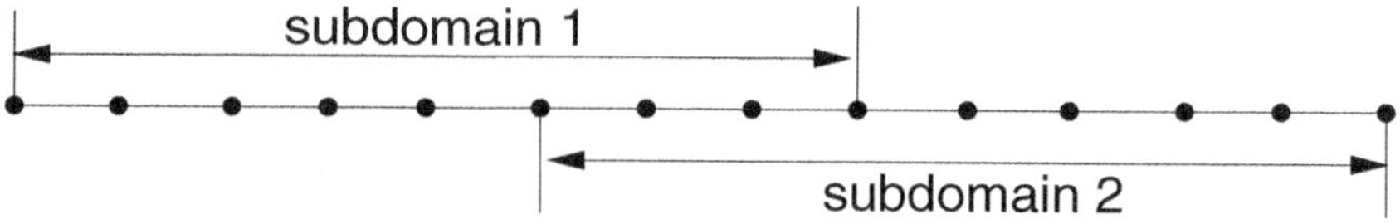

Fig. 11.4. A simple example of two overlapping subdomains.

the correct solution $\boldsymbol{x}|_{\Omega_s}$, the difference between the subdomain solutions in an overlapping zone will eventually disappear.

Mathematically, a subdomain matrix $\boldsymbol{A}_s$ in (11.19) should arise from first building the global matrix $\boldsymbol{A}$ in (11.18) and then cutting out the portion of $\boldsymbol{A}$ that corresponds to the mesh points lying in Ω_s. However, this approach requires unnecessary construction and storage of global matrices, which is not a desired situation during parallel computations. We just make the point that the global matrix $\boldsymbol{A}$ can be conceptually represented by the collection of subdomain matrices $\boldsymbol{A}_s$.

A Layered Design of the Parallel Block Preconditioner

To devise a parallel version of the block preconditioner $\mathcal{D}$ defined in (11.16), we suggest a layered design. First, two separate additive Schwarz iterations (see e.g. [5]) approximately solve the two diagonal blocks in $\mathcal{D}$, i.e., $\chi C_{\mathrm{m}}\mathbf{I}+\theta\Delta t\mathbf{A}_{\mathrm{v}}$ and $\theta\Delta t\mathbf{A}_{\mathrm{u}}$, respectively. Let us demonstrate this for the second diagonal block. For this purpose, one additive Schwarz iteration for approximating the inverse of $\mathbf{A}_{\mathrm{u}}$ can be expressed as

$$\mathbf{A}_{\mathrm{u}}^{-1} \approx \sum_{s=0}^{P} \mathbf{A}_{\mathrm{u},s}^{-1}, \tag{11.21}$$

where it is assumed that the body domain Ω is partitioned into P overlapping subdomains Ω_s, $1 \leq s \leq P$. Thus, $\mathbf{A}_{\mathrm{u},s}$ denotes a subdomain matrix that arises from a discretization restricted to Ω_s. Note that $\mathbf{A}_{\mathrm{u},0}$ in (11.21) is associated with a discretization on a very coarse global grid. The use of $\mathbf{A}_{\mathrm{u},0}^{-1}$, also called *coarse grid correction*, is to ensure convergence independent of the number of subdomains P; see e.g. [30]. The implementation issues for this topic will be discussed in Section 11.4.3.

Second, as an approximate subdomain solver, we use multigrid V-cycles. This is because the complexity of such cycles is linearly proportional to the number of subdomain unknowns. The construction of a required hierarchy of subdomain meshes is described in Section 11.4.2.

In summary, the combination of additive Schwarz iterations on the "global layer" and multigrid V-cycles on the "subdomain layer" constitutes the parallel version of the block preconditioner $\mathcal{D}$ defined in (11.16). Scalability with respect to the number of unknowns is due to the spectral equivalence between the inverse of the two diagonal blocks in $\mathcal{D}$ with the two separate additive Schwarz iterations, such as (11.21) corresponds to the second diagonal block in $\mathcal{D}$. We remark that corresponding scalability in the serial case has been proved in [23]. Moreover, multigrid V-cycles are used as the subdomain solvers in the additive Schwarz iterations. Convergence independent of the number of subdomains is due to using the coarse grid correction; see e.g. [30].

11.4 A Parallel Electro-Cardiac Simulator

Incorporating parallelism into the preceding numerical strategy for electro-cardiac simulations requires parallelization of *each* of the three sub-steps in the time-stepping scheme from Section 11.3.1. The first and third sub-steps are readily parallelizable, because the solution of the ODE system (11.10) at any two mesh points in H can be carried out completely independent of each other. For the second sub-step, preconditioned CG iterations can be parallelized using a subdomain-based approach with a distributed data structure, which will be explained in the following text.

11.4.1 Parallel Computing Based on Subdomains

Let P denote the number of processors; we adopt the approach of explicit domain partitioning that divides the entire computational work among the processors. Recall that the combined global domain Ω consists of the heart domain H and the torso domain T, so we partition both H and T into P pieces; see Figure 11.5. Processor s is thus responsible for the composite subdomain $\Omega_s = H_s \cup T_s$. In addition, we also introduce a certain amount of overlap between the subdomains to enable the additive Schwarz iterations useful in the parallel preconditioner. More details on the issue of domain partitioning will be presented in Section 11.4.2.

For the parallelization of the first and third sub-steps of the time-stepping scheme, i.e., solving the ODE system (11.10), processor s only needs to consider the mesh points inside the heart subdomain H_s. In principle, no communication is needed between the processors. However, to reduce the unnecessary duplicated local computations due to overlap between the subdomains, some additional communication may help to reduce the local computation volume; see Section 11.5.

To achieve parallel CG iterations using the block preconditioner, both the involved linear algebra operations and the block preconditioner need to be executed in parallel. A distributed data structure suits this purpose very well. We recall that the global domains H and Ω are partitioned into a set of subdomains $\{H_s\}_{s=1}^P$ and $\{\Omega_s\}_{s=1}^P$. The global matrices (such as $\mathbf{A}_\mathrm{v}$ and $\mathbf{A}_\mathrm{u}$) and global vectors can thus be represented *collectively* by the subdomain matrices (such as $\mathbf{A}_{\mathrm{v},s}$ and $\mathbf{A}_{\mathrm{u},s}$) and subdomain vectors that are *distributed* on the different processors. There is no need to physically construct the global matrices and vectors, because all the global linear algebra operations involved in a Krylov subspace method (such as the CG method) can be parallelized by doing subdomain linear algebra operations with additional communication between neighboring subdomains. In such a setup, each subdomain becomes an independent working unit, which mostly concentrates on local discretizations within H_s or Ω_s and solving local linear systems. In addition, the subdomains frequently collaborate with each other, in a form that neighboring subdomains exchange local solutions within overlapping zones. A loose synchronization of the work progress on the subdomains also has to be enforced.

During parallel computations for the PDE part, i.e., the second sub-step of the time-stepping scheme, the work on subdomain s consists of local operations that are restricted to H_s and T_s. That is, local finite element discretizations are carried out independently on each processor. No communication between the processors is needed for this task of distributed discretizations. Afterwards, during the parallel CG iterations for solving the global 2×2 block linear system (11.14), which is distributed as a set of subdomain 2×2 block systems, subdomain local operations need to be interleaved with inter-processor communication.

Such a subdomain-based approach also suits very well the parallel version of the block preconditioner $\mathcal{D}$ presented in Section 11.3.3. Subdomain multigrid V-cycles need only the local matrices and vectors to find, e.g., the approximate inverse of $\mathbf{A}_{\mathrm{u},s}$, which is needed in the additive Schwarz iterations. Besides, the distributed data structure is compatible with the communication between neighboring subdomains when all the processors have finished the subdomain multigrid V-cycles. The rules for the communication are described earlier in Section 11.3.3.

11.4.2 A Special Strategy for Mesh Partitioning

A special feature of our numerical strategy is that both the ODE system (11.10) and the linear parabolic PDE (11.9) use H as the solution domain, whereas the composite elliptic PDE (11.15) uses $\Omega = H \cup T$ as the solution domain. Therefore, we need to have $H_s \subset \Omega_s$ on each subdomain in order to form a distributed set of subdomain 2×2 block matrices $\mathcal{A}_s$ $(1 \leq s \leq P)$ for running the parallel CG iterations, where

$$\mathcal{A}_s = \begin{bmatrix} \chi C_{\mathrm{m}} \mathbf{I}_s + \theta \Delta t \mathbf{A}_{\mathrm{v},s} & \theta \Delta t \hat{\mathbf{A}}_{\mathrm{v},s} \\ \theta \Delta t \hat{\mathbf{A}}_{\mathrm{v},s}^T & \theta \Delta t \mathbf{A}_{\mathrm{u},s} \end{bmatrix}. \tag{11.22}$$

In addition, to apply multigrid V-cycles as the subdomain solvers in each additive Schwarz iteration, we also need to have a hierarchy of subdomain meshes for H_s and a hierarchy of subdomain meshes for Ω_s, respectively. To satisfy these two requirements, we use the following scheme for partitioning the global domains H and Ω, while generating desired subdomain mesh hierarchies:

1. For the global domain Ω, we start with a global coarse mesh $\mathcal{G}_{\Omega,0}$. This mesh can be used in connection with coarse grid corrections inside the additive Schwarz iterations; see Section 11.4.3.
2. Then, if necessary, we recursively perform an adaptive mesh refinement $J_g \geq 0$ times, such that we obtain a hierarchy of global meshes: $\mathcal{G}_{\Omega,0}$, $\mathcal{G}_{\Omega,1}, \ldots, \mathcal{G}_{\Omega,J_g}$.
3. The so-far finest global mesh $\mathcal{G}_{\Omega,J_g}$ is then partitioned to give rise to a set of overlapping subdomain meshes: $\{\mathcal{G}_{\Omega_s,0}\}_{s=1}^P$. This overlapping mesh partitioning is achieved by first individually carrying out non-overlapping partitioning of the global meshes $\mathcal{G}_{H,J_g}$ and $\mathcal{G}_{T,J_g}$. (Note

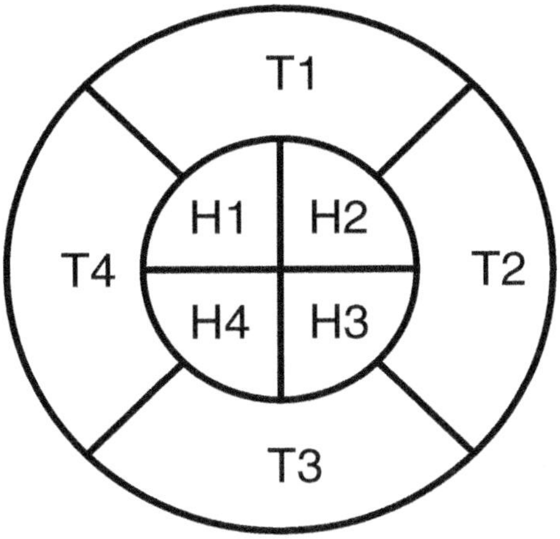

Fig. 11.5. A simplified diagram showing the strategy of domain partitioning for producing subdomains H_s and Ω_s. Here are four heart subdomains and four torso subdomains. Each subdomain Ω_s is composed of H_s and T_s.

that $\mathcal{G}_{T,J_g}$ may contain a considerable number of mesh points, necessary for describing the fine-scale changes of the conductivity tensor M_o due to, e.g., lungs and bones.) Thereafter, a certain amount of overlap is added to the subdomain meshes to give rise to the overlapping subdomain meshes $\{\mathcal{G}_{\Omega_s,0}\}$, where each $\mathcal{G}_{\Omega_s,0}$ contains a subdomain heart mesh $\mathcal{G}_{H_s,0}$. Note that the second subscript '0' in the notation $\mathcal{G}_{\Omega_s,0}$ indicates that $\mathcal{G}_{\Omega_s,0}$ is the coarsest subdomain mesh on subdomain s.

4. Afterwards, we recursively perform an adaptive mesh refinement J_s times on each subdomain independently. The final result is that, on subdomain number s, we obtain a subdomain mesh hierarchy: $\mathcal{G}_{\Omega_s,0}$, $\mathcal{G}_{\Omega_s,1}, \ldots, \mathcal{G}_{\Omega_s,J_s}$, which can be used by the subdomain multigrid V-cycles for solving $\mathbf{A}_{\mathrm{u},s}$. In addition, another subdomain mesh hierarchy is also created for the subdomain H_s, i.e., $\mathcal{G}_{H_s,0}, \mathcal{G}_{H_s,1}, \ldots, \mathcal{G}_{H_s,J_s}$.

The basic idea of domain partitioning in the above partitioning scheme can be depicted by a simplified diagram in Figure 11.5, where we have four heart subdomains and four torso subdomains. Each body subdomain Ω_s is composed of H_s and T_s. A realistic example (which was first reported in connection with [21]) is shown in Figure 11.6.

Remarks.

Balancing the work load on the subdomains depends on an even distribution of the sizes of the subdomain meshes $\mathcal{G}_{\Omega_s,j}$ and $\mathcal{G}_{H_s,j}$ on all the subdomain mesh levels, $0 \le j \le J_s$. This can be roughly achieved by first finding balanced overlapping subdomain meshes $\mathcal{G}_{H_s,0}$ and $\mathcal{G}_{T_s,0}$, for $1 \le s \le P$. (Note that $\mathcal{G}_{\Omega_s,0} = \mathcal{G}_{H_s,0} \cup \mathcal{G}_{T_s,0}$.)

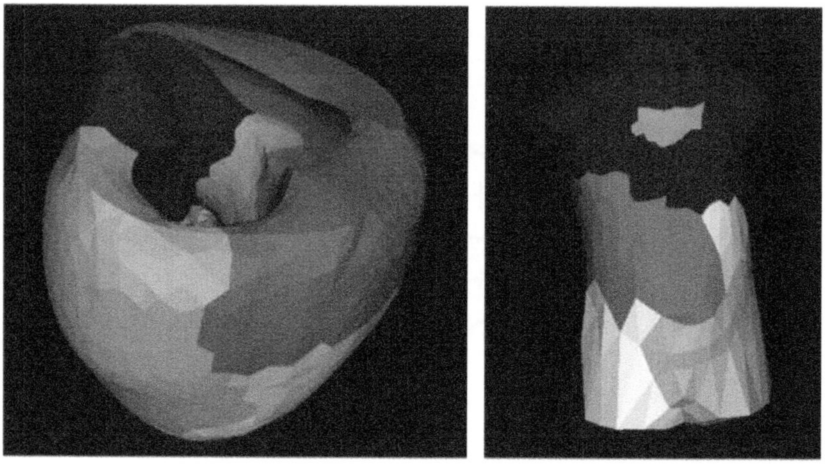

Fig. 11.6. An example of partitioning an unstructured heart mesh (left) and an unstructured torso mesh (right). (For the color version, see Figure A.25 on page 479).

Then, the adaptive mesh refinement on each subdomain has to take care to maintain the load balance, while ensuring that elements in the overlapping regions are refined in exactly the same way between the neighboring subdomains. We remark that the subdomain meshes at the finest level, e.g., $\{\mathcal{G}_{\Omega_s,J_s}\}_{s=1}^{P}$, constitute the finest *virtual* global mesh. In practice, the subdomain meshes normally have a certain degree of load imbalance, which is one of the main obstacles for achieving perfect speedup results. The need for a relatively fine global coarse mesh $\mathcal{G}_{\Omega,J_g}$ (or $\mathcal{G}_{H,J_g}$) may arise when the coarsest global mesh $\mathcal{G}_{\Omega,0}$ (or $\mathcal{G}_{H,0}$) is too coarse to limit the amount of overlap between the subdomains during the overlapping mesh partitioning. More details can be found in [8].

11.4.3 Coarse Grid Corrections

We recall from Section 11.3.3 that $\mathbf{A}_{\mathrm{u},0}^{-1}$ is included in the formula of the additive Schwarz iteration (11.21), which approximately inverts $\mathbf{A}_{\mathrm{u}}$. The action of $\mathbf{A}_{\mathrm{u},0}^{-1}$, which is denoted a coarse grid correction, refers to solving a global problem associated with the coarsest global mesh $\mathcal{G}_{\Omega,0}$. For the block matrix $\mathbf{A}_{\mathrm{u}}$, which arises from discretizing the combined elliptic equation (11.15), the coarse grid correction is particularly important for obtaining constant convergence speed, independently of the number of subdomains P. Otherwise, without the coarse grid correction, the effect of the parallel block preconditioner will deteriorate when P becomes large. The effect of coarse grid corrections are well illustrated in Table 11.3.

In the case where the coarsest global mesh $\mathcal{G}_{\Omega,0}$ has a small number of mesh points, it suffices for every subdomain to solve the same global coarse grid problem using a serial algorithm. Compared with using a parallel coarse grid solver, this serial approach requires fewer inter-processor communications at the cost of a small

Table 11.3. An example of the effect of coarse grid corrections (CGCs), associated with the parallel block preconditioner, where I_{CG} denotes the number of CG iterations needed for solving the 2×2 block system (11.14).

		Without CGCs	With CGCs
$N_H + N_\Omega$	P	I_{CG}	I_{CG}
302,166	2	31	7
	4	41	7
	8	52	7
	16	79	8
1,552,283	2	32	7
	4	44	7
	8	63	8
	16	113	8

increase of subdomain computations. However, if $\mathcal{G}_{\Omega,0}$ has quite many mesh points, the serial approach will become too costly. A parallel coarse grid solver must therefore be adopted. In such a case where the $\mathcal{G}_{\Omega,0}$ mesh is quite fine, we assume that there is no need to refine it before carrying out overlapping mesh partitioning, i.e., $J_g = 0$. Therefore, the subdomain coarsest meshes $\mathcal{G}_{\Omega_s,0}$ constitute a decomposition of $\mathcal{G}_{\Omega,0}$, which can be used to build a distributed data structure on the coarsest global mesh level and run the coarse grid solver in parallel.

11.5 Some Techniques for Overhead Reduction

Recall that the parallel block preconditioner from Section 11.3.3 uses two separate additive Schwarz iterations as approximate solvers for the two diagonal block matrices of $\mathcal{D}$ as defined in (11.16). The mathematical theory of the additive Schwarz method requires overlapping subdomains for ensuring the convergence; see [9, 30, 39]. Therefore, a number of elements and mesh points are shared between each pair of neighboring subdomains. We also remark that sufficient overlap between neighboring subdomains is important for obtaining a rapid convergence.

However, the mathematically required overlap between the subdomains will give rise to some duplicated computations in both the ODE part and the PDE part of the parallel electro-cardiac simulator. Regarding the ODE part, in order to avoid communication in the embarrassingly parallel ODE solver, neighboring subdomains must duplicate computations on the mesh points lying inside the overlap. As the ODE solver may be quite computationally intensive, this duplication gives rise to considerable overhead, which deteriorates the speedup. Similarly, duplication also arises in the PDE part, associated with the overlap between the subdomains. Such duplicated PDE computations are present on each shared mesh point inside the overlap, when the matrix-vector product of the global CG method is distributed as a set of subdomain linear algebra operations. However, we stress that there is no duplicated computation during the parallel preconditioning operations, because neighboring subdomains are meant to compute different values for the shared mesh points.

To remove the overhead due to duplicated computations described above, we need to do a *disjoint* re-distribution of all the mesh points. Such a disjoint re-distribution should be done on the basis of the existing overlapping subdomain meshes. For this purpose, let us denote by N_s the total number of mesh points in subdomain s. Since there is overlap between the subdomains, we have (for $P > 1$)

$$N < \sum_{s=1}^{P} N_s,$$

where N denotes the total number of points in a global mesh. Moreover, let us denote by N_s^O and N_s^I (where $N_s^O + N_s^I = N_s$) the number of overlapping points and the number of interior (non-overlapping) points in subdomain s, respectively. The essence of a disjoint re-distribution is to divide the N_s^O overlapping points on subdomain s into two parts: $N_s^O = N_s^{O_c} + N_s^{O_n}$, where $N_s^{O_c}$ is the number of overlapping points that participate in all the computations, whereas the remaining $N_s^{O_n}$ *non-computational* overlapping points do not participate in either the ODE solver or the global-level CG operations in the PDE solver. The objective of the disjoint re-distribution is to achieve

$$N = \sum_{s=1}^{P} \left(N_s^{O_c} + N_s^I\right),$$

while approximately maintaining a constant value of $N_s^{O_c} + N_s^I$ independent of s.

The removal of the duplicated computations, however, comes at the cost of an increased volume of communication. This is because the values that are needed on the $N_s^{O_n}$ non-computational overlapping points need to be provided by the neighboring subdomains, through message passing. Normally, the increase of the communication overhead can be justified by the removal of the duplicated computations.

Devising a high-quality and efficient re-distribution scheme is not trivial. Research is currently under way to find an optimal scheme which is also executable in parallel, involving collaboration between all the subdomains. For the time being, Tables 11.4 and 11.5 show two examples of applying a relatively simple re-distribution scheme to the mesh points in the global H and Ω domains, respectively. We can see that although the resulting value of $\max_s(N_s^I + N_s^{O_c})$ is dramatically reduced in comparison with $\max_s N_s$, the load balancing quality can be further improved. Regarding Table 11.5, the reason for only a slight decrease in WT_{CG} is because the portion of the duplicated computations inside the PDE part is relatively small. Figure 11.7 depicts the distribution of the values of $N_{\Omega_s}^{O_n}, N_{\Omega_s}^{O_c}, N_{\Omega_s}^I$ after applying the simple re-distribution scheme to the Ω mesh points, where $N_\Omega = 919,851$. We can see that the load balancing situation is considerably improved with respect to $N_{\Omega_s}^{O_c} + N_{\Omega_s}^I$, compared with N_{Ω_s}. However, the disjoint re-distribution can clearly be further improved.

Another possibility of overhead reduction lies in the coarse grid corrections within the additive Schwarz iterations, when the global $\mathcal{G}_{\Omega,0}$ mesh has a relatively large number of points. As mentioned in Section 11.4.3, parallel coarse grid corrections should in such a case replace serial coarse grid corrections. Table 11.6 shows

Table 11.4. An example of applying a disjoint re-distribution to the H domain mesh points, where $N_H = 632,432$ and $WT_{\rm ODE}$ denotes the time consumptions by the ODE solver for one time step on a Linux cluster; see Section 11.6 for more details.

	Before re-distribution		After re-distribution	
P	$\max_s N_{H_s}$	$WT_{\rm ODE}$	$\max_s(N^I_{H_s} + N^{O_c}_{H_s})$	$WT_{\rm ODE}$
2	405,893	120.75	316201	117.40
4	225,296	73.48	162,823	65.31
8	137,446	51.82	89,025	37.20
16	71,643	29.48	49,057	23.08

Table 11.5. An example of applying a disjoint re-distribution to the Ω domain mesh points, where $N_\Omega = 919,851$ and $WT_{\rm CG}$ denotes the time consumptions by the parallel CG solver for one time step on a Linux cluster; see Section 11.6 for more details.

	Before re-distribution		After re-distribution	
P	$\max_s N_{\Omega_s}$	$WT_{\rm CG}$	$\max_s(N^I_{\Omega_s} + N^{O_c}_{\Omega_s})$	$WT_{\rm CG}$
2	600,773	44.30	476,382	42.63
4	332,163	27.16	248,841	26.42
8	200,987	21.57	123,662	20.86
16	110,146	18.40	70,164	17.21

Table 11.6. The effect of using parallel coarse grid corrections (CGCs) instead of serial CGCs. The total number of unknowns at the global fine mesh level is $N_H + N_\Omega = 1,552,283$, and $WT_{\rm CG}$ denotes the time consumptions for one time step on a Linux cluster; see Section 11.6 for more details.

	Serial CGCs	Parallel CGCs
P	$WT_{\rm CG}$	$WT_{\rm CG}$
2	42.63	44.82
4	26.42	26.23
8	20.86	19.42
16	17.21	13.46

the results associated with the parallel PDE solver when $N_H + N_\Omega = 1,552,283$ for the global fine mesh level and $N_{\Omega,0} = 28,283$ for the global coarse mesh level. (We note that coarse grid corrections are not used in the additive Schwarz iterations for treating the $(1,1)$ diagonal block of $\mathcal{D}$.) Both the serial and parallel coarse grid solvers use CG iterations without preconditioning on the level of the global $\mathcal{G}_{\Omega,0}$ mesh. The reason for a deteriorated performance for $P = 2$ in Table 11.6 is that the overhead in the parallel coarse grid solver exceeds the gain from parallelization. The advantage of a parallel coarse grid solver thus becomes visible when P is large.

11.6 Numerical Experiments

In this section, we will report some more numerical experiments of the electro-cardiac simulation, where we have used the realistic global domains H and Ω as

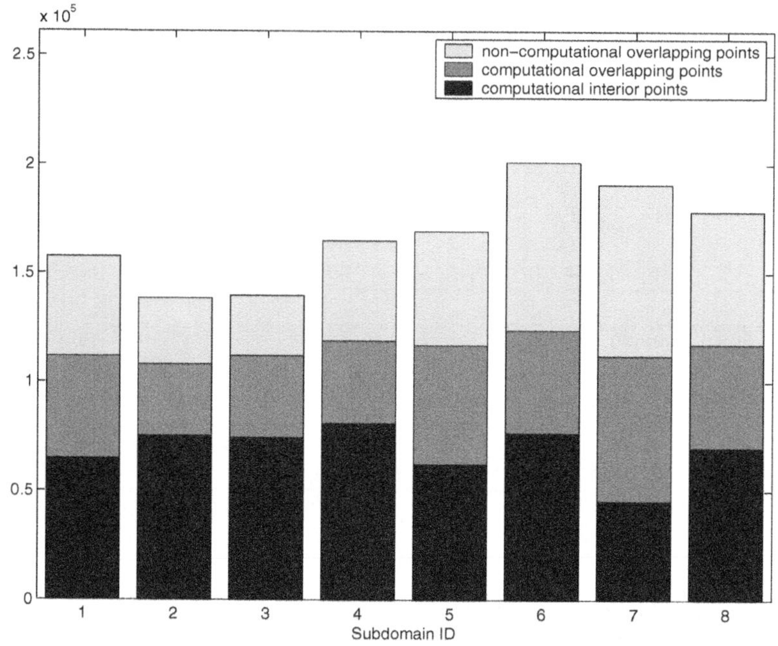

Fig. 11.7. The effect of applying a disjoint re-distribution to the Ω mesh points, where $N_\Omega = 919,851$ and the number of subdomains is 8. (For the color version, see Figure A.26 on page 479).

depicted in Figure 11.1. The coarsest global $\mathcal{G}_{H,0}$ mesh consists of 56,568 tetrahedral elements and 11,306 mesh points, while the coarsest global $\mathcal{G}_{\Omega,0}$ mesh consists of 162,120 tetrahedral elements and 28,283 mesh points. Since both the global coarsest meshes have quite high resolution, we directly partition the $\mathcal{G}_{\Omega,0}$ mesh (without adaptive refinement) into overlapping subdomain $\mathcal{G}_{\Omega_s,0}$ meshes, $1 \le s \le P$. We remark that our mesh partitioning scheme uses the Metis [16] software package to first generate intermediate non-overlapping subdomain meshes, which are then expanded to introduce a certain amount of overlap between neighboring subdomains. Depending on the desired resolution of the finest subdomain meshes, we adaptively refine the $\mathcal{G}_{\Omega_s,0}$ subdomain mesh J_s times on each subdomain s, as described in Section 11.4.2. We remark that the subdomain mesh hierarchy $\{\mathcal{G}_{H_s,j}\}$ $(0 \le j \le J_s)$ is obtained by "cutting out" the portion of the subdomain mesh hierarchy $\{\mathcal{G}_{\Omega_s,j}\}$ that lies inside H.

We have chosen the Winslow cell model (see [38]) for the numerical experiments in this section. Regarding the anisotropic conductivity tensors M_i and M_e, which are defined in (11.4), the following σ values are used:

$$\sigma_i^l = 3.0, \quad \sigma_i^t = 0.31525, \quad \sigma_i^n = 1.0, \tag{11.23}$$

$$\sigma_e^l = 2.0, \quad \sigma_e^t = 1.3514, \quad \sigma_e^n = 1.65. \tag{11.24}$$

Moreover, the vectors $\mathbf{a}_l$ and $\mathbf{a}_n$ are depicted in Figure 11.2. For the surface-to-volume scaling factor used in (11.1), we have chosen $\chi = 5.0 \times 10^{-4}$.

When the global 2×2 block system (11.14) is solved by the preconditioned parallel CG iterations at each time step, we consider that the convergence is reached when the global residual vector is reduced by a factor of 10^4 from its initial value at the beginning of the time step, measured in the L_2-norm. For the multigrid V-cycles that are used as the subdomain solvers in the additive Schwarz iterations, the pre- and post-smoothers are chosen as three SOR iterations for the subdomain mesh levels $1 \leq j \leq J_s$. On the level of the coarsest subdomain mesh $\mathcal{G}_{H_s,0}$ or $\mathcal{G}_{\Omega_s,0}$, 20 SSOR iterations are used. Coarse grid corrections have been used in association with the additive Schwarz iteration handling the second diagonal block in $\mathcal{D}$. The chosen global coarse grid solver on the $\mathcal{G}_{\Omega,0}$ mesh level is also a parallel CG solver, which aims to reduce the associated residual vector by a factor of 10. We remark that for the additive Schwarz iteration handling the first diagonal block in $\mathcal{D}$, experiments show that coarse grid correction is not necessary.

The parallel electro-cardiac simulator is programmed in the scientific computing environment of Diffpack; see [17, 10]. The implementation of the additive-Schwarz block preconditioner uses the object-oriented programming techniques and a generic framework for overlapping domain decomposition algorithms; see e.g. [4, 18].

In Table 11.7, we list the wall-time measurements of four different tasks within one time step:

1. WT_{ODE} denotes the wall-time consumption of the ODE part.
2. WT_{discr} denotes the wall-time consumption of the finite element discretization procedure for building the 2×2 block system (11.14). We note that WT_{discr} is noticeably larger for the first time step (which is reported in Tables 11.7 and 11.8) than for the subsequent time steps, because both $\mathcal{A}$ and the right-hand side vector in (11.14) need to be built in the first time step, whereas only the right-hand side vector needs to be updated in the subsequent time steps.
3. WT_{CG} denotes the wall-time consumption of the preconditioned parallel CG iterations for solving (11.14).
4. WT_{step} denotes the total wall-time consumption of one time step.

The measurements in Table 11.7 are obtained on a Linux cluster that consists of 1.3 GHz Itanium2 processors, inter-connected through a Gigabit ethernet. We can observe that the number of the parallel CG iterations, which is denoted by I_{CG}, remains independent of both the number of degrees of freedom and the number of subdomains (when $P \geq 2$). In fact, the parallel additive-Schwarz block preconditioner has better convergence properties than the serial multigrid block preconditioner. For $J_s = 3$, the finest meshes contain so many points that the simulator can not be run on fewer than four processors. Therefore, measurements for $P = 1$ and $P = 2$ are absent for $J_s = 3$ in Table 11.7.

In Table 11.8, we list the corresponding measurements that are obtained on an SGI Origin 3800 machine. Due to a faster communication network on the SGI machine (and slower processors), the scalability of the measurements should in general be better than that of Table 11.7. However, due to the extremely heavy work load on

Table 11.7. The wall-time measurements (in seconds and obtained on an Itanium-cluster) of different tasks within one time step during parallel full-scale electro-cardiac simulations.

J_s	N_H	N_Ω	P	WT_{ODE}	WT_{discr}	I_{CG}	WT_{CG}	WT_{step}
1	82,768	219,398	1	52.85	41.08	15	28.30	153.19
			2	33.42	24.26	7	13.52	93.44
			4	18.98	13.20	7	8.25	53.64
			8	10.26	7.55	7	4.63	30.55
			16	6.20	4.00	8	3.89	23.14
2	632,432	919,851	1	381.58	246.65	16	134.93	982.11
			2	241.85	153.36	7	44.82	585.88
			4	133.45	84.31	7	26.23	327.59
			8	75.21	49.78	8	19.42	195.79
			16	47.77	26.18	8	13.46	127.79
3	4,942,624	5,762,729	1	N/A	N/A	N/A	N/A	N/A
			2	N/A	N/A	N/A	N/A	N/A
			4	1518.95	676.02	9	196.51	2511.18
			8	942.48	383.61	10	121.36	1534.65
			16	617.41	191.27	10	78.19	952.51

Table 11.8. The wall-time measurements (in seconds and obtained on an SGI Origin 3800 machine) of different tasks within one time step during parallel full-scale electro-cardiac simulations.

J_s	N_H	N_Ω	P	WT_{ODE}	WT_{discr}	I_{CG}	WT_{CG}	WT_{step}
1	82,768	219,398	8	19.91	20.68	7	9.07	55.30
			16	16.12	10.84	8	4.70	33.25
			32	9.13	5.65	8	3.47	21.70
2	632,432	919,851	8	155.63	169.46	8	27.30	385.08
			16	121.67	83.76	8	16.37	242.93
			32	68.59	41.65	8	15.45	142.51
3	4,942,624	5,762,729	8	1233.72	1481.66	9	386.63	3406.51
			16	949.98	755.67	10	286.89	2206.67
			32	534.43	381.01	10	144.98	1197.09

the SGI machine, some of the measurements in Table 11.8 have been "slowed down" quite considerably and are thus not very accurate.

11.7 Concluding Remarks

Building a high-performance parallel electro-cardiac simulator relies mostly on an efficient parallel solver for the involved PDEs. Numerically, the convergence speed of the parallel CG iterations needed in the PDE part is ensured by a matching parallel substitute of the serial multigrid block preconditioner $\mathcal{D}$ defined in (11.16). That is, two separate additive Schwarz iterations act as approximate solvers for the two diagonal blocks in $\mathcal{D}$, while also incorporating coarse grid correction and using multigrid

V-cycles as the subdomain solvers. With respect to the parallel implementation, satisfactory performance can be obtained when attention is paid to a good work load balance and the reduction of unnecessary duplicated computations. However, due to the complicated shape of the realistic three-dimensional heart and torso meshes, finding a well balanced set of subdomain mesh hierarchies while restricting the volume of communication remains a challenging task. This issue should be further investigated in the future work.

Acknowledgement

The work presented in this chapter has been partially supported by the Norwegian Research Council (NFR) through *Programme for Supercomputing* in form of a grant of computing time.

References

1. W. L. Briggs, V. E. Henson, and S. F. McCormick. *A Multigrid Tutorial*. SIAM, 2nd edition, 2000.
2. A. M. Bruaset. *A Survey of Preconditioned Iterative Methods*. Pitman Research Notes In Mathematics Series 328. Longman Scientific & Technical, 1995.
3. M. L. Buist and A. J. Pullan. The effect of torso impedance on epicardial and body surface potentials: A modeling study. *IEEE Trans. Biomed. Eng.*, 50(7):816–824, 2003.
4. X. Cai. Domain decomposition in high-level parallelization of PDE codes. In C.-H. L. et al., editor, *Domain Decompostion Methods in Science and Engineering*, pages 382–389. Domain Decomposition Press, 1999.
5. X. Cai. Overlapping domain decomposition methods. In H. P. Langtangen and A. Tveito, editors, *Advanced Topics in Computational Partial Differential Equations – Numerical Methods and Diffpack Programming*, pages 57–95. Springer, 2003.
6. X. Cai, G. Lines, and A. Tveito. Parallel solution of the bidomain equations with high resolutions. In G. R. Joubert, W. E. Nagel, F. J. Peters, and W. V. Walter, editors, *Parallel Computing: Software Technology, Algorithms, Architectures & Applications*, pages 837–844. Elsevier Science, 2004.
7. X. Cai and G. T. Lines. Enabling numerical and software technologies for studying the electrical activity in human heart. In J. Fagerholm et al., editor, *Applied Parallel Computing - Advanced Scientific Computing, 6th International Conference, PARA 2002*, number 2367 in Lecture Notes in Computer Science, pages 3–17, Espoo, Finland, 2002. Springer-Verlag.
8. X. Cai and K. Samuelsson. Parallel multilevel methods with adaptivity on unstructured grids. *Computing and Visualization in Science*, 3:133–146, 2000.
9. T. F. Chan and T. P. Mathew. Domain decomposition algorithms. In *Acta Numerica 1994*, pages 61–143. Cambridge University Press, 1994.
10. Diffpack Home Page. http://www.diffpack.com.
11. L. Formaggia, M. Sala, and F. Saleri. Domain decomposition techniques. In A. M. Bruaset and A. Tveito, editors, *Numerical Solution of Partial Differential Equations on Parallel Computers*, volume 51 of *Lecture Notes in Computational Science and Engineering*, pages 135–163. Springer-Verlag, 2005.

12. P. C. Franzone and L. F. Pavarino. A parallel solver for reaction-diffusion systems in computational electrocardiology. *Mathematical Models and Methods in Applied Sciences*, 14(6):883–911, 2004.
13. I. L. Grice, P. Hunter, and B. Smaill. Laminar structure of the heart: a mathematical model. *Am. J. Physiol. Heart. Circ. Physiol.*, 272:H2466–H2476, 1997.
14. W. Hackbusch. *Multigrid Methods and Applications*. Springer, Berlin, 1985.
15. F. Hülsemann, M. Kowarschik, M. Mohr, and U. Rüde. Parallel geometric multigrid. In A. M. Bruaset and A. Tveito, editors, *Numerical Solution of Partial Differential Equations on Parallel Computers*, volume 51 of *Lecture Notes in Computational Science and Engineering*, pages 165–208. Springer-Verlag, 2005.
16. G. Karypis and V. Kumar. Metis: Unstructured graph partitioning and sparse matrix ordering system. Technical report, Department of Computer Science, University of Minnesota, Minneapolis/St. Paul, MN, 1995.
17. H. P. Langtangen. *Computational Partial Differential Equations - Numerical Methods and Diffpack Programming*. Texts in Computational Science and Engineering. Springer, 2nd edition, 2003.
18. H. P. Langtangen and X. Cai. A software framework for easy parallelization of PDE solvers. In *Proceedings of the Parallel Computational Fluid Dynamics 2000 Conference*, 2001.
19. G. T. Lines. *Simulating the Electrical Activity in the Heart - A Bidomain Model of the Ventrciles Embedded in a Torso*. PhD thesis, Department of informatics, University of Oslo, 1999.
20. G. T. Lines, M. L. Buist, P. Grøttum, A. J. Pullan, J. Sundnes, and A. Tveito. Mathematical models and numerical methods for the forward problem in cardiac electrophysiology. *Computations and Visualization in Science*, 5:215–239, 2003.
21. G. T. Lines, X. Cai, and A. Tveito. A parallel solution of the bidomain equations modeling the electrical activity of the heart. Technical report, Simula Resserch Laboratory, 2001.
22. C. H. Luo and Y. Rudy. A dynamic model of the cardiac ventricular action potenial. *Circulation Research*, 74:1071–1096, 1994.
23. K.-A. Mardal and A. Tveito. Optimal preconditioners for discrete versions of the bidomain model. *Submitted to: SIAM J. Sci. Comp.*, 2004.
24. M. Murillo and X.-C. Cai. A fully implicit parallel algorithm for simulating the non-linear electrical activity of the heart. *Numerical Linear Algebra with Applications*, 11:261–277, 2004.
25. M. Pennacchio and V. Simoncini. Efficient algebraic solution of reaction–diffusion systems for the cardiac excitation process. *Journal of Computational and Applied Mathematics*, 145:49–70, 2002.
26. J. B. Pormann, C. S. Henriquez, J. A. B. Jr., D. J. Rose, D. M. Harrild, and A. P. Henriquez. Computer simulations of cardiac electrophysiology. In *Proceedings of the 2000 ACM/IEEE conference on Supercomputing*. IEEE Computer Society, 2000.
27. D. Porras, J. M. Rogers, W. M. Smith, and A. E. Pollard. Distributed computing for membrane-based modeling of action potential propagation. *IEEE Trans. Biomed. Eng.*, 47(8):1051–1057, 2000.
28. H. I. Saleheen and K. T. Ng. A new three dimensional finite-differenc bidomain formulation for inhomogeneous anisotropic cardiac tissues. *IEEE Trans. Biomed. Eng.*, 45(1):15–25, 1998.
29. K. Skouibine and W. Krassowska. Increasing the computational efficiency of a bidomain model of defibrillation using a time-dependent activating function. *Ann Biomed Eng.*, 28(7):772–780, 2000.

30. B. F. Smith, P. E. Bjørstad, and W. Gropp. *Domain Decomposition: Parallel Multilevel Methods for Elliptic Partial Differential Equations*. Cambridge University Press, 1996.
31. J. Sundnes, G. Lines, K.-A. Mardal, and A. Tveito. Multigrid block preconditioning for a coupled system of partial differential equations modeling the electrical activity of the heart. *Computer Methods in Biomechanics and Biomedical Engineering*, 5(6):397–409, 2002.
32. J. Sundnes, G. T. Lines, X. Cai, B. F. Nielsen, K.-A. Mardal, and A. Tveito. *Computing the Electrical Activity in the Human Heart*. Book accepted for publication by Springer-Verlag, 2004. 278 pages.
33. J. Sundnes, G. T. Lines, P. Grøttum, and A. Tveito. Electrical activity in the human heart. In H. P. Langtangen and A. Tveito, editors, *Advanced Topics in Computational Partial Differential Equations – Numerical Methods and Diffpack Programming*, volume 33 of *Lecture Notes in Computational Science and Engineering*, pages 401–449. Springer-Verlag, 2003.
34. J. Sundnes, G. T. Lines, and A. Tveito. ODE-solvers for a stiff system arising in the modeling of the electrical activity of the heart. *International Journal of Nonlinear Sciences and Numerical Simulation*, 4(1):67–80, 2003.
35. M.-C. Trudel, B. Dubé, M. Potse, R. M. Gulrajani, and L. J. Leon. Simulation of QRST integral maps with a membrane-based computer heart model employing parallel processing. *IEEE Trans. Biomed. Eng.*, 51(8):1319–1329, 2004.
36. L. Tung. *A Bi-domain model for describing ischemic myocardial D-C potentials*. PhD thesis, MIT, Cambridge, MA, 1978.
37. R. Weber dos Santos, G. Plank, S. Bauer, and E. J. Vigmond. Parallel multigrid preconditioner for the cardiac bidomain model. *IEEE Transactions on Biomedical Engineering*, 51(11):1960–1968, 2004.
38. R. L. Winslow, J. Rice, S. Jafri, E. Marban, and B. O'Rourke. Mechanisms of altered excitation-contraction coupling in canine tachycardia-induced heart failure, II, model studies. *Circulation Research*, 84:571–586, 1999.
39. J. Xu. Iterative methods by space decomposition and subspace correction. *SIAM Review*, 34(4):581–613, December 1992.

12

Developing a Geodynamics Simulator with PETSc

Matthew G. Knepley[1], Richard F. Katz[2], and Barry Smith[1]

[1] Mathematics and Computer Science Division, Argonne National Laboratory, Argonne, IL, USA
[knepley,bsmith]@mcs.anl.gov

[2] Department of Earth and Environmental Sciences, Lamont Doherty Earth Observatory, Palisades, NY, USA
katz@ldeo.columbia.edu

Summary. Most high-performance simulation codes are not written from scratch but begin as desktop experiments and are subsequently migrated to a scalable, parallel paradigm. This transition can be painful, however, because the restructuring required in conversion forces most authors to abandon their serial code and begin an entirely new parallel code. Starting a parallel code from scratch has many disadvantages, such as the loss of the original test suite and the introduction of new bugs. We present a disciplined, incremental approach to parallelization of existing scientific code using the PETSc framework. In addition to the parallelization, it allows the addition of more physics (in this case strong nonlinearities) without the user having to program anything beyond the new pieces of discretization code. Our approach permits users to easily develop and experiment on the desktop with the same code that scales efficiently to large clusters with excellent parallel performance. As a motivating example, we present work integrating PETSc into an existing plate tectonic subduction code.

12.1 Geodynamics of Subduction Zones

Subduction zones, where one of the Earth's surface plates collides with another and sinks into the deep mantle (see Figure 12.1a) [24], are the locus of many of the world's most devastating natural disasters, especially volcanic eruptions and earthquakes. Subduction zone volcanism such as the 1980 eruption of Mount St. Helens is characterized by violent explosions of ash and rock. Despite the relevance of this type of volcanism to problems ranging from public safety to global climate change and mass-extinction events in the geologic record, a detailed understanding of its source is lacking. Clues are abundant, however, in the rocks erupted from subduction zone volcanos which record a history of formation, transport, and eruption in their distinct geochemistry. The complexity of these processes in terms of the governing reactive thermochemical fluid dynamics and non-Newtonian rheology is significant; simplified models have proven inadequate in explaining basic observations [18].

More sophisticated PDE-based computational models are needed to address the sharp nonlinearities typically unexplored in past work. The large separation in length

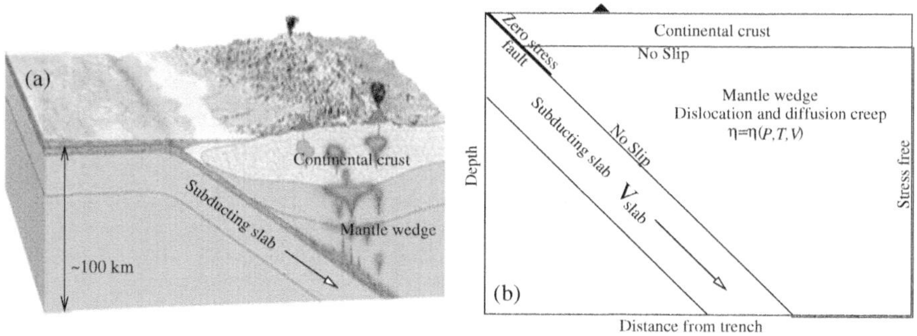

Fig. 12.1. **(a)** Schematic diagram of a subduction zone. Oceanic lithosphere is colliding with continental lithosphere and being subducted. Volatile compounds are released from the subducting slab at depth and enter the mantle wedge, lowering the melting temperature of the rock and causing partial melting. This melt rises to feed volcanos at the surface. **(b)** Schematic diagram of the computational domain. Boundary conditions on the flow field are imposed at the bottom of the crust, the top of the slab, and the intersection of the mantle wedge and the domain boundary. Flow velocity within the mantle wedge is the solution to equation (12.1). Potential temperature throughout the domain is the solution to equation (12.3). The "zero stress fault" represents a frictional sliding surface. On geologic time scales all stress on this boundary is relieved in earthquakes and does not cause deformation.

and time scales of the constituent physics implies a great computational cost due to the need for fine meshes to resolve the small (but relevant) scales. Moreover, the highly nonlinear character of the non-Newtonian, temperature-dependent viscosity demands costly solution algorithms. This high computational complexity coupled with long development times have put such models out of reach for most geoscientists. The Portable Extensible Toolkit for Scientific Computation (PETSc) [5] makes it easier for geoscientists to overcome these barriers. In an abstract sense, PETSc provides a framework for collaboration between geoscientists with complex modeling problems and numerical analysts and software engineers who have the encapsulated numerical methods for solving those problems. In the case described here, the original model was a specialized, single linear solver serial code capable of calculating the thermal structure due to an analytically prescribed isoviscous flow field. By porting this code into the PETSc framework we were able to extend it to solve for fully coupled thermal structure and non-Newtonian flow with a choice of many scalable parallel solvers, flexible boundary conditions, convenient parameter input, real-time code steering, and many other features not available with the serial version. This was done using an incremental process by first replacing the custom linear solver with PETSc's general purpose nonlinear solver, then replacing the sequential data structures with PETSc's parallel ones and then finally adding the additional nonlinear physics (to the portion of the code that discretizes the PDE).

Any model of subduction zone volcanism must be based on the thermal and flow structure of the slowly moving solid mantle wedge, shown in Figure 12.1. While the magnesium and iron-rich mantle rock is solid on human time-scales (as evidenced by the seismic waves it transmits), on geologic time-scales it undergoes two modes of solid-state creep [14]. Its motion can be described by the Stokes equation for steady flow of an incompressible, highly viscous fluid (with zero Reynolds number),

$$\nabla P = \nabla \cdot \left[\eta \left(\nabla \mathbf{V} + \nabla \mathbf{V}^T\right)\right]; \quad s.t. \; \nabla \cdot \mathbf{V} = 0, \tag{12.1}$$

$$\eta = \left(1/\eta_{disl} + 1/\eta_{difn}\right)^{-1}, \tag{12.2}$$

where P is the fluid pressure, $\mathbf{V}$ is the mantle velocity field, η_{difn} is the diffusion creep viscosity and η_{disl} is the dislocation creep viscosity. Both creep mechanisms give an Arrhenius-type dependence on pressure and temperature. Dislocation creep has an additional non-Newtonian strain rate dependence. The strength of the non-linearity of viscosity makes this equation difficult to solve without a good initial guess. In our code the initial guess in provided by a continuation method: we modify equation (12.1) to let $\eta \rightarrow \eta^\alpha$, where α is a number between zero and one. The continuation method, described in section 12.5.1, varies α over a sequence of solves of increasing nonlinearity to reach natural variation in viscosity. The solution at each step is used as a guess for the following solve. The first step in this process is to solve the problem for constant viscosity (α=0), where the system of equations is linear and analytically tractable.

The production of molten rock depends fundamentally on the mantle temperature field. The distribution of heat is governed by the conservation of enthalpy, expressed as an advection-diffusion equation,

$$\frac{\partial \theta}{\partial t} + \mathbf{V} \cdot \nabla \theta = \kappa \nabla^2 \theta, \tag{12.3}$$

where θ is the mantle potential temperature and κ is the thermal diffusivity of the mantle. The mantle has a very low thermal diffusivity compared to materials such as metal or water, and thus the advection term dominates in equation (12.3).

For $0 < \alpha \leq 1$, equations (12.1) and (12.3) form a nonlinear set coupled through the viscosity. The solution of these equations is the first stage in modeling magma genesis in subduction zones. Future work will incorporate equations of porous flow of volatiles and magma through the mantle, reactive melting, and geochemical transport. Even without these complexities, however, we have achieved interesting results with the simple, though highly nonlinear, set of equations given above. Some of these results are presented below, after an in-depth look at the application development process in the PETSc framework.

12.2 Integrating PETSc

PETSc is a set of library interfaces built in a generally hierarchical fashion. A user may decide to use some libraries and disregard others. Figure 12.2 illustrates this hierarchy of dependencies. For instance, a user can use only the PETSc linear algebra (vectors and matrices) libraries, or add linear solvers to those, or add nonlinear solvers to the entire group. Thus, integration may proceed in several stages, which we

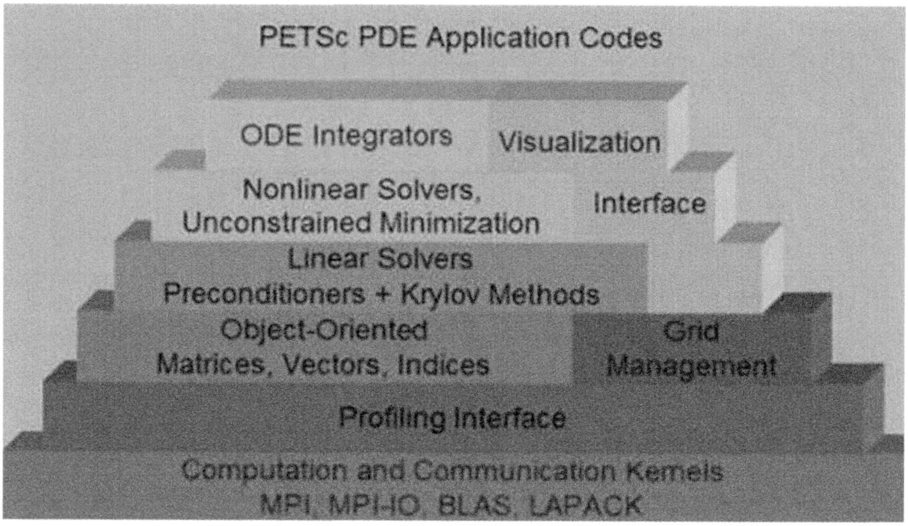

Fig. 12.2. Interface hierarchy in a PETSc application.

```
CFLAGS =
FFLAGS =
CPPFLAGS =
FPPFLAGS =

include ${PETSC_DIR}/bmake/common/base

ex1: ex1.o util.o chkopts
    -${CLINKER} -o $@ $^ ${PETSC_SNES_LIB}
    ${RM} $^

ex1f: ex1f.o phys.o chkopts
    -${FLINKER} -o $@ $^ ${PETSC_FORTRAN_LIB} ${PETSC_SNES_LIB}
    ${RM} $^
```

Fig. 12.3. Typical PETSc makefile.

discuss in the following sections. The first step is to incorporate the PETSc libraries into the existing compile system (i.e., `make`) and to initialize the PETSc runtime. This may be done in two ways.

The simplest approach is to adopt the PETSc makefile structure, which is presented in Figure 12.3 for an application using the nonlinear solver package. The user includes the `bmake/common/base`, which defines both rules for compiling source and variables that are used during the compile and link. Then, only a simple rule for

```
include ${PETSC_DIR}/bmake/common/variables

.c.o:
	massageC.pl $<
	${CC} -c ${MY_CFLAGS} ${COPTFLAGS} ${CFLAGS} ${CCPPFLAGS} $<

.F.o:
	massageFortran.pl $<
	${FC} -c ${MY_FFLAGS} ${FOPTFLAGS} ${FFLAGS} ${FCPPFLAGS} $<

ex1: ex1.o util.o
	-${CLINKER} -o $@ $^ ${PETSC_SNES_LIB}
	${RM} $<

ex1f: ex1f.o phys.o
	-${FLINKER} -o $@ $^ ${PETSC_FORTRAN_LIB} ${PETSC_SNES_LIB}
	${RM} $<
```

Fig. 12.4. Boilerplate custom makefile using PETSc.

the executable is necessary using the appropriate PETSc library variable. Individual compilation can be customized with the variables shown at the top of the makefile.

Users with large existing build systems may choose not to inherit the PETSc make rules but instead use a lower-level interface based only on PETSc make variables. A boilerplate example is given in Figure 12.4. The user now includes only the `bmake/common/variables` file, which defines the make variables but does not prescribe any rules for compilation or linking. The variables provide information about all compilation flags, libraries, and external packages necessary to link with PETSc.

Before any PETSc code can be run, the user must call `PetscInitialize()`, and likewise after all PETSc code has completed the user must call `PetscFinalize()`. This is analogous to the requirements for using MPI. In fact, if the user has not done so already, PETSc will handle the initialization and cleanup of MPI automatically in these routines. A simple C driver is shown in Figure 12.5.

The initialization call provides the command line arguments to PETSc for processing and can take an optional help string for the user (printed when the `-help` option is given). The finalization call frees any resources used by PETSc and provides summary logging and diagnostic information, most notably performance profiling. The equivalent Fortran driver is shown in Figure 12.6.

Notice that the command line arguments are now obtained directly from the Fortran runtime library, rather than the user code. Once these calls are inserted into the application code, the user can verify a successful link and run with the PETSc libraries.

```
static char help[] = "Boilerplate PETSc Example.\n\n";

#include "petsc.h"

int main(int argc, char **argv)
{
  ierr = PetscInitialize(&argc, &argv, (char *) 0, help);CHKERRQ(ierr);

  /* User code */

  return PetscFinalize();
}
```

Fig. 12.5. Boilerplate C driver for PETSc.

```
      program main
      implicit none
#include "include/finclude/petsc.h"

      integer ierr

      call PetscInitialize(PETSC_NULL_CHARACTER, ierr)

!     User code

      call PetscFinalize(ierr)
      end
```

Fig. 12.6. Boilerplate Fortran driver for PETSc.

12.3 Data Distribution and Linear Algebra

The PETSc solvers (discussed in Section 12.4) require the user to provide C or Fortran routines that compute the residual of the equation they wish to solve (after they have discretized the PDE) and optionally the Jacobian of that residual function. We refer to these functions generically as `FormFunction()` and `FormJacobian()`. For nonlinear problems the PETSc solvers use truncated Newton methods, with line searches or trust-regions for robustness, and possibly approximate or incomplete Jacobians for efficiency.

The central objects in any PETSc simulation are the abstractions from linear algebra, vectors and matrices, or in PETSc terms the `Vec` and `Mat` classes. These form the basis of all solver and preconditioner interfaces, as well as providing the link to user-supplied discretization and physics routines. Thus, the most important step in the migration of an application to the PETSc framework is the incorporation of its linear algebra and data distribution abstractions. Two obvious strategies emerge for accomplishing this integration: a gradual approach that seeks to minimize the change to existing code and a more aggressive approach that leverages as much of the PETSc technology as possible. The next two subsections address these complementary paths toward integration. We use the simple example of the solid-fuel ignition problem, or Bratu problem, to illustrate the gradual evolution of a simulation, and then we give an example of mantle subduction for the more aggressive approach.

12.3.1 Gradual Evolution

A new PETSc user with very complicated code, which perhaps was originally written by someone else, may opt to make as few changes as possible when moving to PETSc. PETSc facilitates this approach with low-level interfaces to both `Vec` and `Mat` objects that closely resemble common serial data structures. Vectors are especially easy to migrate because the default PETSc storage format, contiguous arrays on each process, is that most common in serial applications. Matrices are somewhat more complicated, and in order to hide the actual matrix data storage format, PETSc requires the user to access values through a functional interface [6].

Let us examine the Bratu problem as an example of integrating PETSc vectors into an existing simulation [21]. This problem is modeled by the partial differential equation

$$-\Delta u - \lambda e^{u} = 0. \tag{12.4}$$

We take the domain to be the unit square and impose homogeneous Dirichlet boundary conditions on the edges. We discretize the equation using a 5-point stencil finite difference scheme, which results in a set of nonlinear algebraic equations $F(u_h) = 0$, where we use u_h to indicate the discrete solution vector. In Figure 12.7, we present a Fortran 90 routine that calculates the residual F as a function of the input vector u_h. It also takes a user context as input that holds the domain information and problem coefficient.

Notice that the boundary conditions on the residual are of the form $u - u_\Gamma$, where u_Γ are the specified boundary values, because we are driving the residual $F(u)$ to zero. An alternative would be to eliminate those variables altogether, substituting the boundary values in any interior calculation. However, this technique currently imposes a greater burden on the programmer.

When integrating PETSc vectors into this code, we can let Fortran allocate the storage, or we can use PETSc for allocation. In Figure 12.8, we let PETSc manage the memory. The input vectors already have storage allocated by PETSc, which can be accessed as an F90 array by using the `VecGetArrayF90()` function or in C by using `VecGetArray()`. A sample `driver()` routine shows how PETSc allocates vector storage. If we let Fortran manage memory, then we use `VecCreateSeqWithArray()`, or `VecCreateMPIWithArray()` in parallel, to create the `Vec` objects.

During a Newton iteration to solve this equation, we would require the Jacobian J of the mapping F. Rather than being accessed as a multidimensional array, the

```
      module f90module
      type userctx
!       The start, end, and number of vertices in the x- and y-directions
        integer xs,xe,ys,ye,integer mx,my
        double precision lambda
      end type userctx
      contains
      end module f90module

      subroutine FormFunction(u,F,user,ierr)
      use f90module
      type (userctx) user
      double precision u(user%xs:user%xe,user%ys:user%ye)
      double precision F(user%xs:user%xe,user%ys:user%ye)
      double precision two,one,hx,hy,hxdhy,hydhx,sc,uij,uxx,uyy
      integer i,j,ierr

      hx    = 1.0/dble(user%mx-1)
      hy    = 1.0/dble(user%my-1)
      sc    = hx*hy*user%lambda
      hxdhy = hx/hy
      hydhx = hy/hx

      do 20 j=user%ys,user%ye
         do 10 i=user%xs,user%xe
!           Apply boundary conditions
            if (i == 1 .or. j == 1 .or. i == user%mx .or. j == user%my) then
               F(i,j) = u(i,j)
!           Apply finite difference scheme
            else
               uij = u(i,j)
               uxx = hydhx * (2.0*uij - u(i-1,j) - u(i+1,j))
               uyy = hxdhy * (2.0*uij - u(i,j-1) - u(i,j+1))
               F(i,j) = uxx + uyy - sc*exp(uij)
            endif
 10      continue
 20   continue
```

Fig. 12.7. Residual calculation for the Bratu problem.

```
subroutine driver(u,F,user,ierr)
implicit none

Vec u,F
int N
parameter(N=10000)

call VecCreate(PETSC_COMM_WORLD,u,ierr)
call VecSetSizes(u,PETSC_DECIDE,N,ierr)
call VecDuplicate(u,F,ierr)
call SolverLoop(u,F,ierr)
call VecDestroy(u,ierr)
call VecDestroy(F,ierr)
end

subroutine FormFunctionPETSc(u,F,user,ierr)
use f90module
implicit none

Vec u,F
type (userctx) user
double precision,pointer :: u_v(:),f_v(:)

call VecGetArrayF90(u,u_v,ierr)
call VecGetArrayF90(F,f_v,ierr)
call FormFunction(u_v,f_v,user,ierr)
call VecRestoreArrayF90(u,u_v,ierr)
call VecRestoreArrayF90(F,f_v,ierr)
end
```

Fig. 12.8. Driver for the Bratu problem using PETSc.

PETSc `Mat` object provides the `MatSetValues()` function to set logically dense blocks of values into the structure. A function that computes the Jacobian and stores it in a PETSc matrix is shown in Figure 12.9. With this addition, the user can now run serial code using the PETSc solvers (details are given in Section 12.4). Note that the storage mechanism is the only change to user code necessary for its incorporation into the PETSc framework using the wrapper in Figure 12.8.

Although PETSc removes the need for low-level parallel programming, such as direct calls to the MPI library, the user must still identify parallelism inherent in the application and must structure the computation accordingly. The first step is to partition the domain and have each process calculate the residual and Jacobian only over its local piece. This is easily accomplished by redefining the `xs`, `xe`, `ys`, and `ye` variables on each process, converting loops over the entire domain into loops over the local domain. However, this method uses nearest-neighbor information, and thus

```
      subroutine FormJacobian(u,jac,user,ierr)
      use f90module
      type (userctx) user
      Mat jac
      double precision x(user%xs:user%xe,user%ys:user%ye)
      double precision two,one,hx,hy,hxdhy,hydhx,sc,v(5)
      integer row,col(5),i,j,ierr

      one   = 1.0
      two   = 2.0
      hx    = one/dble(user%mx-1)
      hy    = one/dble(user%my-1)
      sc    = hx*hy*user%lambda
      hxdhy = hx/hy
      hydhx = hy/hx
      do 20 j=user%ys,user%ye
         row = (j - user%ys)*user%xm - 1
         do 10 i=user%xs,user%xe
            row = row + 1
!           boundary points
            if (i == 1 .or. j == 1 .or. i == user%mx .or. j == user%my) then
               col(1) = row
               v(1)   = one
               call MatSetValues(jac,1,row,1,col,v,INSERT_VALUES,ierr)
!           interior grid points
            else
               v(1) = -hxdhy
               v(2) = -hydhx
               v(3) = two*(hydhx + hxdhy) - sc*exp(x(i,j))
               v(4) = -hydhx
               v(5) = -hxdhy
               col(1) = row - user%xm
               col(2) = row - 1
               col(3) = row
               col(4) = row + 1
               col(5) = row + user%xm
               call MatSetValues(jac,1,row,5,col,v,INSERT_VALUES,ierr)
            endif
 10      continue
 20   continue
```

Fig. 12.9. Jacobian calculation for the Bratu problem using PETSc.

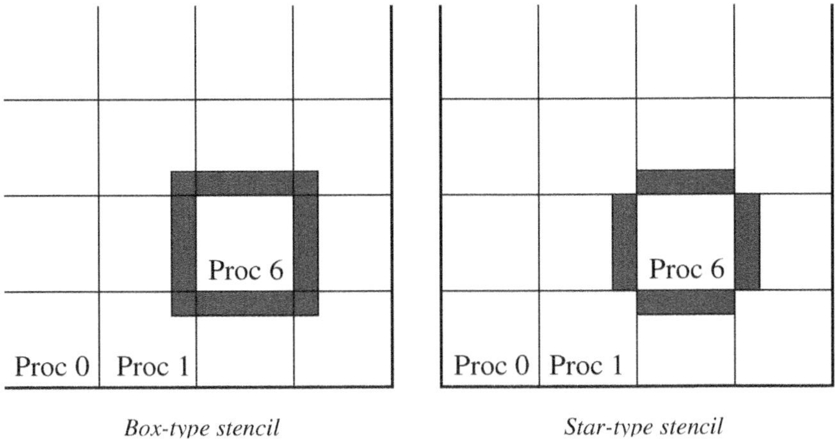

Fig. 12.10. Star and box stencils using a DA.

we must have access to some values not present locally on the process, as shown in Figure 12.10.

We must store values for these *ghost* points locally so that they may be used for the computation. In addition, we must ensure that these values are identical to the corresponding values on the neighboring processes. At the lowest level, PETSc provides *ghosted vectors*, created using `VecCreateGhost()`, with storage for some values owned by other processes. These values are explicitly indicated during construction. The `VecGhostUpdateBegin()` and `VecGhostUpdateEnd()` functions transfer values between ghost storage and the corresponding storage on other processes. This method, however, can become complicated for the user. Therefore, for the common case of a logically rectangular grid, PETSc provides the `DA` object to manage the determination, allocation, and coherence of ghost values. The `DA` object is discussed fully in Section 12.3.2, but we give here a short introduction in the context of the Bratu problem.

The `DA` object represents a distributed, structured (possibly staggered) grid and thus has more information than our serial grid. If we augment our user context to include information about ghost values, we can obtain all the grid information from the `DA`. Figure 12.11 shows the creation of a `DA` when one initially know only the total number of vertices in the x and y directions. Alternatively, we could have specified the local number of vertices in each direction.

Now we need only modify our PETSc residual routine to update the ghost values, as shown in Figure 12.12.

Notice that the ghost value scatter is a two-step operation. This allows the communication to overlap with local computation that may be taking place while the messages are in transit. The only change necessary for the residual calculation itself is the correct declaration of the input array, `double precision u(user%gxs:user%gxe,user%gys:user%gye)`, which now includes ghost values,

```
      type userctx
         DA da
!        The start, end, and number of vertices in the x-direction
         integer xs,xe,xm
!        The start, end, and number of ghost vertices in the x-direction
         integer gxs,gxe,gxm
!        The start, end, and number of vertices in the y-direction
         integer ys,ye,ym
!        The start, end, and number of ghost vertices in the y-direction
         integer gys,gye,gym
!        The number of vertices in the x- and y-directions
         integer mx,my
!        The MPI rank of this process
         integer rank
!        The coefficient in the Bratu equation
         double precision lambda
       end type userctx

       call DACreate2d(PETSC_COMM_WORLD,DA_NONPERIODIC,                 &
     &        DA_STENCIL_STAR,user%mx,user%my,PETSC_DECIDE,             &
     &        PETSC_DECIDE,1,1,PETSC_NULL_INTEGER,                      &
     &        PETSC_NULL_INTEGER,user%da,ierr)
       call DAGetCorners(user%da,user%xs,user%ys,PETSC_NULL_INTEGER,    &
     &        user%xm,user%ym,PETSC_NULL_INTEGER,ierr)
       call DAGetGhostCorners(user%da,user%gxs,user%gys,                &
     &        PETSC_NULL_INTEGER,user%gxm,user%gym,                     &
     &        PETSC_NULL_INTEGER,ierr)
!   Here we shift the starting indices up by one so that we can easily
!   use the Fortran convention of 1-based indices, rather than 0-based.
       user%xs  = user%xs+1
       user%ys  = user%ys+1
       user%gxs = user%gxs+1
       user%gys = user%gys+1
       user%ye  = user%ys+user%ym-1
       user%xe  = user%xs+user%xm-1
       user%gye = user%gys+user%gym-1
       user%gxe = user%gxs+user%gxm-1
```

Fig. 12.11. Creation of a DA for the Bratu problem.

The changes to `FormJacobian()` are similar, resizing the input array and changing slightly the calculation of row and column indices, and can be found in the PETSc example source. We have now finished the initial introduction of PETSc linear algebra, and the simulation is ready to run in parallel.

```
subroutine FormFunctionPETSc(u,F,user,ierr)
implicit none

Vec u,F
integer ierr
type (userctx) user

double precision,pointer :: lu_v(:),lf_v(:)
Vec uLocal

call DAGetLocalVector(user%da,uLocal,ierr)
call DAGlobalToLocalBegin(user%da,u,INSERT_VALUES,uLocal,ierr)
call DAGlobalToLocalEnd(user%da,u,INSERT_VALUES,uLocal,ierr)

call VecGetArrayF90(uLocal,lu_v,ierr)
call VecGetArrayF90(F,lf_v,ierr)

!      Actually compute the local portion of the residual
call FormFunctionLocal(lu_v,lf_v,user,ierr)

call VecRestoreArrayF90(uLocal,lu_v,ierr)
call VecRestoreArrayF90(F,lf_v,ierr)

call DARestoreLocalVector(user%da,uLocal,ierr)
```

Fig. 12.12. Parallel residual calculation for the Bratu problem using PETSc.

```
DASetLocalFunction(user.da, (DALocalFunction1) FormFunctionLocal);
```

Fig. 12.13. Using the DA to form a function.

12.3.2 Rapid Evolution

The user who is willing to raise the level of abstraction in a code can start by employing the `DA` to describe the problem domain and discretization. The data interface mimics a multidimensional array and is thus ideal for finite difference, finite volume, and low-order finite element schemes on a logically rectangular grid. In fact, after the definition of a `FormFunction()` routine, the simulation is ready to run because the nonlinear solver provides approximate Jacobians automatically, as discussed in Section 12.4.

We begin by examining the Bratu problem from the last section, only this time in C. We can now provide our local residual routine to the `DA`; see Figure 12.13.

The grid information is passed to `FormFunctionLocal()` in a `DALocalInfo` structure, as shown in Figure 12.14.

```
int FormFunctionLocal(DALocalInfo *info,double **u,double **f,AppCtx *user){
  double  two = 2.0,hx,hy,hxdhy,hydhx,sc;
  double  uij,uxx,uyy;
  int     i,j;

  hx     = 1.0/(double)(info->mx-1);
  hy     = 1.0/(double)(info->my-1);
  sc     = hx*hy*user->lambda;
  hxdhy  = hx/hy;
  hydhx  = hy/hx;
  /* Compute function over the locally owned part of the grid */
  for (j=info->ys; j<info->ys+info->ym; j++) {
    for (i=info->xs; i<info->xs+info->xm; i++) {
      if (i == 0 || j == 0 || i == info->mx-1 || j == info->my-1) {
        f[j][i] = u[j][i];
      } else {
        uij     = u[j][i];
        uxx     = (two*uij - u[j][i-1] - u[j][i+1])*hydhx;
        uyy     = (two*uij - u[j-1][i] - u[j+1][i])*hxdhy;
        f[j][i] = uxx + uyy - sc*exp(uij);
      }
    }
  }
}
```

Fig. 12.14. `FormFunctionLocal()` for the Bratu problem.

```
DASetLocalJacobian(user.da, (DALocalFunction1) FormJacobianLocal);
```

Fig. 12.15. Using the DA to form a Jacobian.

The fields are passed directly as multidimensional C arrays, complete with ghost values. An application scientist can use a boilerplate example, such as the Bratu problem, merely altering the local physics calculation in `FormFunctionLocal()`, to rapidly obtain a parallel, scalable simulation that runs on the desktop as well as on massively parallel supercomputers.

If the user has an expression for the Jacobian of the residual function, then this matrix can also be computed by using the `DA` interface with the call, as in Figure 12.15, and the corresponding routine for the local calculation in Figure 12.16.

We have so far presented simple examples involving a perturbed Laplacian, but this development strategy extends far beyond toy problems. We have developed large-scale, parallel geophysical simulations [20] that are producing new results in the field. We began by replacing `FormFunctionLocal()` in the Bratu example with one containing the linear physics of the previous sequential linear subduction

```
int FormJacobianLocal(DALocalInfo *info,double **x,Mat jac,AppCtx *user){
  MatStencil   col[5],row;
  double       lambda,v[5],hx,hy,hxdhy,hydhx,sc;
  int          i,j;

  lambda = user->param;
  hx     = 1.0/(double)(info->mx-1);
  hy     = 1.0/(double)(info->my-1);
  sc     = hx*hy*lambda;
  hxdhy  = hx/hy;                    hydhx  = hy/hx;

  for (j=info->ys; j<info->ys+info->ym; j++) {
    for (i=info->xs; i<info->xs+info->xm; i++) {
      row.j = j; row.i = i;
      /* boundary points */
      if (i == 0 || j == 0 || i == info->mx-1 || j == info->my-1) {
        v[0] = 1.0;
        MatSetValuesStencil(jac,1,&row,1,&row,v,INSERT_VALUES);
      } else {
      /* interior grid points */
        v[0] = -hxdhy; col[0].j = j - 1; col[0].i = i;
        v[1] = -hydhx; col[1].j = j; col[1].i = i-1;
        v[2] = 2.0*(hydhx + hxdhy) - sc*exp(x[j][i]); col[2].j = j; col[2].i = i;
        v[3] = -hydhx; col[3].j = j; col[3].i = i+1;
        v[4] = -hxdhy; col[4].j = j + 1; col[4].i = i;
        MatSetValuesStencil(jac,1,&row,5,col,v,INSERT_VALUES);
      }
    }
  }
  MatAssemblyBegin(jac,MAT_FINAL_ASSEMBLY);
  MatAssemblyEnd(jac,MAT_FINAL_ASSEMBLY);
  MatSetOption(jac,MAT_NEW_NONZERO_LOCATION_ERR);
}
```

Fig. 12.16. `FormJacobianLocal()` for the Bratu problem.

code (dropping the previous code's linear solve). Then the PETSc solver results were compared, for correctness, with the complete prior subduction code. Next the new code was immediately run in parallel for correctness and efficiency tests. Finally the nonlinear terms from the more complicated physics of the mantle subduction problem, discussed in Section 12.1 and shown in Figures 12.17-12.21 and better quality discretizations of the boundary conditions were added. In fact, since the structure of the `FormFunction()` changed significantly, the final `FormFunction()` looks nothing like the original, but the process of obtaining through the incremental approach reduced the learning curve for PETSc and make correctness checking at the various stages much easier.

```
int FormFunctionLocal(DALocalInfo *info,Field **x,Field **f,void *ptr){
  AppCtx      *user = (AppCtx*)ptr;
  Parameter   *param = user->param;
  GridInfo    *grid  = user->grid;
  double      mag_w, mag_u;
  int         ilim = info->mx-1, jlim = info->my-1, i, j;

  for (j=info->ys; j<info->ys+info->ym; j++) {
    for (i=info->xs; i<info->xs+info->xm; i++) {
      calculateXMomentumResidual(i, j, x, f, user);
      calculateZMomentumResidual(i, j, x, f, user);
      calculatePressureResidual(i, j, x, f, user);
      calculateTemperatureResidual(i, j, x, f, user);
    }
  }
}
```

Fig. 12.17. `FormFunctionLocal()` for the mantle subduction problem.

The details of the subduction code are not as important as the recognition that a very complicated physics problem need not involve any more complication in our simulation infrastructure. To give more insight into the actual calculation, we show in Figure 12.22 the explicit calculation of the residual from the continuity equation.

Here we have used the `DA` for a multicomponent problem on a staggered mesh without alteration because the structure of the discretization remains logically rectangular.

Although the `DA` manages communication during the solution process, data post-processing, often necessary for visualization or analysis, also requires this functionality. For the mantle subduction simulation, we want to calculate both the second invariant of the strain rate tensor and the viscosity over the grid (at both cell centers and corners). Using the `DAGlobalToLocalBegin()` and `DAGlobalToLocalEnd()` functions, we transferred the global solution data to local ghosted vectors and proceeded with the calculation. We can also transfer data from local vectors to a global vector using `DALocalToGlobal()`. The entire viscosity calculation is given in Figure 12.23.

Note that this framework allows the user to manage arbitrary fields over the domain. For instance, a porous flow simulation might manage material properties of the medium in this fashion.

12.4 Solvers

The `SNES` object in PETSc is an abstraction of the "inverse" of a nonlinear operator. The user provides a `FormFunction()` routine, as seen in Section 12.3, which computes the action of the operator on an input vector. Linear problems can also be

```
int calculateXMomentumResidual(int i, int j, Field **x,Field **f,AppCtx *user){
  Parameter *param = user->param;
  GridInfo   *grid  = user->grid;
  int        ilim = info->mx-1, jlim = info->my-1;

  if (i<j) {
    f[j][i].u = x[j][i].u - SlabVel('U',i,j,user);
  } else if (j<=grid->jlid || (j<grid->corner+grid->inose &&
                               i<grid->corner+grid->inose)) {
    /* in the lithospheric lid */
    f[j][i].u = x[j][i].u - 0.0;
  } else if (i==ilim) { /* on the right side boundary */
    if (param->ibound==BC_ANALYTIC) {
      f[j][i].u = x[j][i].u - HorizVelocity(i,j,user);
    } else {
      f[j][i].u = XNormalStress(x,i,j,CELL_CENTER,user) - EPS_ZERO;
    }
  } else if (j==jlim) { /* on the bottom boundary */
    if (param->ibound==BC_ANALYTIC) {
      f[j][i].u = x[j][i].u - HorizVelocity(i,j,user);
    } else if (param->ibound==BC_NOSTRESS) {
      f[j][i].u = XMomentumResidual(x,i,j,user);
    }
  } else { /* in the mantle wedge */
    f[j][i].u = XMomentumResidual(x,i,j,user);
  }
}
```

Fig. 12.18. X-momentum residual for the mantle subduction problem.

solved by using `SNES` with no loss of efficiency compared to using the underlying `KSP` object (PETSc linear solver object) directly. Thus `SNES` seems the appropriate framework for a general-purpose simulator, providing the flexibility to add or subtract nonlinear terms from the equation at will.

The `SNES` solver does not require a user to implement a Jacobian. Default routines are provided to compute a finite difference approximation using coloring to account for the sparsity of the matrix. The user does not even need the PETSc `Mat` interface but can initially interact only with `Vec` objects, making the transition to PETSc nearly painless. However, these approximations to the Jacobian can have numerical difficulties and are not as efficient as direct evaluation. Therefore, the user has the option of providing a routine or of using the ADIC [15] or similar system for automatic differentiation. The finite difference approximation can be activated by using the `-snes_fd` and `-mat_fd_coloring_freq` options or by providing the `SNESDefaultComputeJacobianColor()` function to `SNESSetJacobian()`.

```
int calculateZMomentumResidual(int i, int j, Field **x,Field **f,AppCtx *user){
  Parameter *param = user->param;
  GridInfo  *grid  = user->grid;
  int       ilim = info->mx-1,jlim = info->my-1;

  if (i<=j) {
    f[j][i].w = x[j][i].w - SlabVel('W',i,j,user);
  } else if (j<=grid->jlid || (j<grid->corner+grid->inose &&
                               i<grid->corner+grid->inose)) {
    /* in the lithospheric lid */
    f[j][i].w = x[j][i].w - 0.0;
  } else if (j==jlim) { /* on the bottom boundary */
    if (param->ibound==BC_ANALYTIC) {
      f[j][i].w = x[j][i].w - VertVelocity(i,j,user);
    } else {
      f[j][i].w = ZNormalStress(x,i,j,CELL_CENTER,user) - EPS_ZERO;
    }
  } else if (i==ilim) { /* on the right side boundary */
    if (param->ibound==BC_ANALYTIC) {
      f[j][i].w = x[j][i].w - VertVelocity(i,j,user);
    } else if (param->ibound==BC_NOSTRESS) {
      f[j][i].w = ZMomentumResidual(x,i,j,user);
    }
  } else { /* in the mantle wedge */
    f[j][i].w =    ZMomentumResidual(x,i,j,user);
  }
}
```

Fig. 12.19. Z-momentum residual for the mantle subduction problem.

Through the `SNES` object, the user may also access the great range of direct and iterative linear solvers and preconditioners provided by PETSc. Sixteen different Krylov solvers are available including GMRES, BiCGStab, and LSQR, along with interfaces to popular sparse direct packages, such as MUMPS [3] and SuperLU [9]. In addition to a variety of incomplete factorization preconditioners, including PILUT [16], PETSc supports additive Schwartz preconditioning, algebraic multigrid through BoomerAMG [13], and the geometric multigrid discussed below. The full panoply of solvers and preconditioners available is catalogued on the PETSc Website [22].

The modularity of PETSc allows users to easily customize each solver. For instance, suppose the user wishes to increase the number of levels in ILU(k) preconditioning on one block of a block-Jacobi scheme. The code fragment in Figure 12.24 will set the number of levels of fill to `levels`.

PETSc provides an elegant framework for managing geometric multigrid in combination with a `DA`, which is abstracted in the `DMMG` object. In the same way that

```
int calculatePressureResidual(int i, int j, Field **x,Field **f,AppCtx *user){
  Parameter *param = user->param;
  GridInfo   *grid  = user->grid;
  int         ilim = info->mx-1, jlim = info->my-1;

  if (i<j || j<=grid->jlid || (j<grid->corner+grid->inose &&
                                 i<grid->corner+grid->inose)) {
    /* in the lid or slab */
    f[j][i].p = x[j][i].p;
  } else if ((i==ilim || j==jlim) && param->ibound==BC_ANALYTIC) {
    /* on an analytic boundary */
    f[j][i].p = x[j][i].p - Pressure(i,j,user);
  } else {
    /* in the mantle wedge */
    f[j][i].p = ContinuityResidual(x,i,j,user);
  }
}
```

Fig. 12.20. Pressure, or continuity, residual for the mantle subduction problem.

any linear solve can be performed with `SNES`, any preconditioner or solver combination is available in `DMMG` by using a single level. That is, the same user code supports both geometric multigrid as well as direct methods, Krylov methods etc. When creating a `DMMG`, the user specifies the number of levels of grid refinement and provides the coarse-grid information (see Figure 12.25), which is always a `DA` object at present, although unstructured prototypes are being developed. Then the `DMMG` object calculates the intergrid transfer operators, prolongation and restriction, and allocates objects to hold the solution at each level. The user must provide the action of the residual operator F and can optionally provide a routine to compute the Jacobian matrix (which will be called on each level) and a routine to compute the initial guess, as in Figure 12.26.

With this information, the user can now solve the equation and retrieve the solution vector, as in Figure 12.27.

12.5 Extensions

PETSc is not a complete environment for simulating physical phenomena. Rather it is a set of tools that allow the user to assemble such an environment tailored to a specific application. As such, PETSc will never provide every facility appropriate for a given simulation. However, the user can easily extend PETSc and supplement its capabilities. We present two examples in the context of the mantle subduction simulation.

```
int calculateTemperatureResidual(int i, int j, Field **x,Field **f,AppCtx *user){
  Parameter *param = user->param;
  GridInfo   *grid   = user->grid;
  int          ilim = info->mx-1, jlim = info->my-1;

  if (j==0) { /* on the surface */
    f[j][i].T = x[j][i].T + x[j+1][i].T + PetscMax(x[j][i].T,0.0);
  } else if (i==0) { /* slab inflow boundary */
    f[j][i].T = x[j][i].T - PlateModel(j,PLATE_SLAB,user);
  } else if (i==ilim) { /* right side boundary */
    mag_u = 1.0 - pow((1.0-PetscMax(PetscMin(x[j][i-1].u/param->cb,1.0),0.0)),5.0);
    f[j][i].T = x[j][i].T - mag_u*x[j-1][i-1].T -
                        (1.0-mag_u)*PlateModel(j,PLATE_LID,user);
  } else if (j==jlim) { /* bottom boundary */
    mag_w = 1.0 - pow((1.0-PetscMax(PetscMin(x[j-1][i].w/param->sb,1.0),0.0)),5.0);
    f[j][i].T = x[j][i].T - mag_w*x[j-1][i-1].T - (1.0-mag_w);
  } else { /* in the mantle wedge */
    f[j][i].T = EnergyResidual(x,i,j,user);
  }
}
```

Fig. 12.21. Temperature, or energy, residual for the mantle subduction problem.

```
double precision ContinuityResidual(Field **x, int i, int j, AppCtx *user)
{
  GridInfo               *grid = user->grid;
  double precision uE,uW,wN,wS,dudx,dwdz;

  uW = x[j][i-1].u; uE = x[j][i].u; dudx = (uE - uW)/grid->dx;
  wS = x[j-1][i].w; wN = x[j][i].w; dwdz = (wN - wS)/grid->dz;
  return dudx + dwdz;
}
```

Fig. 12.22. Calculation of the continuity residual for the mantle subduction problem.

12.5.1 Simple Continuation

Because of the highly nonlinear dependence of viscosity on temperature and strain rate in equation (12.1), the iteration in Newton's method can fail to converge without a good starting guess of the solution. On the other hand, for the isoviscous case where temperature is coupled to velocity only through advection, a solution is easily reached. It is therefore natural to propose a continuation scheme in the viscosity beginning with constant viscosity and progressing to full variability. A simple adaptive continuation model was devised, in which the viscosity was raised to a power between zero and one, $\eta \rightarrow \eta^{\alpha}$, where α=0 corresponds to constant viscosity and α=1

```
/* Compute both the second invariant of the strain rate tensor and the viscosity */
int ViscosityField(DA da, Vec X, Vec V,AppCtx *user){
  Parameter    *param = user->param;
  GridInfo     *grid  = user->grid;
  Vec          localX;
  Field        **v, **x;
  double       eps, dx, dz, T, epsC, TC;
  int          i,j,is,js,im,jm,ilim,jlim,ivt;

  ivt          = param->ivisc;
  param->ivisc = param->output_ivisc;

  DACreateLocalVector(da, &localX);
  DAGlobalToLocalBegin(da, X, INSERT_VALUES, localX);
  DAGlobalToLocalEnd(da, X, INSERT_VALUES, localX);
  DAVecGetArray(da,localX,(void**)&x);
  DAVecGetArray(da,V,(void**)&v);

  /* Parameters */
  dx   = grid->dx;   dz   = grid->dz;
  ilim = grid->ni-1; jlim = grid->nj-1;

  /* Compute real temperature, strain rate and viscosity */
  DAGetCorners(da,&is,&js,PETSC_NULL,&im,&jm,PETSC_NULL);
  for (j=js; j<js+jm; j++) {
    for (i=is; i<is+im; i++) {
      T  = param->potentialT * x[j][i].T * exp( (j-0.5)*dz*param->z_scale );
      if (i<ilim && j<jlim) {
        TC = param->potentialT * TInterp(x,i,j) * exp( j*dz*param->z_scale );
      } else {
        TC = T;
      }
      /* Compute the values at both cell centers and cell corners */
      eps  = CalcSecInv(x,i,j,CELL_CENTER,user);
      epsC = CalcSecInv(x,i,j,CELL_CORNER,user);
      v[j][i].u = eps;
      v[j][i].w = epsC;
      v[j][i].p = Viscosity(T,eps,dz*(j-0.5),param);
      v[j][i].T = Viscosity(TC,epsC,dz*j,param);
    }
  }
  DAVecRestoreArray(da,V,(void**)&v);
  DAVecRestoreArray(da,localX,(void**)&x);
  param->ivisc = ivt;
}
```

Fig. 12.23. Calculation of the second invariant of the strain tensor and viscosity field for the mantle subduction problem.

```
KSP ksp, *subksp;
PC bjacobi, ilu;

SNESGetKSP(snes, &ksp);
KSPGetPC(ksp, &bjacobi);
PCBJacobiGetSubKSP(bjacobi, PETSC_NULL, PETSC_NULL, &subksp);
KSPGetPC(subksp[0], &ilu);
PCILUSetLevels(ilu, levels);
```

Fig. 12.24. Customizing the preconditioner on a single process.

```
DMMGCreate(comm, grid.mglevels, user, &dmmg);
DMMGSetDM(dmmg, (DM) da);
```

Fig. 12.25. Creating a DMMG.

```
DMMGSetSNESLocal(dmmg, FormFunctionLocal, FormJacobianLocal, 0, 0);
DMMGSetInitialGuess(dmmg, FormInitialGuess);
```

Fig. 12.26. Providing discretization and assembly routines to DMMG.

```
DMMGSolve(dmmg);
soln = DMMGGetx(dmmg);
```

Fig. 12.27. Solving the problem with DMMG.

corresponds to natural viscosity variation. In the continuation loop, α was increased towards unity by an amount dependent on the rate of convergence of the previous `SNES` solve.

PETSc itself provides no special support for continuation, but it is sufficiently modular that a continuation loop was readily constructed in the application code, using repeated calls to `DMMGSolve()`, and found to work quite well.

12.5.2 Simple Steering

No rigorous justification had been given for the continuation strategy above, and consequently it was possible that, for certain initial conditions, Newton would fail to converge, thereby resulting in extremely long run times. An observant user could detect this situation and abort the run, but all the potentially useful solution information up to that point would be lost. A clean way to asynchronously abort the continuation loop was thus needed. On architectures where OS signals are available, PETSc provides an interface for registering signal handlers. Thus we were able to define a handler. The user wishing to change the control flow of the simulation simply sends the appropriate signal to the process. This sets a flag in the user data that causes the change at the next iteration.

12.6 Simulation Results

A long-standing debate has existed between theorists and observationalists over the thermal structure of subduction zones. Modelers, using constant viscosity flow simulations to compute thermal structure, have predicted relatively cold subduction zones [19]. Conversely, observationalists, who use heat flow measurements and mineralogical thermobarometry to estimate temperatures at depth, have long claimed that subduction zones are hotter than predicted. They have invoked the presence of strong upwelling of hot mantle material to supply the heat. This upwelling was never predicted by isoviscous flow models, however.

The debate remained unresolved until recently when several groups, including our own, succeeded in developing simulations with realistically variable viscosity (for other examples [12, 10, 8, 18, 25]). A comparison of flow and temperature fields for variable and constant viscosity simulations generated with our code is shown in Figure 12.28. Results from these simulations are exciting because they close the gap between models and observations: they predict hotter mantle temperatures, steeper surface thermal gradients, and upwelling mantle flow.

Furthermore, these simulations allow for quantitative predictions of variation of observables with subduction parameters. A recent study of subduction zone earthquakes has identified an intriguing trend: the vertical distance from subduction zone volcanoes to the surface of the subducting slab is anti-correlated with descent rate of the slab [11]. Preliminary calculations with our model are consistent with this trend, indicating that flow and thermal structure may play an important role in determining not only the quantity and chemistry of magmas but also their path of transport to the surface. Further work is required to resolve this issue.

Simulating the geodynamics of subduction is an example of our success in porting an existing code into the PETSc framework, simultaneously parallelizing the code and increasing its functionality. Using this code as a template, we rapidly developed a related simulation of another tectonic boundary, the mid-ocean ridge. This work was done to address a set of observations of systematic morphological asymmetry in the global mid-ocean ridge system [7]. Our model confirms the qualitative mechanism that had been proposed to explain these observations. Furthermore, it shows a quantitative agreement with trends in the observed data [17]. As with our models of subduction, the key to demonstrating the validity of the hypothesized dynamics was simulating them with the strongly nonlinear rheology that characterizes mantle rock on geologic timescales. The computational tools provided by PETSc enabled us easily handle this rheology, reducing development time and allowing us to focus on model interpretation.

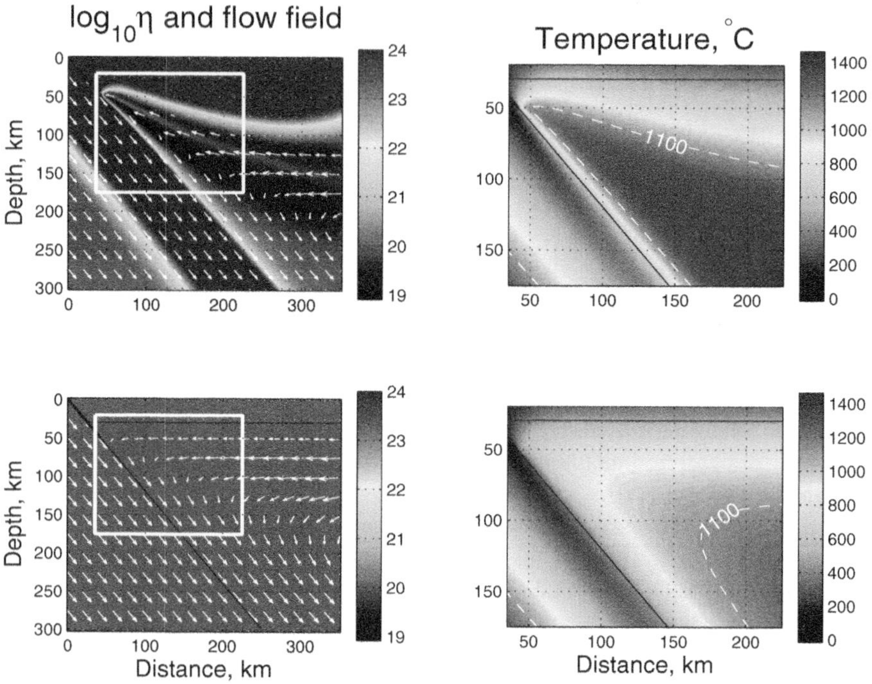

Fig. 12.28. 2D viscosity and potential temperature fields from simulations on 8 processors with 230,112 degrees of freedom. Panels in the top row are from a simulation with α=1 in equation (12.1). Panels in bottom row have α=0. The white box in panels (a) and (c) shows the region in which temperature is plotted in panels (b) and (d). **(a)** Colors show $\log_{10}$ of the viscosity field. Note that there are more than five orders of magnitude variation in viscosity. Arrows show the flow direction and magnitude (the slab is being subducted at a rate of 6 cm/year). Upwelling is evident in the flow field near the wedge corner. **(b)** Temperature field from the variable viscosity simulation; 1100°C isotherm is shown as a dashed line. **(c)** (Constant) viscosity and flow field from the isoviscous simulation. Strong flow is predicted at the base of the crust despite the low-temperature rock there. No upwelling flow is predicted. **(d)** Temperature field from isoviscous simulation. Note that the mantle wedge corner is much colder than in (b). (For the color version, see Figure A.27 on page 480).

PETSc has been used in a large number of parallel applications, [23], including three Gordon Bell Special Prize winning codes, [1, 2, 4]. The largest nonlinear problem solved with PETSc involved 500,000,000 unknowns, [1]. In this calculation, on an unstructured grid, the average time, per linear solve, was one and one-half minutes, corresponding to almost six million degrees of freedom per second. This demonstrates that well implemented software libraries, like PETSc, can be used on the most challenging application problems; in this case micro finite element analysis of a thoracic vertebral body (spinal disk).

References

1. M. F. Adams, H. H. Bayraktar, T. M. Keaveny, and P. Papadopoulos. Ultrascalable implicit finite element analyses in solid mechanics with over a half a billion degrees of freedom. In *Proceedings of SC04*, 2004. Winner of Gordon Bell Special Prize at SC2004: Large scale trabecular bone finite element modeling.
2. V. Akcelik, J. Bielak, G. Biros, I. Epanomeritakis, A. F. O. Ghattas, E. J. Kim, D. O'Hallaron, and T. Tu. High resolution forward and inverse earthquake modeling on terascale computers. In *Proceedings of SC2003*, 2003. A winner of the Gordon Bell Prize for special achievement at SC2003.
3. P. R. Amestoy, I. S. Duff, J.-Y. L'Excellent, and J. Koster. A fully asynchronous multifrontal solver using distributed dynamic scheduling. *SIAM Journal on Matrix Analysis and Applications*, 23(1):15–41, 2001.
4. W. K. Anderson, W. D. Gropp, D. K. Kaushik, D. E. Keyes, and B. F. Smith. Achieving high sustained performance in an unstructured mesh CFD application. In *Proceedings of SC 99*, 1999.
5. S. Balay, K. Buschelman, V. Eijkhout, W. D. Gropp, D. Kaushik, M. G. Knepley, L. C. McInnes, B. F. Smith, and H. Zhang. PETSc Web page. `http://www.mcs.anl.gov/petsc`.
6. S. Balay, W. D. Gropp, L. C. McInnes, and B. F. Smith. Efficient management of parallelism in object oriented numerical software libraries. In E. Arge, A. M. Bruaset, and H. P. Langtangen, editors, *Modern Software Tools in Scientific Computing*, pages 163–202. Birkhäuser Press, 1997.
7. S. Carbotte, C. Small, and K. Donnelly. The influence of ridge migration on the magmatic segmentation of mid-ocean ridges. *Nature*, 429:743–746, 2004.
8. J. Conder, D. Wiens, and J. Morris. On the decompression melting structure at volcanic arcs and back-arc spreading centers. *Geophys. Res. Letts.*, 29, 2002.
9. J. W. Demmel, J. R. Gilbert, and X. S. Li. SuperLU user's guide. Technical Report LBNL-44289, Lawrence Berkeley National Laboratory, October 2003.
10. M. Eberle, O. Grasset, and C. Sotin. A numerical study of the interaction of the mantle wedge, subducting slab, and overriding plate. *Phys. Earth Planet. In.*, 134:191–202, 2002.
11. P. England, R. Engdahl, and W. Thatcher. Systematic variation in the depth of slabs beneath arc volcanos. *Geophys. J. Int.*, 156(2):377–408, 2003.
12. Y. Furukawa. Depth of the decoupling plate interface and thermal structure under arcs. *J. Geophys. Res.*, 98:20005–20013, 1993.
13. V. E. Henson and U. M. Yang. BoomerAMG: A parallel algebraic multigrid solver and preconditioner. Technical Report UCRL-JC-133948, Lawrence Livermore National Laboratory, 2000.
14. G. Hirth and D. Kohlstedt. Rheology of the upper mantle and the mantle wedge: A view from the experimentalists. In *Inside the Subduction Factory*, volume 138 of *Geophysical Monograph*. American Geophysical Union, 2003.
15. P. Hovland, B. Norris, and B. Smith. Making automatic differentiation truly automatic: Coupling PETSc with ADIC. In *Proceedings of ICCS2002*, 2002.
16. D. Hysom and A. Pothen. A scalable parallel algorithm for incomplete factor preconditioning. *SIAM Journal on Scientific Computing*, 22:2194–2215, 2001.
17. R. Katz, M. Spiegelman, and S. Carbotte. Ridge migration, asthenospheric flow and the origin of magmatic segmentation in the global mid-ocean ridge system. *Geophys. Res. Letts.*, 31, 2004.

18. P. Kelemen, J. Rilling, E. Parmentier, L. Mehl, and B. Hacker. Thermal structure due to solid-state flow in the mantle wedge beneath arcs. In *Inside the Subduction Factory*, volume 138 of *Geophysical Monograph*. American Geophysical Union, 2003.
19. S. Peacock and K. Wang. Seismic consequences of warm versus cool subduction metamorphism: Examples from southwest and northeast Japan. *Science*, 286:937–939, 1999.
20. PETSc SNES Example 30. `http://www.mcs.anl.gov/petsc/petsc-2/snapshots/petsc-current/src/snes/examples/tutorials/ex30.c.html`.
21. PETSc SNES Example 5. `http://www.mcs.anl.gov/petsc/petsc-2/snapshots/petsc-current/src/snes/examples/tutorials/ex5f90.F.html`.
22. PETSc Solvers. `http://www.mcs.anl.gov/petsc/petsc-2/documentation/linearsolvertable.html`.
23. B. Smith et al. Scientific Applications Using PETSc. `http://www.mcs.anl.gov/petsc/petsc-2/publications`.
24. R. Stern. Subduction zones. *Rev. Geophys.*, 40(4), 2002.
25. P. van Keken, B. Kiefer, and S. Peacock. High-resolution models of subduction zones: Implications for mineral dehydration reactions and the transport of water into the deep mantle. *Geochem. Geophys. Geosys.*, 3(10), 2003.

13

Parallel Lattice Boltzmann Methods for CFD Applications

Carolin Körner[1], Thomas Pohl[2], Ulrich Rüde[2], Nils Thürey[2], and Thomas Zeiser[3]

[1] Lehrstuhl Werkstoffkunde und Technologie der Metalle, Martensstraße 5, 91058 Erlangen, Germany
carolin.koerner@ww.uni-erlangen.de
[2] Lehrstuhl für Systemsimulation, Cauerstraße 6, 91058 Erlangen, Germany
tom@thomas-pohl.info,
ulrich.ruede@informatik.uni-erlangen.de,
nils@thuerey.de
[3] Regionales Rechenzentrum Erlangen, Martensstraße 1, 91058 Erlangen, Germany
thomas.zeiser@rrze.uni-erlangen.de

Summary. The lattice Boltzmann method (LBM) has evolved to a promising alternative to the well-established methods based on finite elements/volumes for computational fluid dynamics simulations. Ease of implementation, extensibility, and computational efficiency are the major reasons for LBM's growing field of application and increasing popularity. In this paper we give a brief introduction to the involved theory and equations for LBM, present various techniques to increase the single-CPU performance, outline the parallelization of a standard LBM implementation, and show performance results. In order to demonstrate the straightforward extensibility of LBM, we then focus on an application in material science involving fluid flows with free surfaces. We discuss the required extensions to handle this complex scenario, and the impact on the parallelization technique.

13.1 Introduction

Our approach to CFD applications is based on the lattice Boltzmann method (LBM) which belongs to the class of cellular automata (CA). Cellular automata represent a physical system in an idealized way where space and time are discrete, i.e., a fully discrete universe made up of identical cells. CA are defined by a regular lattice of cells characterized by a set of boolean state variables. The evolution rule, which is a function of the state of the neighboring cells, is the same for all cells and updating of the cells occurs simultaneously in discrete time steps.

A special class of CA [45], the lattice gas automata (LGA) [36], describe the dynamics of point particles moving and colliding in a discrete space-time universe. Lattice gas models with an appropriate choice of the lattice symmetry in fact represent numerical solutions of the Navier-Stokes equations and are therefore able to describe macroscopic hydrodynamic problems [17]. Besides a simple implementation, the main advantages of lattice gas techniques are their stability, easy introduction of boundary conditions and intrinsic parallel structure allowing high performance computing. However, lattice gas models suffer from some drawbacks: Statistical noise, non-Galilean invariance[4], a velocity dependent pressure and spurious invariants. Particularly, the statistical noise requires time and/or space averaging procedures to extract macroscopic quantities like the density or the velocity. This intrinsic property of LGAs is the reason why they were not able to compete with conventional numerical methods of hydrodynamics.

The lattice Boltzmann method [12] historically developed from lattice gas automata. McNamara and Zanetti [30] were the first who extended the boolean dynamics of the automaton to real numbers, the particle distribution functions, representing the probability for a cell to have a given state. The philosophy behind this procedure is that it is more efficient to average the micro dynamics before than after simulation. That is, the discrete nature of the fluid particles vanishes on the macroscopic level of observation.

The LBM is characterized by a much higher numerical efficiency than the Boolean dynamics. In addition, lattice Boltzmann methods maintain the intuitive microscopic level of interpretation belonging to the CA. These properties make LBMs promising approaches to model complex physical systems, especially fluid flow. Though it will not be a replacement for well-established CFD technology, it may have advantages in certain application areas. After a short introduction to the lattice Boltzmann method, the implementation of the standard algorithm is explained. Its parallelization and extension to free surface flows for simulating foaming processes are presented in Section 13.4 and 13.5, respectively.

13.2 Basics of the Lattice Boltzmann Method

The lattice Boltzmann method [39, 44, 9] can not only be seen as a successor of lattice gas automata (LGA). The equations can also be derived rigorously from the underlying physical model, the *Boltzmann equation*, and it can be shown that Navier-Stokes flow behavior is recovered in the macroscopic limit [23, 44]. The Boltzmann equation is a partial differential equation (PDE) describing the evolution of the single particle distribution function f in phase space. This distribution function is defined in such a way that $f(\mathbf{x}, \boldsymbol{\xi}, t)$ is the probability for particles to be located within a phase space control element $d\mathbf{x}\, d\boldsymbol{\xi}$ about $\mathbf{x}$ and $\boldsymbol{\xi}$ at time t where $\mathbf{x}$ and $\boldsymbol{\xi}$ are the spatial position vector and the particle velocity vector, respectively. The macroscopic quantities, such as the density ρ and the momentum $\rho\mathbf{u}$, can then be obtained by evaluating the first moments of the distribution function f.

In the following paragraphs, we outline the major steps of the rigorous derivation of the relevant equations and relations of the lattice Boltzmann method.

Neglecting external forces, the transport equation for $f(\mathbf{x}, \boldsymbol{\xi}, t)$ can be expressed by the Boltzmann equation as

[4] Galilean invariance means that the behavior of a system is not influenced by rotation or translation.

Fig. 13.1. Discretized distribution functions f_i for the D2Q9 model: eight distribution functions associated with the particles moving to the neighboring cells and one distribution function corresponding to the resting particles.

$$\frac{\partial f}{\partial t} + \xi \frac{\partial f}{\partial \mathbf{x}} = \mathcal{Q}(f, f) \,. \tag{13.1}$$

The collision term $\mathcal{Q}(f, f)$ is quadratic in f, consisting of a complex integro-differential expression. A suitable simplification of the collision integral for the near-equilibrium state of low Mach number hydrodynamics is the single relaxation time approximation, the so-called Bhatnagar-Gross-Krook (BGK) model [5],

$$\mathcal{Q}(f, f) = -\frac{1}{\lambda}\left(f - f^{(0)}\right) , \tag{13.2}$$

where $f^{(0)}$ is the Maxwell-Boltzmann equilibrium distribution function, and λ is the relaxation time which controls the rate of approaching equilibrium, or in other words the viscosity of the fluid. The BGK relaxation still fulfills Boltzmann's H-theorem and locally conserves mass and momentum.

To solve for f numerically, (13.1) is first discretized in the velocity space using a finite set of velocity vectors $\mathbf{e}_i$ ($i = 0, \ldots, N$) leading to the velocity discrete Boltzmann equation,

$$\frac{\partial f_i}{\partial t} + \mathbf{e}_i \frac{\partial f_i}{\partial \mathbf{x}} = -\frac{1}{\lambda}(f_i - f_i^{eq}) \,, \qquad i = 0, \ldots, N \,, \tag{13.3}$$

where $f_i(\mathbf{x}, t)$ is equivalent to $f(\mathbf{x}, \mathbf{e}_i, t)$ (see Figure 13.1).

For simulating two-dimensional flows, the 9-velocity D2Q9 model ($i = 0, \ldots, 8$), and for three-dimensional simulations, both, the 15-velocity D3Q15 ($i = 0, \ldots, 14$) and the 19-velocity D3Q19 model ($i = 0, \ldots, 18$) are widely used (cf. Table 13.1). Surprisingly, the discretization with such a low number of collocation points is sufficient to describe the fluid in the near-equilibrium state of low Mach number hydrodynamics. For all these models, a suitable equilibrium distribution function f_i^{eq} is of the form

$$f_i^{eq} = \rho w_i \left[1 + \frac{3}{c^2}\mathbf{e}_i \cdot \mathbf{u} + \frac{9}{2c^4}\left(\mathbf{e}_i \cdot \mathbf{u}\right)^2 - \frac{3}{2c^2}\mathbf{u} \cdot \mathbf{u}\right] , \tag{13.4}$$

with $c = \Delta x / \Delta t$ and the discrete particle velocity vectors $\mathbf{e}_i$[5]. The weighting factors w_i depend only on the lattice model [35] and are given in Table 13.2 for the three

[5]In contrast to the common notation h for the spatial discretization length, we use Δx here in conformance with the standard LBM literature. Δx is normalized to 1 for 2D and 3D.

Table 13.1. Discrete velocity vectors $\mathbf{e}_i$ for three different LB models. The order of the discrete velocity vectors does not matter. Thus any permutation of the given columns can be used in practice.

model	discrete velocity vectors $\mathbf{e}_i$
D2Q9	$\begin{pmatrix} 0 & +1 & 0 & -1 & 0 & +1 & -1 & -1 & +1 \\ 0 & 0 & +1 & 0 & -1 & +1 & +1 & -1 & -1 \end{pmatrix}$
D3Q15	$\begin{pmatrix} 0 & +1 & 0 & 0 & -1 & 0 & 0 & +1 & -1 & -1 & +1 & +1 & -1 & -1 & +1 \\ 0 & 0 & +1 & 0 & 0 & -1 & 0 & +1 & +1 & -1 & -1 & +1 & +1 & -1 & -1 \\ 0 & 0 & 0 & +1 & 0 & 0 & -1 & +1 & +1 & +1 & +1 & -1 & -1 & -1 & -1 \end{pmatrix}$
D3Q19	$\begin{pmatrix} 0 & +1 & 0 & 0 & -1 & 0 & 0 & +1 & -1 & -1 & +1 & +1 & 0 & -1 & 0 & +1 & 0 & -1 & 0 \\ 0 & 0 & +1 & 0 & 0 & -1 & 0 & +1 & +1 & -1 & -1 & 0 & +1 & 0 & -1 & 0 & +1 & 0 & -1 \\ 0 & 0 & 0 & +1 & 0 & 0 & -1 & 0 & 0 & 0 & 0 & +1 & +1 & +1 & +1 & -1 & -1 & -1 & -1 \end{pmatrix}$

Table 13.2. Weighting factors w_i for two different LB models for the discrete velocity vectors $\mathbf{e}_i$ (with $i = 0 \ldots N$).

model	$\lvert\mathbf{e}_i\rvert^2 = 0$	$\lvert\mathbf{e}_i\rvert^2 = 1$	$\lvert\mathbf{e}_i\rvert^2 = 2$	$\lvert\mathbf{e}_i\rvert^2 = 3$
D2Q9	$w_i = 4/9$	$w_i = 1/9$	$w_i = 1/36$	
D3Q15	$w_i = 2/9$	$w_i = 1/9$		$w_i = 1/72$
D3Q19	$w_i = 1/3$	$w_i = 1/18$	$w_i = 1/36$	

models. This discrete equilibrium distribution function f_i^{eq} has been derived from the Maxwell-Boltzmann equilibrium distribution function $f^{(0)}$ in such a way that the velocity moments up to fourth order are identical with those of $f^{(0)}$.

The (macroscopic) values of density ρ, momentum $\rho\mathbf{u}$ and the momentum flux tensor $\Pi_{\alpha\beta}$ can be evaluated as

$$\rho = \int_{-\infty}^{\infty} f \, d\boldsymbol{\xi} = \sum_{i=0}^{N} f_i = \sum_{i=0}^{N} f_i^{eq} , \tag{13.5}$$

$$\rho\mathbf{u} = \int_{-\infty}^{\infty} \boldsymbol{\xi} f \, d\boldsymbol{\xi} = \sum_{i=0}^{N} \mathbf{e}_i f_i = \sum_{i=0}^{N} \mathbf{e}_i f_i^{eq} , \tag{13.6}$$

$$\Pi_{\alpha\beta} = \int_{-\infty}^{\infty} \xi_\alpha \xi_\beta f \, d\boldsymbol{\xi} = \sum_{i=0}^{N} \mathbf{e}_{i\alpha} \mathbf{e}_{i\beta} f_i . \tag{13.7}$$

The speed of sound in these models is $c_s = c/\sqrt{3}$ and the pressure is given by the equation of state of an ideal gas,

$$p = \rho c_s^2 . \tag{13.8}$$

To obtain the main equation of the lattice Boltzmann approach, (13.3) is discretized numerically in a very special manner. The discretization of space and time is accomplished by an explicit finite difference approximation. By scaling the lattice spacing, the time step and the discrete velocities appropriately, the discretized equations take the following explicit form:

$$f_i(\mathbf{x} + \mathbf{e}_i \Delta t, t + \Delta t) - f_i(\mathbf{x}, t) = -\frac{1}{\tau} \left[f_i(\mathbf{x}, t) - f_i^{eq}(\mathbf{x}, t) \right], \tag{13.9}$$

where $\tau = \lambda / \Delta t$ is the dimensionless relaxation time and $\mathbf{x}$ is a point in the discretized physical space.

The right hand side of (13.9) is usually called *collision step* and the left hand side *streaming step*. For the collision step, the equilibrium distribution function has to be calculated at each cell and at each time step from the local density ρ using (13.5) and the local macroscopic flow velocity $\mathbf{u}$ using (13.6).

In particular when dealing with complicated formulations of the boundary conditions (see, e.g., Section 13.5), it can be advantageous or even necessary to split the update process into the following two equations

$$f_i^{out}(\mathbf{x},\, t) = f_i^{in}(\mathbf{x},\, t) - \frac{1}{\tau} \left[f_i^{in}(\mathbf{x},\, t) - f_i^{eq}(\mathbf{x},\, t) \right] \tag{13.10}$$

$$f_i^{in}(x + \mathbf{e_i},\, t + \Delta t) = f_i^{out}(\mathbf{x},\, t), \tag{13.11}$$

where f_i^{out} denotes the distribution values after collision (but before propagation), and f_i^{in} are the values after collision and propagation, thus the values entering the neighboring cell as data for the next time step.

The Navier-Stokes equations (up to second order accuracy in space and time) can be derived formally from the lattice Boltzmann equation through the Chapman-Enskog expansion by a standard multi-scale expansion with time and space rescaled and the distribution function f_i expanded up to second order [17, 23, 7].

The relation between the relaxation time τ and the kinematic shear viscosity ν, including a correction for the truncation error due to the discretization, can be obtained from the result of the Chapman-Enskog expansion. As the discretization error is known a priori from this analysis, it can be corrected and thus, the lattice Boltzmann method does not suffer from numerical diffusion as many other finite difference methods do. The final relation for the kinematic viscosity is

$$\nu = (\tau - 1/2)\, c_s^2 \Delta t\,. \tag{13.12}$$

As the lattice Boltzmann method is a kinetic method, macroscopic boundary conditions do not have direct equivalents. They have to be replaced by appropriate microscopic rules which induce the desired macroscopic behavior. The easiest solution for introducing solid walls (i.e. a no-slip boundary condition) is the introduction of the *bounce back rule* on wall nodes

$$f_i^{out}(\mathbf{x},\, t) = f_{\bar{i}}^{in}(\mathbf{x},\, t) \qquad \text{with } x \in \text{wall}, \tag{13.13}$$

with $\mathbf{e}_{\bar{i}} = -\mathbf{e}_i$ and $f_{\bar{i}}(\mathbf{x}, t) = f(\mathbf{x}, \mathbf{e}_{\bar{i}}, t) = f(\mathbf{x}, -\mathbf{e}_i, t)$. This rule can be seen as a replacement of (13.10). It rotates the distribution functions on the wall node and thus they return back to the fluid with opposite momentum in the next time step. This results in zero velocities at the wall (which is located half-way between the last fluid cell and the first wall node) and ensures that there is no flux across the wall. This is equivalent to the macroscopic no-slip boundary condition.

As a computational tool, the lattice Boltzmann method differs from methods which are directly based on the Navier-Stokes equations in various aspects. The major differences are summarized according to Yu et al. [46] as follows:

1. The Navier-Stokes equations are second-order partial differential equations (PDEs); the discrete velocity Boltzmann equation from which the lattice Boltzmann model is derived, consists of a set of first order PDEs.
2. Navier-Stokes solvers inevitably need to treat the nonlinear convective term $\mathbf{u} \cdot \nabla \mathbf{u}$. The lattice Boltzmann method totally avoids the nonlinear convective term, because the convection becomes a simple advection (uniform data shift). The non-linearity of the Navier-Stokes equations is hidden in the quadratic velocity terms of the equilibrium distribution function (13.4).
3. CFD solvers for the incompressible Navier-Stokes equations need to solve the Poisson equation for the pressure. This involves global data communication, while in the lattice Boltzmann method data communication is always local and the pressure is obtained through an equation of state.
4. In the lattice Boltzmann method, the Courant-Friedrichs-Lewy (CFL) number is proportional to $\Delta t/\Delta x$, in other words, the grid CFL number is equal to unity based on the lattice units of $\Delta x = \Delta t = 1$. Consequently, the time dependent lattice Boltzmann method is inefficient for solving steady-state problems, because its speed of convergence is dictated by acoustic propagation, which is very slow.[6]
5. Boundary conditions involving complicated geometries require a careful treatment in both Navier-Stokes and lattice Boltzmann solvers. In Navier-Stokes solvers, normal and shear stress components require appropriate handling of geometric estimates of normals and tangents, as well as one-sided extrapolations. In lattice Boltzmann solvers, the boundary condition issue [11, 19, 20, 6] arises because the continuum framework, such as the no-slip condition at the wall, does not have a direct counterpart.
6. Since the Boltzmann equation is kinetic-based, the physics associated with molecular level interactions can be incorporated more easily. Hence, the lattice Boltzmann model might be fruitfully applied to micro-scale fluid flow problems.
7. The spatial discretization in the lattice Boltzmann method is dictated by the discretization of the particle velocity space. This coupling between discretized velocity space and configuration space leads to regular square grids. This is a limitation of the lattice Boltzmann methods,

[6] However, especially in the case of complex geometries, the lattice Boltzmann methods can still be competitive or even faster than Navier-Stokes solvers, see e.g. [3, 4].

especially for aerodynamic applications where both the far field boundary condition and the near wall boundary layer need to be carefully implemented. Local grid refinement techniques [15, 16, 47] cannot completely solve this issue.

However, the algorithm of the LBM is similar to the structure of Jacobi's method for the iterative solution of linear systems on structured meshes. In particular, the time loop in the LBM corresponds to the iteration loop in Jacobi's method [27].

To compare the performance of different computational methods is always a difficult task. Since established finite volume/difference/element methods are the result of an evolution over many decades, one might expect that the simple LBM cannot compete. However, the work presented in [3, 4, 39] or the recent extensive comparison in [18] demonstrate that LBM are competitive – although there is still room for improvements. From the current point of view, LBM based solvers are in particular suited for problems involving complex geometries, complex physics or weakly compressible transient flows with Mach numbers up to about 0.2.

13.3 General Implementation Aspects and Optimization of the Single CPU Performance

For realistic applications, the LBM is computationally very demanding, since it needs fine spatial resolution and small time steps. The natural answer is parallelization, but before a parallel LBM can be designed successfully, it is necessary to discuss the efficient implementation of the computational kernel on current CPUs and node architectures. As with many scientific applications, the LBM has high memory requirements and in many cases, the memory access can be more time consuming than performing the arithmetic operations.

The lattice Boltzmann algorithm as outlined in Section 13.2 can be implemented easily. The major components of the algorithm together with the relevant equations are summarized in Figure 13.2. Through a simple analysis, the number of required floating point operations (Flops) can be significantly reduced by taking into account that many components of $\mathbf{e}_i$ are zero, as well as by precomputing common subexpressions. This results in less than 200 Flops per cell update for the D3Q19 BGK model which we used in our simulations. Memory access is the other determining factor for the speed of execution.

A straightforward implementation would consist of three nested loops over the three spatial dimensions and treat the collision step separately from the propagation process (cf. (13.10) and (13.11)). This algorithm would read the values of the current time step from the local cell, execute the relaxation and write the results back to a temporary array, as this can be done independently for all cells. In a separate nested loop which would only contain copy operations, these values would then be propagated back to adjacent cells in the original array.

Numerous improvements are possible starting from this implementation and have been studied carefully in [42, 34, 25, 14, 33, 41, 14]. In the following subsections we will summarize some of the core ideas.

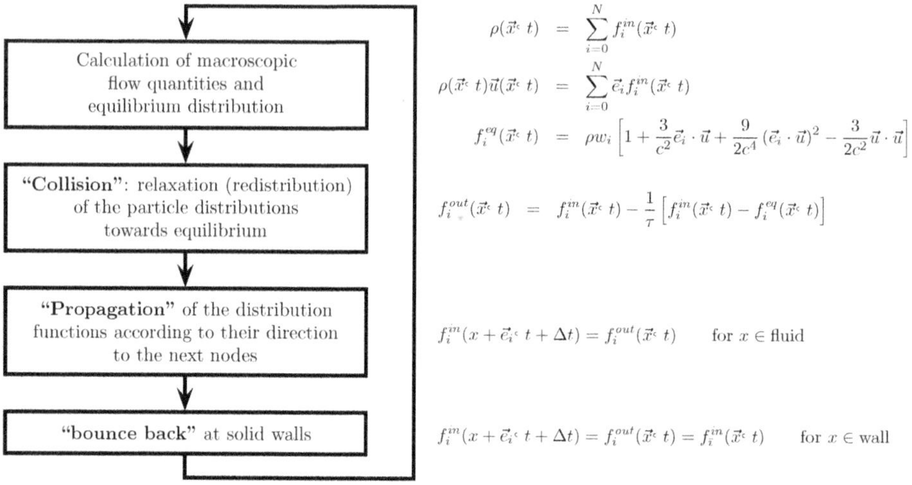

Fig. 13.2. Major components of the LB algorithm together with the relevant equations. Each "box" is executed for all cells during each time step. One sweep through all "boxes" represents one time step.

13.3.1 Combined Collision and Propagation Step

Naturally, the collision process can be combined with the propagation step. To keep the implementation simple and to be able to use any order for cell updates in the propagation step, two copies of the f_i array can be kept in memory (i.e. to ignore data dependencies). During an update, values are read from the "*old*" array and written to the other ("*new*") including the propagation step (within the reading or writing step). At the end of each time step the two arrays are swapped, i.e. the source becomes the destination and vice versa. Depending on the implementation, the propagation step is realized as first or last step of the iteration loop as depicted in Figure 13.3, resulting in a *pull* or *push* scheme of the update process.

Obviously, the main difference consists in non-local read operations (gather data) in the first case compared to non-local write operations (scatter data) in the second.

The data for the values of the distribution functions can be stored in one array of dimension five: three spatial coordinates, the discrete-velocity direction $i \in \{0 \ldots 18\}$ and an index distinguishing between the two arrays ("old" and "new"). The order of the indices controls the actual layout of the data in memory and can therefore have a substantial performance impact [14, 41]. Here, we only compare the index orders $(x, y, z, i, old/new)$ *(propagation optimized)* and $(i, x, y, z, old/new)$ *(collision optimized)* with the first index moving fastest as the benchmark kernel was implemented in Fortran.

13.3.2 Vector Optimization

For vector architectures, long inner loops can be advantageous. In the case of the LBM, the three nested spatial loops with index (x, y, z) (see Figure 13.4) can be

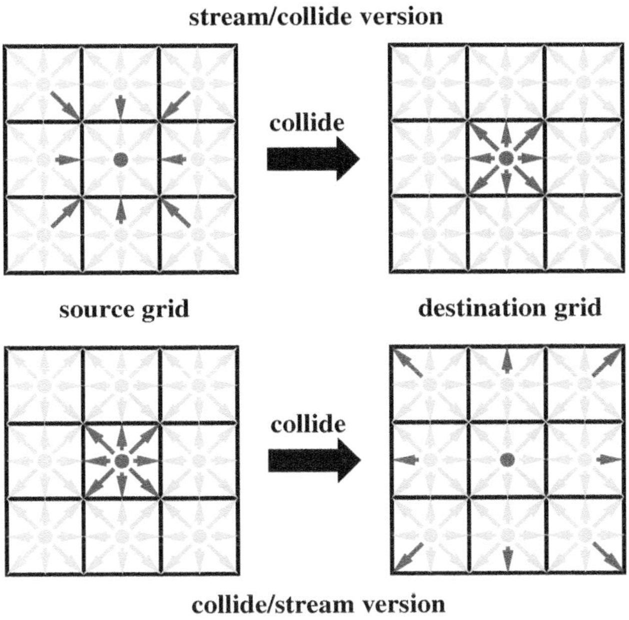

Pull: stream/collide

- read distribution functions $f_i^{out}(t)$ from *adjacent* cells of the "old" array (propagation)
- calculate ρ, $\mathbf{u}$ and f_i^{eq}
- write updated $f_i^{out}(t + \Delta t)$ values to *current* cell of the "new" array

Push: collide/stream

- read distribution functions $f_i^{in}(t)$ from *current* cell of the "old" array
- calculate ρ, $\mathbf{u}$ and f_i^{eq}
- write updated $f_i^{in}(t + \Delta t)$ values to *adjacent* cells of the "new" array (propagation)

Fig. 13.3. Comparison of the data access for the stream/collide (pull) and the collide/stream (push) version of LBM.

fused into just one large loop (m) sweeping over the full 3D domain. This may improve the efficiency of vectorization (Figure 13.5). However, to allow correct updates at the boundaries of the computational domain, an additional ghost layer must be added around the actual computational box together with an appropriate mask which blocks out the ghost cells from computing the collision and streaming process. The idea of ghost cells will reappear in Section 13.4 for purposes of parallelization.

13.3.3 Cache Optimization

On cache-based architectures, the data path between CPU and main memory is a serious bottleneck for memory intensive applications like LBM codes. Therefore, a major aim of an efficient implementation is the increase of cache reuse — if necessary even at the cost of additional arithmetic operations.

```
real (kind=8),dimension(0:xE+1,0:yE+1,0:zE +1,0:18,0:1):: f
logical , dimension(1:xE,1:yE,1:zE ) :: fluidCell
real (kind=8)  :: dens, ne (0:18), ...
do z=1,zE; do y=1,yE; do x=1,xE
  if ( fluidCell (x,y,z) then
    ! read distributions from local cell and
    ! calculate moments
    dens=f(x,y,z,0, old)+f(x,y,z,1, old)+f(x,y,z,2, old )+...
    ...
    ! compute non-equilibrium parts
    ne (0)=...
    ...
    ! write updates to neighboring cells
    f(x  ,y  ,z  , 0, new)=f(x,y,z , 0, old)*ImOmega+ne(0)
    f(x+1,y+1,z  , 1, new)=f(x,y,z , 1, old)*ImOmega+ne(1)
    ...
    f(x  ,y-1,z-1,18,new)=f(x,y,z,18, old)*ImOmega+ne(18)
  endif
enddo; enddo; enddo
```

Fig. 13.4. Layout of the "standard propagation optimized version".

A data layout with the 19 discrete-velocity directions i as first index (collision optimized) results in the distribution functions of a cell being stored contiguously in memory and therefore only few cache lines are required to access the data of a cell. While this is beneficial for the read access, the results have to be stored to non-contiguous memory areas in the streaming phase and therefore involve many different cache lines. Additionally, the index x of the innermost loop does "move fastest", i.e. it does not result in a stride 1 memory access, which it ideally should.

Using $(x, y, z, i, old/new)$ as data layout (propagation optimized), makes the x index the fastest. In contiguous memory areas, complete x lines for the different discrete velocity directions i can be found. To exploit this fact even better, the computational work within the x loop is divided into several parts. First of all, current distribution values of a complete x line are read and the different moments which are required later on for the calculation of the equilibrium distribution and the non-equilibrium part are precomputed and stored in temporary arrays. Then in another loop, the non-equilibrium parts are calculated. Finally, in separate loops for all discrete velocity directions i, the relaxation is executed and the results are written back to the adjacent cells for the next time step. By writing back the data in a line-wise fashion, contiguous memory areas are accessed. The basic code structure of this implementation is outlined in Figure 13.6.

When the propagation optimized data layout is used, powers of two should be avoided for the leading dimension as this will result in severe cache trashing [41].

Starting at a certain architecture-dependent x size, blocking of the inner loops [27] (x) can be advantageous in order to ensure that all temporary data remains in

```
real(kind=8), dimension(0:(1+xE)*(1+yE)*(1+zE ),0:18,0:1) :: f
logical , dimension(0:(xE+1),0:(yE+1),0:(zE +1)) :: fluidCell
real(kind=8)  :: dens, ne (0:18), ...
do m=0, (1+xE)*(1+yE)*(1+zE)
  if ( fluidCell (m) ) then
    ! read distributions from local cell and
    ! calculate moments
    dens=f(m,0,old)+f(m,1,old)+f(m,2,old )+...
    ...
    ! compute non-equilibrium parts based on local moments
    ne (0)=...
    ...
    ! write updates to neighboring cells
    f(m                     , 0, new)=f(m, 0,old)*ImOmega+ne(0)
    f(m+1+(xE+1)            , 1, new)=f(m, 1,old)*ImOmega+ne(1)
    ...
    f(m  -(xE+1)-(xE+1)*(yE+1),18,new)=f(m,18,old)*ImOmega+ne(18)
  endif
enddo
```

Fig. 13.5. Layout of the "vector propagation optimized version".

the cache. The additional modifications of the code are outlined in Figure 13.7. Of course, the blocking can be easily extended to three dimensional 3-way blocking [27].

13.3.4 Further Optimization Strategies

In literature, some other recent optimization strategies can be found which aim at reducing the memory consumption and/or improving performance.

Pohl *et al.* [34, 42, 43, 25] have extended the idea of blocking and demonstrated the effect of n-way blocking. In particular, a 4-way blocking (three-fold spatial blocking and additional blocking in time) can provide additional performance improvements on certain architectures.

Pohl *et al.* [34] and Schulz *et al.* [37] have presented two different "compressed grid" approaches. Here, a carefully designed update order is used when executing the propagation step of the individual cells. In this way the array can be reused to store the new distribution values overwriting old ones as soon as they are no longer needed. This strategy reduces the total memory consumption almost by a factor of two. However, in particular for the method suggested in [37] it is not clear yet whether the cache reuse is really improved as the propagation step has to be done direction dependent.

The compressed grid technique can be combined with a 4-way blocking. As mentioned above, blocking in the fourth dimension (time) can again only be done if the

```
real (kind =8), dimension(1:xE,1:yE,1:zE ,0:18,0:1) :: f
real (kind =8), dimension(1:xE ) :: dens , ...,
real (kind =8), dimension(1:xE ,0:18) :: ne
do z=1,zE; do y=1,yE
   ! read distributions from local cell and calculate moments
   do x=1,xE
      dens(x)=f(x,y,z,0, old)+f(x,y,z,1, old)+f(x,y,z,2, old )+...
      ...
   enddo
   ! compute non-equilibrium parts based on local moments
   do x=1,iE
      ne(x ,0) = ...
      ...
   enddo
   ! write updates to neighboring cells ;
   ! separate loops for all directions
   do x=1,xE
      if ( fluidCell (x,y,z ) ) then
         f(x ,y ,z , 0, new)=f(x,y,z , 0, old)*ImOmega+ne(x, 0)
      endif
   enddo
   ...
   do x=1,xE
      if ( fluidCell (x,y,z ) ) then
         f(x ,y-1,z-1,18,new)=f(x,y,z,18, old)*ImOmega+ne(x,18)
      endif
   enddo
enddo; enddo
```

Fig. 13.6. Cache-optimized RISC version.

data dependencies are carefully taken into account. This results in more complex code, and in particular in "short" loops which may be difficult to execute efficiently. For the details the reader is referred to [34, 42, 43, 25].

Here, we only want to point out that a blocking in the fourth dimension (time) is the only technique that can fundamentally improve the *temporal* cache-reuse. Any implementation short of a 4-way blocking will have to transfer the complete set of distribution functions through the memory hierarchy at least once per time step and is thus fundamentally limited by the bandwidth of each CPU to main memory. The other techniques discussed here were primarily aimed at improving the *spatial* locality by re-using the contents of each cache line efficiently.

On the other hand, the 4-way blocking algorithm and the compressed grids make it much more difficult to incorporate advanced boundary conditions or models for more complicated physics as both might depend on pre- as well as post-collision values of more than just the local cell itself.

```
do xx=1,xE, BLOCKSIZE
  do z=1,zE; do y=1,yE
    ! read  distributions  from local  cell  ...
    do x=xx,min(xE,xx+BLOCKSIZE-1)
      dens(x)=f(x,y,z,0,old)+f(x,y,z,1,old)+...
       ...
    enddo
    ! compute non-equilibrium parts  ...
    do x=xx,min(xE,xx+BLOCKSIZE-1)
      ne(x ,0)=...
       ...
    enddo
    ! write  updates  to  neighboring  cells ;
    ! separate  loops  for  all  directions
    do x=xx,min(xE,xx+BLOCKSIZE-1)
      if (  fluidCell (x,y,z ) )  then
        f(x  ,y  ,z  , 0, new)=f(x,y,z , 0, old)*ImOmega+ne(x, 0)
      endif
    enddo
     ...
    do x=xx,min(xE,xx+BLOCKSIZE-1)
      if (  fluidCell (x,y,z ) )  then
        f(x  ,y-1,z-1,18,new)=f(x,y,z,18,old)*ImOmega+ne(x,18)
      endif
    enddo
  enddo; enddo
enddo
```

Fig. 13.7. Cache-optimized version with blocking.

Argentini *et al.* [2] use the 3D non-BGK model of Ladd [28] and thus succeed in storing only 9 moments of the distribution functions instead of all the distribution values. However, they admit that an application of this idea to the common BGK model is not (transparently) possible.

Pan *et al.* [32] and Schulz *et al.* [37] presented data structures — in particular for porous media applications with low porosities (i.e. with a low ratio of fluid to solid nodes) — which abandon the "full matrix representation" and instead use several lists with the data of the distribution functions and the connectivity. Thus, only memory for the data of fluid cells consisting of the distribution functions and pointers to all N neighbors are required. For low porosities, a considerable fraction of memory can be saved in this way. However, the memory access patterns get much more complicated and include vast indirect addressing. As long as vectorization is not prevented, vector systems probably can do rather well despite the indirect memory access [37]. However, on RISC machines, cache reuse is significantly reduced by the scattered memory access and thus a considerable performance loss is expected.

Careful data ordering in the lists (e.g. Morton ordering a shown in [32]) can lessen the performance impact to some extent. However, as the optimal reordering process is probably np-complete, some heuristics will always be required.

13.4 Parallelization of a Simple Full-Grid LBM Code

As for many other scientific computing algorithms, the most natural approach to parallelize the lattice Boltzmann method is by *domain decomposition*, or technically more precise (see [24]) *domain partitioning*. For domain partitioning, the entire computational domain is divided into several subdomains. An example of such a domain decomposition is depicted in Figure 13.8. Each subdomain is then assigned to a processing unit (PU) which can be a single CPU or a group of CPUs accessing the same memory. If every PU only had to read and write data in its own subdomain, then no communication would be necessary. Unfortunately, the streaming step of LBM needs to access data from neighboring cells which might be located in adjacent subdomains. Assuming a distributed memory environment, this cannot be accomplished without *explicit communication* and the question is how this communication can be organized both conveniently and efficiently.

A common way to approach the necessary data exchange is the introduction of a so-called *halo* (also known as *ghost cell layer*) at the subdomain interfaces, i.e. where two subdomains are adjacent. Halos hold copies of the values of neighboring domains, and naturally they must be updated at specific times when the original value has been changed. Depending on the type of data dependencies the halo must consist of a single or more layers. For the streaming step of the LBM, only nearest neighbor communication is necessary (corresponding to a compact stencil), and thus a single layer of cells is sufficient, but certain extensions of the LBM may need wider halos.

With the halo, the algorithm can proceed in each subdomain concurrently and can perform a time step in parallel. However, between time steps, the halos must be exchanged with the neighboring subdomains. This technique simplifies the parallelization, since the exact details of the data dependencies between neighboring cells need not be considered, and, since data is exchanged in larger blocks, the startup overhead for communication on cluster parallel machines is better amortized.

If communication is so slow that the exchange of the halos becomes a significant contribution to the computing time, it can be overlapped with the computation if supported by the architecture. This can be accomplished when the cell updates in each subdomain are computed first for those cells which are halos of neighboring subdomains. The data in the halo can then be sent simultaneously while computing the remaining cells in the interior of each subdomain. Again, these are standard strategies, well-known from other parallel grid or mesh-based algorithms.

Figure 13.8 shows a simple one-dimensional domain partitioning for a two dimensional rectangular grid of cells. In the light of the extensions required for our application, as discussed in Section 13.5.1, we will primarily use this one-dimensional partitioning.

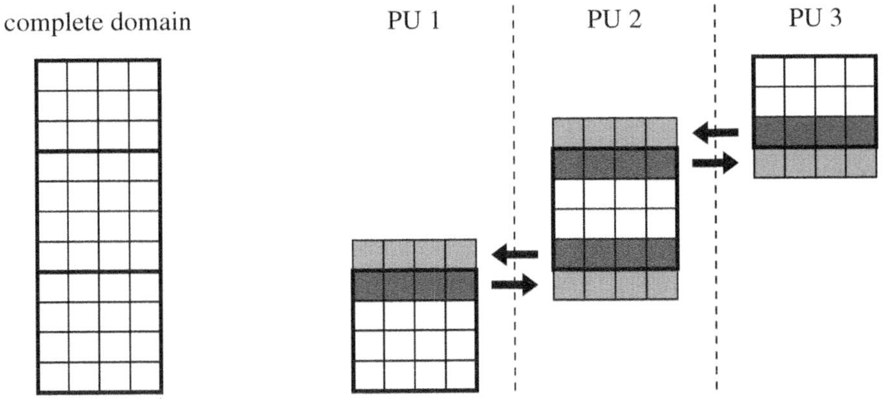

Fig. 13.8. Example of a one dimensional domain decomposition with 3 PUs. Before a new iteration can start, each PU has to send required cells (*dark grey cells*) to other PUs. Additionally, each PU receives an update of its halo (*light grey cells*).

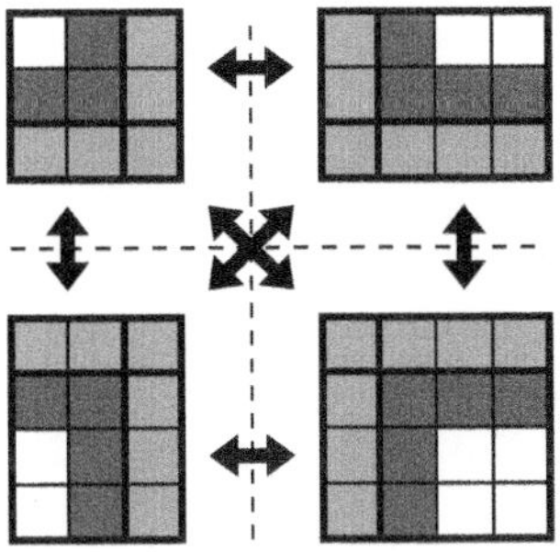

Fig. 13.9. For a two dimensional domain decomposition more messages have to be sent.

Of course it is also possible to use a more complicated two dimensional domain partitioning as illustrated in Figure 13.9. This will asymptotically result in a better surface-to-volume ratio of each subdomain and thus in less communication. This, however, comes at the price of the communication being performed in smaller message packages and therefore increasing startup cost. Note that for a 3D grid, a subdomain in the interior will not only have to communicate with the six subdomains adjacent to its faces, but also to the twelve neighboring subdomains with which it shares an edge (for the D3Q19 model).

To alleviate the cost of these communication steps, sophisticated schemes can be devised. The communication of data across an edge can be accomplished by, e.g., sending the data as part of the messages between face neighbors and using one intermediate neighbor as a mailman. This, however, can become quite complicated if the dependencies between neighboring cells are more complex (and therefore the halos are wider), and when communication and processing need to be overlapped.

Figure 13.10 shows the scalability (performance on an increasing number of computational units) of a standard parallel LBM implementation for different architectures (see Table 13.3 for details). The scale-up performance (fixed CPU load by scaling the domain size) is almost linear for all three systems, but the speed-up (fixed domain size) performance typically degrades with an increasing number of CPUs due to the load shift from computation to communication.

13.5 Free Surfaces

13.5.1 Motivation and Application

Free surface flow is omnipresent in nature, in everyday life as well as in technological processes. Examples are river flow, rain, and filling of cavities with liquids or foams. The advantage of the LBM approach becomes apparent when certain complex boundary conditions must be implemented, since the microscopic interpretation of the dynamics allows a more natural treatment of these boundary conditions compared to the description based on differential equations. Thus, the LB approach seems to be especially suited for modeling complex fluid mechanical problems in complex time-dependent domains. For example foams which develop by interacting and growing gas bubbles in a melt, are challenging due to their large and highly dynamic internal surface, as illustrated in Figure 13.11.

In the following, our approach to handle free surface flow within the LB framework is described. In principle, there are two difficulties to be resolved. The first one is the description of the movement of the interface. Besides the prevention of the spreading of the interface, one has to ensure that the interface movement preserves mass. The second challenge is to fulfill the pressure boundary conditions at the interface.

Modeling of foaming processes leads to additional challenges if parallelization is considered. Information with respect to interface movement has to be transferred between two domains when an interface crosses the partition boundary. In addition, foam expansion makes some kind of dynamical load balance indispensable in order to reach a high performance.

13.5.2 Free Surface and Fluid Advection

The description of the liquid–gas interface is very similar to that of the marker and cell or the volume of fluid methods. An additional variable, the volume fraction of fluid ϵ, defined as the portion of the area of the cell filled with fluid, is assigned to each interface cell. The representation of liquid–gas interfaces is illustrated in Figure 13.12.

Gas cells are separated from liquid cells by a layer of interface cells. These interface cells form a "completely closed" boundary in the sense that no distribution function is directly advected from fluid to gas cells and vice versa. This is a crucial

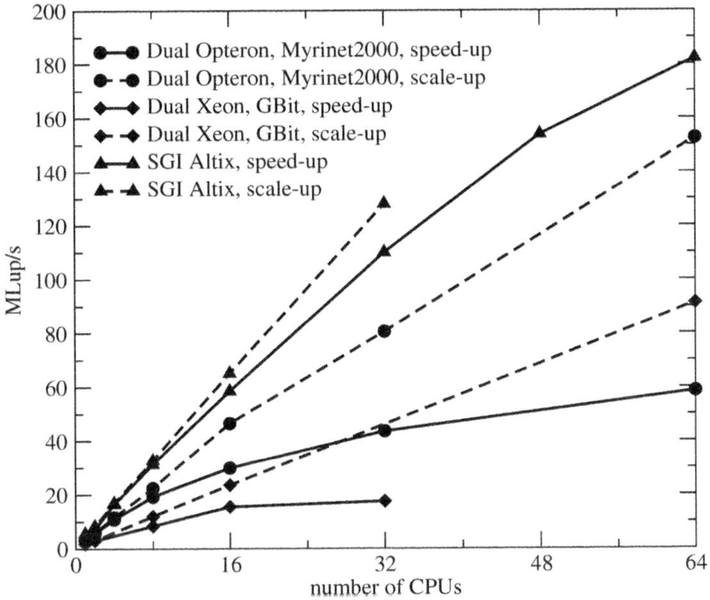

Fig. 13.10. Scalability tests for modern cluster configurations. The domain size is $256 \times 129 \times 128$ for speed-up (fixed domain size) and 128^3 per processor for scale-up tests (fixed CPU load by scaling the domain size). For reference the corresponding results of a shared memory system (SGI Altix 3000) are given. The common performance unit MLup/s (million lattice site updates per second) has been used. (For the color version, see Figure A.28 on page 481).

Table 13.3. Characteristics of architectures which have been used for benchmarking.

System	Xeon/GBit	Opteron/Myrinet	SGI Altix 3000
Basic building block (BB)	2-way SMP node with 1 memory path	2-way SMP node with 2 memory paths	4-way SMP node with 2 memory paths
CPU	Intel Xeon 2.66 GHz	AMD Opteron 2.0 GHz	Intel Itanium2 1.3 GHz, 3 MB L3 cache
Peak performance per CPU	5.3 GFlop/s	4.0 GFlop/s	5.2 GFlop/s
Memory bandwidth per BB	4.3 GByte/s	2 × 5.4 GByte/s	2 × 6.4 GByte/s
Interconnect	Cisco 4503 GBit Ethernet switch	Myrinet2000	SGI NUMAlink 3 2 × 1.6 GByte/s bidirect.
Operating system	Debian Linux 3.0	SuSE SLES 8 Linux	Redhat AS2.1 & SGI Propack 2.4
Compiler	Intel ifc 7.1	PGI 5.0	Intel efc 7.1

Fig. 13.11. 3D foam: The bubbles grow and coalescence occurs. The disjoining pressure Π stabilizes the foam and eventually a polygonal structure develops (initial number of bubbles: 1000; system size: $120 \times 120 \times 140$; $\tau = 0.8$; g = 0; $\sigma = 0.01$; $c_\Pi = 0.006$). (For the color version, see Figure A.29 on page 482).

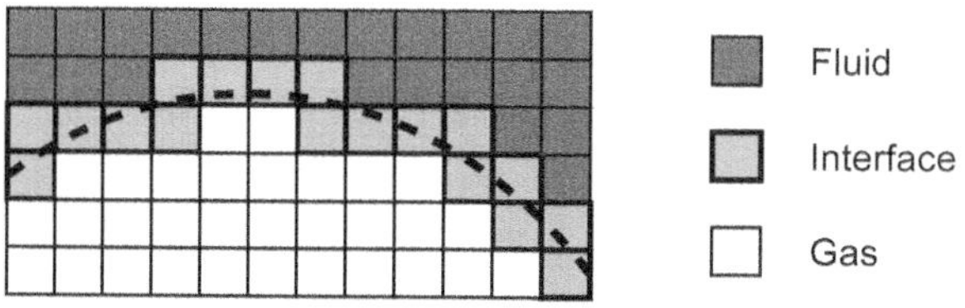

Fig. 13.12. 2D representation of a free liquid–gas interface by interface cells. The real interface (*dashed line*) is captured by assigning the interface cells their liquid fraction.

point to assure mass conservation since mass coming from the liquid or mass transfered to the liquid always passes through the interface cells where the total mass is balanced. Hence, global conservation laws are fulfilled if mass and momentum conservation is ensured for interface cells. The cell types and their state variables and possible state transformations are listed in Table 13.4.

Per definition, the volume fraction ϵ of fluid and gas cells is 1 and 0, respectively. The fluid mass content of a cell is denoted with $M = M(\mathbf{x}, t)$. The mass content is a function of the volume fraction and the density. For a gas cell the fluid mass content M is zero whereas that of a fluid cell is given by its density ρ and the cell volume ΔV: $M(\mathbf{x}, t) = \rho(\mathbf{x}, t) \cdot \Delta V$ for $\mathbf{x} \in F$. Fluid cells gain and lose mass due to streaming of the f_i. For fluid cells M and ρ are equivalent. If interface cells

Table 13.4. Cell types: state variables and possible state transformations. Applicable variables for the concerning cell type are marked with "•"; with "-" otherwise.

cell type	distr. func. f_i	volume fraction ϵ	gas pressure p^G	change of state
fluid F	•	-	-	$\rightarrow$ I
gas G	-	-	•	$\rightarrow$ I
interface I	•	•	•	$\rightarrow$ G, $\rightarrow$ F

are considered, M and ρ are not equivalent and we have to account for the partially filled state by the volume fraction $\epsilon = \epsilon(\mathbf{x}, t)$. The fluid mass content M, the volume fraction ϵ and the fluid density ρ are related by $M(\mathbf{x}, t) = \rho(\mathbf{x}, t) \cdot \epsilon \cdot \Delta V$ for $\mathbf{x} \in I$.

All cells are able to change their state. It is important to notice that direct state changes from fluid to gas and vice versa are not possible. Hence, fluid and gas cells are only allowed to transform into interface cells whereas interface cells can be transformed into both gas and fluid cells. A fluid cell is transformed into an interface cell if a direct neighbor is transformed into a gas cell. At the moment of transformation the fluid cell contains a certain amount of fluid mass M which is stored. During further development the interface cell may gain mass from or lose mass to the neighboring cells. These mass currents are calculated and lead to a temporal change of M. If M drops below zero, the interface cell is transformed into a gas cell. It is important to pronounce that mass and density are completely decoupled for interface cells. For gas cells the density is given by the volume and mass of a bubble, while for fluid cells, the density can be determined using (13.5). As they are always completely filled, their volume fraction is one, and their mass is equal to their density. Interface cells, on the other hand, contain the fluid interface, and thus have changing values for mass and volume fraction. The change of mass during each time step is calculated by computing the mass exchange of the current interface cell with all surrounding fluid and interface cells.

The mass exchange $\Delta M_i(\mathbf{x}, t)$ between an interface cell at lattice site $\mathbf{x}$ and its neighbor in $\mathbf{e}_i$-direction at $\mathbf{x} + \mathbf{e}_i$ (with $\mathbf{e}_{\bar{i}} = -\mathbf{e}_i$ and $f_{\bar{i}}(\mathbf{x}, t) = f(\mathbf{x}, \mathbf{e}_{\bar{i}}, t) = f(\mathbf{x}, -\mathbf{e}_i, t)$) is calculated as

$$\Delta M_i(\mathbf{x}, t) = \begin{cases} 0 \\ f_{\bar{i}}(\mathbf{x} + \mathbf{e}_i, t) - f_i(\mathbf{x}, t) \\ \frac{1}{2}\left[\epsilon(\mathbf{x}, t) + \epsilon(\mathbf{x} + \mathbf{e}_i, t)\right]\left[f_{\bar{i}}(\mathbf{x} + \mathbf{e}_i, t) - f_i(\mathbf{x}, t)\right], \end{cases} \tag{13.14}$$

where the three clauses are used for $\mathbf{x} + \mathbf{e}_i \in G$, F and I, respectively. There is no mass transfer between gas cells and interface cells. The interchange between an interface cell and a fluid cell should be the same as that of two fluid cells since the cell boundary is completely covered with liquid. In this case, the mass exchange can be directly calculated from the particle distribution functions. The interchange between two interface cells is approximated by assuming that the mass current is weighted by the mean occupied volume fraction. It is crucial to note that mass is explicitly conserved in (13.14):

$$\Delta M_{\bar{i}}(\mathbf{x} + \mathbf{e}_i, t) = -\Delta M_i(\mathbf{x}, t) \,. \tag{13.15}$$

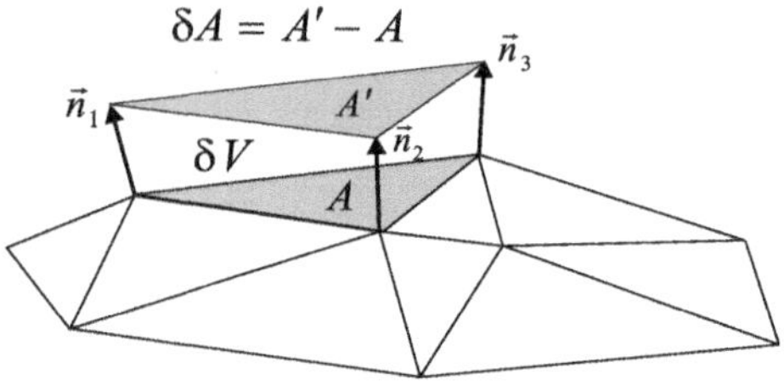

Fig. 13.13. Calculation of the curvature.

That is, the mass which a certain cell receives from a neighboring cell is automatically lost there and vice versa. The temporal evolution of the mass content of an interface cell is thus given by

$$M(\mathbf{x}, t + \Delta t) = M(\mathbf{x}, t) + \sum_{i=0}^{N} \Delta M_i(\mathbf{x}, t) \,. \tag{13.16}$$

An interface cell is transformed into a gas or fluid cell if $M < 0$ or $M > \rho \Delta V$, respectively. At the same moment, new interface cells emerge in order to guarantee the continuity of the interface. The initial distribution functions of these new interface cells are extrapolated from the cells in normal direction towards the fluid.

The calculation of the local curvature of the interface is complex and time consuming, an overview of our algorithm is shown in Figure 13.13. In a first step, the marching cube algorithm is used to generate a triangulation of the interface. Secondly, the curvature κ belonging to each triangle is estimated by $\kappa = \frac{1}{2}\frac{\delta A}{\delta V}$ where δA denotes the rate of change of the triangle area when its vertices are infinitesimally shifted in normal direction. The covered volume is denoted by δV. In the last step, the curvature of an interface cell is estimated by averaging the curvature of the triangles belonging to it.

13.5.3 Boundary Conditions

Interface cells separate gas cells from fluid cells. After streaming, only distribution functions from fluid and interface cells are defined. Distribution functions arriving from gas cells are not defined (see Figure 13.14, left).

The symmetry between known and unknown distribution functions, i.e., if f_i is known $f_{\bar{i}}$ is unknown, is essential to fulfill the boundary conditions. We demand force balance for opposite lattice directions. In addition, we make use of the fact that the forces exerted by the gas are known and are given by the gas pressure and the velocity at the interface. Hence, the missing distribution functions are reconstructed as

$$f_i^{out}(\mathbf{x} - \mathbf{e}_i, t) = \left(f_i^{eq}(\rho^G, \mathbf{v}) + f_{\bar{i}}^{eq}(\rho^G, \mathbf{v})\right)(1 + \kappa\sigma) - f_{\bar{i}}^{out}(\mathbf{x}, t) \tag{13.17}$$

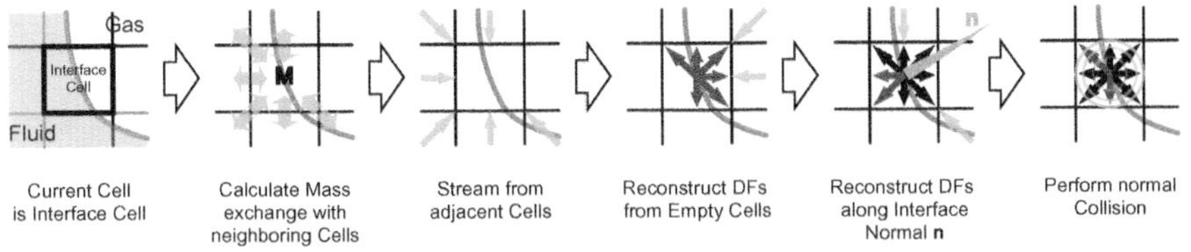

Fig. 13.14. An overview over the steps for an interface cell are shown. Note that not only distribution function from empty cells, but also those along the interface normal are reconstructed. (For the color version, see Figure A.30 on page 482).

using the gas density $\rho^G = \frac{1}{c_s^2}\, p^G$ and the velocity $\mathbf{v}$ of the interface cell. The reconstructed equilibrium distribution functions are scaled according to the curvature and the strength of the surface tension of the fluid σ. Depending on the sign of the curvature, this results in a force normal to the interface.

It is important to note that not only the missing distribution functions are reconstructed but all distribution functions with $\mathbf{e}_i \cdot \mathbf{n} \geq 0$ (see Figure 13.14). After completion of the whole set of distribution functions, the new density ρ and velocity $\mathbf{v}$ can be calculated. The outgoing distribution functions of interface cells are calculated as

$$f_i^{out}(\mathbf{x},t) = f_i^{in}(\mathbf{x},t) - \frac{1}{\tau}\left(f_i^{in}(\mathbf{x},t) - f_i^{eq}(\rho,\mathbf{v})\right) + \epsilon(\mathbf{x})\, w_i\, \rho\, \mathbf{e}_i\, \mathbf{g}\,, \qquad (13.18)$$

where g is the gravity constant, now weighted according to the volume fraction of the interface cell.

13.5.4 Visualization

To visualize the complex surfaces generated by the foam simulations we use the ray tracing algorithm [21]. Ray tracing is capable of producing high quality visualizations even for many layers of transparent surfaces like those of a foamed liquid, as can be seen in Figure 13.11.

The ray tracing algorithm determines the color of each pixel of an image by shooting a ray through an imaginary camera into a scene, and computing the amount of light reflected into the camera at the first intersection of the ray with a surface. Effects like transparency can be by included by tracing the ray further into the scene after the first point of intersection. For our ray tracer implementation we use the triangulated surface from the marching cubes algorithm, as the ray intersection can be easily computed for triangles. To prevent intersection tests with all generated triangles for each ray, we use a binary space partitioning tree [38]. Another important effect for the visualization of free surface liquids is the focusing of light on nearby objects. These patches of stronger light are called caustics, and can be generated by tracing rays into the scene from the light sources during an additional pass before the actual ray tracing.

Still, the number of rays to be traced for a typical scene like Figure 13.11 can be high due to the large amount of surface layers. But, as the memory requirements of the algorithm are usually not problematic, the image generation process can be conveniently parallelized by having each CPU compute a separate image. Our implementation writes the surface data of each image to the hard disk, while a small script running on each workstation in our network checks whether a new set of surface data is available. If this is the case and the computer is currently idle, the script locks the surface data and initiates the rendering process. The ray tracing can be done without further communication with other workstations. The generated image is stored in a central repository.

On the other hand, the aim of the parallelization may sometimes not be the total speedup, but to get a single picture as soon as possible. Some scenes that need to be visualized may also be too large to be handled by a single workstation. In both cases, a more sophisticated way of parallelization can be used to have multiple computers generate parts of the same image, which, however, requires more communication between the involved computers. But as shown in [40, 29] this distributed ray tracing algorithm can be fast enough to produce complex visualizations interactively and in real time.

13.5.5 Concepts for Parallelization

In a standard LBM code with fused stream/collide step, a time step can be performed in a single sweep over the computational domain. For the more complex handling of free surfaces with the described method, however, our current implementation requires five sweeps.

1. The first sweep performs the usual streaming step augmented by the reconstruction of missing distribution functions from adjacent gas cells taking into account the gas pressure of the concerned bubble. After that, the mass exchange with neighboring cells is computed. For interface cells the new mass M leads to an updated fill value ϵ. This value determines whether an interface cell is now completely filled or emptied resulting in a conversion to a gas or fluid cell, respectively. Finally, the collision step according to the BGK approximation with an additional force term representing gravity is performed.
2. The second sweep checks and reestablishes the strict division of fluid and gas cells by the interface cell layer. Therefore, it might be necessary to undo cell conversions that have been scheduled in the first sweep.
3. In the third sweep converted cells have be to initialized. Depending on the new cell type, a set of distribution functions (only for former gas cells) and the fill value ϵ and mass value M are computed.
4. Interface cells converting to gas or fluid cells normally still contain or miss a certain amount of mass. This excess mass, which can also be negative for emptied interface cells, is distributed among the adjacent interface cells. In rare cases, no such cells can be found and the mass cannot be redistributed locally. To deal with this exception, the mass could be distributed in a wider neighborhood.

5. During the last sweep, for all former and new interface cells, the change in the fill value ϵ is calculated and the volume of the corresponding bubble is updated.

In addition to the communication of the grid data, the changes in the volume of the gas bubbles have to be collected and added for each subdomain to obtain the global volume change for each bubble. Therefore, the uppermost process starts to send its volume changes to its lower neighbor which adds its changes and sends this updated data to its lower neighbor. This procedure is continued until the lowermost process receives the volume update. Meanwhile the same chain of communication has been started at the lowermost process traveling upwards. When both chains reach the opposite end of the domain all processes have a global update of the gas bubbles' volumes. Instead of this procedure an "all to all" communication could be initiated using a dedicated MPI call. For a large number of subdomains, however, this would result in a large amount of send/receive events.

Load Balancing

The time required for updating a cell strongly depends on its type: For gas cells nothing has to be done; fluid cells do the normal stream/collide step, whereas interface cells are treated like fluid cells with several additional checks and calculations that also depend on the neighboring cell types. This makes an ab initio domain partitioning with an even load on each processing unit (PU) difficult. Apart from that, a static partitioning would not help, since a typical foam simulation involves large topological changes that shift the fluid and therefore the workload from one computational domain to an other.

The only way to distribute the changing workload evenly on all PUs is to implement one of the available load balancing schemes which can be classified in two major groups:

Global Load Balancing: By taking the performance and load data of all PUs into account, a single (master) node decides when and how to perform a global re-partitioning of the domain. This allows for fast adaptation to large-scale load changes at the cost of a considerable overhead for global communication and re-partitioning.

Local Load Balancing: Each PU relies on its neighboring PUs to send updated halos. If one PU has to wait for this update, it might be advantageous to shift a fraction of the domain of the neighbor with capacity overload to the waiting PU. This scheme is applicable for slow and "diffusive" load changes and does not require global communication or costly re-partitioning of the entire domain.

By construction, cell types can only travel the distance of one cell in each time step and much less than this in practice. Likewise, the computational load moves slowly from a PU to its neighbors which can adequately be compensated by a local load balancing scheme.

Implementation Details of Load Balancing

The initial domain partitioning can be chosen according to simple heuristics like an equal distribution of fluid and interface cells on all PUs. Any deviation from a balanced computational load will be compensated within a few time steps.

During the communication phase in the simulation, each PU measures the waiting time for a halo update from neighboring PUs. If the waiting time exceeds a certain threshold for a given number of time steps in succession, the waiting PU requests a layer of cells from the overloaded PU. This prevents unnecessary re-partitioning due to negligible fluctuations in the load.

The simple 1D domain partitioning in combination with an adapted memory layout where all data for one cell layer is located contiguously in memory allows for efficient re-partitioning since a cell layer can be sent without repacking from one PU to its neighbor.

If the fluid simulation is running exclusively on the computers, it is beneficial to allocate as much memory as possible on each PU during initialization to avoid the overhead of allocating new memory for the smaller/larger new domain, copying data, and freeing the memory for the old domain in the case of re-partitioning.

13.6 Summary and Outlook

Since the core algorithm of LBM is simple compared the other CFD codes based on finite volume/element methods, it is easy to optimize and to extend. We presented techniques to improve the single CPU performance, applied domain partitioning to perform the computations in parallel, and introduced an extension to handle free surface flows. Many other extensions are being developed by other groups to improve boundary conditions [16, 15, 20], capture turbulence [8, 1], multi-phase flows [10, 22, 26], or solidification processes [31, 13].

Notation

$f = f(\mathbf{x}, \boldsymbol{\xi}, t)$	$[kg/m^3]$	particle distribution function
$f_i = f_i(\mathbf{x}, t)$	$[kg/m^3]$	velocity discrete particle distribution function
$f^{(0)}$	$[kg/m^3]$	Maxwell-Boltzmann equilibrium distribution function
f_i^{eq}	$[kg/m^3]$	discretized equilibrium distribution function
$\mathbf{x}$	$[\{m\}]$	spatial position vector (to continuous or discrete position)
t	$[s]$	continuous or discrete time
$\boldsymbol{\xi}$	$[\{m/s\}]$	microscopic particle velocity vector
$\mathbf{e}_i$	$[\{m/s\}]$	discrete particle velocity vector
Δt	$[s]$	discrete time increment ("time spacing")
Δx	$[m]$	discrete position increment ("lattice spacing")
c	$[m/s]$	unit velocity
ρ	$[kg/m^3]$	(number) density
$\mathbf{u}$	$[\{m/s\}]$	macroscopic fluid velocity vector
τ	$[-]$	dimensionless relaxation time
λ	$[s]$	relaxation time
c_s	$[m/s]$	speed of sound
w_i	$[-]$	weighting factor
ν	$[m^2/s]$	kinematic viscosity
p	$[kg/ms^2]$	pressure

Subscripts and superscripts

i	discrete velocity direction, $i = 0, \ldots, N$
α, β	velocity component in the Cartesian directions x, y and z
$f^{(0)}$	Maxwell-Boltzmann equilibrium distribution
f^{eq}	discretized equilibrium distribution
f^{in}	incoming particle distribution function, i.e. before collision
f^{out}	outgoing particle distribution function, i.e. after collision
old/new	source / destination array (see Section 13.3)

References

1. S. Ansumali, I. V. Karlin, and S. Succi. Kinetic Theory of Turbulence Modeling: Smallness Parameter, Scaling and Microscopic Derivation of Smagorinsky Model. *Physica A*, 338(3):379–394, 2004.
2. R. Argentini, A. Bakker, and C. Lowe. Efficiently Using Memory in Lattice Boltzmann Simulations. *Future Generation Computer Systems*, 20(6):973–980, 2004.
3. J. Bernsdorf, F. Durst, and M. Schäfer. Comparison of Cellular Automata and Finite Volume Techniques for Simulation of Incompressible Flows in Complex Geometries. *Int. J. Numer. Meth. Fluids*, 29(3):251–264, 1999.
4. V. Bhandari. Detailed Investigations of Transport Properties in Complex Reactor Components. Master thesis, Lehrstuhl für Strömungsmechanik, Universität Erlangen-Nürnberg, Erlangen, Germany, 2002.

5. P. Bhatnagar, E. P. Gross, and M. K. Krook. A Model for Collision Processes in Gases. I. Small Amplitude Processes in Charged and Neutral One-Component Systems. *Phys. Rev.*, 94(3):511–525, 1954.
6. M. Bouzidi, M. Firdaouss, and P. Lallemand. Momentum Transfer of a Boltzmann Lattice Fluid with Boundaries. *Phys. Fluids*, 13(11):3452–3459, 2001.
7. S. Chapman and T. G. Cowling. *The Mathematical Theory of Non-Uniform Gases*. Cambridge University Press, 1995.
8. H. Chen, S. Kandasamy, S. Orszag, R. Shock, S. Succi, and V. Yakhot. Extended Boltzmann Kinetic Equation for Turbulent Flows. *Science*, 301(5644):633–636, 2003.
9. S. Chen and G. D. Doolen. Lattice Boltzmann Method for Fluid Flows. *Annu. Rev. Fluid Mech.*, 30:329–364, 1998.
10. S. Chen, G. D. Doolen, and K. G. Eggert. Lattice-Boltzmann Fluid Dynamics: A Versatile Tool for Multiphase and Other Complicated Flows. *Los Alamos Science*, 22:98–111, 1994.
11. S. Chen and D. Martinez. On Boundary Conditions in Lattice Boltzmann Methods. *Phys. Fluids*, 8(9):2527–2536, 1996.
12. B. Chopard and M. Droz. *Cellular Automata Modeling of Physical Systems*. Cambridge University Press, 1998.
13. C. Denniston, E. Orlandini, and J. Yeomans. Lattice Boltzmann Simulations of Liquid Crystal Hydrodynamics. *Phys. Rev. E*, 63:056702_10, 2001.
14. S. Donath. On Optimized Implementations of the Lattice Boltzmann Method on Contemporary High Performance Architectures. Bachelor thesis, Lehrstuhl für Informatik 10 (Systemsimulation), Universität Erlangen-Nürnberg, 2004.
15. A. Dupuis and B. Chopard. Theory and Applications of an Alternative Lattice Boltzmann Grid Refinement Algorithm. *Phys. Rev. E*, 67(6):066707_7, 2003.
16. O. Filippova and D. Hänel. Grid Refinement for Lattice BGK Models. *J. Comput. Phys.*, 147:219–228, 1998.
17. U. Frisch, D. d'Humières, B. Hasslacher, P. Lallemand, Y. Pomeau, and J.-P. Rivet. Lattice Gas Hydrodynamics in Two and Three Dimensions. *Complex Systems*, 1:649–707, 1987.
18. S. Geller, M. Krafczyk, J. Tölke, S. Turek, and J. Hron. Benchmark Computations based on Lattice Boltzmann, Finite Element, and Finite Volume Methods for Laminar Flows. *Submitted to Computers and Fluids*, 2004. Also available as *Ergebisbericht 274 des Fachbereichs Mathematik der Universtität Dortmund*, `http://www.mathematik.uni-dortmund.de/lsiii/german/preprintfb.html`.
19. I. Ginzburg and P. M. Adler. Boundary Flow Condition Analysis for Three-Dimensional Lattice Boltzmann Model. *J. Phys. II France*, 4:191–214, 1994.
20. I. Ginzburg and D. d'Humières. Multireflection Boundary Conditions for Lattice Boltzmann Models. *Phys. Rev. E*, 68(6):066614_30, 2003.
21. A. S. Glassner. *An Introduction to Ray Tracing*. Harlekijn, 1989.
22. X. He and G. D. Doolen. Thermodynamic Foundations of Kinetic Theory and Lattice Boltzmann Models for Multiphase Flows. *J. Stat. Phys.*, 107(1-2):309–328, 2002.
23. X. He and L.-S. Luo. A Priori Derivation of the Lattice Boltzmann Equation. *Phys. Rev. E*, 55(6):R6333–R6336, 1997.
24. F. Hülsemann, M. Kowarschik, M. Mohr, and U. Rüde. Parallel geometric multigrid. In A. M. Bruaset and A. Tveito, editors, *Numerical Solution of Partial Differential Equations on Parallel Computers*, volume 51 of *Lecture Notes in Computational Science and Engineering*, pages 165–208. Springer-Verlag, 2005.
25. K. Iglberger. Cache Optimizations for the Lattice Boltzmann Method in 3D. Studienarbeit, Lehrstuhl für Informatik 10 (Systemsimulation), Universität Erlangen-Nürnberg, 2003.

26. D. Kehrwald. *Numerical Analysis of Immiscible Lattice BGK*. PhD thesis, Fachbereich Mathematik, Universität Kaiserslautern, 2002.
27. M. Kowarschik. *Data Locality Optimizations for Iterative Numerical Algorithms and Cellular Automata on Hierarchical Memory Architectures*. PhD thesis, Universität Erlangen-Nürnberg, Technische Fakultät, 2004.
28. A. J. C. Ladd. Numerical Simulations of Particulate Suspensions via a Discrete Boltzmann Equation. Part 1. Theoretical Foundation. *J. Fluid Mech.*, 271:285–309, 1994.
29. G. Marmitt, A. Kleer, I. Wald, H. Friedrich, and P. Slusallek. Fast and Accurate Ray-Voxel Intersection Techniques for Iso-Surface Ray Tracing. In *Vision, Modelling, and Visualization 2003 (VMV), November 16-18, Stanford (CA), USA*, 2004.
30. G. R. McNamara and G. Zanetti. Use of the Boltzmann Equation to Simulate Lattice Gas Automata. *Phys. Rev. Lett.*, 61:2332–2335, 1988.
31. W. Miller and S. Succi. A Lattice Boltzmann Model for Anisotropic Crystal Growth from Melt. *J. Stat. Phys.*, 107(1-2):173–186, 2002.
32. C. Pan, J. Prins, and C. T. Miller. A High-Performance Lattice Boltzmann Implementation to Model Flow in Porous Media. *Comp. Phys. Com.*, 158(1):89–105, 2004.
33. T. Pohl, F. Deserno, N. Thürey, U. Rüde, P. Lammers, G. Wellein, and T. Zeiser. Performance Evaluation of Parallel Large-Scale Lattice Boltzmann Applications on Three Supercomputing Architectures. In *Supercomputing Conference*, 2004.
34. T. Pohl, M. Kowarschik, J. Wilke, K. Iglberger, and U. Rüde. Optimization and Profiling of the Cache Performance of Parallel Lattice Boltzmann Codes. *Parallel Processing Letters*, 13(4):549–560, 2003.
35. Y. H. Qian, D. d'Humières, and P. Lallemand. Lattice BGK Models for Navier-Stokes Equation. *Europhys. Lett.*, 17(6):479–484, 1992.
36. D. H. Rothman and S. Zaleski. *Lattice Gas Cellular Automata. Simple models of Complex Hydrodynamics*. Cambridge University Press, 1997.
37. M. Schulz, M. Krafczyk, J. Tölke, and E. Rank. Parallelization Strategies and Efficiency of CFD Computations in Complex Geometries Using Lattice Boltzmann Methods on High Performance Computers. In M. Breuer, F. Durst, and C. Zenger, editors, *High Performance Scientific and Engineering Computing*, pages 115–122, Berlin, 2001. Springer.
38. P. Shirley and K. Sung. *Graphics Gems III*. Morgan Kaufmann, 1994.
39. S. Succi. *The Lattice Boltzmann Equation – For Fluid Dynamics and Beyond*. Clarendon Press, 2001.
40. I. Wald, C. Benthin, M. Wagner, and P. Slusallek. Interactive Rendering with Coherent Ray Tracing. In A. Chalmers and T.-M. Rhyne, editors, *Computer Graphics Forum (Proceedings of EUROGRAPHICS 2001)*, volume 20. Blackwell Publishers, Oxford, 2001.
41. G. Wellein, T. Zeiser, S. Donath, and G. Hager. On the Single Processor Performance of Simple Lattice Boltzmann Kernels. *Computers & Fluids*, accepted, 2004.
42. J. Wilke. Cache Optimizations for the Lattice Boltzmann Method in 2D. Studienarbeit, Lehrstuhl für Informatik 10 (Systemsimulation), Universität Erlangen-Nürnberg, 2002.
43. J. Wilke, T. Pohl, M. Kowarschik, and U. Rüde. Cache Performance Optimization for Parallel Lattice Boltzmann Code in 2D. Technical Report 03-3, Lehrstuhl für Informatik 10 (Systemsimulation), Universität Erlangen-Nürnberg, 2003.
44. D. A. Wolf-Gladrow. *Lattice Gas Cellular Automata and Lattice Boltzmann Models*, volume 1725 of *Lecture Notes in Mathematics*. Springer, Berlin, 2000.
45. S. Wolfram. Cellular Automaton Fluids 1: Basic Theory. *J. Stat. Phys.*, 3/4:471–526, 1986.

46. D. Yu, R. Mei, L.-S. Luo, and W. Shyy. Viscous Flow Computations with the Method of Lattice Boltzmann Equation. *Progr. Aero. Sci.*, 39:329–367, 2003.
47. D. Yu, R. Mei, and W. Shyy. A Multiblock Lattice Boltzmann Method for Viscous Fluid Flows. *Int. J. Numer. Meth. Fluids*, 39(2):99–120, 2002.

A

Color Figures

Partitioning and Dynamic Load Balancing for the Numerical Solution of Partial Differential Equations

J. D. Teresco, K. D. Devine, and J. E. Flaherty

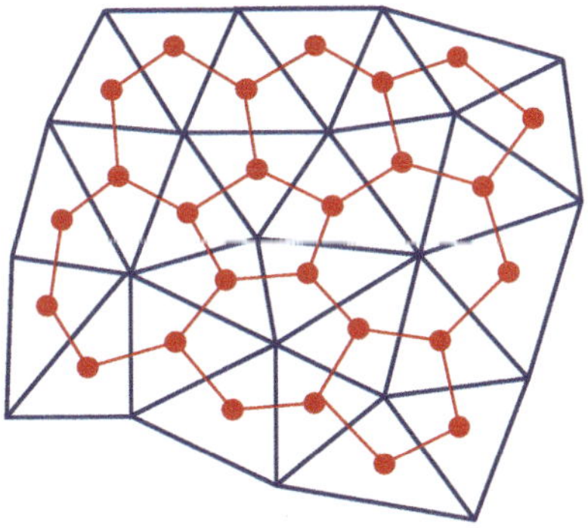

Fig. A.1. Example two-dimensional mesh from Figure 2.1 (left) with its induced graph. (This is a color version of Figure 2.6 on page 64).

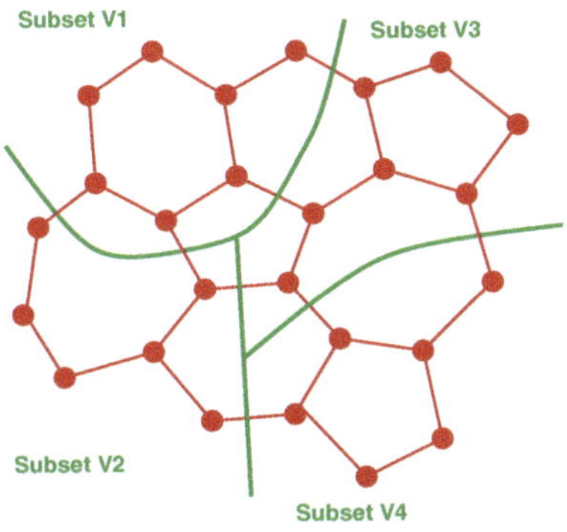

Fig. A.2. Four-way partitioning of the graph from Figure 2.6. (This is a color version of Figure 2.7 on page 64).

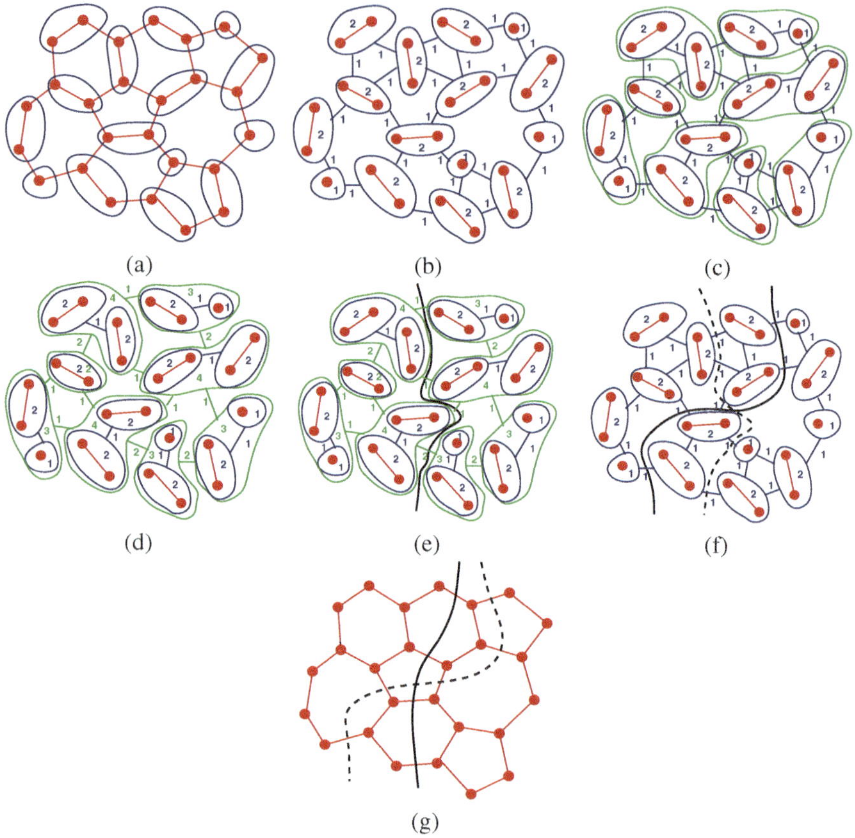

Fig. A.3. Multilevel partitioning of the induced graph of Figure 2.6. Vertex matching in (a) leads to the coarse graph in (b). A second round of vertex matching in (c) produces the coarse graph in (d). This coarsest graph is partitioned in (e). The graph is uncoarsened one level and the partitioning is optimized in (f). The second level of uncoarsening, and another round of local optimization on this partitioning produces the final two-way partitioning shown in (g). (This is a color version of Figure 2.9 on page 67).

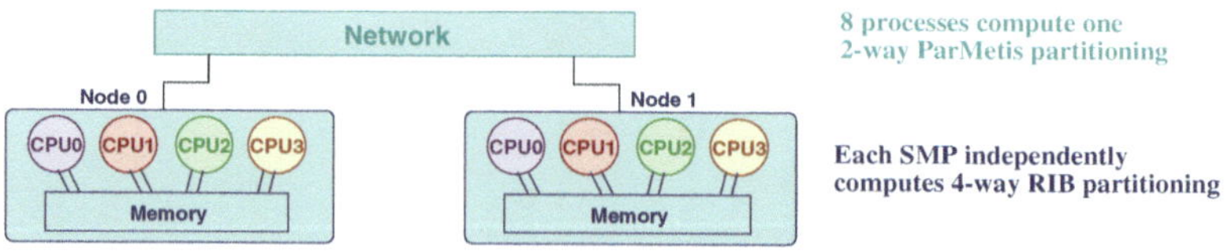

Fig. A.4. Hierarchical balancing algorithm selection for two 4-way SMP nodes connected by a network. (This is a color version of Figure 2.10 on page 79).

Domain Decomposition Techniques

L. Formaggia, M. Sala, and F. Saleri

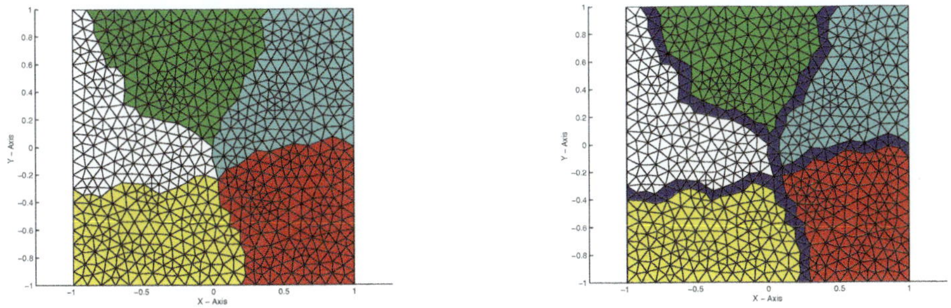

Fig. A.5. Example of element-oriented (left) and vertex-oriented (right) decomposition in the case of a partition of Ω into several subdomains. (This is a color version of Figure 4.2 on page 140).

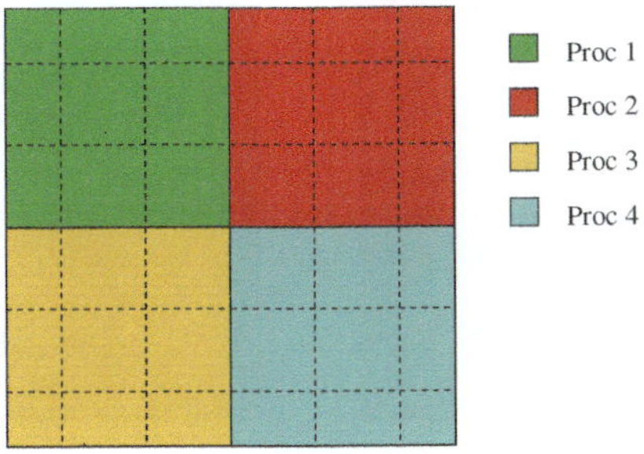

Fig. A.6. Example of two-level decomposition. First level in continuous line, and second level in dashed line. Typically, each first-level decomposition subdomain is given to a different processor. (This is a color version of Figure 4.4 on page 152).

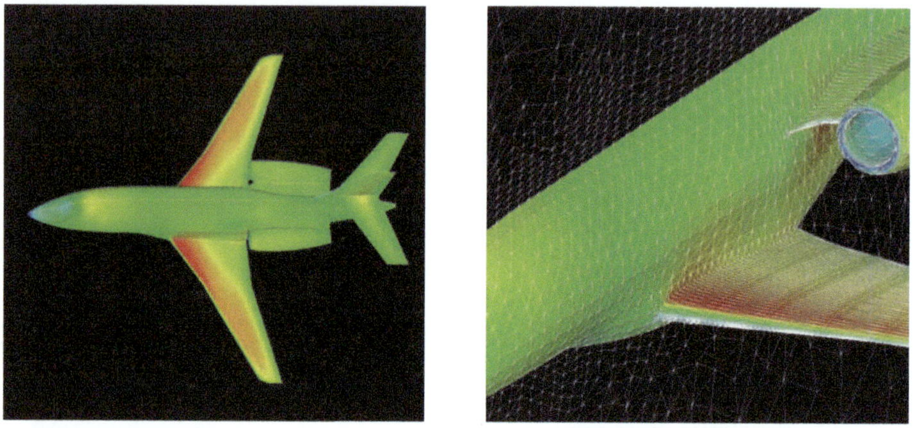

Fig. A.7. Pressure coefficient contours for FALCON_45k. (This is a color version of Figure 4.5 on page 156).

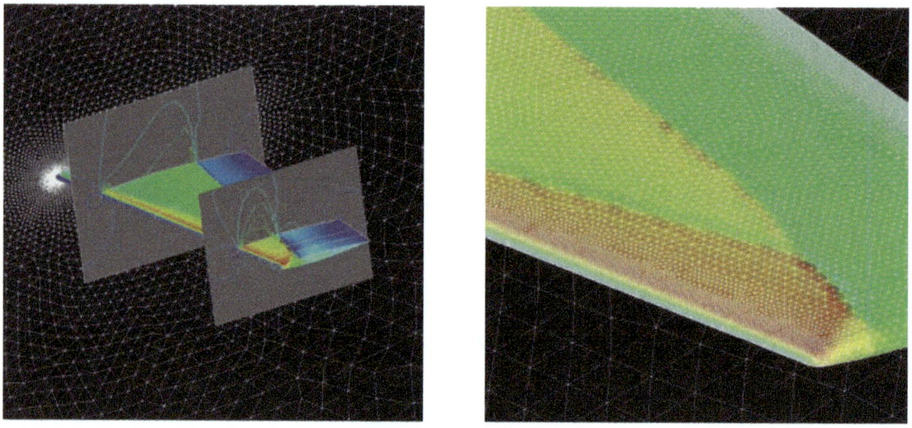

Fig. A.8. Mach number contours for M6_316K. (This is a color version of Figure 4.6 on page 156).

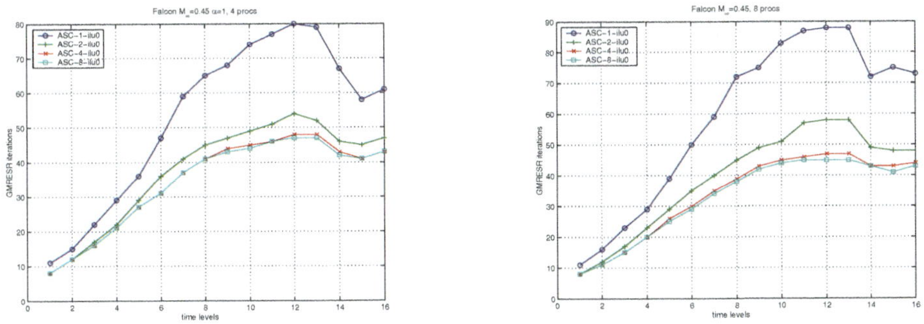

Fig. A.9. FALCON_45k. Convergence history with 4 (left) and 8 processors (right) with the ASC preconditioner, for different values of L. (This is a color version of Figure 4.7 on page 157).

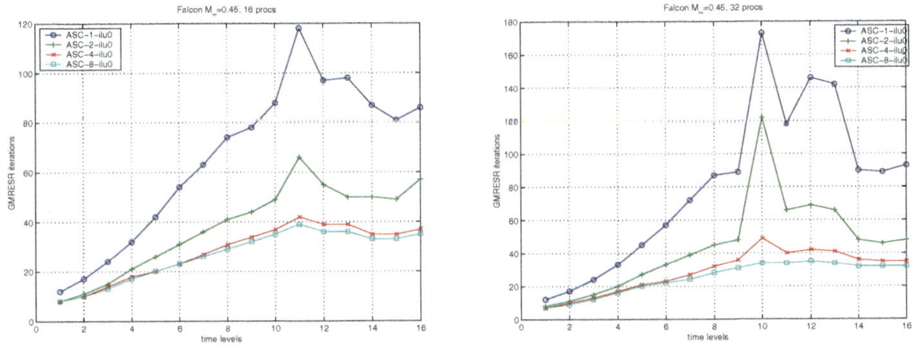

Fig. A.10. FALCON_45k. Convergence history with 16 processors (left) and 32 processors (right) with the ASC preconditioner, for different values of L. (This is a color version of Figure 4.8 on page 158).

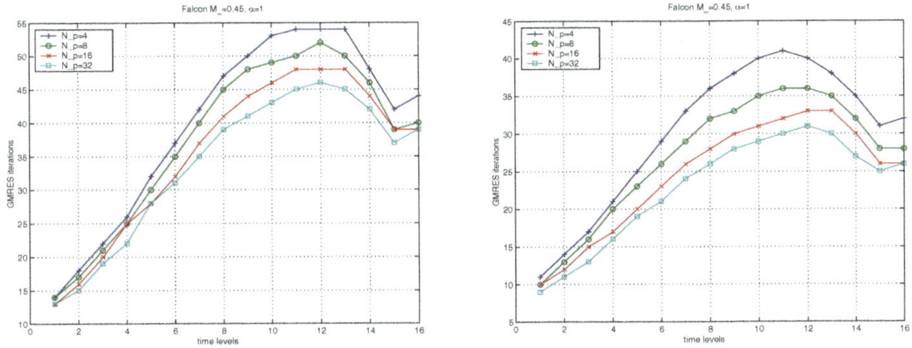

Fig. A.11. FALCON_45k. Iterations to converge with different values of N_p for $P_{\mathrm{ACM},1}$ (left) and $P_{\mathrm{ACM},2}$ (right). (This is a color version of Figure 4.9 on page 158).

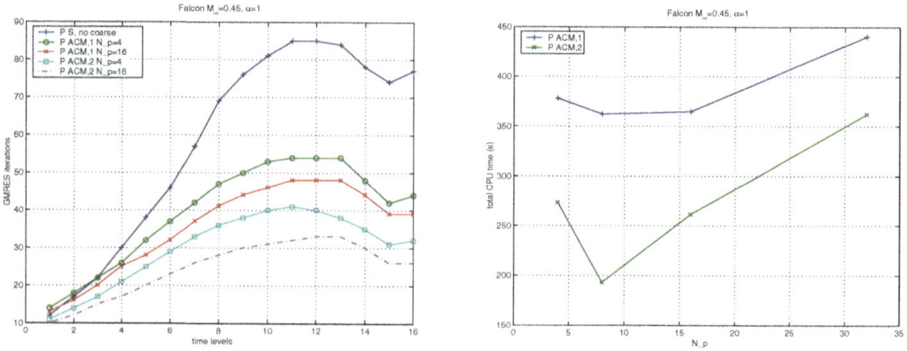

Fig. A.12. FALCON_45k. Comparison among different preconditioners (left) and CPU time, in seconds (right). 16 SGI-Origin3000 processors. (This is a color version of Figure 4.10 on page 159).

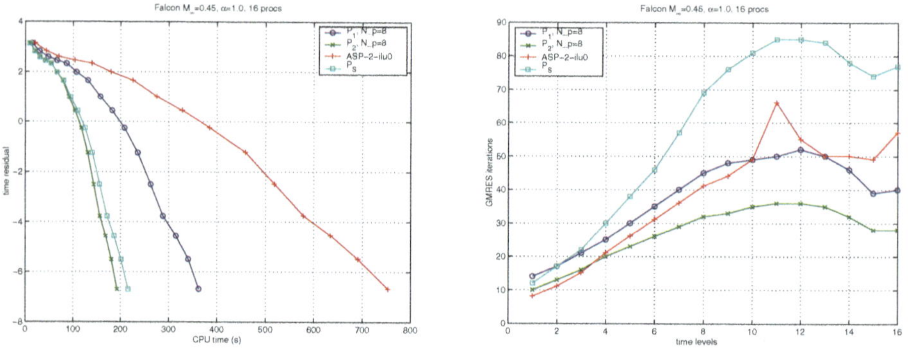

Fig. A.13. FALCON_45k. Residual versus CPU-time (left) and iterations to converge at each time level (right), using 16 SGI-Origin3000 processors. (This is a color version of Figure 4.11 on page 159).

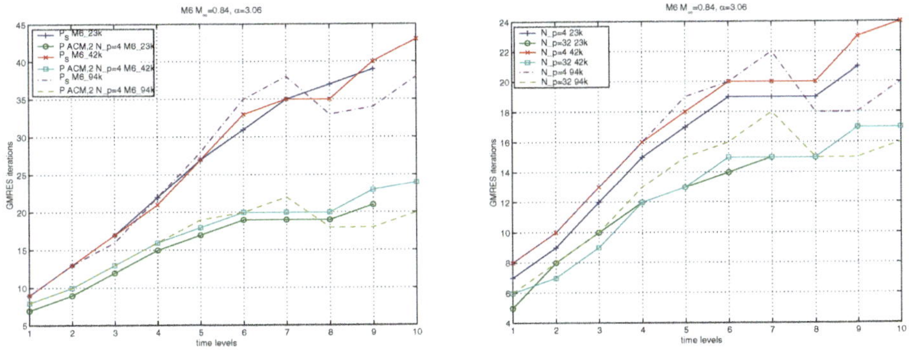

Fig. A.14. M6_94k. Iterations to converge with P_S and $P_{\mathrm{ACM},2}$ (left), and iterations to converge with $P_{\mathrm{ACM},2}$ (right) using two different values of $N_p$and 16 processors. (This is a color version of Figure 4.12 on page 160).

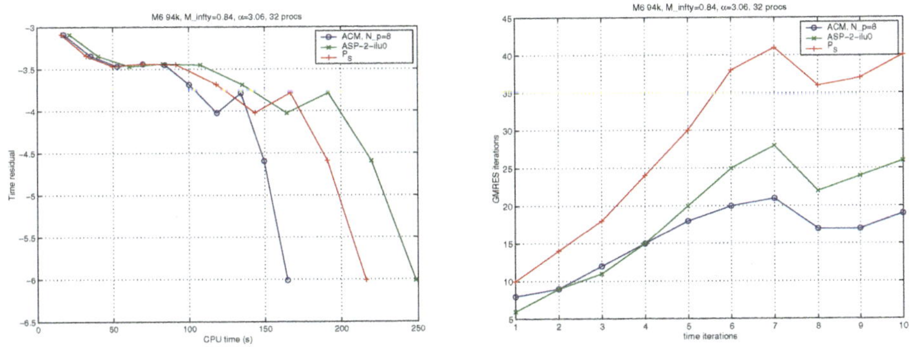

Fig. A.15. M6_94k. Residual versus CPU-time (right) and iterations to converge at each time level (right), using 32 processors. (This is a color version of Figure 4.13 on page 160).

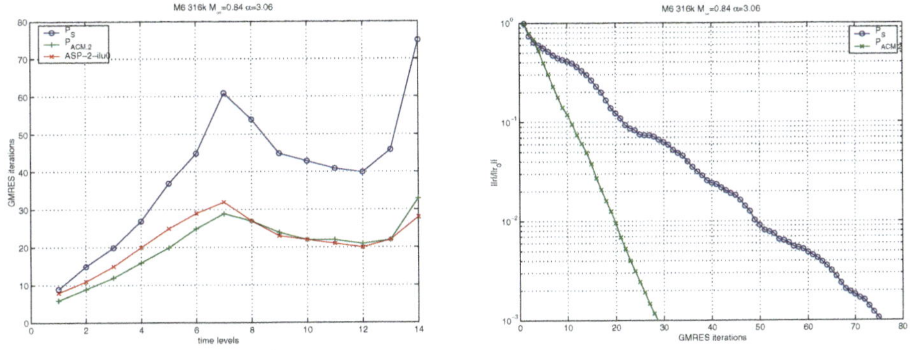

Fig. A.16. M6_316k. Iterations at each time level (left) and converge history at the 14th time step (right). (This is a color version of Figure 4.14 on page 161).

Parallel Mesh Generation

N. Chrisochoides

Fig. A.17. Domain decomposition of the continuous geometry [52] and the discrete geometry [17] of a cross section of a rocket pipe. (This is a color version of Figure 7.1 on page 239).

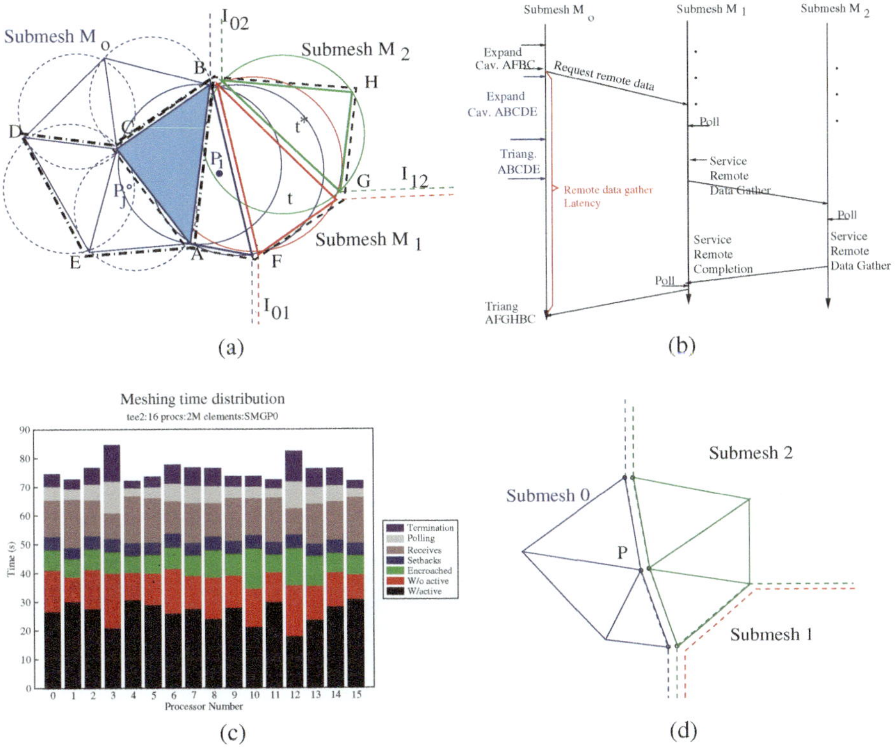

Fig. A.18. a) cavity extension beyond submesh interfaces, b) time diagram with concurrent point insertion, c) a breakdown of execution time for PODM, and finally d) the refinement of a cavity with simultaneous distribution of the newly created elements. (This is a color version of Figure 7.4 on page 243).

Parallel PDE-Based Simulation Using the Common Component Architecture

L. C. McInnes et al.

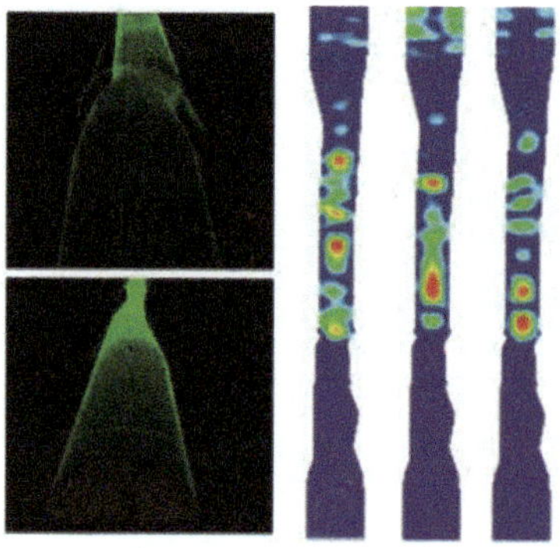

Fig. A.19. State-of-the-art simulation tools are used to help design the next generation of accelerator facilities. *(Left)*: Mesh generated for the PEP-II interaction region using the CUBIT mesh generation package. Image courtesy of Tim Tautges of Sandia National Laboratories. *(Right)*: Excited fields computed using Tau3P. Image courtesy of the numerics team at SLAC. (This is a color version of Figure 10.2 on page 330).

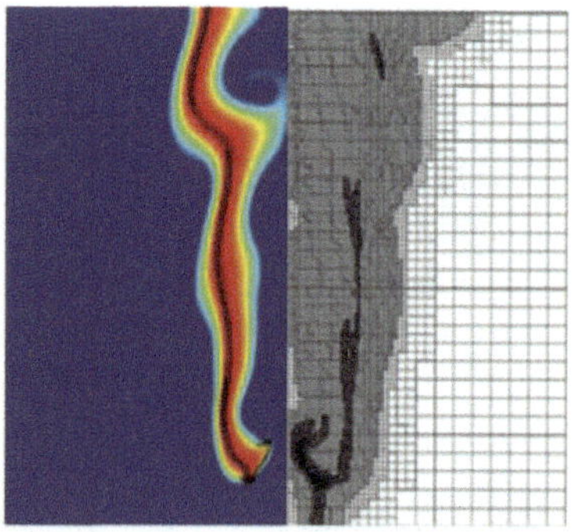

Fig. A.20. A 10-cm-high pulsating methane-air jet flame, computed on an adaptive mesh. On the left is the temperature field with a black contour showing regions of high heat release rates. On the right is the adaptive mesh, in which regions corresponding to the jet shear layer are refined the most. (This is a color version of Figure 10.4 on page 333).

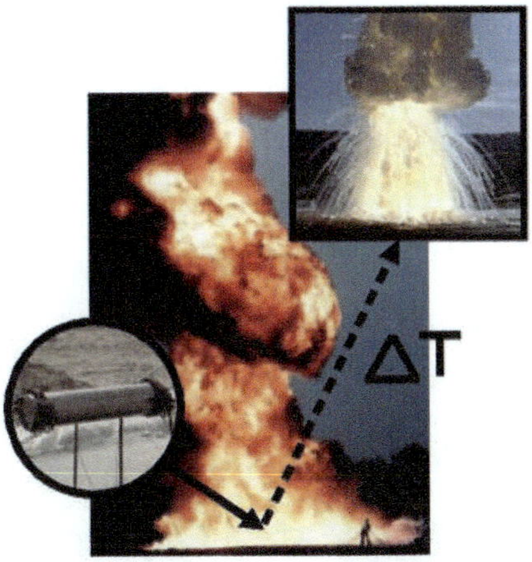

Fig. A.21. A typical C-SAFE problem involving hydrocarbon fires and explosions of energetic materials. This simulation involves fluid dynamics, structural mechanics, and chemical reactions in both the flame and the explosive. Accurate simulations of these events can lead to a better understanding of high-energy explosives, can help evaluate the design of shipping and storage containers for these materials, and can help officials determine a response to various accident scenarios. The fire image is courtesy of Schonbucher Institut for Technische Chemie I der Universitat Stuttgart, and the images of the container and explosion are courtesy of Eric Eddings of the University of Utah. (This is a color version of Figure 10.5 on page 334).

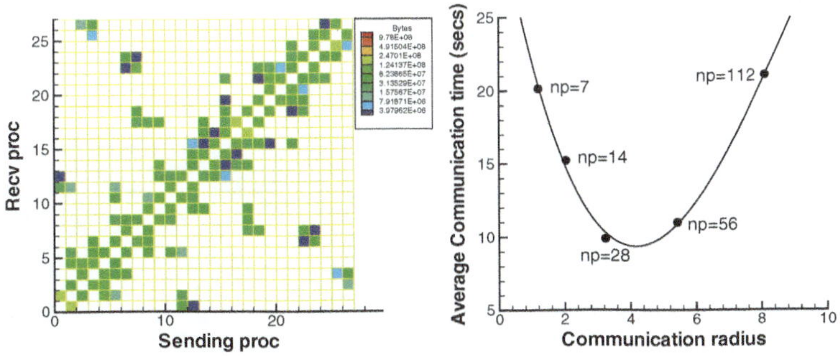

Fig. A.22. *(Left)*: Communication patterns for 28 processors at timestep 40. *(Right)*: Communication costs as a function of the communication radius at timestep 40. (This is a color version of Figure 10.17 on page 367).

Full-Scale Simulation of Cardiac Electrophysioology on Parallel Computers

X. Cai and G. T. Lines

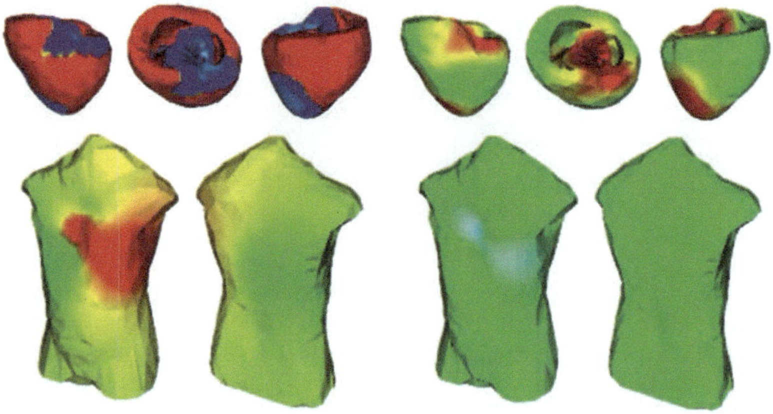

t=30ms t=200ms

Fig. A.23. Snapshots from two time levels of a simulation of the electrical field in the human heart and torso. At each time level, the electrical potential distribution on the heart surface is shown at three different angles, while the distribution on the torso surface is shown at two different angles. (This is a color version of Figure 11.1 on page 387).

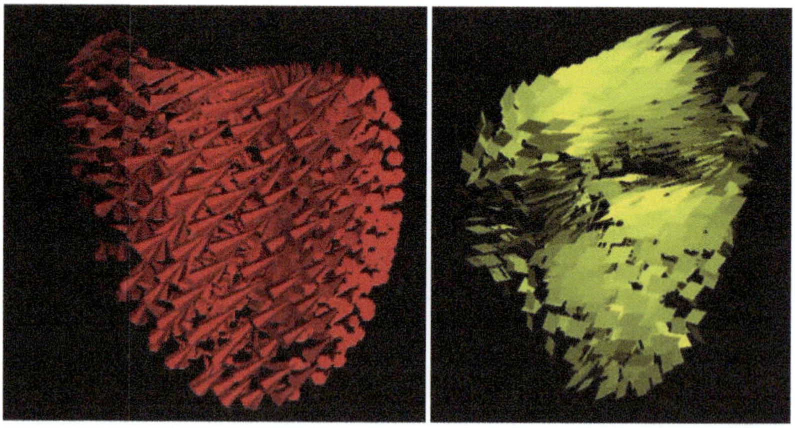

Fig. A.24. The orientation of the muscle fibers (left) and sheet layers (right) in the heart. (This is a color version of Figure 11.2 on page 391).

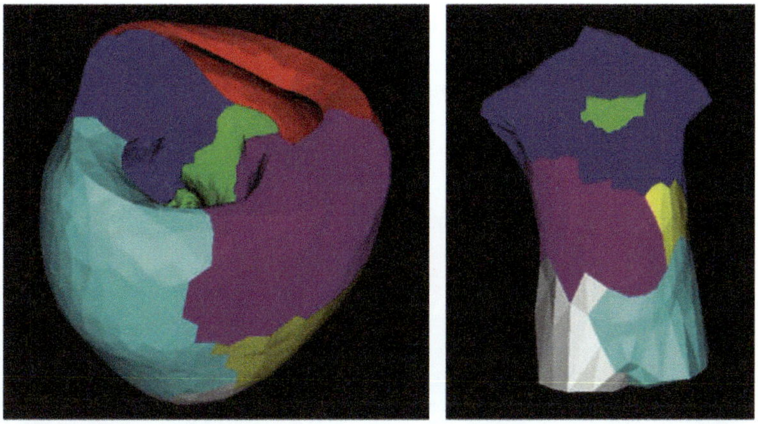

Fig. A.25. An example of partitioning an unstructured heart mesh (left) and an unstructured torso mesh (right). (This is a color version of Figure 11.6 on page 402).

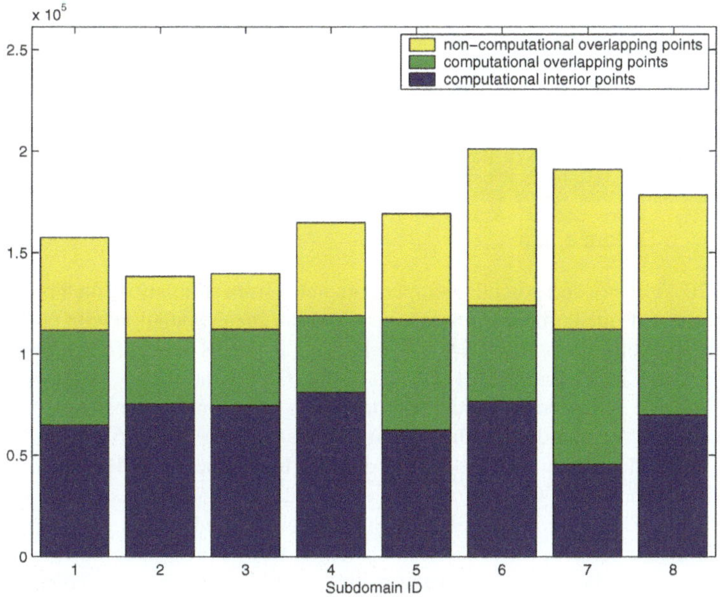

Fig. A.26. The effect of applying a disjoint re-distribution to the Ω mesh points, where $N_{\Omega} = 919,851$ and the number of subdomains is 8. (This is a color version of Figure 11.7 on page 406).

Developing a Geodynamics Simulator with PETSc

M. G. Knepley, R. F. Katz, and B. Smith

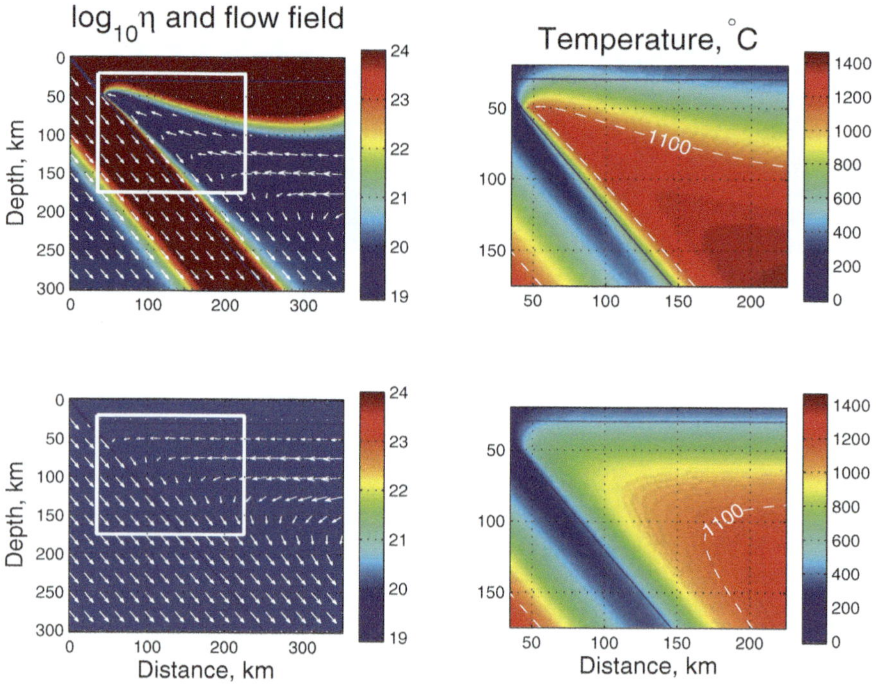

Fig. A.27. 2D viscosity and potential temperature fields from simulations on 8 processors with 230,112 degrees of freedom. Panels in the top row are from a simulation with α=1 in equation (12.1). Panels in bottom row have α=0. The white box in panels (a) and (c) shows the region in which temperature is plotted in panels (b) and (d). **(a)** Colors show $\log_{10}$ of the viscosity field. Note that there are more than five orders of magnitude variation in viscosity. Arrows show the flow direction and magnitude (the slab is being subducted at a rate of 6 cm/year). Upwelling is evident in the flow field near the wedge corner. **(b)** Temperature field from the variable viscosity simulation; 1100°C isotherm is shown as a dashed line. **(c)** (Constant) viscosity and flow field from the isoviscous simulation. Strong flow is predicted at the base of the crust despite the low-temperature rock there. No upwelling flow is predicted. **(d)** Temperature field from isoviscous simulation. Note that the mantle wedge corner is much colder than in (b). (This is a color version of Figure 12.28 on page 436).

Parallel Lattice Boltzmann Methods for CFD Applications

C. Körner, T. Pohl, U. Rüde, N. Thürey, and T. Zeiser

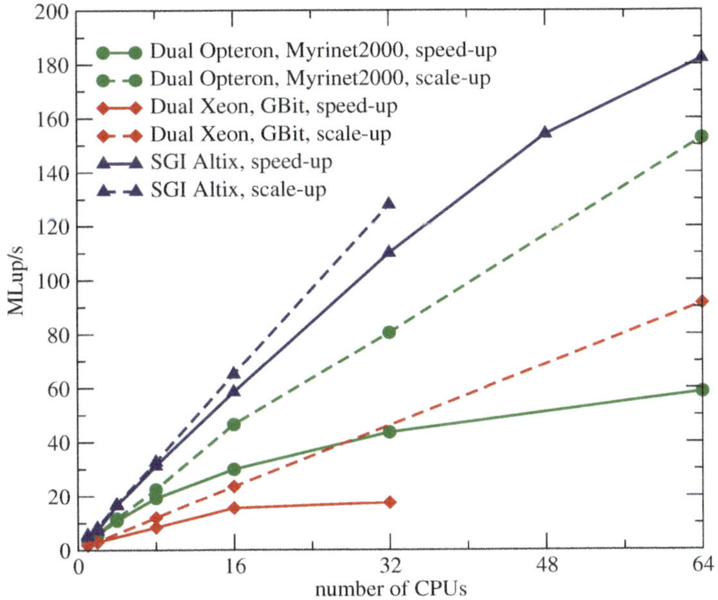

Fig. A.28. Scalability tests for modern cluster configurations. The domain size is $256 \times 129 \times 128$ for speed-up (fixed domain size) and 128^3 per processor for scale-up tests (fixed CPU load by scaling the domain size). For reference the corresponding results of a shared memory system (SGI Altix 3000) are given. The common performance unit MLup/s (million lattice site updates per second) has been used. (This is a color version of Figure 13.10 on page 455).

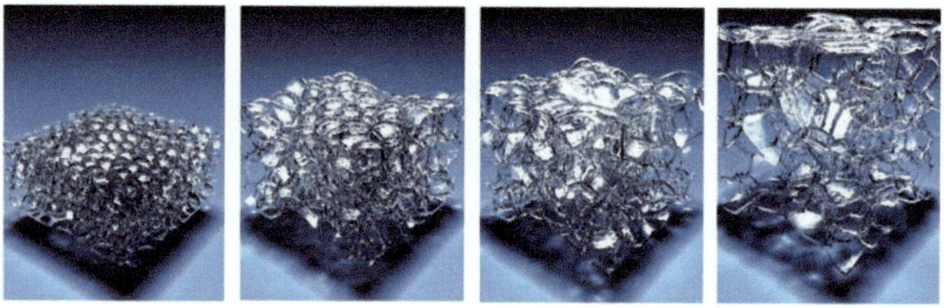

Fig. A.29. 3D foam: The bubbles grow and coalescence occurs. The disjoining pressure Π stabilizes the foam and eventually a polygonal structure develops (initial number of bubbles: 1000; system size: $120 \times 120 \times 140$; $\tau = 0.8$; g = 0; $\sigma = 0.01$; $c_\Pi = 0.006$). (This is a color version of Figure 13.11 on page 456).

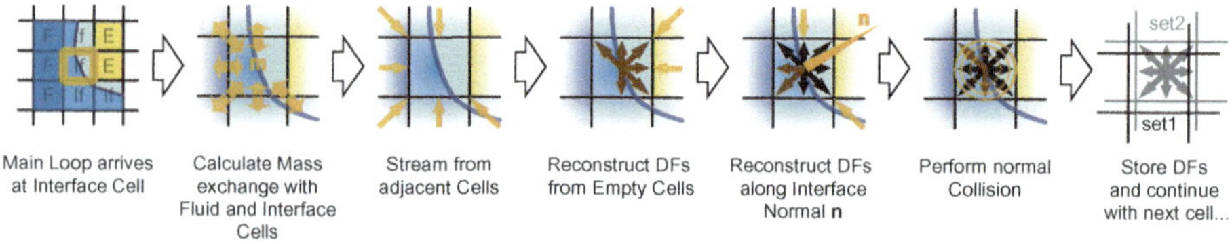

Fig. A.30. An overview over the steps for an interface cell are shown. Note that not only distribution function from empty cells, but also those along the interface normal are reconstructed. (This is a color version of Figure 13.14 on page 459).

Editorial Policy

§1. Volumes in the following three categories will be published in LNCSE:

i) Research monographs
ii) Lecture and seminar notes
iii) Conference proceedings

Those considering a book which might be suitable for the series are strongly advised to contact the publisher or the series editors at an early stage.

§2. Categories i) and ii). These categories will be emphasized by Lecture Notes in Computational Science and Engineering. **Submissions by interdisciplinary teams of authors are encouraged.** The goal is to report new developments – quickly, informally, and in a way that will make them accessible to non-specialists. In the evaluation of submissions timeliness of the work is an important criterion. Texts should be well-rounded, well-written and reasonably self-contained. In most cases the work will contain results of others as well as those of the author(s). In each case the author(s) should provide sufficient motivation, examples, and applications. In this respect, Ph.D. theses will usually be deemed unsuitable for the Lecture Notes series. Proposals for volumes in these categories should be submitted either to one of the series editors or to Springer-Verlag, Heidelberg, and will be refereed. A provisional judgment on the acceptability of a project can be based on partial information about the work: a detailed outline describing the contents of each chapter, the estimated length, a bibliography, and one or two sample chapters – or a first draft. A final decision whether to accept will rest on an evaluation of the completed work which should include

- at least 100 pages of text;
- a table of contents;
- an informative introduction perhaps with some historical remarks which should be accessible to readers unfamiliar with the topic treated;
- a subject index.

§3. Category iii). Conference proceedings will be considered for publication provided that they are both of exceptional interest and devoted to a single topic. One (or more) expert participants will act as the scientific editor(s) of the volume. They select the papers which are suitable for inclusion and have them individually refereed as for a journal. Papers not closely related to the central topic are to be excluded. Organizers should contact Lecture Notes in Computational Science and Engineering at the planning stage.

In exceptional cases some other multi-author-volumes may be considered in this category.

§4. Format. Only works in English are considered. They should be submitted in camera-ready form according to Springer-Verlag's specifications.
Electronic material can be included if appropriate. Please contact the publisher.
Technical instructions and/or TeX macros are available via
http://www.springer.com/sgw/cda/frontpage/0,11855,5-40017-2-71391-0,00.html
The macros can also be sent on request.

General Remarks

Lecture Notes are printed by photo-offset from the master-copy delivered in camera-ready form by the authors. For this purpose Springer-Verlag provides technical instructions for the preparation of manuscripts. See also *Editorial Policy*.

Careful preparation of manuscripts will help keep production time short and ensure a satisfactory appearance of the finished book.

The following terms and conditions hold:

Categories i), ii), and iii):
Authors receive 50 free copies of their book. No royalty is paid. Commitment to publish is made by letter of intent rather than by signing a formal contract. Springer-Verlag secures the copyright for each volume.

For conference proceedings, editors receive a total of 50 free copies of their volume for distribution to the contributing authors.

All categories:
Authors are entitled to purchase further copies of their book and other Springer mathematics books for their personal use, at a discount of 33,3 % directly from Springer-Verlag.

Addresses:

Timothy J. Barth
NASA Ames Research Center
NAS Division
Moffett Field, CA 94035, USA
e-mail: barth@nas.nasa.gov

Michael Griebel
Institut für Numerische Simulation
der Universität Bonn
Wegelerstr. 6
53115 Bonn, Germany
e-mail: griebel@ins.uni-bonn.de

David E. Keyes
Department of Applied Physics
and Applied Mathematics
Columbia University
200 S. W. Mudd Building
500 W. 120th Street
New York, NY 10027, USA
e-mail: david.keyes@columbia.edu

Risto M. Nieminen
Laboratory of Physics
Helsinki University of Technology
02150 Espoo, Finland
e-mail: rni@fyslab.hut.fi

Dirk Roose
Department of Computer Science
Katholieke Universiteit Leuven
Celestijnenlaan 200A
3001 Leuven-Heverlee, Belgium
e-mail: dirk.roose@cs.kuleuven.ac.be

Tamar Schlick
Department of Chemistry
Courant Institute of Mathematical Sciences
New York University
and Howard Hughes Medical Institute
251 Mercer Street
New York, NY 10012, USA
e-mail: schlick@nyu.edu

Mathematics Editor at Springer: Martin Peters
Springer-Verlag, Mathematics Editorial IV
Tiergartenstrasse 17
D-69121 Heidelberg, Germany
Tel.: *49 (6221) 487-8185
Fax: *49 (6221) 487-8355
e-mail: martin.peters@springer.com

Lecture Notes in Computational Science and Engineering

Vol. 18 U. van Rienen, M. Günther, D. Hecht (eds.), *Scientific Computing in Electrical Engineering.* Proceedings of the 3rd International Workshop, August 20-23, 2000, Warnemünde, Germany. 2001. XII, 428 pp. Softcover. 3-540-42173-4

Vol. 19 I. Babuška, P. G. Ciarlet, T. Miyoshi (eds.), *Mathematical Modeling and Numerical Simulation in Continuum Mechanics.* Proceedings of the International Symposium on Mathematical Modeling and Numerical Simulation in Continuum Mechanics, September 29 - October 3, 2000, Yamaguchi, Japan. 2002. VIII, 301 pp. Softcover. 3-540-42399-0

Vol. 20 T. J. Barth, T. Chan, R. Haimes (eds.), *Multiscale and Multiresolution Methods.* Theory and Applications. 2002. X, 389 pp. Softcover. 3-540-42420-2

Vol. 21 M. Breuer, F. Durst, C. Zenger (eds.), *High Performance Scientific and Engineering Computing.* Proceedings of the 3rd International FORTWIHR Conference on HPSEC, Erlangen, March 12-14, 2001. 2002. XIII, 408 pp. Softcover. 3-540-42946-8

Vol. 22 K. Urban, *Wavelets in Numerical Simulation.* Problem Adapted Construction and Applications. 2002. XV, 181 pp. Softcover. 3-540-43055-5

Vol. 23 L. F. Pavarino, A. Toselli (eds.), *Recent Developments in Domain Decomposition Methods.* 2002. XII, 243 pp. Softcover. 3-540-43413-5

Vol. 24 T. Schlick, H. H. Gan (eds.), *Computational Methods for Macromolecules: Challenges and Applications.* Proceedings of the 3rd International Workshop on Algorithms for Macromolecular Modeling, New York, October 12-14, 2000. 2002. IX, 504 pp. Softcover. 3-540-43756-8

Vol. 25 T. J. Barth, H. Deconinck (eds.), *Error Estimation and Adaptive Discretization Methods in Computational Fluid Dynamics.* 2003. VII, 344 pp. Hardcover. 3-540-43758-4

Vol. 26 M. Griebel, M. A. Schweitzer (eds.), *Meshfree Methods for Partial Differential Equations.* 2003. IX, 466 pp. Softcover. 3-540-43891-2

Vol. 27 S. Müller, *Adaptive Multiscale Schemes for Conservation Laws.* 2003. XIV, 181 pp. Softcover. 3-540-44325-8

Vol. 28 C. Carstensen, S. Funken, W. Hackbusch, R. H. W. Hoppe, P. Monk (eds.), *Computational Electromagnetics.* Proceedings of the GAMM Workshop on "Computational Electromagnetics", Kiel, Germany, January 26-28, 2001. 2003. X, 209 pp. Softcover. 3-540-44392-4

Vol. 29 M. A. Schweitzer, *A Parallel Multilevel Partition of Unity Method for Elliptic Partial Differential Equations.* 2003. V, 194 pp. Softcover. 3-540-00351-7

Vol. 30 T. Biegler, O. Ghattas, M. Heinkenschloss, B. van Bloemen Waanders (eds.), *Large-Scale PDE-Constrained Optimization.* 2003. VI, 349 pp. Softcover.
3-540-05045-0

Vol. 31 M. Ainsworth, P. Davies, D. Duncan, P. Martin, B. Rynne (eds.), *Topics in Computational Wave Propagation.* Direct and Inverse Problems. 2003. VIII, 399 pp. Softcover. 3-540-00744-X

Vol. 32 H. Emmerich, B. Nestler, M. Schreckenberg (eds.), *Interface and Transport Dynamics.* Computational Modelling. 2003. XV, 432 pp. Hardcover. 3-540-40367-1

Vol. 33 H. P. Langtangen, A. Tveito (eds.), *Advanced Topics in Computational Partial Differential Equations.* Numerical Methods and Diffpack Programming. 2003. XIX, 658 pp. Softcover. 3-540-01438-1

Vol. 34 V. John, *Large Eddy Simulation of Turbulent Incompressible Flows.* Analytical and Numerical Results for a Class of LES Models. 2004. XII, 261 pp. Softcover. 3-540-40643-3

Vol. 35 E. Bänsch (ed.), *Challenges in Scientific Computing - CISC 2002.* Proceedings of the Conference *Challenges in Scientific Computing*, Berlin, October 2-5, 2002. 2003. VIII, 287 pp. Hardcover. 3-540-40887-8

Vol. 36 B. N. Khoromskij, G. Wittum, *Numerical Solution of Elliptic Differential Equations by Reduction to the Interface.* 2004. XI, 293 pp. Softcover. 3-540-20406-7

Vol. 37 A. Iske, *Multiresolution Methods in Scattered Data Modelling.* 2004. XII, 182 pp. Softcover. 3-540-20479-2

Vol. 38 S.-I. Niculescu, K. Gu (eds.), *Advances in Time-Delay Systems.* 2004. XIV, 446 pp. Softcover. 3-540-20890-9

Vol. 39 S. Attinger, P. Koumoutsakos (eds.), *Multiscale Modelling and Simulation.* 2004. VIII, 277 pp. Softcover. 3-540-21180-2

Vol. 40 R. Kornhuber, R. Hoppe, J. Périaux, O. Pironneau, O. Wildlund, J. Xu (eds.), *Domain Decomposition Methods in Science and Engineering.* 2005. XVIII, 690 pp. Softcover. 3-540-22523-4

Vol. 41 T. Plewa, T. Linde, V.G. Weirs (eds.), *Adaptive Mesh Refinement – Theory and Applications.* 2005. XIV, 552 pp. Softcover. 3-540-21147-0

Vol. 42 A. Schmidt, K.G. Siebert, *Design of Adaptive Finite Element Software.* The Finite Element Toolbox ALBERTA. 2005. XII, 322 pp. Hardcover. 3-540-22842-X

Vol. 43 M. Griebel, M.A. Schweitzer (eds.), *Meshfree Methods for Partial Differential Equations II.* 2005. XIII, 303 pp. Softcover. 3-540-23026-2

Vol. 44 B. Engquist, P. Lötstedt, O. Runborg (eds.), *Multiscale Methods in Science and Engineering.* 2005. XII, 291 pp. Softcover. 3-540-25335-1

Vol. 45 P. Benner, V. Mehrmann, D.C. Sorensen (eds.), *Dimension Reduction of Large-Scale Systems.* 2005. XII, 402 pp. Softcover. 3-540-24545-6

Vol. 46 D. Kressner (ed.), *Numerical Methods for General and Structured Eigenvalue Problems.* 2005. XIV, 258 pp. Softcover. 3-540-24546-4

Vol. 47 A. Boriçi, A. Frommer, B. Joó, A. Kennedy, B. Pendleton (eds.), *QCD and Numerical Analysis III.* 2005. XIII, 201 pp. Softcover. 3-540-21257-4

Vol. 48 F. Graziani (ed.), *Computational Methods in Transport.* 2006. VIII, 524 pp. Softcover. 3-540-28122-3

Vol. 49 B. Leimkuhler, C. Chipot, R. Elber, A. Laaksonen, A. Mark, T. Schlick, C. Schütte, R. Skeel (eds.), *New Algorithms for Macromolecular Simulation.* 2006. XVI, 376 pp. Softcover. 3-540-25542-7

Vol. 50 M. Bücker, G. Corliss, P. Hovland, U. Naumann, B. Norris (eds.), *Automatic Differentiation: Applications, Theory, and Implementations.* 2006. XVIII, 362 pp. Softcover. 3-540-28403-6

Vol. 51 A.M. Bruaset, A. Tveito (eds.), *Numerical Solution of Partial Differential Equations on Parallel Computers* 2006. XII, 482 pp. Softcover. 3-540-29076-1

For further information on these books please have a look at our mathematics catalogue at the following URL: `www.springer.com/series/3527`

Texts in Computational Science and Engineering

Vol. 1 H. P. Langtangen, *Computational Partial Differential Equations.* Numerical Methods and Diffpack Programming. 2nd Edition 2003. XXVI, 855 pp. Hardcover. ISBN 3-540-43416-X

Vol. 2 A. Quarteroni, F. Saleri, *Scientific Computing with MATLAB.* 2003. IX, 257 pp. Hardcover. ISBN 3-540-44363-0

Vol. 3 H. P. Langtangen, *Python Scripting for Computational Science.* 2nd Edition 2006. XXIV, 736 pp. Hardcover. ISBN 3-540-29415-5

For further information on these books please have a look at our mathematics catalogue at the following URL: `www.springer.com/series/5151`

GPSR Compliance

The European Union's (EU) General Product Safety Regulation (GPSR) is a set of rules that requires consumer products to be safe and our obligations to ensure this.

If you have any concerns about our products, you can contact us on ProductSafety@springernature.com

In case Publisher is established outside the EU, the EU authorized representative is:

Springer Nature Customer Service Center GmbH
Europaplatz 3
69115 Heidelberg, Germany

Batch number: 10755090

Printed by Printforce, the Netherlands